ANALYSIS IN
EUCLIDEAN SPACE

Essential Textbooks in Mathematics

ISSN: 2059-7657

The *Essential Textbooks in Mathematics* explores the most important topics in Mathematics that undergraduate students in Pure and Applied Mathematics are expected to be familiar with.

Written by senior academics as well lecturers recognised for their teaching skills, they offer in around 200 to 250 pages a precise, introductory approach to advanced mathematical theories and concepts in pure and applied subjects (e.g. Probability Theory, Statistics, Computational Methods, etc.).

Their lively style, focused scope and pedagogical material make them ideal learning tools at a very affordable price.

Published:

Analysis in Euclidean Space
by Joaquim Bruna (Universitat Autònoma de Barcelona, Spain & Barcelona Graduate School of Mathematics, Spain)

Introduction to Number Theory
by Richard Michael Hill (University College London, UK)

A Friendly Approach to Functional Analysis
by Amol Sasane (London School of Economics, UK)

A Sequential Introduction to Real Analysis
by J M Speight (University of Leeds, UK)

Essential Textbooks in Mathematics

ANALYSIS IN EUCLIDEAN SPACE

Joaquim Bruna
Universitat Autònoma de Barcelona
Barcelona Graduate School of Mathematics

World Scientific

NEW JERSEY • LONDON • SINGAPORE • BEIJING • SHANGHAI • HONG KONG • TAIPEI • CHENNAI • TOKYO

Published by

World Scientific Publishing Europe Ltd.
57 Shelton Street, Covent Garden, London WC2H 9HE
Head office: 5 Toh Tuck Link, Singapore 596224
USA office: 27 Warren Street, Suite 401-402, Hackensack, NJ 07601

Library of Congress Cataloging-in-Publication Data
Names: Bruna, Joaquim, author.
Title: Analysis in Euclidean space / Joaquim Bruna, Universitat Autònoma de Barcelona, Barcelona Graduate School of Mathematics.
Description: New Jersey : World Scientific, [2023] | Series: Essential textbooks in mathematics, 2059-7657 | Includes bibliographical references and index.
Identifiers: LCCN 2021054632 | ISBN 9781800611719 (hardcover) | ISBN 9781800611726 (ebook for institutions) | ISBN 9781800611733 (ebook for individuals)
Subjects: LCSH: Mathematical analysis. | Calculus.
Classification: LCC QA300 .B755 2023 | DDC 515--dc23/eng/20220104
LC record available at https://lccn.loc.gov/2021054632

British Library Cataloguing-in-Publication Data
A catalogue record for this book is available from the British Library.

For any available supplementary material, please visit
https://www.worldscientific.com/worldscibooks/10.1142/Q0345#t=suppl

Desk Editors: Jayanthi Muthuswamy/Shi Ying Koe

Typeset by Stallion Press
Email: enquiries@stallionpress.com

To my wife Anna Maria

About the Author

Joaquim Bruna obtained his PhD in Mathematics in 1978, at Universitat Autònoma de Barcelona (UAB), and was a postdoctoral researcher at Université de Paris-Sud (Orsay). Since 1984 he has been full Professor at the Departament de Matemàtiques de la UAB, having visiting professor positions at University of Wisconsin–Madison and University of New York at Albany. He has been Editor-in-Chief of *Publicacions Matemàtiques* and has served as editor of *Revista Matemàtica Iberoamericana.* His research interests are classical real analysis, harmonic and complex analysis, several complex variables and signal analysis. He has published 65 research papers, 11 dissemination articles and 3 textbooks, and has mentored 10 PhD students. He has participated in 25 research projects, 11 of which as a PI. He was the Director of the Centre de Recerca Matemàtica (CRM) from 2008 to 2015.

Contents

Introduction

This book is mostly based on Catalan hand-written notes I have accumulated during many years teaching at *Universitat Autònoma de Barcelona*, the main topic being Differentiation and Integration theory in several variables. Besides developing this main topic, this text provides glimpses into a number of closely related areas, including measure theory, differential geometry, classical theory of curves and surfaces, complex analysis, ordinary and partial differential equations, geometric measure theory, integral geometry, probability, and others.

The text is organized in 21 chapters, each with an introduction summarizing its contents, and an additional chapter containing other miscellaneous exercises. By choosing among the chapters, lecturers might use this book for different undergraduate courses in analysis. The only prerequisites are a basic course in linear algebra and a standard first-year calculus course in differentiation and integration in one real variable. As the text progresses, the level increases. Some of the last sections might be classified to be at graduate level and are indicated with an asterisk (*). The chapters are structured in sections and subsections, the rule being that a subsection starts a new issue.

There is some emphasis on concepts and rigorous proofs. I have tried to introduce concepts in the most natural way possible and to supplement the proofs with intuition, best if geometric. While writing it I have been thinking that the potential reader is an undergraduate student, probably working independently, interested in acquiring a solid footing in Analysis and expanding his/her background. Somehow, I hope that by studying this book the student not only learns the content but also how to think in a mathematical way. There are many examples and exercises inserted in the text for the student to work on independently. As a result, the text is especially suited for undergraduate students in Mathematics and Physics.

The approach to some topics is original, and in fact some results are new and unpublished. In particular, a multi-dimensional version of the fundamental theorem of calculus which surprisingly enough is new to the best of my knowledge. This result is exploited systematically to prove theorems like the changes of variables formula and the basic theorems in vector analysis.

The list in the bibliography is minimal, as Wikipedia makes it unnecessary to provide references for most of the keywords being introduced. It contains a number of other books that I have used to prepare my courses or that I used as a student. I want to mention particularly the excellent monograph [5], unfortunately a forgotten one.

I have not checked the source of all examples and presentations explained to me by colleagues and reproduced here, so a credit to some authors might be missing and apologies are due. Somehow, the front cover image is related to that. Building human towers (*castellers*) is an old Catalan tradition. Such towers can only be accomplished by the join effort of hundreds of people, in the same way that all textbooks in mathematics benefit from the contribution of many authors along the centuries.

Finally, I must acknowledge the valuable comments of my colleagues at the Departament de Matemàtiques on parts of this book, particularly Agustí Reventós and Julià Cufí. Thanks are due to Robert Ferréol for allowing me to use some figures in his outstanding website [11] and to Gregori Guasp and Rosa Rodríguez for their help in LaTex matters.

Joaquim Bruna
Camprodon, Catalonia, July 2022

Chapter 1

Euclidean Space

This chapter introduces preliminary concepts and notations regarding the linear space and metric structure of Euclidean space $\mathbb{E}^n$ as well as the maps that preserve these structures, the *rigid motions*. The final section relates volumes of parallelepipeds with determinants and introduces the k-th Jacobians of linear maps.

1.1 Linear and Metric Structure

1.1.1 We denote by $\mathbb{E}^3$ the *Euclidean space* of dimension three and by $\mathbb{E}^2$ the plane of dimension two and more generally the abstract Euclidean space $\mathbb{E}^n$ of dimension n. In general, we deal with $\mathbb{E}^n$, but the reader should imagine $\mathbb{E}^2$ and $\mathbb{E}^3$ to achieve a geometrical intuition of some of the concepts. Elements of $\mathbb{E}^n$ are points, which will hereafter be denoted by letters $p, q, \ldots$. Two points p and q determine a vector $\overrightarrow{pq}$ from p to q.

How do we manipulate points? Since there are n degrees of freedom in $\mathbb{E}^n$, we need an ordered family of n numbers to identify each point p. The set of such families is the Cartesian product of n copies of the set $\mathbb{R}$ of real numbers:

$$\mathbb{R}^n = \{x = (x_1, x_2, \ldots, x_n), x_i \in \mathbb{R}\}.$$

The classical way to locate each point using n numbers is through the *affine coordinates* in a *reference system*, consisting of a point 0, named the *origin*, and n axes through 0 with linearly independent direction vectors $e_1, e_2, \ldots, e_n$, hence a *basis* of n-dimensional vectors. For each point p, we consider the vector $\overrightarrow{0p}$ and write it in terms of the basis,

$$\overrightarrow{0p} = \sum_{i=1}^{n} x_i e_i.$$

The numbers $x = (x_1, x_2, \ldots, x_n)$ are the *coordinates* of p in this reference system, and each $x \in \mathbb{R}^n$ defines a unique point.

Thus, we have the abstract concept of point $p \in \mathbb{E}^n$ on one hand and the coordinates of this point in a reference system, which can be arbitrary, on the other. At certain points, for instance, when discussing general coordinate systems, it will be important to make this distinction. For the time being, we will assume that we have fixed a *canonical* Cartesian reference system (for instance, with origin at Perpignan's train station, the center of the universe according to Salvador Dalí). We identify $\mathbb{E}^n$ with $\mathbb{R}^n$ and p with the vector $\overrightarrow{0p}$ which in turn is identified with the coordinates, $p = x = (x_1, \ldots, x_n)$. When convenient, we use the notation X as well to denote the column vector with x_i as entries. The origin of coordinates is $0 = (0, 0, \ldots, 0)$, and the vectors $e_i = (0, \ldots, 1, \ldots, 0)$ constitute the *canonical basis* of $\mathbb{R}^n$ and $(x_1, \ldots, x_n)$ are called the *canonical Cartesian coordinates.* The sets $R = \prod [a_i, b_i]$ are *rectangles* with faces parallel to axes, for which each coordinate x_i freely varies in an interval $I_i = [a_i, b_i]$ independent of the other variables. We call them *n-dimensional intervals* or simply *intervals.* When all sizes $b_i - a_i$ are equal, we call them *cubes* or *squares* if $n = 2$.

In $\mathbb{R}^3$, we sometimes use the notation $p = (x, y, z)$ instead of (x_1, x_2, x_3) and $p = (x, y)$ in the plane.

1.1.2 In $\mathbb{R}^n$, there is a *vector space* structure defined by vector sum and scalar multiplication, with standard rules that we use, for instance, when changing coordinates. Consider another reference system consisting of another origin q and vectors $v_1, \ldots, v_n$, whose expressions in the canonical system are

$$q = (b_1, \ldots, b_n), \quad v_j = \sum_{i=1}^{n} a_{ij} e_i.$$

We organize the numbers a_{ij} in a matrix $A = (a_{ij})$, where i is for rows and j is for columns, so that the column vectors are v_j. A point p having coordinates $p = (x_1, x_2, \ldots, x_n)$ in the canonical system has now different coordinates $(y_1, \ldots, y_n)$. The relationship between them is obtained by equalizing the components in

$$(x_1 - b_1, \ldots, x_n - b_n) = \overrightarrow{qp} = \sum_{j=1}^{n} y_j v_j,$$

that is,

$$x_i = b_i + \sum_{j=1}^{n} y_j a_{ij}, \quad i = 1, \ldots, n.$$

At this point, the following matrix notation is useful: If X is the column vector with x_i, Y the one with y_i, and B the one with b_i, then

$$X = B + AY.$$

The inverse transformation is $Y = A^{-1}(X - B)$, where A^{-1} denotes the inverse matrix of A.

In Euclidean space, we have *linear subspaces* V of dimension $k \leq n$, which can be described in two ways. First, with $n - k$ linearly independent equations:

$$\sum_{j=1}^{n} m_{ij} x_j = 0, \quad i = 1, \ldots, n - k \tag{1.1}$$

(that is, as an intersection of $n - k$ different hyperplanes). With matrix notation, the above is written as $MX = 0$ with $M = (m_{ij})$ a $(n - k) \times n$ matrix of rank $n - k$. Second, we can describe V *parametrically*, choosing k linearly independent vectors in V, $v_1, \ldots, v_k$, that is, a basis of V, and setting

$$\overrightarrow{0p} = \sum_{j=1}^{k} \lambda_j v_j.$$

Then, points $p \in V$ are in one-to-one correspondence with k-tuples $(\lambda_1, \ldots, \lambda_k)$. If, in terms of the canonical basis,

$$v_j = \sum_{i=1}^{n} a_{ij} e_i = (a_{1j}, \ldots, a_{nj}),$$

then with matrix notation, points in V are described by

$$X = A\Lambda,$$

where A is an $n \times k$ matrix of rank k and Λ is the column vector with λ_j as entries. This is called a (global) *parametrization* of V; $\Lambda = (\lambda_1, \ldots, \lambda_k)$

are called the *parameters* of $p \in V$. Choosing a different basis of V leads to a different set of parameters for the same point p.

Together with the linear subspace V, which contains 0, we have the *affine sub-manifold* $p+V$ parallel to V through $p \notin V$. This is described by setting $X = p + A\Lambda$ or else with $n-k$ equations like those in (1.1) but not homogeneous.

1.1.3 In $\mathbb{R}^n$, there is also a *metric structure* given by the scalar product among vectors: if v, w are given in the canonical basis by $v = (v_1, \ldots, v_n), w = (w_1, \ldots, w_n)$, respectively, then their scalar product is defined as

$$\langle v, w\rangle = v_1 w_1 + \cdots + v_n w_n. \tag{1.2}$$

The number $|v|$ defined by $|v|^2 = \langle v, v\rangle = v_1^2 + \cdots + v_n^2$ is called the length or *Euclidean norm* of v.

If $\langle v, w\rangle = 0$, we say that v, w are *perpendicular* or *orthogonal.* An *orthonormal basis* of a subspace V is a basis of V consisting of unit vectors (of length one) and pairwise orthogonal. We assume that every subspace has an orthonormal basis (for instance, starting from an arbitrary basis and applying the so-called *Gram–Schmidt orthonormalization process*). Coordinates with respect to orthonormal basis are called *Cartesian coordinates.*

It is straightforward to check that

$$|v+w|^2 = |v|^2 + |w|^2 + 2\langle v, w\rangle, \tag{1.3}$$

and so, $|v+w|^2 = |v|^2 + |w|^2$ when v, w are orthogonal. If given two vectors v, w, we consider the linear combination $w + tv$ and apply (1.3), we get

$$t^2|v|^2 + 2t\langle w, v\rangle + |w|^2 = |w+tv|^2 \geq 0.$$

This implies the following:

(a) The discriminant of the left-hand binomial must be negative or zero, and we find

$$(\langle w, v\rangle)^2 \leq |w|^2|v|^2,$$

which amounts to

$$|\langle w, v\rangle)| \leq |w||v|. \tag{1.4}$$

This is called the *Cauchy–Schwarz* inequality.

(b) There is a value t for which $w + tv = 0$ if and only if the discriminant is zero, that is, v, w are linearly dependent if and only if the equality holds in the Cauchy–Schwarz inequality.

The Cauchy–Schwarz inequality is non-trivial only if both vectors are non-zero, and in this case, it can be written as

$$-1 \leq \frac{\langle w, v \rangle}{|w||v|} \leq 1.$$

The middle quantity is called the *correlation* between w and v. In geometric terms, it is the cosine of the angle between w and v, and so, it may be considered as quantifying the degree of linear dependence between both vectors. So, the angle between two non-zero vectors v, w is

$$\arccos \frac{\langle w, v \rangle}{|w||v|} \in [0, \pi].$$

Note that this is always a number in $[0, \pi]$. If we draw a unit circle C in the plane spanned by w, v, this is the absolute, non-signed measure of the shortest arc determined by w, v. Given two unit vectors w, v, this does not serve to fix the position of w with that of v known. Instead, *oriented angles* are used. Once an orientation of C is chosen, the oriented angle takes values in $[-\pi, +\pi]$.

Combining (1.4) with (1.3), we get

$$\begin{aligned} |w+v|^2 = |w|^2 + |v|^2 + 2\langle w, v \rangle &\leq |w|^2 + |v|^2 + 2|\langle w, v \rangle| \\ &\leq |w|^2 + |v|^2 + 2|w||v| = (|w| + |v|)^2, \end{aligned}$$

whence

$$|w + v| \leq |w| + |v|.$$

This is called the *triangle inequality.*

Given two points $x = (x_1, \ldots, x_n), y = (y_1, \ldots, y_n)$ in $\mathbb{R}^n$, their *Euclidean distance* is defined as

$$d(x, y) = d(y, x) = |y - x| = \sqrt{(y_1 - x_1)^2 + \cdots + (y_n - x_n)^2}.$$

Given three points x, y, z, since $z - x = (z - y) + (y - x)$, the triangle inequality gives

$$d(x, z) \leq d(x, y) + d(y, z). \tag{1.5}$$

Permuting points, we also have $d(x, y) \leq d(x, z) + d(z, y)$, and so,

$$|d(x, z) - d(x, y)| \leq d(y, z). \tag{1.6}$$

Thus, we have a quantification of the notion of *proximity* among points, and we say in an informal way that x, y are *close* if $d(x, y)$ is small.

The set $B(x, r) = \{y : d(x, y) < r\}$ is called the *open ball centered at x with radius r*.

Given a linear subspace V of dimension $k, 0 < k < n$, the linear subspace

$$V^{\perp} = \{w : \langle v, w\rangle = 0, v \in V\}$$

is the *orthogonal* of V. Since $V^{\perp}$ is the subspace of solutions of the homogeneous linear system $MX = 0$, where M is a matrix having as columns a basis of V (hence of rank k), it follows that $V^{\perp}$ has dimension $n - k$. Obviously, $V \cap V^{\perp} = 0$. We claim that given any x, there is a unique decomposition

$$x = y + z, \quad y \in V, \quad z \in V^{\perp}.$$

To check unicity, assume $x = y_1 + z_1 = y_2 + z_2$; then, $y_1 - y_2 = z_2 - z_1 \in V \cap V^{\perp}$, and so, $y_1 = y_2, z_1 = z_2$. To check existence, let $v_1, v_2, \dots, v_k$ be an orthonormal basis of V; then, *a fortiori*

$$\lambda_i = \langle y, v_i\rangle = \langle x - z, v_i\rangle = \langle x, v_i\rangle,$$

and so, $y = \sum_i \langle x, v_i\rangle v_i$. This also shows that $x - y \in V^{\perp}$. This decomposition is called *the orthogonal decomposition* of x, and $y \in V$ is called the *orthogonal projection* of x onto V.

If $w \in V$, then $x - w = y - w + z$ with $y - w \in V, z \in V^{\perp}$, and so, $|x - w|^2 = |y - w|^2 + |z|^2$. It follows that $d(x, w), w \in V$ is minimal when $w = y$ and in this case equals $|z|$, that is, y is the point in V closest to x. The mapping defined by $P_V x = y$ (which is the identity on V) is linear and is called the *projection map* on V.

In $\mathbb{R}^3$, there is the notion of *cross product* $v \times w$ of two vectors defined as

$$v \times w = \det\begin{pmatrix} e_1 & e_2 & e_3 \\ v_1 & v_2 & v_3 \\ w_1 & w_2 & w_3 \end{pmatrix} = (v_2 w_3 - v_3 w_2, v_3 w_1 - v_1 w_3, v_1 w_2 - v_2 w_1).$$

It is straightforward to see that $v \times w$ is zero iff v, w are linearly dependent, and otherwise, it is orthogonal to both v, w and thus orthogonal to the plane v, w they span. The direction of $v \times w$ is given by the right-hand rule: if the index finger points in the direction of v and the others in the direction of w, then $v \times w$ points in the thumb direction. For further use, note that $\det(v, w, v \times w) \geq 0$.

If u is the third vector, a computation shows that

$$u \times (v \times w) = \langle u, w \rangle v - \langle u, v \rangle w.$$

If v is a unit vector, in the orthogonal decomposition $w = \lambda v + w_2$, $\langle v, w_2 \rangle = 0$,

$$\lambda = \langle w, v \rangle, \quad w_2 = w - \langle w, v \rangle v = -v \times (v \times w).$$

Thus,

$$w = \langle w, v \rangle v - v \times (v \times w)$$

is the orthogonal decomposition.

In paragraph 1.3.2, the definition of cross product is extended to $\mathbb{R}^n$, $n > 3$ and a geometrical interpretation of $|v \times w|$ is given.

1.1.4 When $n = 1$, the real line has an *order relation* and also a multiplicative structure satisfying natural compatibility properties. When $n = 2$, it is not possible to define a compatible order relation, but still one has a multiplicative structure coming from the identification with the complex field. For $n > 2$, it is not possible to define in $\mathbb{R}^n$ a field structure extending that of $\mathbb{R}^2$. However, in $\mathbb{R}^4$, a non-commutative field structure can be defined, i.e., the field of *quaternions*. One may also define in $\mathbb{R}^8$ a non-associative field structure, i.e., the field of *octonions*.

1.2 Linear Transformations: Rigid Motions

1.2.1 In relation to the vector space structure of $\mathbb{R}^n$, we can consider the maps $T : \mathbb{R}^n \to \mathbb{R}^m$ that preserve this structure:

$$T(x + y) = T(x) + T(y), \quad T(\lambda x) = \lambda T(x).$$

These are called *linear maps*. Such maps are completely determined by the vectors $T(e_j), j = 1, \ldots, n$, which can be chosen arbitrarily. If $T(e_j) = \sum_{i=1}^{m} m_{ij} e_i, j = 1, \ldots, n$ (here, we use the same notation for the canonical basis of $\mathbb{R}^n, \mathbb{R}^m$), we consider the matrix $M = (m_{ij})$ whose jth column is $T(e_j)$; this is called the *matrix of T in the canonical basis.* In matrix notation, the linear mapping is $X \to Y = MX$, where MX denotes matrix multiplication.

We recall that T has two distinguished associate subspaces:

$$N(T) = \{x \in \mathbb{R}^n : Tx = 0\}, \quad R(T) = \{y \in \mathbb{R}^m : y = Tx, x \in \mathbb{R}^n\},$$

called respectively the *kernel* and the *range space*. By definition, the linear system $Tx = y$ has a solution if and only if $y \in R(T)$, in which case the general solution is an affine sub-manifold $x_0+N(T)$. If k, r are the respective dimensions of $N(T), R(T)$, one has $n = k+r, r \leq m$. The number r is called the *rank* of T (or M). The map T is injective if and only if $k = 0$ and is onto if and only if $r = m$, and so, it is bijective if and only if $r = m = n$.

Recall from linear algebra that a linear map T from $\mathbb{R}^n$ into itself is bijective if and only if $\det M \neq 0$. The $n \times n$ matrices M with $\det M \neq 0$ form a group called *the linear group* $GL(n)$.

When $m = n$, maps of the form $X \to q + TX$, the composition of a linear map with a translation, are called *affine maps*.

If we use other coordinates X', Y' related to the canonical ones X, Y by $X = AX', Y = BY'$, from $BY' = Y = MX = MAX'$, we see that $M' = B^{-1}MA$ is the matrix of T in the new coordinates. Recall too from linear algebra that by choosing convenient coordinates, that is, choosing conveniently A, B, it is possible to get simpler M', for instance, upper-triangular or in *Jordan form*.

It is worth pointing out two different interpretations of an invertible matrix with linearly independent column vectors $v_1, \ldots, v_n$ expressed in the canonical basis. This is the matrix in the canonical basis of the linear transformation mapping e_i to $v_i, i = 1, \ldots, n$ (which indeed moves points); it is also the matrix of the identity map (which does not move points) when using the basis $v_1, \ldots, v_n$ and the canonical one, that is, the matrix that converts the coordinates X' into the coordinates X.

1.2.2 Looking at Euclidean space with its metric structure, it is natural to relate linear maps with the scalar product and metric.

If $T : \mathbb{R}^n \to \mathbb{R}^m$ is linear, the scalar map $x \to \langle Tx, y\rangle$ is linear, and so, $\langle Tx, y\rangle = \langle x, T^t y\rangle$ for a linear map $T^t : \mathbb{R}^m \to \mathbb{R}^n$ called the *transpose* of T. If M is the matrix in the canonical basis of $\mathbb{R}^n, \mathbb{R}^m$ (or more generally in orthonormal basis of $\mathbb{R}^n, \mathbb{R}^m$), then T^t has matrix M^t in the same basis. If $m = n$ and $T = T^t, T$ is called *self-adjoint.*

Now, we consider maps that preserve distances. We say that a map $T : \mathbb{R}^n \to \mathbb{R}^m$ is a *rigid motion* if $d(Tx, Ty) = d(x, y)$ for all x, y pairs. The following theorem states that they are affine maps.

Theorem 1.1. *Every rigid motion from $\mathbb{R}^n$ to $\mathbb{R}^m$ has the form $X \to p + TX$ for some linear map T. In particular, the range of T is an affine sub-manifold of dimension n, whence T is onto in case $m = n$.*

Proof. We define $p = T(0)$; by replacing $T(x)$ by $T(x) - p$, we must show that a rigid motion T fixing the origin is linear. Note that since $T(0) = 0$, one has $|Tx| = |x|$; from (1.3), it then follows that T preserves the scalar product, $\langle Tx, Ty\rangle = \langle x, y\rangle$. Now, in

$$|T(x+y) - T(x) - T(y)|^2 = \langle T(x+y) - T(x) - T(y), T(x+y)$$
$$- T(x) - T(y)\rangle = |T(x+y)|^2 + |T(x)|^2 + |T(y)|^2 - 2\langle T(x+y), T(x)\rangle$$
$$- 2\langle T(x+y), T(y)\rangle + 2\langle T(x), T(y)\rangle,$$

one can drop off the T's to get $|x + y - x - y|^2 = 0$. A similar argument shows that $T(\lambda x) = \lambda T(x)$. □

Thus, when $m = n$, every rigid motion fixing the origin is a linear map $T : \mathbb{R}^n \to \mathbb{R}^n$ preserving the scalar product,

$$\langle x, y\rangle = \langle Tx, Ty\rangle = \langle x, T^tTy\rangle.$$

Thus, $T^tT = I$, $TT^tT = T$, and $TT^t(Tx) = Tx$. Since T is one-to-one and a bijection, Tx is a general vector, whence $TT^t = I$. Thus, $T^{-1} = T^t$. The columns of the corresponding matrices form an orthonormal basis. We call T and the corresponding matrices M satisfying $M^t = M^{-1}$ *orthogonal*. The set of $n \times n$ orthogonal matrices is a group under matrix multiplication (corresponding to composition of maps), called the *orthogonal group* and denoted as $O(n)$. From $M^tM = I, (\det M)^2 = 1, \det M = \pm 1$.

The rigid motions T fixing the origin with $\det M = 1$ is a subgroup, the *special orthogonal group* or the group of *rotations*, denoted as $SO(n)$. The group $SO(2)$ is commutative, identified with the unit circle in the plane and $\mathbb{R}/2\pi\mathbb{Z}$:

$$M_\theta = \begin{pmatrix} \cos\theta & -\sin\theta \\ \sin\theta & \cos\theta \end{pmatrix}.$$

To understand intuitively why this group is important, assume we have a rigid body C in space containing the origin from where we start to move with our hands keeping the origin fixed, that is, T_t is a rigid motion for every time t and T_0 is the identity map. If $p = (\varepsilon, 0, 0), q = (0, \varepsilon, 0)$ are in C, it is clear that the position $T_t C$ is known as soon as $T_t p, T_t q$ are known; T_t being linear means $T_t e_1, T_t e_2$ are known. Then, necessarily $T_t e_3 = T_t e_1 \times T_t e_2$, which amounts to $\det T_t = 1$. More formally, moving with our hands means that $T_t x$ must be continuous in t, x (see Chapter 2), which implies that $\det T_t$ must be continuous in t. Since $\det T_0 = 1$, necessarily $\det T_t = 1$.

Thus, $SO(3)$ can be interpreted as the class of all possible positions of C in space. Considering $T_t e_1, T_t e_2$ as a point in $\mathbb{R}^6$, whose coordinates must satisfy three equations $\langle T_t e_i, T_t e_j \rangle = \delta_{ij}, i, j = 1, 2$, we see that, on an intuitive basis, there are three degrees of freedom in $SO(3)$. If we also allow the origin to move, we conclude that there are six degrees of freedom for the general position of a rigid body.

A formal proof is based on *Euler's theorem.*

Theorem 1.2. *Every $T \in O(n)$ with n odd has an invariant line.*

Proof. We prove that if $T \in SO(n), Tv = v$ has a non-zero solution, that is, $\det(M - I) = 0$. This follows from

$$\begin{aligned}\det(M - I) &= \det(M^t - I) = \det(M^{-1} - I) = \det[M^{-1}(I - M)] \\ &= \det(M^{-1}) \det(I - M) = \det(I - M) = -\det(M - I).\end{aligned}$$

Analogously, if $\det M = -1$, then there is a unit vector with $Tv = -v$. $\square$

If $T \in SO(3), Tv = v$ and V is orthogonal to the line L spanned by v, then T fixes every point of L and leaves V invariant. Thus, it consists of a rotation around L by a certain angle θ. Analytically, in terms of the orthogonal decomposition

$$w = \langle w, v \rangle v + w_2, \quad w_2 = -v \times (v \times w),$$

one has

$$\begin{aligned}&Tv = v, Tw_2 = \cos\theta\, w_2 + \sin\theta\, (v \times w_2) = \cos\theta\, w_2 + \sin\theta\, (v \times w), \\ &Tw = \langle w, v \rangle v + \cos\theta\, w_2 + \sin\theta\, (v \times w).\end{aligned}$$

Inserting $w_2 = w - \langle w, v \rangle v$, we get a description $T = T(v, \theta)$:

$$Tw = w + (1 - \cos\theta)\, [v \times (v \times w)] + \sin\theta\, (v \times w). \tag{1.7}$$

This is called *the Olinde Rodrigues representation.* In terms of matrices, the map $w \to v \times w, v = (v_1, v_2, v_3)$ has matrix

$$A_v = \begin{pmatrix} 0 & -v_3 & v_2 \\ v_3 & 0 & -v_1 \\ -v_2 & v_1 & 0 \end{pmatrix},$$

so

$$M = I + \sin\theta A_v + (1 - \cos\theta) A_v^2.$$

Altogether, elements in $SO(3)$ depend on $v, |v| = 1$ (two degrees of freedom) and one angle (the third degree of freedom), it has *dimension* three. Later, in Section 3.2, we formalize this concept of dimension.

The group $SO(3)$ is not commutative, and so, neither is $SO(n)$ if $n \geq 3$. If $T \in O(3)$ and $\det M = -1, Tv = -v$, then T consists of a reflection through V followed by a rotation.

1.2.3 A different parametrization of $SO(3)$ is the one in terms of the *Euler angles.* If $T \in SO(3)$ and we denote by X, Y, Z the axes defined by $Te_i, i = 1, 2, 3$, the Euler angles are defined as in Figure 1.1.

First, β is the absolute unsigned angle between e_3 and Te_3. The *line of nodes* is the intersection line of the planes xy and XY, spanned by $e_3 \times Te_3$. In the plane xy, α is the oriented angle from e_1 to $e_3 \times Te_3$ in the counterclockwise direction. In the plane XY, γ is the oriented angle from

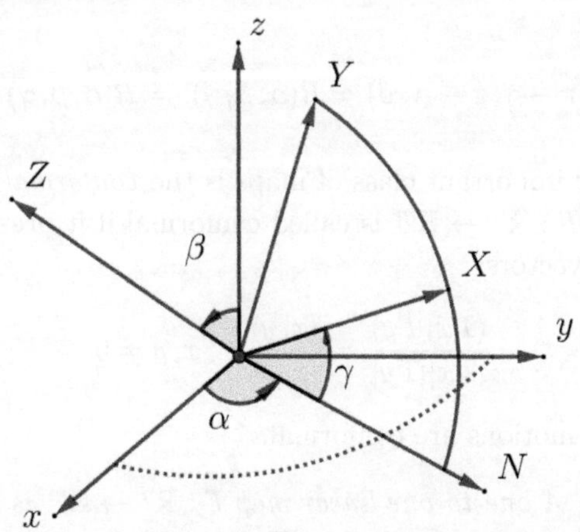

Figure 1.1. Euler angles.

$e_3 \times Te_3$ to Te_1, also in the counterclockwise direction. Thus, α, γ must be understood as modulo 2π and $0 \leq \beta \leq \pi$.

Exercise 1.1. Check geometrically that $T = T_\alpha T_\beta T_\gamma$, where T_α, T_γ are the rotations around the z-axis:

$$T_\alpha = \begin{pmatrix} \cos\alpha & -\sin\alpha & 0 \\ \sin\alpha & \cos\alpha & 0 \\ 0 & 0 & 1 \end{pmatrix}, \quad T_\gamma = \begin{pmatrix} \cos\gamma & -\sin\gamma & 0 \\ \sin\gamma & \cos\gamma & 0 \\ 0 & 0 & 1 \end{pmatrix},$$

and T_β is the rotation around the x-axis:

$$T_\beta = \begin{pmatrix} 1 & 0 & 0 \\ 0 & \cos\beta & -\sin\beta \\ 0 & \sin\beta & \cos\beta \end{pmatrix}.$$

A solution for the exercise can be found in [13]. Using this, we find the following general expression of a rotation $R(\alpha, \beta, \gamma)$ in $\mathbb{R}^3$:

$$\begin{pmatrix} \cos\alpha\cos\gamma - \cos\beta\sin\alpha s, \gamma & -\cos\alpha\sin\gamma - \cos\beta\sin\alpha\cos\gamma & \sin\alpha\sin\beta \\ \sin\alpha\cos\gamma + \cos\beta\cos\alpha\sin\gamma & -\sin\alpha\sin\gamma + c\,\beta\cos\alpha\cos\gamma & -\cos\alpha\sin\beta \\ \sin\gamma\sin\beta & \cos\gamma\sin\beta & \cos\beta \end{pmatrix}. \tag{1.8}$$

The expression is unique if α, β are not integer multiples of 2π and $0 < \beta < \pi$. Note that

$$R(\pi - \gamma, \pi - \alpha, \beta) = R(\alpha, \beta, \gamma)^t = R(\alpha, \beta, \gamma)^{-1}. \tag{1.9}$$

1.2.4 Another important class of maps is the *conformal* class. A one-to-one linear map $T : \mathbb{R}^n \to \mathbb{R}^m$ is called conformal if it preserves the size of angles between vectors:

$$\frac{\langle Tx, Ty \rangle}{|Tx||Ty|} = \frac{\langle x, y \rangle}{|x||y|}, \quad x, y \neq 0.$$

Of course, rigid motions are conformal.

Theorem 1.3. *A one-to-one linear map $T : \mathbb{R}^n \to \mathbb{R}^m$ is conformal if and only if there exists $\lambda > 0$ such that $|Tx| = \lambda|x|$ for all x. In other words, T is a multiple of a rigid motion.*

Proof. We must show that for $x, y \neq 0$,

$$\frac{|Tx|}{|x|} = \frac{|Ty|}{|y|}.$$

Composing with rigid motions on both sides of T, we may assume that $n = m = 2$. If e_1, e_2 is the canonical basis, by composing again with a dilation and a rigid motion in the plane, we may assume that $Te_1 = e_1$ and must show that $T = \pm I$. Since T is conformal, $Te_2 = \lambda e_2$. If $e_\theta = (\cos\theta, \sin\theta), Te_\theta = (\cos\theta, \lambda\sin\theta)$ forms angle θ with e_1 iff $|Te_\theta| = 1$, that is, $\lambda = \pm 1$. □

The following exercise is analogous to Theorem 1.1.

Exercise 1.2. Prove that if $T : \mathbb{R}^n \to \mathbb{R}^m$ is a one-to-one map such that the size of the angle between $Ty - Tx, Tz - Tx$ equals that of $y - x, z - x$ for all x, y, z, then $Tx = p + \lambda Ux$ for a (linear) rigid motion U.

1.2.5 The self-adjoint operators are particularly simple to handle using the following result.

Theorem 1.4. *If A is a $n \times n$ symmetric matrix, there is an orthogonal matrix M such that M^tAM is a diagonal matrix $D(\lambda_1, \ldots, \lambda_n)$, with $\lambda_i \in \mathbb{R}$. The numbers $\lambda_i, i = 1, \ldots, n$ (which may occur with multiplicities) are the eigenvalues of A and hence unique.*

Proof. For a better understanding of this result, it is convenient to analyze *a fortiori* the meaning of the λ_i. Let us consider A as a linear transformation, the one having matrix A in the canonical basis. The symmetry of A means that

$$\langle Aw, v\rangle = \langle w, Av\rangle.$$

Now, $M^tAM = D(\lambda_1, \ldots, \lambda_n)$ amounts to $M^tAM(e_i) = \lambda_i e_i$, and since $M^t = M^{-1}$, this is equivalent to

$$AM(e_i) = \lambda_i M(e_i).$$

Recall from linear algebra that if $Aw = \lambda w$, with $w \neq 0$, w is said to be an *eigenvector* with *eigenvalue* λ. Therefore, $v_i = M(e_i)$ (the ith column of M) is an eigenvector with eigenvalue λ_i. Besides, M is orthogonal if and only if $v_1, \ldots, v_n$ is an orthonormal basis. Altogether, the statement is equivalent to the fact that there is an orthonormal basis consisting of eigenvectors of A.

Now, w is an eigenvector with eigenvalue λ if and only if $(A - \lambda I)(w) = 0$, whence the eigenvectors are the roots of the *characteristic polynomial*, $P_A(\lambda) = \det(A - \lambda I)$. Each eigenvalue λ has an associated eigensubspace $E_\lambda = \ker(A - \lambda I)$, whose dimension is the multiplicity of λ.

Next, we point out two facts: First, if λ, μ are different eigenvalues, then E_λ, E_μ are orthogonal subspaces because if $w \in E_\lambda, v \in E_\mu$,

$$\lambda\langle w, v\rangle = \langle Aw, v\rangle = \langle w, Av\rangle = \mu\langle w, v\rangle.$$

Second, the same argument shows that each orthogonal subspace $E_\lambda^\perp$ is invariant by A.

With all these preliminaries, we can now prove the theorem using induction on n; as we will see, it is sufficient to prove that all eigenvalues are real. For $n = 2$, assume

$$A = \begin{pmatrix} a & b \\ b & c \end{pmatrix}.$$

The characteristic polynomial is $(a-\lambda)(c-\lambda)-b^2$ with positive discriminant $(a+c)^2 - 4(ac - b^2) = (a - c)^2 + b^2$. Hence, if $b = 0, a = c$, and A is already diagonal, or else there exist two different real eigenvalues. In the latter case, choosing a unit eigenvector in each eigenspace, we are done.

Assume by induction that every symmetric transformation A acting on a subspace of dimension $n' < n$ diagonalizes in an orthonormal basis. Let now A be a $n \times n$ symmetric matrix, and let us prove that it has a real eigenvalue (*a fortiori* all are real). By the fundamental theorem of algebra, we know that there are complex eigenvalues and complex eigenvectors, that is, working in $\mathbb{C}^n$ instead, we see that there are complex vectors X such that $AX = \lambda X$. From this, we get

$$(\overline{X})^t AX = \lambda|X|^2.$$

But the left-hand term is $\sum_{ij} \overline{x_i} a_{ij} x_j$, a real value because $a_{ij} = a_{ji}$, and therefore, λ is real.

Now, we consider the eigenspace E_λ, in which we can choose an arbitrary orthonormal basis. If E_λ is the whole space, we are done. Otherwise, we consider the restriction A' of A to $E_\lambda^\perp$ that has dimension $n' < n$ and apply the induction hypothesis. □

The theorem can thus be stated that self-adjoint operators are exactly those that diagonalize in an orthonormal basis.

1.2.6 Next, we introduce the *norm of a linear transformation* $T : \mathbb{R}^n \longrightarrow \mathbb{R}^m$. Using the notations above, $y = Tx$ has components $y_j = \sum_{i=1}^n m_{ji}x_i$, $j = 1, \ldots, m$. Therefore, as $|x_i| \leq |x|$,

$$|y_j| \leq \left(\sum_i |m_{ji}|\right)|x|,$$

and so,

$$|y| = \left(\sum_j |y_j|^2\right)^{\frac{1}{2}} \leq \left(\sum_j \left(\sum_i |m_{ji}|\right)^2\right)^{\frac{1}{2}} |x|.$$

This means that $\frac{|Tx|}{|x|}$ remains bounded if $x \neq 0$, and we may define the *norm* of T by

$$|T| = \sup\left\{\frac{|Tx|}{|x|}, x \neq 0\right\}.$$

Hence, one has $|Tx| \leq |T||x|$, and $|T|$ is the smallest constant C for which $|Tx| \leq C|x|$ holds for all x. Note that if $y = \frac{x}{|x|}$, then $|y| = 1$ and by linearity, $\frac{|Tx|}{|x|} = |Ty|$, and so,

$$|T| = \sup\{|Ty| : |y| = 1\} = \sup\{|Ty| : |y| \leq 1\}.$$

Since this equals the supremum in $|x|, |y| \leq 1$ of

$$\langle Ty, x\rangle = \langle y, T^t x\rangle,$$

it follows that $|T^t| = |T|$.

The explicit computation of $|T|$ is easy in case $m = 1$. In this case, $T(x) = t_1x_1 + \cdots + t_mx_n$; with $v = (t_1, \ldots, t_m)$, one has $T(w) = \langle w, v\rangle$, i.e., T consists of scalar product with v; by the Cauchy–Schwarz inequality,

$$|T(w)| \leq |w||v|.$$

Therefore, $|T| \leq |v|$; but for $w = v$, the above is an equality, and so, $|T| = |v|$. In dimension $n = 2$, we can describe all unit vectors by $y = (\cos\theta, \sin\theta), 0 \leq \theta \leq 2\pi$; since $|Ty|$ is then a continuous function of θ, this supremum is attained at some y. We will see later that this holds true for an arbitrary dimension.

For self-adjoint maps, the computation of $|T|$ is trivial. By Theorem 1.4, there exists an orthonormal basis $v_1, \ldots, v_n$ of eigenvectors of T, that is, $T(v_j) = \lambda_j v_j$. Now, if $x = \sum y_j v_j$, then

$$|x|^2 = \sum y_j^2, \quad Tx = \sum \lambda_j y_j u_j, \quad |Tx|^2 = \sum |\lambda_j|^2 |y_j|^2,$$

from which it follows that $|T|$ equals $\max|\lambda_j|$.

For our purposes later in this book, we do not need in general the optimal value $|T|$ and the estimate $|Tx| \leq C|x|$ for some C will suffice.

Exercise 1.3. Prove that $|T|^2 = |T^tT|$. Since T^tT is symmetric, $|T|^2$ equals the largest eigenvalue of T^tT.

Note that if T is invertible with inverse S, then $|Sx| \leq |S||x|$ may be written as $|x| \leq |S||Tx|$, and therefore,

$$m|x| \leq |Tx| \leq M|x| \tag{1.10}$$

holds for some constants m, M.

Exercise 1.4.

(a) Prove that if $T \in O(n)$ and v, w are eigenvectors with different eigenvalues, v, w are orthogonal.

(b) Working in $\mathbb{C}^n$ and using the same proof as in Theorem 1.4, prove that there is an orthogonal decomposition in lines L and planes Π:

$$\mathbb{R}^n = L_1 \oplus \cdots \oplus L_m \oplus \Pi_1 \oplus \cdots \oplus \Pi_k, m + 2k = n,$$

such that $T = \pm I$ in each L and T is a rotation in each Π. That is, the matrix of T in this basis is

$$\begin{pmatrix} 1 &&&&&&&&&&& \\ & \ddots &&&&&&&&&& \\ && 1 &&&&&&&&& \\ &&& -1 &&&&&&&& \\ &&&& \ddots &&&&&&& \\ &&&&& -1 &&&&&& \\ &&&&&& \cos\theta_1 & -\sin\theta_1 &&&& \\ &&&&&& \sin\theta_1 & \cos\theta_1 &&&& \\ &&&&&&&& \ddots &&& \\ &&&&&&&&& \cos\theta_k & -\sin\theta_k \\ &&&&&&&&& \sin\theta_k & \cos\theta_k \end{pmatrix}.$$

This includes Euler's theorem in case n is odd.

1.3 Volume and Determinants

1.3.1 We have an intuitive idea of what length, area, and volume mean. However, what is the precise definition of the length of a set $A \subset \mathbb{R}$, the area of $A \subset \mathbb{R}^2$, and the volume of $A \subset \mathbb{R}^3$? How can one define length of curves and area of surfaces?

In this book we address these questions. In this section, we consider for now only parallelepipeds. By a *parallelepiped*, we understand the set determined by a point p and k linearly independent vectors $v_1, \ldots, v_k$:

$$P(p, v_1, \ldots, v_k) = \left\{ q \in R^n : q = p + \sum_i t_i v_i, 0 \leq t_i \leq 1 \right\}.$$

When the vectors are orthogonal, we use the term *rectangle.*

We find the k-dimensional volume $m_k(P)$ of a parallelepiped. We assume that $m_k(E)$ is defined for polygons E lying on k-dimensional linear subspaces, and we assume the following three properties to be true.

First, the measure of a rectangle is the product of the lengths of its edges:

$$m_k(P(p, v_1, \ldots, v_k)) = |v_1| \cdots |v_k|.$$

Second, it is invariant by rigid motions, implying $m_k(T(P)) = m_k(P)$ when T is a rigid motion. Finally, it is finitely additive, that is,

$$m_k(E_1 \cup \cdots \cup E_N) = m_k(E_1) + \cdots + m_k(E_N),$$

whenever the sets are disjoint.

It is elementary that in dimension two, the area of the parallelepiped determined by two linearly independent vectors v_1, v_2 (we assume hereafter that $p = 0$) is the length of the basis, say $|v_1|$, by the height. Since the projection of v_2 on the line spanned by v_1 has length $|v_2||\cos\theta|$, where θ is the angle between them, the height is $|v_2||\sin\theta|$.

Proceeding analogously, we see that the three-dimensional volume of a parallelepiped in space determined by three linearly independent vectors v_1, v_2, v_3 is the product of the area of its basis, which is a two-dimensional parallelepiped, and its height. In general,

$$m_k(P(v_1,\dots,v_k)) = m_{k-1}(P(v_1,\dots,v_{k-1}))h,$$

with h the height with respect to the subspace spanned by $v_1,\dots,v_{k-1}$.

Exercise 1.5. Prove *Heron's formula* for the area A of a triangle with sides a, b, c:

$$A = \sqrt{s(s-a)(s-b)(s-c)},$$

where $s = \frac{1}{2}(a+b+c)$ is the semi-perimeter.

Theorem 1.5. *If V is the $n \times k$ matrix with column vectors $v_1,\dots,v_k$ (expressed in the canonical basis or in any other orthonormal basis), then*

$$m_k(P(v_1,\dots,v_k))^2 = \det V^tV.$$

In particular, if $k = n$, $m_n(P(v_1,\dots,v_n)) = |\det V|$.

Proof. We prove first the case $k = n$ by induction. For $n = 2$, the area is, as mentioned before, $A = |v_1||v_2||\sin\theta|$. Set $V = (v_{ij})$, that is, $v_j = v_{1j}e_1 + v_{2j}e_2, j = 1, 2$. Then,

$$\begin{aligned} A^2 &= |v_1|^2|v_2|^2\sin^2\theta = |v_1|^2|v_2|^2(1-\cos^2\theta) = |v_1|^2|v_2|^2 - |\langle v_1, v_2\rangle|^2 \\ &= (v_{11}^2 + v_{21}^2)(v_{12}^2 + v_{22}^2) - (v_{11}v_{12} + v_{21}v_{22})^2 = (v_{11}v_{22} - v_{12}v_{21})^2. \end{aligned}$$

Assuming that the result holds up to dimension $n-1$, let T be an orthogonal transformation with matrix B in the canonical basis mapping the subspace spanned by $v_1,\dots,v_{n-1}$ to $\mathbb{R}^{n-1}$, that is, the matrix BV has the last row of the form $(0,\dots,0,c)$ and above the zeros, it has the matrix V' having as column vectors $T(v_1),\dots,T(v_{n-1})$ viewed in $\mathbb{R}^{n-1}$. If P' is the parallelepiped defined by the vectors $v_1,\dots,v_{n-1}$, by the induction hypothesis,

$$m_n(P) = m_n(T(P)) = m_{n-1}(T(P'))|h| = |\det V'|h,$$

with h the height of $T(P)$ over $\mathbb{R}^{n-1}$, the last component c of Tv_n. Therefore,

$$m_n(P) = |\det V'||c| = |\det BV| = |\det V|$$

because $|\det B| = 1$.

Now, we prove the case $k < n$. Let $w_1, w_2, \ldots, w_k$ be an orthonormal basis of the subspace spanned by $v_1, \ldots, v_k$ and write $v_j = \sum_i b_{ij} w_i$. By what has just been seen, $m_k(P) = |\det B|$. But the entry (i, j) of $V^t V$ is the scalar product $\langle v_i, v_j \rangle$. If we compute this scalar product using $v_j = \sum_i b_{ij} w_i$ and the fact that w_j are orthogonal, we find that $V^t B = B^t B$, whence $\det(V^t V) = (\det B)^2 = m_k(P)^2$. □

The $k \times k$ matrix $V^t V$ is called the *Gram matrix* or *Gramian* of the vectors $v_1, \ldots, v_k$, denoted as $G(v_1, \ldots, v_k)$. The proof also shows that

$$m_k(P(v_1, \ldots, v_k)) \leq |v_k| m_{k-1}(P(v_1, \ldots, v_{k-1})),$$

and that the equality holds if and only if v_k is orthogonal to $v_1, \ldots, v_{k-1}$. As a consequence, we can state the following obvious intuitive fact.

Corollary 1.1. *$m_k(P(v_1, \ldots, v_k)) = |v_1| \cdots |v_k|$ if and only if v_i are pairwise orthogonal. For an invertible $n \times n$ matrix M with column vectors v_i, $|\det M| = |v_1| \cdots |v_n|$ if and only if the v_i are pairwise orthogonal.*

1.3.2 At this point, we quote a nice result from the matrix theory, the *Cauchy–Binnet* formula.

Let B be a $k \times n$ matrix and A an $n \times k$ matrix. To obtain square matrices, we select k columns in B. This choice is coded by a set S of k ordered indexes, and we call B_S the resulting square matrix. Then, we take the k rows in A corresponding to the same indexes and call A_S the resulting square matrix. The statement is then

$$\det BA = \sum_S \det B_S \det A_S.$$

This can also be viewed as a generalization of the rule $\det BA = \det A \det B$ for square matrices.

Exercise 1.6. Prove the above formula.

If we apply this to $A = V, B = V^t$ and use Theorem 1.5, we get a sort of Pythagorean theorem for volumes of parallelepipeds.

Theorem 1.6. *For a k-dimensional parallelepiped in R^n, the square of its measure equals the sum of the squares of the measures of its projection on each of the $\binom{n}{k}$ coordinate k planes.*

An appropriate language is the exterior algebra one. The minors are the coefficients of $v_1 \wedge \cdots \wedge v_k$ in the basis

$$e_I = e_{i_1} \wedge \cdots \wedge e_{i_k}, \quad I = i_1 < \cdots < i_k$$

of $\Lambda^k\mathbb{R}^n$. Considering e_I as an orthonormal basis in $\Lambda^k\mathbb{R}^n$, we may write

$$m_k(P(v_1, \ldots, v_k)) = |v_1 \wedge \cdots \wedge v_k|.$$

In case $k = n - 1$, we may generalize the cross product in $\mathbb{R}^3$. There is an isometry of $\Lambda^{n-1}\mathbb{R}^n$ with $\mathbb{R}^n$,

$$v_1 \wedge \cdots \wedge v_{n-1} \to v_1 \times \cdots \times v_{n-1},$$

obtained as follows: Given $n-1$ vectors $v_1, \ldots, v_{n-1}$ in $\mathbb{R}^n$, row vectors of M, its *cross product* denoted as $v_1 \times v_2 \times \cdots \times v_{n-1}$ is the vector whose expression in the canonical basis is the determinant of the matrix N obtained by completing M with a first row $e_1, \ldots, e_n$. Using this first row to compute the determinant, the coefficients of the cross product are thus the n minors of order $(n-1) \times (n-1)$ of M with the corresponding signs. Evidently, this vector is zero if the vectors are linearly dependent. Let us find out what this vector is otherwise. Note that for an arbitrary vector v, the scalar product $\langle v_1 \times v_2 \times \cdots \times v_{n-1}, v\rangle$ is the determinant of the matrix obtained by completing M with a first row with the components of v in the canonical basis. In particular, $\langle v_1 \times v_2 \times \cdots \times v_{n-1}, v_i\rangle, i = 1, 2, \ldots, n-1$ equals the determinant of a matrix with two equal rows, which is zero. Therefore, $v_1 \times v_2 \times \cdots \times v_{n-1}$ is orthogonal to each v_i and so to the subspace of dimension $n-1$ they span. Applied to $v = v_1 \times v_2 \times \cdots \times v_{n-1}$, it follows that $|v_1 \times v_2 \times \cdots \times v_{n-1}|^2$ is the sum of the squares of the n minors of order $n-1$ of M. In conclusion, $v_1 \times v_2 \times \cdots \times v_{n-1}$ is zero iff the vectors are linearly dependent; otherwise, it is perpendicular to all of them and has length

$$|v_1 \times v_2 \times \cdots \times v_{n-1}| = m_{n-1}(P(v_1, \ldots, v_{n-1})).$$

1.3.3 For further use, we analyze how a linear map $A : \mathbb{R}^n \to \mathbb{R}^m$ with matrix A in the canonical basis transforms volumes of parallelepipeds. The quantity,

$$J_k(A) = \sup \frac{m_k(P(Av_1, \ldots, Av_k))}{m_k(P(v_1, \ldots, v_k))} = \sup \frac{|Av_1 \wedge \cdots \wedge Av_k|}{|v_1 \wedge \cdots \wedge v_k|},$$

where the supremum is among all linearly independent $v_1, \ldots, v_k$, is called the *k-th Jacobian of A*. In terms of the linear map

$$\Lambda^k A : \Lambda^k\mathbb{R}^n \to \Lambda^k\mathbb{R}^m, \quad (\Lambda^k A)(v_1 \wedge \cdots \wedge v_k) = Av_1 \wedge \cdots \wedge Av_k,$$

$J_k(A)$ is the norm of $\Lambda^k A$, in particular the supremum is achieved.

Note that the above ratio among measures only depends on the linear space spanned by v_i, for if $v_1, \ldots, v_k$ and $w_1, \ldots, w_k$ span the same subspace and $v_j = \sum_i b_{ij} w_i, V = WB$ in matrix notation, then

$$G(v_1, \ldots, v_k) = BG(w_1, \ldots, w_k)B^t,$$

and thus,

$$m_k(P(v_1, \ldots, v_k)) = |\det B|\, m_k(P(w_1, \ldots, w_k)).$$

But the same linear relation holds between $Av_1, \ldots, Av_k$ and $Aw_1, \ldots, Aw_k$. Or using the exterior algebra notation,

$$v_1 \wedge \cdots \wedge v_k = \lambda w_1 \wedge \cdots \wedge w_k.$$

Thus, we may think that each k-dimensional subspace V has a distortion factor, given, for instance, by $m_k(P(Aw_1, \ldots, Aw_k))$ with $w_1, \ldots, w_k$ an orthonormal basis of V, and $J_k(A)$ is the maximum of those achieved at some V.

If r is the rank of A, of course $m_k(P(Av_1, \ldots, Av_k)) = 0$ if $k > r$. For $k \leq r \leq \min(n, m)$, the matrix of $\Lambda^k A$ in the orthonormal basis e_I, e_J is an $\binom{n}{k} \times \binom{m}{k}$ matrix whose I, J entry is the minor of A consisting of the I columns and J rows. To compute $J_k(A)$, one would need to use the solution of Exercise 1.3.

In case $k = r = n \leq m$, A is a one-to-one map, the matrix of $\Lambda^k A$ is a column vector, and there is just one distortion factor equal to the square root of the sum of the squares of the $k \times k$ minors of A, that is,

$$J_k(A) = m_k(P(Ae_1, \ldots, Ae_k)) = |\det(A^t A)|^{\frac{1}{2}}.$$

In case $k = m = r < n$, A has a $(n-k)$-dimensional kernel N, so there is just one distortion factor in $N^\perp$, the matrix of $\Lambda^k A$ is a row vector and

$$J_k(A) = |\det(AA^t)|^{\frac{1}{2}}.$$

For a general parallelepiped P, clearly $A(P) = A(P')$, where P' is the orthogonal projection of P onto the orthogonal complement $N^\perp$ of N. From this, it follows that $J_k(A)$ is in this case the distortion factor in $N^\perp$.

So, we have completed the proof of the following.

Theorem 1.7. *Let $A : \mathbb{R}^n \to \mathbb{R}^m$ be a linear map, r its rank.*

(a) *If $r = n, m_r(A(P)) = |\det(A^t A)|^{\frac{1}{2}} m_r(P)$ for an r-dimensional parallelepiped P.*
(b) *If $r = m$, $m_r(A(P)) = |\det(AA^t)|^{\frac{1}{2}} m_r(P)$ for an r-dimensional parallelepiped P lying on the orthogonal of the kernel of A.*
(c) *If $r = n = m$, both distortion factors agree with $|\det A|$.*

Chapter 2

Continuous Functions

In this chapter, we first introduce two basic concepts, *completeness* and *compactness*. Both are the basis of *existence theorems* in Analysis, as illustrated by the fixed point theorem. Then we start dealing with functions defined in $\mathbb{E}^n$ and the way to visualize them, via *level sets*. The last section introduces *continuous functions*, the maps $y = f(x)$ for which small increments of x translate to small increments of y, and their basic properties.

2.1 Topological Aspects of Euclidean Space

2.1.1 Recall that $B(x,r) = \{y : d(x,y) < r\}$ is called the *open ball* centered at x of radius r. The set $S(x,r) = \{y : d(x,y) = r\}$ is called *the sphere* of center x and radius r and $\overline{B}(x,r) = \{y : d(x,y) \leq r\}$ is called the *closed ball*.

The distance $d(x,y)$ quantifies proximity among points. However, there are other ways to define the length of a vector, the distance and the balls. In general, a *norm* ρ in $\mathbb{R}^n$ is a map $\rho : \mathbb{R}^n \to [0,+\infty)$ satisfying

(1) $\rho(w+v) \leq \rho(w) + \rho(v)$.
(2) $\rho(\lambda v) = |\lambda| \rho(v)$.
(3) $\rho(v) = 0$ if and only if $v = 0$.

Note that this implies as before

$$|\rho(w) - \rho(v)| \leq \rho(w - v).$$

Associated to ρ there is a distance $d_\rho(x,y) = \rho(x-y)$ and a family of balls $B_\rho(x,r) = \{y : d_\rho(x,y) < r\}$. Examples of norms are

$$|v|_1 = |v_1| + |v_2| + \cdots + |v_n|, \quad |v|_\infty = \max(|v_1|, |v_2|, \ldots, |v_n|).$$

In fact, there is a whole scale of norms depending on a parameter $p \geq 1$

$$|v|_p = (|v_1|^p + |v_2|^p + \cdots + |v_n|^p)^{\frac{1}{p}}.$$

Exercise 2.1.

(a) If $a, b > 0, 1 < p, q < \infty, 1/p + 1/q = 1$, prove *Young's inequality*

$$ab \leq \frac{a^p}{p} + \frac{b^q}{q}.$$

(b) If $v = (v_1, \ldots, v_n), w = (w_1, \ldots, w_n) \in \mathbb{R}^n$, prove *Holder's inequality*

$$\sum_1^n |v_i w_i| \leq |v|_p |w|_q.$$

(c) Prove *Minkowski's inequality*

$$|v + w|_p \leq |v|_p + |w|_p,$$

using Holder's inequality for $|v_i + w_i|^{p-1}, |v_i|, |w_i|$.

A set E is called *convex* if with every two points $x, y \in E$ the segment joining x, y is contained in E, that is, $tx + (1-t)y \in E, 0 \leq t \leq 1$. As a consequence of the triangle inequality, balls associated to norms are convex. The balls defined by $|v|_p$ with $p < 1$ are not convex, as shown in Figure 2.1, and hence $|v|_p$ is not a norm if $p < 1$.

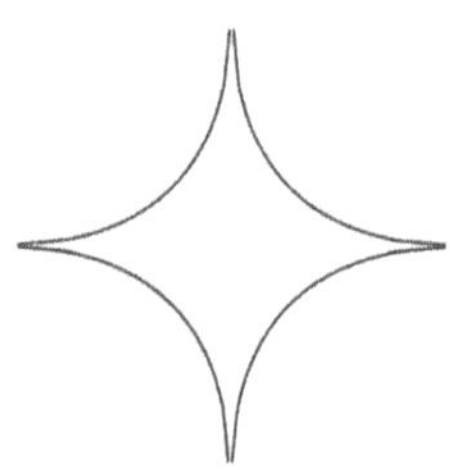

Figure 2.1. Ball for the $|v|_p$-norm, $p < 1$.

Note that if ρ is a norm and L is a one-to-one linear map, then $\rho(Lv)$ is also a norm. In particular,

$$\left(\sum_i \lambda_i |v_i|^q\right)^{\frac{1}{q}}$$

is a norm for all choices of $\lambda_i > 0$.

One may ask why among all possible norms we choose working with $|v|_2 = |v|$. A possible explanation is that this norm is associated with a scalar product, and that computations with this norm are easier. But in certain situations a different norm might be more suitable. When displacements are allowed only in vertical or horizontal directions (as it is the case in some quarters of Barcelona and many other towns), the norm $|v|_\infty$ and the corresponding distance might be more appropriate to compute distances. In spite of this apparent diversity, the fact is that as a measure of proximity all these norms and distances are equivalent as shown by the following theorem to be proved later.

Theorem 2.1. *If ρ is a norm, there are two constants $m, M > 0$ such that $m|v| \le \rho(v) \le M|v|$ for all v, and so*

$$m\, d(x, y) \le \rho(x - y) \le M\, d(x, y). \tag{2.1}$$

As a consequence, two arbitrary norms ρ_1, ρ_2 are equivalent, meaning that

$$m\, \rho_1(v) \le \rho_2(v) \le M\, \rho_1(v)$$

for some constants $m, M > 0$.

We can check this directly for the norms $|v|_p$. It is obvious that $|v_i| \le |v|_p$ and so for $p_1, p_2 \ge 1$,

$$|v|_{p_1} \le n^{\frac{1}{p_1}} |v|_{p_2},$$

and interchanging p_1, p_2,

$$n^{-\frac{1}{p_2}} \le \frac{|v|_{p_1}}{|v|_{p_2}} \le n^{\frac{1}{p_1}}.$$

The following exercise is to be compared with Theorem 1.1.

Exercise 2.2. Prove that the maps $T : \mathbb{R}^2 \to \mathbb{R}^2$ such that $|T(v)|_1 = |v|_1$ or $|T(v)|_\infty = |v|_\infty$ for all v form a finite group and identify it.

2.1.2 A useful concept in analysis is that of a sequence and *convergent sequence*. The concept of sequence arises in iterative processes, for instance, when approximating the root of an equation by Newton's method. It is also useful when proving certain results by contradiction. Formally, a sequence in $\mathbb{R}^n$ is a map $\mathbb{N} \longrightarrow \mathbb{R}^n$, written simply as an ordered infinite string $x^0, x^1, x^2, \ldots, x^k, \ldots$ of points in $\mathbb{R}^n$. It is important to distinguish between the set of points $\{x^k\}$, where the order is irrelevant, and the sequence (x^k). Put differently, a sequence consists of the points and the *order* in which they appear.

A sequence (x^k) is said to be *convergent* with limit point x, written $x^k \to x$, if for arbitrary $\varepsilon > 0$ (that we should think very small) there exists $k_0 \in \mathbb{N}$ such that $d(x^k, x) < \varepsilon$ for $k > k_0$. Of course, if the limit exists, it is unique. Informally, one says that the points x^k are *arbitrarily close* to x. Since

$$|x_i^k - x_i| \leq d(x^k, x) \leq \sum_i |x_i^k - x_i|,$$

$x^k \to x$ iff the coordinates (x_i^k) have limit point x_i in $\mathbb{R}$, for each $i = 1, \ldots, n$. A sequence (x^k) is said to *escape to infinity* if $|x^k| \to +\infty$, that is, for every ball $B = B(0, R)$ there is k_0 such that $x^k \notin B$ for $k \geq k_0$.

A *partial sequence* or *subsequence* (y^m) of (x^k) consists in selecting the terms corresponding to an increasing set of indexes $k_1 < k_2 < \cdots < k_m < \cdots$, $y^m = x^{k_m}$. It is evident that if x^k has limit x, so does a partial sequence.

All distances defined by norms being equivalent, the notion of convergence does not depend on which one is used. This notion has all natural properties regarding sums and multiplication by scalars,

$$x^k \to x, y^k \to y, \lambda_k \to \lambda \implies x^k + y^k \to x + y, \lambda_k x^k \to \lambda x.$$

2.1.3 A fundamental property of the Euclidean space is its *completeness*, which is the base of many existence proofs. One may define this concept in terms of *Cauchy sequences*; these are the sequences (x^k) such that $d(x^k, x^m) \to 0$ when $k, m \to \infty$, that is, terms in the sequence are asymptotically arbitrarily close. Completeness means that every Cauchy sequence has a limit point. Since (x^k) is a Cauchy sequence if and only if each coordinate sequence (x_i^k) is, completeness of $\mathbb{R}^n$ amounts to completeness of the real line $\mathbb{R}$.

Using this concept we may prove the following useful existence theorem:

Theorem 2.2. *Assume that* $T : \mathbb{R}^n \to \mathbb{R}^n$ *is* contractive, *i.e., it satisfies*

$$d(T(x), T(y)) \leq c\, d(x, y), \quad x, y \in \mathbb{R}^n, \quad 0 \leq c < 1.$$

Then there is a unique point p *fixed by* $T, T(p) = p$. *Moreover,* p *is an* attractor point, *meaning that for an arbitrary starting point* x, *the sequence of iterates* $x^k, x^k = T(x^{k-1}), x^0 = x$ *converges to* p. *Also, if some iterate* T^k *of* T *is contractive,* T *has a unique fixed point.*

Proof. Uniqueness of p is clear: if $T(p) = p, T(q) = q$, then

$$d(p, q) = d(T(p), T(q)) \leq c\; d(p, q), \quad c < 1,$$

implies $d(p, q) = 0$. Now we prove that the sequence of iterates x^k is a Cauchy sequence; from

$$d(x^k, x^{k+1}) = d(T(x^{k-1}), T(x^k)) \leq c\, d(x^{k-1}, x^k),$$

by iteration it follows that

$$d(x^k, x^{k+1}) \leq c\, d(x^{k-1}, x^k) \leq c^2 d(x^{k-2}, x^{k-1}) \leq \cdots \leq c^k d(x^0, x^1),$$

and therefore, if $m > k$

$$\begin{aligned} d(x^k, x^m) &\leq (d(x^k, x^{k+1}) + d(x^{k+1}, x^{k+2}) + \cdots + d(x^{m-1}, x^m) \\ &\leq d(x^0, x^1) \sum_{i=k}^{m-1} c^i \leq \frac{c^k}{1-c} d(x^0, x^1) \to 0. \end{aligned}$$

If p is a limit point of x^k, it follows that

$$d(x^{k+1}, T(p)) = d(T(x^k), T(p)) \leq c\; d(x^k, p),$$

whence $T(p) = p$. If T^k is contractive, it has a unique fixed point p; then $T^k(T(p)) = T(T^k(p)) = T(p)$, so $T(p)$ is also fixed by T^k, whence $T(p) = p$ and p is a fixed point of T. Every point fixed by T is fixed by T^k, so unicity is obvious. □

For $n = 1$, by the mean-value theorem, if T is differentiable and $|T'| \leq c$, then T is contractive. In Section 4.6, we will extend the mean-value theorem to functions of several variables.

Example 2.1. Assume T is contractive and define $F(x) = x - T(x)$. Then we claim that F is a bijection from $\mathbb{R}^n$ into itself. First, if $F(x) = F(y)$, then $x - y = T(x) - T(y), d(x, y) = d(T(x), T(y)) \leq cd(x, y)$ implies $x = y$. Now we must see that $F(p) = y$ has a solution p for every y. This means that $p = y + T(p)$, that is p is a fixed point for the map $T'(x) = y + T(x)$. Since the later is also contractive, the result follows. For instance, if $g(t)$ is contractive in one variable, we may take in $n = 2, T(x, y) = (g(y), g(x))$ and conclude that the system

$$x + g(y) = a, \quad y + g(x) = b,$$

has a unique solution (x, y) for every (a, b).

In case T is a linear map in the above example with $|T| < 1$, the iterates of T' are, say starting at 0,

$$T'0 = y, y + T(y), y + T(y) + T^2(y), \ldots, \sum_{j=0}^{k} T^j(y),$$

leading to the *Neumann series* $\sum_{j=0}^{\infty} T^j(y)$. Thus,

$$(I - T)^{-1} = \sum_j T^j.$$

A *metric space* is a set X endowed with a distance d, that is a map

$$d : X \times X \to [0, +\infty),$$

satisfying the triangle inequality

$$d(x, y) \leq d(x, z) + d(z, y), \quad x, y, z \in X,$$

and $d(x, y) = d(y, x), d(x, y) = 0$ if and only if $x = y$. Cauchy and convergent sequences are defined in the same way. A metric space is called *complete* if every Cauchy sequence is convergent to a point in the space.

The previous theorem and its proof goes over to a general complete metric space and is the basis of important existence and uniqueness theorems, for instance, on solutions of Cauchy problem for ordinary differential equations, see Section 8.3.

2.1.4 For a given non-empty set $A \subset \mathbb{R}^n$, we shall consider the following classification of points:

(a) The *interior points* are those $x \in A$ for which there is a ball $B(x,r) \subset A$. The set of interior points is called the *interior* of A and denoted $\mathring{A}$.
(b) The *exterior points* are the interior points of the complement of A, that is those x for which there exists a ball not meeting A. They constitute the *exterior* of A, denoted A^e.
(c) The points that are neither interior nor exterior are the *boundary points*; for those, every ball $B(x,r)$ meets both A and $\mathbb{R}^n \setminus A = A^c$. They constitute the *boundary* of A, denoted $b(A)$.

Of course, in all definitions one may replace the ball $B(x,r)$ by the cube $Q(x,r)$ centered at x of size r, or the balls for any other norm. These three sets are pairwise disjoint. The complement of the exterior, that is the union of the interior and the boundary, consists of points x for which every ball $B(x,r)$ meets A. They constitute the *closure* of A, denoted $\overline{A}$. Among the points of $\overline{A}$, the *accumulation points* x are those for which every ball $B(x,r)$ contains points of A different from x itself; equivalently, every ball $B(x,r)$ contains infinitely many points in A. The set of accumulation points is denoted by A'. It follows that the points of $\overline{A}$ that are not accumulation points are those points $x \in A$ for which there is a ball $B(x,r)$ such that x is the only point of A in $B(x,r)$. These are called *isolated* in A.

The set A is called *open* if $\mathring{A} = A$, and *closed* if $\overline{A} = A$ (equivalently, $A' \subset A$). Thus, A is open if whenever $x \in A$ there is a cube $Q(x,r) \subset A$. The meaning of this property is that around each point the set A has n degrees of freedom, because the coordinates of the points in A may vary freely and with no relationship among them in an interval. The interior $\mathring{A}$ of a set A is open, it is in fact the largest open set included in A. Similarly, the closure $\overline{A}$ is the smallest closed set containing A. Since

$$\overline{A} = \mathring{A} \cup b(A) = \mathbb{R}^n \setminus A^e, \quad \overline{\mathbb{R}^n \setminus A} = \mathbb{R}^n \setminus \mathring{A},$$

A is open if and only if A^c is closed. It is straightforward to check that the union $\cup_i A_i$ of open sets A_i is open, while the intersection $\cap_i A_i$ of a finite number of open sets is open. Equivalently, the intersection of closed sets is closed, and a finite union of closed sets is closed.

A set A consisting of just isolated points is called *discrete*. This is the opposite situation of open sets, around each point $x \in A$ there are zero degrees of freedom within A.

As an example, every set A defined in terms of a strict inequality like

$$A = \{x : f(x_i) < g(x_j), x = (x_1, \ldots, x_n)\},$$

where f, g are continuous functions of one variable, is open. To prove this, suppose that $x \in A$ and let $\tau = g(x_j) - f(x_i)$. By the continuity of f, g there exists δ such that $|f(t) - f(x_i)| < \frac{\tau}{2}, |g(t) - g(x_j)| < \frac{\tau}{2}$ if $|t - x_i| < \delta$. It follows that if $y \in \mathbb{R}^n$ satisfies $|y_i - x_i| < \delta$ for all i, then $f(y_i) < g(y_j)$ and so $y \in A$. As a consequence, every set defined by an arbitrary number of inequalities of type $f_i(x_i) \leq f_j(x_j), x = (x_1, \ldots, x_n)$, is closed. A point $x \in \overline{A}$ satisfies $f(x_i) \leq g(x_j)$. If f, g are, say, strictly monotone, then conversely, if $f(x_i) \leq g(x_j)$, then $x \in \overline{A}$.

Closed sets can be described using sequences; namely, it is straightforward to prove that $x \in \overline{A}$ if and only if there is a sequence (x^k) of points in A such that $x^k \to x$, and x is an accumulation point if and only if the points x^k can be chosen all different. Therefore, A is closed if and only if $x^k \in A, x^k \to x$ implies $x \in A$, a property that somehow explains the terminology.

The closure of a set A can be defined too using the *distance function* $d(x, A)$,

$$d(x, A) = \inf\{d(x, y), y \in A\}.$$

In terms of this function, note that $x \in \overline{A}$ means exactly that $d(x, A) = 0$. It is trivial that $d(x, A) = d(x, \overline{A})$. Therefore, when A is closed, $d(x, A) > 0$ if and only if $x \notin A$; we will see later in Theorem 2.7 that there exists $y \in A$ such that $d(x, A) = d(x, y)$. Taking the infimum in $y \in A$ in the triangle inequality (1.5), we get that for arbitrary x, y one has

$$d(x, A) \leq d(x, y) + d(y, A),$$

and so

$$|d(x, A) - d(y, A)| \leq d(x, y). \tag{2.2}$$

Exercise 2.3. Check that for fixed $A, r > 0$ the sets

$$B = \{x : d(x, A) < r\}, \quad C = \{x : d(x, A) \leq r\}$$

are, respectively, open and closed. Prove that $\overline{B} = C$ and that $b(B) = \{x : d(x, A) = r\}$.

On an intuitive basis, if we think of countries as sets, the mathematical boundary is the political boundary; in most examples the boundary $b(A)$ is "thin", consisting of lines, arcs, etc. But in general the boundary need not be so, it can be a "fat" set. For instance, if $A = \mathbb{Q}^n$, the set of points with rational coordinates, then both the interior and the exterior are empty, and $b(A) = \mathbb{R}^n$.

If $B \subset A$ satisfies $A \subset \overline{B}$ (that is, all points in A can be approached by points of B), one says that B is *dense* in A. Thus, the sets B dense in the whole space $\mathbb{R}^n$ are those that intersect every ball $B(x, r)$. The typical example is the set $\mathbb{Q}^n$ of points with rational coordinates.

2.1.5 Another concept that we will need is that of *bounded set.* A set A is said to be bounded if it is contained in a ball that may be assumed centered at the origin, e.g., there exists M such that $|x| \leq M$ for all $x \in A$.

Finally, an open set U is said to be *connected* if it cannot be written as union of two non-empty open sets. The meaning of this will be made clear in paragraph 3.4.4. Connected open sets are also called *domains.* The domains in case $n = 1$ are exactly the open intervals. If we declare that two points $x, y \in U$ are related whenever there exists a domain $V \subset U$ containing both x, y, the equivalence classes are the *connected components of* U. These are open and disjoint, fill up the whole of U and the component containing x is the largest domain containing x and included in U. Since $\mathbb{Q}^n$ is a countable dense set, an open set U has a countable number of connected components.

More generally, a set $A \subset \mathbb{R}^n$ is called *connected* if whenever $A = A_1 \cup A_2$, with $A_i = A \cap U_i$, with U_i open, $U_1 \cap U_2 = \emptyset$, either A_1 or A_2 is empty.

2.2 Compact Sets

2.2.1 In dimension $n = 1$, a sequence (x^k) either is unbounded, e.g., it has a subsequence with limit ∞, or else is bounded, and in this case it has a monotone convergent subsequence. As a consequence, a bounded sequence has a convergent subsequence. This implies that every infinite bounded set $A \subset \mathbb{R}$ has finite accumulation points. We state this for arbitrary dimension:

Theorem 2.3. *Every bounded sequence* (x^k) *in* $\mathbb{R}^n$ *has a convergent subsequence.*

Proof. We consider the sequence (x_1^k) of first components of (x^k); being bounded it has a convergent subsequence $(x_1^{k_m})$. Thus, $y^m = x^{k_m}$ is a subsequence of (x^k) and the sequence of first components is convergent. Now we consider (y_2^m), the sequence of second components of (y^m), also bounded with a convergent subsequence $(y_2^{m_p})$ and so on. In n steps we get a subsequence (x^{k_m}) of (x^k) such that all components $(x_i^{k_m}), i = 1, 2, \ldots, n$ are convergent, and hence it is convergent. □

As a consequence, every infinite bounded set $A \subset \mathbb{R}^n$ has accumulation points. Intuitively, this means that if one is supposed to mark with a cross, with a fountain pen, a point in a bounded set, then a different one, then another and so on, the different crosses will accumulate somewhere.

In certain parts of the course we will use the following concept. We say that a set K is *sequentially compact* if every sequence (x^k) of points in K has a partial sequence convergent to a point in K.

Theorem 2.4. *A set K is sequentially compact if and only if it is bounded and closed.*

Proof. Assume K bounded and closed and let (x^k) be a sequence of points in K. By the previous theorem, there is a subsequence convergent to some point x; but K being closed, necessarily $x \in K$.

Conversely, assume K sequentially compact and let us prove that K is closed and bounded by contradiction. If it were not bounded, for each $k \in \mathbb{N}$ there would exist $x^k \in K$ such that $|x^k| > k$, and it is evident that the sequence (x^k) cannot have a convergent subsequence. Assume now that $x \in \overline{K}$, that is, there is a sequence (x^k) of points in K convergent to x. By hypothesis, (x^k) has a subsequence (x^{k_m}) convergent to some point $p \in K$. But being a subsequence of (x^k), it has limit point x, and so $x = p \in K$. □

2.2.2 There is another useful characterization of sequentially compact sets in terms of *open coverings*. An open covering of a set A is a collection U_i of open sets such that $A \subset \cup_i U_i$.

A set K is called *compact* if every open covering of K by open sets U_i has a finite sub-covering, meaning that already a finite number of them cover K. We will now state this property in terms of subsets of K, for which purpose we introduce a new definition. A subset $F \subset K$ is called *closed in K* if $K \cap \overline{F} = F$; this amounts to saying that it can be written as $F = K \cap G$ with G closed. Then, if U_i is open, $G_i = U_i^c$ is closed and $F_i = G_i \cap K$ is closed in K. The fact that U_i cover K is equivalent to $\cap_i F_i = \emptyset$. The fact that a finite number of U_i cover K means that a finite number of F_i are disjoint. Hence, K is compact if and only if for every collection F_i of closed sets in K with empty intersection, there exists a finite sub-collection with empty intersection. Or equivalently, K is compact if $\cap_i F_i \neq \emptyset$ for every collection F_i of closed sets in K satisfying $\cap_{j\in J} F_j \neq \emptyset$ for all finite $J \subset I$.

Our purpose is to prove that a set is compact if and only if it is sequentially compact. First we will see that all open coverings can be assumed to be countable.

Theorem 2.5. *If (U_i) is an arbitrary family of open sets, there is a countable subfamily (U_{i_k}) with the same union.*

Proof. We consider the family of all balls having centers with rational coordinates and rational radius. It is countable, and hence we may put all of them in a list $(B_k), k \in \mathbb{N}$. If $x \in U_i$, there is a ball $B(x,r) \subset U_i$. We claim that there is a ball $B(q,s)$ with rational center and rational radius such that $x \in B(q,s) \subset B(x,r) \subset U_i$: indeed, take first $q \in \mathbb{Q}^n$ such that $|x-q| < \frac{r}{4}$ and then a rational number $s, \frac{r}{4} < s < \frac{r}{2}$. Therefore, we can associate to each $x \in \cup U_i$ a natural number $k(x) \in \mathbb{N}$ such that $x \in B_{k(x)}$, and $B_{k(x)}$ is included in some U_i. □

Theorem 2.6. *For a non-empty set K in $\mathbb{R}^n$, the following are equivalent:*

(a) *K is closed and bounded.*
(b) *K is sequentially compact.*
(c) *If F_k is a decreasing sequence, $F_{k+1} \subset F_k$, of non-empty closed sets in K, then $\cap_k F_k$ is non-empty. Equivalently, if $\cap_k F_k$ is empty, some of the F_k is empty, too.*
(d) *K is compact.*

Proof. We already know that (a), (b) are equivalent. Let us check that (b) implies (c). Assume that K is sequentially compact and that F_k are as above. We construct a sequence by choosing $x^k \in F_k$; in particular, $x^m \in F_k, m > k$. This sequence has a subsequence convergent to a point $x \in K$; since x is the limit of the sequence (x^m), $m > k$, and F_k is closed in K, one has $x \in F_k$ for all k.

To check that (c) implies (d), consider an open covering of K; by Theorem 2.5 we may assume that it is a countable covering (U_k); by setting $V_k = U_1 \cup \cdots \cup U_k$ we obtain an increasing sequence of open sets. Then $F_k = K \cap V_k^c$ are closed sets in K with empty intersection. Therefore, some F_k is empty, which means that $K \subset V_k$ for some k.

Finally, we see that (d) implies (a). The collection of balls $B(0,k)$ covers K and so a finite family, say $B(0,k_1),\dots,B(0,k_p)$ cover, too, and therefore $K\subset B(0,\max(k_1,\dots,k_p))$ is bounded. Assume it is not closed and consider $x\in\overline{K}, x\notin K$; we consider the family of balls $B(y,d(x,y)/2), y\in K$, which obviously cover K. A finite number of them $B(y_j,d(x,y_j)/2)$ cover K. Set $r=\min\{d(x,y_j)/2\}$; then if $y\in K$, $y\in B(y_j,d(x,y_j)/2)$ for some j and hence

$$d(x,y_j)\le d(x,y)+d(y,y_j)\le d(x,y)+\frac{1}{2}d(x,y_j),$$

implying $d(x,y)\ge\frac{1}{2}d(x,y_j)\ge r$ for all $y\in K$. This means that the ball $B(x,r)$ does not meet K, in contradiction with $x\in\overline{K}$. □

Note that compact discrete sets are necessarily finite. If A is isolated and closed, then $A\cap\overline{B}(0,N)$ is finite and so A is at most countable.

Exercise 2.4. Prove that if K,L are compact sets in $\mathbb{R}^n,\mathbb{R}^m$, respectively, $K\times L$ is compact in $\mathbb{R}^{n+m}$.

2.2.3 In this paragraph we see a couple of consequences. First, we prove that the distance from a point to a closed set F is attained at some point. In the second statement, the *distance between two sets* A,B is defined as

$$d(A,B)=\inf_{x\in A,y\in B}d(x,y)=\inf_{x\in A}d(x,B)=\inf_{y\in B}d(y,A).$$

Note that $d(A,B)$ may be zero even if A,B are disjoint (for instance if A is the positive x-axis and B is the hyperbola $y=\frac{1}{x}$).

Theorem 2.7.

(a) *If F is a closed set and $x\notin F$, there exists $y\in F$ such that $d(x,y)=d(x,F)$.*
(b) *If F is closed and K is compact, $F\cap K=\emptyset$, there are $x\in F, y\in K$ such that $d(F,K)=d(x,y)>0$.*

Proof. Since F is closed and $x\notin F$, $d(x,F)=d>0$. The set $K=F\cap\overline{B}(x,2d)$ is closed and bounded, whence compact and $d(x,F)=d(x,K)$. By definition of $d(x,K)$, there are $x^k\in K$ such that $d(x,x^k)\to d$. The sequence (x^k) has a subsequence (x^{k_m}) convergent to some point $y\in K$, and by (1.6) $d(x,x^{k_m})\to d(x,y)$, so that $d(x,y)=d$. For the second statement, if $d=d(F,K)$, by definition there are $x^k\in K$ such that $d(x^k,F)\to d$. Again, if $x\in K$ is the limit of some subsequence of (x^{k_m}), by (2.2) one has $d(x,F)=\lim d(x^{k_m},F)=d$. □

The number appearing in the next theorem is called the *Lebesgue number* of the covering.

Theorem 2.8. *If (U_j) is an open covering of a compact set K, there is $\varepsilon > 0$ such that every ball $B(x, \varepsilon), x \in K$ is included in some U_j.*

Proof. For each $y \in K$ let $r(y) > 0$ and $k(y)$ such that $B(y, r(y)) \subset U_{k(y)}$. The family of balls $B(y, r(y)/2), y \in K$ covers K, whence a finite number $B(y_j, r(y_j)/2)$ cover K, too. Then $\varepsilon = \frac{1}{2}\min\{r(y_j)\}$ has the desired property: if $x \in K$ is in $B(y_j, r(y_j)/2$, then $B(x, \varepsilon) \subset B(y_j, r(y_j)) \subset U_{k(y_j)}$. □

2.3 Functions of Several Variables, Level Sets

2.3.1 A real (or scalar) function of n variables is a rule that assigns a real number $f(p)$ to each point $p \in \mathbb{E}^n$ or in a certain subset U, called the *domain of definition* of f. The value $f(p)$ may be given by a formula involving elementary functions and the coordinates $p = (x_1, x_2, \dots, x_n)$ in the canonical reference system or in any other reference system, for instance

$$f(p) = x_1 + x_2^2 + \cdots + x_n^2, \quad f(p) = \sin xy, \quad f(p) = \frac{xyz}{\sin xyz}.$$

Evidently, this expression will change when we change the reference system. In those cases, it is implicitly assumed that the domain of definition is the set of points for which the formula makes sense; for instance, in the last example the domain is $xyz \neq 0$. Later on we will discuss how to extend the definition to other points.

But a function may be given not by a formula but through an empiric observation. For instance, $f(x, y, z)$ may be the height above sea level of the point with geographical coordinates x, y or the atmospheric pressure at height z over this point, etc. A particularly important case in applications is when one of the variables is time; in this case we often use the notation $u(x, t), x = (x_1, \dots, x_n)$ to denote the value of some physical quantity at time t and at point x, involved in some physical phenomena. In these cases, in fact, we do not have in practice the value of $u(x, t)$ for all values of x, t, but for *sampled* values of x, t obtained at some sampling rate. Still, when modeling the phenomena we consider $u(x, t)$ defined for all values. There are two examples that are worth keeping in mind.

The first one is the description of a *vibrating string*, for instance that of a violin. Assume it has length L and that at time $t = 0$ it is forced to have the profile of a certain function $f_0(x), 0 \le x \le L$, that is, $(x, f_0(x))$ is the initial position, $f_0(x)$ being the vertical displacement from the equilibrium position. When free, the string vibrates and we define $u(x,t)$ as the vertical displacement, at time t, of the point initially at $(x, f_0(x))$. Said differently, the position of the string at time t is the graphic of $u(x,t), 0 \le x \le L$. Thus, one has $u(x,0) = f_0(x)$ and, on an intuitive basis, the initial position completely determines $u(x,t)$ for all t.

A second important example is the description of *heat flow*. Assume that a metallic bar of length L is heated in some way so that the initial distribution of temperature is $f_0(x), 0 \le x \le L$; keeping both ends isolated, we ask ourselves what will be the distribution of temperatures $u(x,t)$ at point x at time t. Again, on an intuitive basis, the initial temperature $f_0(x) = u(x,0)$ should determine $u(x,t)$ for all t.

Of course, empirical functions are not known exactly at all points. For instance, the position $u(x,t)$ is not known at all times and for all positions x. What we know are *samples* $u(x_i, t_i)$ taken at uniformly spaced points x_i or instants t_i. Still, in mathematics we consider that an *idealized* analogical function $u(x,t)$ exists, and make models with it.

Coming back to functions defined by formulas, the simplest functions are, besides constants, the linear functions in the coordinates

$$f(x_1, x_2, \ldots, x_n) = c_1x_1 + c_2x_2 + \cdots + c_nx_n.$$

Next, we may consider the *quadratic functions*, polynomials of degree two, in the coordinates

$$f(x) = \sum_{i,j=1}^{n} c_{ij}x_ix_j + \sum_{i=1}^{n} b_ix_i + d.$$

2.3.2 In dimension $n = 1$, it is usual to visualize f using its *graph*, the set of points $(x, f(x))$ in the plane, that we simply denote by $y = f(x)$. In the same way, if f is a function of two variables, the graph is the set of points $(x, y, f(x,y))$ in the space, denoted $z = f(x,y)$. Figure 2.2 is the graph of the sinus of a radial function.

An alternative way to visualize a function f of n variables defined in a set U is by means of the *level sets*. If f is a scalar function defined in U, the set

$$L_c = \{x \in U : f(x) = c\},$$

when non-empty (that is if c is a value of f) is called a *level set*. Obviously, U is then the disjoint union of all level sets of f; each point $p \in U$ belongs

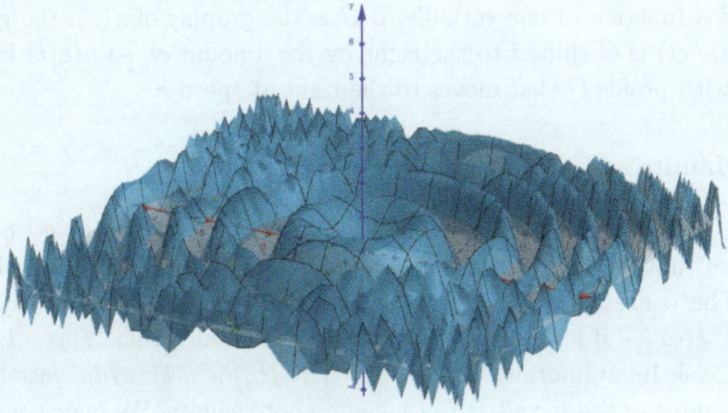

Figure 2.2. Graph of the sinus of a radial function.

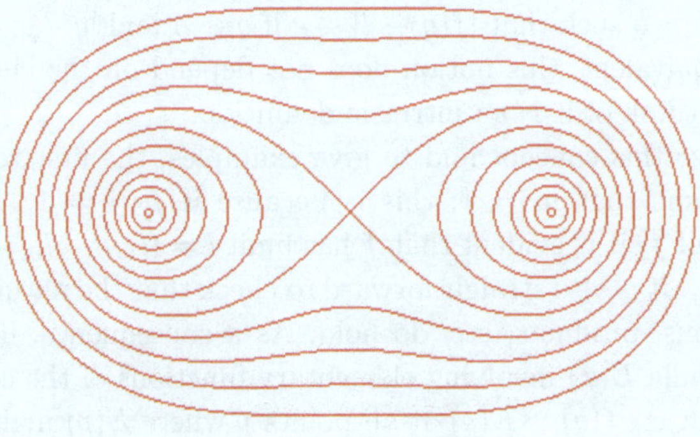

Figure 2.3. The Cassini curve.

to one and only one of the level sets, namely the one with $c = f(p)$, the set of points where f takes the same value as in p. A typical example is when f indicates the height above sea level of a point in a certain geographical region; the level sets are then the curves that we can see in old maps. When $n = 3$, we call them *level surfaces* and when $n = 2$, *level curves.* Figure 2.3 shows the *Cassini curves*, the loci of points such that the product of distances to two given points is constant.

For functions $u(x, t)$ depending on time, it is more intuitive to visualize $u(x, t)$ for fixed t, and then vary t so that we have not a still picture but a motion. The simplest example is

$$u(x, t) = f(x - ct), \quad x \in \mathbb{R},\ t \geq 0,$$

with f a function of one variable. If G is the graphic of $f(x)$, the graphic of $f(x-ct)$ is G shifted to the right by the amount ct, so $u(x,t)$ is like a wave with profile G that moves to the right at speed c.

2.4 Limits and Continuity

2.4.1 In this section, we use the notation $B'(p,r) = \{q : 0 < |p-q| < r\}$. Assume that f is defined in $B'(p,r)$ (not necessarily in p) with values in $\mathbb{R}^m$. The concept of limit is the same as in one variable: one says that $\lim_{q\to p} f(q) = l$ if for all $\varepsilon > 0$ there exists $\delta > 0$ such that $|f(q) - l| < \varepsilon$ if $|q - p| < \delta$. In an informal way, we say that $f(q)$ is *arbitrarily close* (this is the meaning of the *for all* ε) to l when q approaches p. We may say as well that l is the constant function that best approximates f near p.

More generally, if f is defined on a general set $A \subset \mathbb{R}^n$ and p is an accumulation point of A, we say that $\lim_{q\to p, q\in A} f(q) = l$ if for all $\varepsilon > 0$ there exists $\delta > 0$ such that $|f(q) - l| < \varepsilon$ if $q \in A$ and $|q-p| < \delta$. Since all norms are equivalent, this notion does not depend on the choice of norm, and we may think of it is an intrinsic definition.

To analyze this concept and to give examples, the first remark is that we may assume that $m = 1$; this is because if $f_j, j = 1, \ldots, m$ are the components of f, it is evident that f has limit $l = (l_1, \ldots, l_m)$ if and only if f_j has limit l_j. It is also straightforward to check that the natural properties regarding sums, products, etc. do hold. As a consequence, if f is defined by some formula $E(x)$ involving elementary functions of the coordinates of x, one has $\lim_{q\to p} f(q) = E(p)$ at all points p where $E(p)$ makes sense. For instance, in $\mathbb{R}^3$, with $q = (x,y,z)$

$$\lim_{q\to 0} \frac{1 + \cos xyz}{1 + x^2y^2 + z^2} = 2.$$

More interestingly, it may happen that $E(p)$ does not make sense and still the limit l exists, for instance

$$\lim_{q\to 0} \frac{\sin xyz}{xyz} = 1, \quad q = (x,y,z).$$

In the following, we assume $m = 1$ and f defined in $B'(p,r)$. The case $n > 1$ presents significative differences with respect to $n = 1$. Namely, in dimension $n = 1$ the variable q can approach p only in two ways, from the right or from the left; that's why it is customary to talk about *lateral limits*, and the existence of the limit amounts to the existence and equality of both lateral limits. In dimension $n > 1$, the variable q may approach p in an infinite number of ways. It may follow a line through p, or an arc of a parabola, in general moving in any set B having p as an accumulation point. The existence of a limit l means that *independently* of the way q approaches p, $f(q)$ must approach l. Given that there are infinite ways to approach, in some sense the existence of a limit when $n > 1$ is something richer. Another way of expressing this same idea is using sequences:

Proposition 2.1. *One has* $\lim_{q \to p} f(q) = l$ *if and only if* $f(q^k) \to l$ *for every sequence* $(q^k), q^k \neq p$ *with* $q^k \to p$.

Proof. That the condition is necessary is obvious. The proof of its sufficiency is by contradiction, a method that we will often use. If l is not the limit, some $\varepsilon > 0$ exists for which no $\delta > 0$ works. Thus, for $\delta = \frac{1}{k}$ we may find q^k with $0 < |q^k - p| < \frac{1}{k}$ and $|l - f(q^k)| \geq \varepsilon$. Then $q^k \to p$, but $f(q^k)$ does not converge to l. □

A case when the way q approaches p is irrelevant is when f satisfies

$$|f(q) - l| \leq \phi(|q - p|),$$

with ϕ a function of a real variable $t > 0$. If $\lim_{t \to 0} \phi(t) = 0$, then $\lim_{q \to p} f(q) = l$. For instance, in the plane, the limit of

$$f(x, y) = \frac{x^4 + y^3}{x^2 + y^2},$$

at the origin is zero: using polar coordinates we see that $f(x, y) = r^2 \cos^4 \theta + r \sin^3 \theta$, and therefore $|f(x, y)| \leq r^2 + r$. This example also shows that it might be useful to work in a convenient coordinate system (see section 3.1).

Just as in the one-variable case, we may define *directional limits.* If v is a unit vector, we may approach p through the semi-axis of origin p and direction v: $q = p + tv, t > 0$, and consider

$$\lim_{t \to 0} f(p + tv).$$

Obviously, if $\lim_{q \to p} f(q) = l$, all these limits must exist and be equal to l. For instance, consider in $n = 2$

$$f(x, y) = \frac{x^2 - y^2 + x^3}{x^2 + y^2}.$$

On $y = \alpha x$, the function has limit equal to $\frac{1-\alpha^2}{1+\alpha^2}$, so depending on α, and therefore the limit does not exist at the origin. But the limit may fail to exist even if all these limits are equal; for instance define

$$f(x,y) = \frac{x^2y}{x^4+y^2}.$$

All v-directional limits at the origin are zero, but the limit does not exist because on the parabola $y = cx^2$ the function is constant equal to $\frac{c}{1+c^2}$. This example illustrates the meaning of the proposition.

Exercise 2.5. Give an example of a function $f(x,y)$ for which all limits along lines and parabolas through the origin are zero, but yet the limits do not exist.

To deal with a limit in two variables at $p = (a,b)$ it seems natural to consider one of the *iterated limits*

$$\lim_{y\to b}(\lim_{x\to a} f(x,y)), \quad \lim_{x\to a}(\lim_{y\to b} f(x,y)).$$

The existence of the first limit means that $\lim_{x\to a} f(x,y) = \phi(y)$ exists for y close to b, and that $\lim_{y\to b}\phi(y)$ exists. To distinguish them from the limit $\lim_{(x,y)\to(a,b)} f(x,y)$, the latter is sometimes called the *double limit.* It is straightforward to prove that if both the double limit and one iterated limit exist, they must be equal. In particular, if both iterated limits exist and are different, the limit does not exist. This is the case for

$$f(x,y) = \frac{x^4-y^4}{x^4+y^4},$$

at the origin. One has

$$\lim_{x\to 0}\left(\lim_{y\to 0}\frac{x^4-y^4}{x^4+y^4}\right) = \lim_{x\to 0} 1 = 1, \quad \lim_{y\to 0}\left(\lim_{x\to 0}\frac{x^4-y^4}{x^4+y^4}\right) = \lim_{y\to 0}(-1) = -1.$$

Along the same lines, the existence of the limit does not imply the existence of an iterated limit. For instance,

$$\lim_{(x,y)\to(0,0)} x\sin\frac{1}{y} = 0,$$

yet trivially $\lim_{y\to 0} x\sin\frac{1}{y}$ does not exist if $x \neq 0$. Neither the existence nor the equality of the iterated limits implies the existence of the limit, as shown by the example $\frac{xy}{x^2+y^2}$. The property to be added for this implication to hold is that the limit $\lim_{x\to p} f(x,y) = \phi(y)$ be *uniform* in y, but we will not deal with this here.

2.4.2 If $f : A \longrightarrow \mathbb{R}^m, A \subset \mathbb{R}^n, f = (f_1, \ldots, f_m)$ and $p \in A$ is not an isolated point of A, we say that f is *continuous* at p if $\lim_{q \to p, q \in A} f(q) = f(p)$, that is,

$$\forall \varepsilon > 0 \, \exists \delta > 0 : |f(q) - f(p)| < \varepsilon, \quad q \in A, \quad |q - p| < \delta.$$

Of course, this amounts to saying that each component f_j is continuous at p. It is also equivalent to what might be called *continuity through sequences*: for every sequence $(q^k), q^k \in A$ convergent to p one has $f(q^k) \to f(p)$. It is evident that if f is continuous at p and g is continuous at $f(p)$, then $g(f(q))$ is continuous at p. Sums, products, etc. of continuous functions are continuous, and so in particular every formula $E(x_1, \ldots, x_n)$ in terms of the coordinates and the usual elementary functions is continuous in its natural domain of definition, that is, at points p where $E(p)$ makes sense. For instance, with $q = (x, y, z)$,

$$f(x, y, z) = \frac{\sin(xyz) + \log(2 + x^2)}{1 + x^2 y^4}$$

is a continuous function in the whole plane.

The points p where $E(p)$ does not make sense will be, generally speaking, isolated. If at some such point p the limit $\lim_{q \to p} E(q) = l$ exists, assigning the value l at p we have a continuous function at p. This is the case for instance with $f(x, y) = \frac{\sin xy}{xy}$; defining $f(0, 0) = 1$ it becomes a continuous function at $(0, 0)$. In case $A' = A$ and f is continuous at all points $p \in A$, we say that f is *continuous on* A.

Exercise 2.6. Prove that $f : A \to \mathbb{R}^m$ is continuous on A if and only if for every open set V in $\mathbb{R}^m$, the set $U = f^{-1}(V) = \{x \in A : f(x) \in V\}$ is *open in* A, meaning that for each $x \in U$ there is an open ball $B(x, r)$ such that $A \cap B(x, r) \subset U$. In particular, if f, g are real valued continuous functions in $\mathbb{R}^n$, the set $U = \{x : f(x) < g(x)\}$ is open and $F = \{x : f(x) \leq g(x)\}$ is closed.

2.4.3 Next, we study the relationship between compactness and continuity. What follows is Weierstrass' theorem:

Theorem 2.9. *A continuous function* $f : K \longrightarrow \mathbb{R}$ *in a compact set* K *of* $\mathbb{R}^n$ *has an (absolute) maximum and minimum, that is, there is at least one* $p \in K$ *and one* $q \in K$ *such that*

$$f(q) \leq f(x) \leq f(p), \quad x \in K.$$

Proof. The proof is the same as for one variable function. We first prove that f is bounded: $|f(x)| \leq C$, $x \in K$ for some C. If this were not the case, for each $k \in \mathbb{N}$ there would exist $x^k \in K$ such that $|f(x^k)| > k$. By compactness of K, there exists a subsequence (x^{k_m}) convergent to a point $x \in K$. Then, $f(x^{k_m}) \to f(x)$ by the continuity of f at x, then $|f(x^{k_m})| \to |f(x)|$, but $|f(x^{k_m})| > k_m$ escapes to $+\infty$, a contradiction. Being bounded, we may consider $M = \sup_{x \in K} f(x)$, and must see that this is in fact a maximum. By the very definition of M, for each $k \in \mathbb{N}$ there exists $x^k \in K$ such that $M - \frac{1}{k} < f(x^k) \leq M$, so that $f(x^k) \to M$. The sequence (x^k) has a subsequence (x^{k_m}) convergent to some $p \in K$, and again by continuity of f it follows that $f(p^{k_m}) \to f(p)$. But being a subsequence of $f(x^k)$ the later subsequence has limit M, whence $f(p) = M$. The same argument is used for the infimum and minimum. □

As a consequence, the distance from p to a compact set K,

$$d(p, K) = \inf_{y \in K} d(p, y),$$

is attained at some $y \in K$, as already shown in Theorem 2.7.

A natural question that arises is how to compute in practice the extreme values of f in K. In one variable, if we wish to compute the absolute maximum M and minimum m of a nice function f in a closed interval $[a, b]$, we know that the points where these values are attained are either the end-points a, b, or else points in the interior (a, b) where f is not differentiable or else points in the interior where f is differentiable with zero derivative. Typically, there is a finite number of such points, and by evaluating f we can conclude. Later on in this course we will explain a similar strategy for $n > 1$.

The general version $m \geq 1$ of Theorem 2.9 is, with the same proof:

Exercise 2.7. If $f : K \longrightarrow \mathbb{R}^m$ is continuous on a compact K, the image $f(K)$ is a compact set in $\mathbb{R}^m$.

2.4.4 With Theorem 2.9 we can now prove that all norms are equivalent in $\mathbb{R}^n$:

Theorem 2.10. *If ρ_1, ρ_2 are two norms in $\mathbb{R}^n$, there exist two constants $m, M > 0$ such that*

$$m \leq \frac{\rho_1(x)}{\rho_2(x)} \leq M, \quad x \neq 0. \tag{2.3}$$

Proof. It is sufficient to prove that an arbitrary norm $\rho(x)$ is equivalent to the Euclidean norm $|x|$. We first prove that $\rho(x) \leq M|x|$ for some constant M. If $e_1, \ldots, e_n$ is the canonical basis, $x = \sum_i x_i e_i$, then by the properties of ρ,

$$\rho(x) = \rho\left(\sum_i x_i e_i\right) \leq \sum_i \rho(x_i e_i) = \sum_i |x_i|\rho(e_i) \leq |x| \sum_i \rho(e_i) = M|x|.$$

Now, again by the properties of ρ,

$$|\rho(x) - \rho(y)| \leq \rho(x-y) \leq M|x-y|,$$

which proves that ρ is continuous. Since the unit sphere $\{x : |x| = 1\}$ is compact, ρ has an absolute minimum here, in particular

$$\rho(x) \geq m > 0, \quad |x| = 1.$$

If $x \neq 0$, applying this to $x/|x|$ we find that $\rho(x) \geq m|x|$. □

Note that in (2.3) M could be replaced by a bigger constant and m by a smaller one. What is natural is to ask for the optimal values of these constants,

$$m = \inf \frac{\rho_2(x)}{\rho_1(x)}, \quad M = \sup \frac{\rho_2(x)}{\rho_1(x)}.$$

In specific cases this is not an easy computation.

The same kind of proof shows that in the definition of the norm of a linear map,

$$|T| = \sup_{x \neq 0} \frac{|Tx|}{|x|} = \sup_{|x|=1} |Tx|,$$

this supremum is attained and is in fact a maximum,

$$|T| = \max_{x \neq 0} \frac{|Tx|}{|x|} = \max_{|x|=1} |Tx|.$$

2.4.5 Assume now that E is a linear space over the real numbers, of finite dimension n. For instance, the space of polynomials in one real variable t with real coefficients of degree at most $n-1$. A *norm* ρ on E is defined in the same way,

(a) $\rho(v) \geq 0, \rho(v) = 0$ if and only if $v = 0$;
(b) $\rho(\lambda v) = |\lambda|\rho(v)$;
(c) $\rho(v + w) \leq \rho(v) + \rho(w)$.

By exhibiting a basis $v_1, v_2, \dots, v_n$ of E, we establish a linear isomorphism with $\mathbb{R}^n$ and we may transport the norm to $\mathbb{R}^n$: if $x = (x_1, \dots, x_n) \in \mathbb{R}^n$

$$\hat{\rho}(x) = \rho(x_1 v_1 + \cdots + x_n v_n),$$

and conversely. This means that we can transport to an arbitrary linear space of finite dimension all results having to do with norms, distances and concepts related to those. In particular, all norms in a finite dimensional linear space are equivalent and their closed balls are compact.

As an application we consider the linear space of polynomials. We change a bit the notation and write $x = (c_0, c_1, \dots, c_{n-1})$ for the components of $x \in \mathbb{R}^n$; we associate to x the polynomial $P_x(t)$ of degree $\leq (n-1)$ with these coefficients

$$P_x(t) = c_0 + c_1 t + c_2 t^2 + \cdots + c_{n-1} t^{n-1}, \quad t \in \mathbb{R}.$$

We define now

$$\rho(x) = \max\{|P_x(t)| : 0 \leq t \leq 1\}.$$

It is easy to check that ρ is indeed a norm; the last two properties are trivial. If $\rho(x) = 0$, then P_x is identically zero in $[0,1]$, implying $P = 0, c_i = 0$ because $c_i = \frac{P^{(i)}(0)}{i!}$. So we can state:

Proposition 2.2. *Given a degree n and $1 \leq p \leq +\infty$, there exist constants $A_n(p), B_n(p) > 0$ such that for all polynomials $P(t) = c_0 + c_1 t + \cdots + c_n t^n$ of degree $\leq n$ one has*

$$A_n(p) \left(\sum_i |c_i|^p \right)^{\frac{1}{p}} \leq \max\{|P(t)| : 0 \leq t \leq 1\} \leq B_n(p) \left(\sum_i |c_i|^p \right)^{\frac{1}{p}},$$

$$A_n(\infty) \max_i |c_i| \leq \max\{|P(t)| : 0 \leq t \leq 1\} \leq B_n(\infty) \max_i |c_i|.$$

The same can be stated with the norm

$$\int_0^1 |P(t)|\, dt$$

replacing $\max\{|P(t)| : 0 \leq t \leq 1\}$. As pointed out before, the computation of the optimal constants $A_n(p), B_n(p)$ is not easy.

As a second application we consider the linear space of all linear transformations $T : \mathbb{R}^n \to \mathbb{R}^m$. We have already considered a norm in this space

$$|T| = \sup_{x \neq 0} \frac{|Tx|}{|x|} = \max_{|x|=1} |Tx|.$$

On the other hand, it is evident that this space is isomorphic to $\mathbb{R}^{nm}$ through the matrix $A = (a_{i,j})$ of T in the canonical basis (or any other basis). Another norm in this space is $\sqrt{\sum_{i,j} a_{i,j}^2}$. Therefore, there are positive constants m, M such that

$$m^2 \sum_{i,j} a_{i,j}^2 \leq |T|^2 \leq M^2 \sum_{i,j} a_{i,j}^2.$$

2.4.6 Assume now that E, F are linear spaces of finite dimension in which we have chosen corresponding norms $||_E, ||_F$ and let $T : E \to F$ be a linear map. Then $|Tx|_F \leq C|x|_E$ for some constant C; also

$$\sup\left\{\frac{|Tx|_F}{|x|_E}, x \neq 0\right\} = \sup\{|Tx|_F, |x|_E \leq 1\},$$

is attained in the unit sphere of E.

Example 2.2. For instance, consider the space $E(n)$ of polynomials P in t of degree at most $\leq n$, fix $0 \leq t \leq 1$ and define

$$T(P) = P'(t),$$

as a linear map. Choosing the norm $\max\{|P(t)| : 0 \leq t \leq 1\}$ in $E(n)$ and the absolute value in $\mathbb{R}$, we reach the conclusion that for each t, there exists a constant $C_n(t)$ such that

$$|P'(t)| \leq C_n(t) \max\{|P(t)| : 0 \leq t \leq 1\},$$

for any polynomial P of degree at most n. The considerations above show that the optimal value of $C_n(t)$

$$C_n(t) = \sup\{|P'(t)| : |P(t)| \leq 1, 0 \leq t \leq 1\},$$

is in fact a maximum, that is, there exists a polynomial P_t of degree $\leq n$ such that $|P_t'(t)|$ is maximum among all polynomials of degree at most n bounded by 1 in $[0, 1]$.

Considering the linear map $T(P) = P'$ from $E(n)$ to $E(n-1)$, we see that there exists as well an optimal constant C_n such that

$$\max\{|P'(t)| : 0 \leq t \leq 1\} \leq C_n \max\{|P(t)| : 0 \leq t \leq 1\},$$

for any polynomials of degree at most n. Obviously, $C_n = \sup\{C_n(t), 0 \leq t \leq 1\}$. Inequalities of this kind are called *Bernstein's inequalities.*

Exercise 2.8. Compute $C_n(t), C_n$ for $n = 1, 2, 3$.

2.4.7 At certain points in the course we will need as well the concept of *uniformly continuous function.* A function f defined on $A \subset \mathbb{R}^n$, which we may assume real valued, is said to be *uniformly continuous* if it is continuous at each point in a uniform way, meaning that given $\varepsilon > 0$ there exists $\delta > 0$, only depending on ε, such that for all couples of points $x, y \in A$, with $|x - y| < \delta$ one has $|f(x) - f(y)| \leq \varepsilon$. Thus, besides requiring that f be continuous at every point x of A, it is required that the number δ in the definition of continuity be chosen independently of x. An example of a continuous function which is not uniformly continuous is x^2 in the real line.

Proposition 2.3. *Every continuous function f on a compact set K is uniformly continuous.*

Proof. Assume it is not: there exists $\varepsilon > 0$ and couple of points $x^k, y^k \in K, |x^k - y^k| \leq \frac{1}{k}$, with $|f(x^k) - f(y^k)| > \varepsilon$. By compactness of K, there exists a partial sequence (x^{k_m}) convergent to a point $p \in K$; as $|x^k - y^k| \leq \frac{1}{k}$, also $y^{k_m} \to p$. By the continuity of f at p, both $f(x^{k_m})$ and $f(y^{k_m})$ have limit $f(p)$, so their difference has zero limit, in contradiction with $|f(x^k) - f(y^k)| > \varepsilon$. □

2.4.8 We end this section with the concept of *homeomorphism.* If $A \subset \mathbb{R}^n, B \subset \mathbb{R}^m$, we say that a map f from A to B is a homeomorphism if it is a bijection, continuous, with continuous inverse. This means that if $p \in A, f(p) \in B$, a variable $x \in A$ approaches p if and only if $f(x)$ approaches $f(p)$: $|p - x| \to 0$ is equivalent to $|f(p) - f(x)| \to 0$. We also say that A, B are *homeomorphic.*

Exercise 2.9. Prove that if $f : A \to \mathbb{R}^m$ is continuous and one-to-one and A compact, then f is an homeomorphism between A and the compact $f(A)$.

The *invariance of domain theorem* by Brouwer states that if U is an open set in $\mathbb{R}^n$ and $f : U \to \mathbb{R}^n$ is continuous and one-to-one, then $f(U)$ is open and f is a homeomorphism between U and $f(U)$. This result is elementary when $n = 1$ but much harder to prove in dimension $n > 1$. Later on we will prove an analogue of this result in the differentiable category (the inverse function theorem). A consequence of this theorem is that open sets $A \subset \mathbb{R}^n, B \subset \mathbb{R}^m$ with $n \neq m$ cannot be homeomorphic. However, they may exist bijective maps between A, B, they may have the same *cardinality*.

Another important theorem is *Brouwer's fixed point theorem* stating that a continuous self-map $f : B \to B$ in the unit ball $B = \{|x| \leq 1\}$ of $\mathbb{R}^n$ has at least one fixed point. Again, this is trivial if $n = 1$ and much harder if $n > 1$. Brouwer's theorems are central in the field of *Algebraic Topology* [18]. Their common proofs rely on techniques in *singular homology*.

Chapter 3

Coordinate Systems, Curves and Surfaces

The goal of this chapter is to become familiar with the concept of *dimension* of a subset M in Euclidean space, intuitively the number $k \leq n$ of degrees of freedom within M. This is not done for all subsets M, just for those admitting *local coordinate systems*, which establish a one-to-one correspondence with open cubes in $\mathbb{R}^k$. First we deal with the case $k = n$, general coordinate systems in $\mathbb{R}^n$. We have chosen introducing parametrizations of k-dimensional sets M already at this stage, with continuous functions. Later, in Chapter 7, *smooth k-dimensional sets* and their tangent spaces will be considered. The last section deals with *parametrized arcs*, closely related to one-dimensional sets, instrumental for next chapters.

3.1 General Coordinate Systems

3.1.1 Besides affine coordinates with respect to an origin and n axes, there are other ways to locate points using n numbers. There are *non-affine coordinates.* For instance, on the plane a point $p = (x, y)$ is located by giving its distance r to the origin and the angle θ between the position vector and the x axis:

$$x = r\cos\theta, \quad y = r\sin\theta.$$

These are the so-called *polar coordinates* of p. Strictly speaking, $\theta(p)$ is for $p \neq 0$ a coset in $\mathbb{R}/2\pi\mathbb{Z}$. See Figure 3.1.

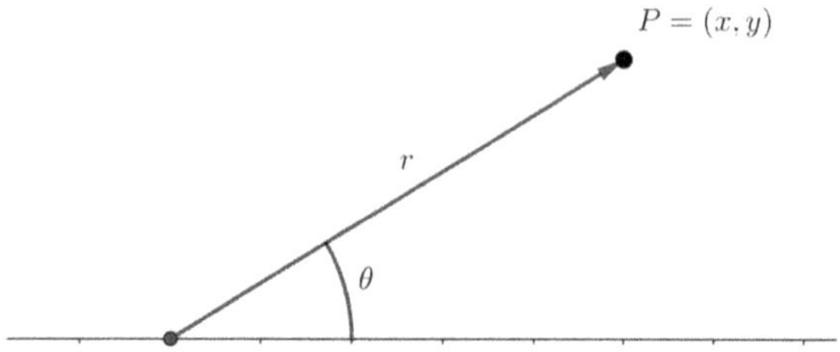

Figure 3.1. Polar coordinates.

In complete generality, by a *coordinate system* in $\mathbb{E}^n$ or in a subset $U \subset \mathbb{E}^n$ we mean a map

$$\Phi : U \longrightarrow \mathbb{R}^n,$$

having components $\Phi = (u_1, \ldots, u_n)$, and assigning to each point $p \in U$ an n-tuple $(u_1(p), \ldots, u_n(p))$ (the new coordinates of p) with two properties:

(a) It is one-to-one (bijective onto its image), that is, different points have different coordinates.
(b) The map Φ is a homeomorphism, that is, two points p, q are close iff their coordinates are

$$\lim_{d(p,q)\to 0} d(\Phi(p), \Phi(q)) = 0,$$

and conversely.

To properly define polar coordinates according to this, we must set $r(p) = \sqrt{x^2 + y^2}$ and we must choose one $\theta(p)$ among all possible angles that differ one from another in integer multiples of 2π. The most common choice is $\theta \in [0, 2\pi)$; that is, for p different from the origin, $\theta(p)$ is the angle between $\vec{0p}$ and the positive x axis, measured counterclockwise. Analytically,

$$\theta(p) = \arccos \frac{x}{r},\ y \geq 0, \quad \theta(p) = 2\pi - \arccos \frac{x}{r},\ y < 0,$$

where $\arccos$ is the inverse function of $\cos : [0, \pi] \to (-1, 1]$. Choosing $\theta(p) \in (-\pi, \pi]$ means that $\theta(p)$ is the angle between the $\vec{0p}$ and the positive x-axis, measured counterclockwise (positive angles) when $y \geq 0$ and clockwise (negative angles) when $y < 0$. Analytically, in this case

$$\theta(p) = \arccos \frac{x}{r},\ y \geq 0, \quad \theta(p) = -\arccos \frac{x}{r},\ y < 0.$$

Strictly speaking, to meet the continuity condition we must delete some points. In the first case, $\theta(p)$ is not continuous at points $p = (a, 0), a > 0$ because it takes the value zero at p and values close to 2π at $(a, -\varepsilon)$ if ε is a small number. Obviously, it is not defined at zero either. So in the first case the polar coordinates are defined in the complement of the positive x-axis. Similarly, the second choice is well defined in the complement of the negative x-axis. Note that we might define polar coordinates in the complement of any ray through the origin. Still, by abuse of notation, it is usual to speak about polar coordinates in the whole plane.

A similar example is that with the *spherical coordinates* in space, consisting in $\rho = \sqrt{x^2 + y^2 + z^2}$, the distance to the origin, and two angles ϕ, θ so that

$$x = \rho \sin\phi \cos\theta, \quad y = \rho \sin\phi \sin\theta, \quad z = \rho \cos\phi.$$

The angle $\phi \in [0, \pi]$ between $\vec{0p}$ and e_3 is called the *latitude*; the projection of p on the plane xy has polar coordinates $\rho \sin\phi, \theta$. Thus, keeping θ constant and varying ϕ, we get a meridian; the north pole corresponds to $\phi = 0$ and the south pole to $\phi = \pi$, while keeping ϕ constant and varying θ we describe a parallel. Again, in order that the spherical coordinates be a coordinate system in the strict sense, we need to choose $\theta \in [0, 2\pi)$ and delete the discontinuities lying in the meridian $\theta = 0$. See Figure 3.2.

Another classical example is the *cylindrical coordinates* (r, θ, z) in the space, consisting in z, and the polar coordinates of the projection q of p on the plane xy. They are a coordinate system in the strict sense in the complement of the half-space $y = 0, x \geq 0$.

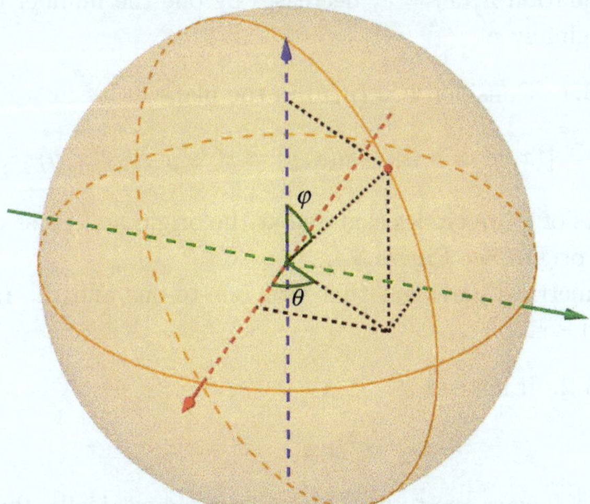

Figure 3.2. Spherical coordinates.

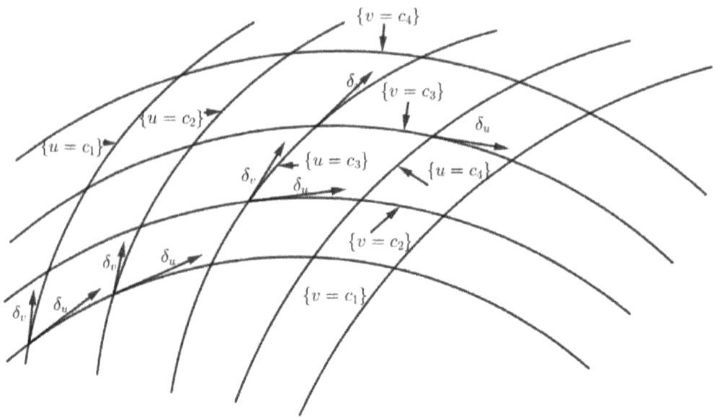

Figure 3.3. Level curves of a plane coordinate system.

3.1.2 We explain the geometrical meaning of a general coordinate system Φ in U, in terms of the components $u_1, \ldots, u_n$. The map Φ being injective means that n level sets of the form

$$\{x \in U : u_j(x) = c_j\}, \quad j = 1, \ldots, n,$$

if not empty, they intersect at a unique point. See in Figure 3.3 a coordinate system u, v for $n = 2$; the level curves $u = c, v = c$ are the new axis.

So, level sets defined by the new coordinates $u_j, j = 1, \ldots, n$ intersect at single points. The intuitive idea, to be formalized rigorously later on, is that each equation $u_j(x) = c_j$ decreases by one the number of degrees of freedom, originally n.

Example 3.1. Consider $\Phi = (u, v)$ in the plane

$$U = \{(x, y), x > 0\}, \quad u(x, y) = x^2 + y^2, \quad v(x, y) = \frac{y}{x}.$$

The level sets of u are circles centered at the origin and those of v are lines through the origin. See Figure 3.4.

It is geometrically evident that Φ is one-to-one and the range of Φ is $V = (0, +\infty) \times \mathbb{R}$.

Example 3.2. If instead

$$u = x^2 + y^2, \quad v = xy,$$

then Φ is no longer one-to-one, as we see geometrically that the circle $x^2 + y^2 = u$ and the hyperbola $xy = v$ meet in general at two different

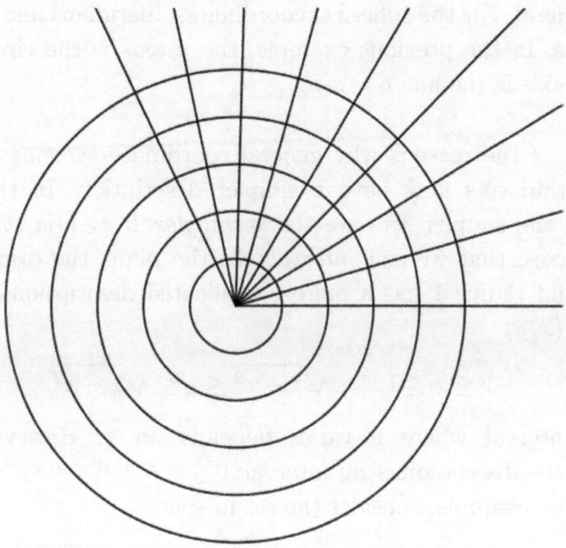

Figure 3.4. Coordinates axes for Example 3.1.

points. Solving analytically for x, y in terms of u, v leads to, with $s = x^2$,

$$s = \frac{1}{2}\left(u \pm \sqrt{u^2 - 4v^2}\right).$$

From this we see that the intersection is a single point only when $u = 2|v|$ or $v = 0$. To get a one-to-one map we must choose one of these two roots, say $s = \frac{1}{2}(u + \sqrt{u^2 - 4v^2}), x = \sqrt{s}$; this means replacing U by $V = \{(x, y), |y| < x\}$, and then Φ is one-to-one from V onto $\{(u, v) : |v| < 2u\}$.

Generally speaking, we will not be interested in the range V of Φ. We should not think of Φ as effectively moving points, we should rather think that points do not move and Φ associates to each point its new coordinates in terms of the canonical coordinates. As in the case with polar and spherical coordinates, often we define the coordinate system by defining the inverse mapping Ψ instead.

The set

$$L_i(p) = \{x \in U : u_j(x) = u_j(p), j \neq i, \},$$

obtained by keeping all new coordinates of p but $u_i(p)$, can be thought as the new i-th axis. If $U = \mathbb{R}^n$ and Φ is linear, these are lines through p, but not so in general. For the spherical coordinates, meridians and parallels are the new axis. In the previous example, the v axis is the circle $u = u(p)$ while the u axis is the line $v = v(p)$.

3.1.3 One of the reasons why general coordinate systems are useful is because certain sets may have a simpler description. In the Cartesian coordinates, the simpler sets are the *rectangles* $R = \prod[a_i, b_i]$ with faces parallel to axes, that we call *intervals*. In the plane the disc centered at the origin and radius 1 has a more complicated description in Cartesian coordinates (x, y)

$$-1 \leq x \leq 1, \quad -\sqrt{1-x^2} \leq y \leq \sqrt{1-x^2}.$$

Here, the interval where y varies depends on x. However, in polar coordinates the disc becomes an interval: $0 \leq r \leq 1, 0 \leq \theta \leq 2\pi$.

As another example, consider the set in space

$$U = \left\{(x, y, z) : x^2 + y^2 + z^2 \leq 1, z \geq \sqrt{x^2 + y^2}\right\}.$$

Its description in cylindrical coordinates is

$$0 \leq r \leq \frac{1}{\sqrt{2}}, \quad 0 \leq \theta \leq 2\pi, \quad r \leq z \leq \sqrt{1-r^2},$$

while in spherical coordinates it is an interval

$$0 \leq \rho \leq 1, \quad 0 \leq \phi \leq \frac{\pi}{4}, \quad 0 \leq \theta \leq 2\pi.$$

Also some functions may take simpler forms in certain coordinates. Generally speaking, when faced with a specific problem, it may be useful to choose the coordinate system in which the problem has a simpler expression. That is the same idea as in linear algebra when solving linear systems; we look for linear coordinates in which the matrix system becomes simpler, say triangular or diagonal.

3.2 Curves, Surfaces and Sub-Manifolds

3.2.1 Up to now we have considered sets U in $\mathbb{R}^n$ that need n coordinates, meaning that intuitively there are n degrees of freedom within U. To formalize this better, we review the concept of *open set*:

Definition 3.1. We say that U is an open set in $\mathbb{R}^n$ if for every $p \in U$ there is a cube $Q(p, r)$ centered at p included in U.

This is because every ball contains a cube and conversely. The intuitive meaning is that around each point there is no restriction on the coordinates of points of U, each can vary freely in an interval. Hence, open sets are the n-dimensional sets in $\mathbb{R}^n$.

But there are sets with fewer degrees of freedom where we do not need n-coordinates. One example is a linear subspace V of dimension k defined by (1.1), where the points are described by

$$X = M\Lambda, \quad \Lambda \in \mathbb{R}^k.$$

There are as many points in V as parameters Λ, so that we may consider the λ as coordinates of points in V.

Another intuitive way to introduce sets with fewer degrees of freedom is by imposing some constraint. Consider a set of points in space defined by one single equation

$$S = \{(x, y, z) : f(x, y, z) = c\}.$$

Intuitively, in $\mathbb{R}^3$ there are three degrees of freedom, so imposing $f(x, y, z) = c$ should decrease them in one, and S should have two degrees of freedom. For instance, a linear subspace V of dimension k in $\mathbb{R}^n$ is also defined by $m = n - k$ linearly independent equations

$$f_j(x_1, \dots, x_n) = \sum_i c_{ji} x_i = 0, \quad j = 1, \dots, m.$$

But not always does $f = 0$ define an object with two degrees of freedom. For instance, if a line is the intersection of two planes of equations $f_1(x, y, z) = 0, f_2(x, y, z) = 0$, it is also defined by the single equation $f = f_1^2 + f_2^2 = 0$.

A precise way of stating that there are k degrees of freedom in some set M is as in the definition that follows.

Definition 3.2. We say that M is a *topological sub-manifold of dimension* $k, k = 1, \dots, n-1$ *of* $\mathbb{R}^n$ if for each $p \in M$ there is a ball $B(p, r)$ such that $M \cap B(p, r)$ is parametrizable with k parameters $u_1, u_2, \dots, u_k$, meaning that there exists a map $\Phi : V \longrightarrow \mathbb{R}^n$, defined in an open cube (or ball) $V, 0 \in V$, in $\mathbb{R}^k$ such that Φ is a homeomorphism from V to $M \cap B(p, r), \Phi(0) = p$.

The $u = (u_1, u_2, \dots, u_k)$ are thought as the coordinates of $\Phi(u)$. The pair (V, Φ) is called a *local coordinate system*, a *local parametrization* or a *local chart*. An *atlas* of S is a collection of local coordinate systems (V_i, Φ_i) such that $M = \cup_i \Phi_i(V_i)$.

The other extreme case is $k = 0$, which corresponds to discrete sets. In those, there are zero degrees of freedom around every isolated point.

For $k = 1$ we use the term *curve* and notation $M = \Gamma$, t for parameters; for $k = 2$ we use the term *surface* and notation $M = S$, s, t for parameters, $\Phi(s,t) = (x(s,t), y(s,t), z(s,t))$ for the parametrization. Note that these curves need not have tangents at all points, vertices and cusps are admissible. For instance, the boundary of a polygon is a curve. The same comment applies to surfaces that might have edges, like the boundary of a cube.

In most cases, M is *given* as the set of points satisfying $m = n - k$ equations

$$f_j(x_1, \dots, x_n) = 0, \quad j = 1, \dots, m.$$

As pointed above, this does not guarantee that M has dimension k. In all examples we check it by inspection and find explicit parametrizations. Later on, when discussing the implicit function theorem and *regular* sub-manifolds, we will analyze when a set defined by m equations $f_j(x) = 0$ is automatically a regular sub-manifold of dimension $k = n - m$. At the present continuous category we do not have such type of statement.

In some cases there is a *global parametrization* $\Phi : V \to M$ of the whole of M, as it is the case in linear subspaces. Another example is a graphic $y = f(x)$ of a function f of one variable, or $z = g(x,y)$ of two variables, having the global parametrization

$$x = x, y = f(x); \quad x = x, y = y, z = g(x,y).$$

Example 3.3. The unit circle T in the plane is defined by the equation $x^2 + y^2 = 1$. The map

$$\Phi_1(t) = (\cos t, \sin t), \quad 0 < t < 2\pi,$$

serves as a local parametrization around every $p \in T, p \neq (1,0)$. Together with

$$\Phi_2(t) = (\cos t, \sin t), \quad -\pi < t < \pi,$$

which is a parametrization around $p = (1,0)$, it constitutes an atlas for T. The sphere centered at the origin of radius r

$$x^2 + y^2 + z^2 = r^2,$$

can be parametrized using the spherical coordinates

$$x = \rho \sin\phi \cos\theta, \quad y = \rho \sin\phi \sin\theta, \quad z = \rho \cos\phi, \quad 0 < \phi < \pi, \ 0 < \theta < 2\pi.$$

This parametrization omits a meridian from the north-pole to the south-pole. To get an atlas, we would need three such carts with other poles.

In practice, though, when a parametrization omits a subset of M of fewer dimensions, we will consider it as a global parametrization.

Example 3.4. An example of a completely different nature is $SO(3)$. Identifying T with its matrix in the canonical basis, we can view $SO(3)$ as a subset of R^9. In paragraph 1.2.3, a parametrization with three angles has been exhibited.

From the definition it follows that if M is a topological sub-manifold and Φ an homeomorphism defined in an open set containing M, then $\Phi(M)$ is a sub-manifold as well.

Evidently, sub-manifolds can be defined using general coordinate systems; in fact, a main use of general coordinate systems is to describe curves and surfaces in a simpler way. For instance, in the plane with polar coordinates (r, θ), the curve described parametrically by $r = g(\theta)$ with g nondecreasing is a spiral that in cartesian coordinates is

$$x = g(\theta)\cos\theta, \quad y = g(\theta)\sin\theta.$$

Again in polar coordinates, $r = 1 - \cos\theta$ defines a curve, the *cardioid.* Its parametrization in the usual coordinates is

$$x = (1 - \cos\theta)\cos\theta, \quad y = (1 - \cos\theta)\sin\theta,$$

while its equation in cartesian coordinates is

$$\sqrt{x^2 + y^2} = 1 - \frac{x}{\sqrt{x^2 + y^2}},$$

or $(x + x^2 + y^2)^2 = x^2 + y^2$. In space, with spherical coordinates, $\rho = 1, 2\phi = \theta$ is a spiral connecting the poles on the unit sphere.

In this textbook, we will be mainly interested in *regular sub-manifolds* (Chapter 7). In fact, all examples that follow, conics and quadrics, fall within this category. Yet, it is convenient to get acquainted with local parametrizations of sub-manifolds already at this stage.

3.2.2 In the next paragraph, we study curves and surfaces given by quadratic equations. Before that we point out some general facts about *surfaces of revolution*. Suppose that Γ is a curve in the plane yz defined by $f(y, z) = 0$, with $y > 0$. Then

$$f(\sqrt{x^2 + y^2}, z) = 0$$

is the equation of the surface S in $\mathbb{R}^3$ obtained by rotating Γ around the z axis, because $\sqrt{x^2 + y^2}$ is the distance from (x, y, z) to this axis. If $y = h(t), z = g(t)$ is a parametrization of Γ, then

$$x = h(t)\cos\theta, \quad y = h(t)\sin\theta, \quad z = g(t),$$

is a parametrization of S. Analogously, $f(y, \sqrt{x^2 + z^2}) = 0$ is the surface in $\mathbb{R}^3$ obtained rotating Γ around the y axis, with parametrization $x = g(t)\cos\theta, y = h(t), z = g(t)\sin\theta$.

For instance, the *torus* S in $\mathbb{R}^3$ is obtained by revolving a circle Γ around an axis in its plane not meeting it, see Figure 3.5. If Γ is

$$(y - R)^2 + z^2 = r^2, \quad r < R, \quad y = R + r\cos t, \quad z = r\sin t, \quad 0 \le t \le 2\pi,$$

then S has the equation

$$(\sqrt{x^2 + y^2} - R)^2 + z^2 = r^2,$$

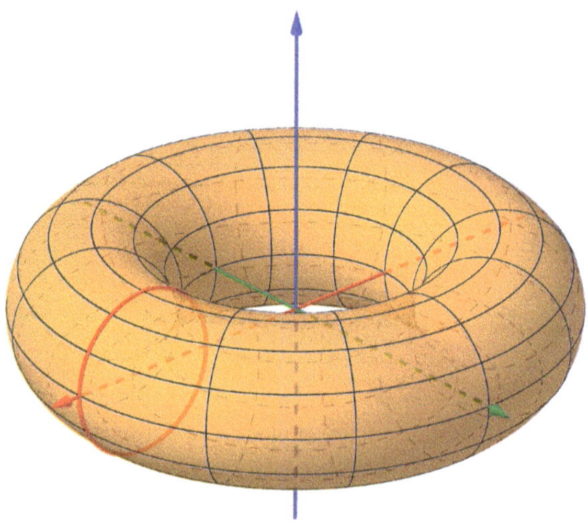

Figure 3.5. A torus in space.

(and so has a polynomial equation of degree four) and parametrization

$$x = (R+r\cos t)\cos\theta, \quad y = (R+r\cos t)\sin\theta, \quad z = r\sin t, \quad 0 \le t, \theta \le 2\pi. \tag{3.1}$$

3.3 Conics and Quadrics

3.3.1 Besides linear functions, the simplest functions are the quadratic ones, polynomials of degree two in the cartesian coordinates $x_i, i = 1, 2, \ldots, n$. These functions have three components, $f = f_1 + f_2 + c$, where c is constant, f_1 is linear

$$f_1(x) = \sum_i b_i x_i,$$

and f_2 is purely quadratic or *homogeneous*

$$f_2(x) = \sum_{i,j=1}^{n} a_{ij} x_i x_j.$$

The term homogeneous is associated to the property $f_2(\lambda x) = \lambda^2 f_2(x)$). If $i \neq j$, replacing a_{ij} by $\frac{1}{2}(a_{ij} + a_{ji})$ it can be assumed that $a_{ij} = a_{ji}$. An alternative way of writing f_2 is then

$$f_2(x) = \sum_{i=1}^{n} a_{ii} x_i^2 + 2\sum_{i<j} a_{ij} x_i x_j.$$

It is very convenient to use matrix notation. If A stands for the symmetric matrix with entries a_{ij}, and B for the row matrix with entries b_i, we write

$$f(x) = X^t A X + BX + c.$$

Suppose that y_j denote new linear coordinates related to X by $X = MY, Y = M^{-1}X$, with M invertible. Then the expression of f in the new coordinates is

$$f(y) = Y^t M^t A M Y + BMY + c,$$

so that A is changed to $M^t A M$. Then, the natural question is whether there are coordinates — that is an M — such that $M^t A M$ is as simple as possible, diagonal for instance. In this case, one would have just the terms y_i^2.

This is indeed possible, and in dimension $n = 2$ can be seen, in an elementary way, completing squares; in this case, we start from

$$Ax^2 + 2Bxy + Cy^2 + Dx + Ey + F = 0.$$

First we get rid of the term in xy; if $A \neq 0$, write

$$Ax^2 + 2Bxy + Cy^2 = A\left(x + \frac{B}{A}y\right)^2 + \left(C - \frac{B^2}{A}\right)y^2,$$

and analogously if $C \neq 0$; if $A = C = 0$, write

$$xy = \frac{1}{4}((x+y)^2 - (x-y)^2).$$

We see that in all cases there are new coordinates x', y' such that $Ax^2 + 2Bxy + Cy^2 = A'x'^2 + B'y'^2$. In dimension $n = 3$, there are as well coordinates x', y', z' such that

$$Ax^2 + By^2 + Cz^2 + 2Dxy + 2Exz + 2Fyz = A'x'^2 + B'y'^2 + C'z'^2.$$

Indeed, assume that one of the three coefficients A, B, C is not zero, say $A \neq 0$; then the above equals

$$A\left(x + \frac{D}{A}y + \frac{E}{A}z\right)^2 + \Phi(y, z),$$

where $\Phi(y, z)$ is quadratic in y, z; setting $x' = x + \frac{D}{A}y + \frac{E}{A}z$ and applying the case $n = 2$ to Φ we are done. If $A = B = C = 0$ and say $D \neq 0$, we write $xy = \frac{1}{4}((x+y)^2 - (x-y)^2), x' = x + y, y' = x - y$ so that we have terms in x'^2, y'^2 and proceed as before.

However, these simplified expressions are not unique. For instance,

$$3x^2 + 3y^2 - 2xy = 3\left(x - \frac{y}{3}\right)^2 + \frac{8}{9}y^2 = (x+y)^2 + 2(x-y)^2.$$

There is an alternative intrinsic procedure leading to an essentially unique reduced form using Theorem 1.4. With the notations there, in the coordinates Y defined by $X = MY$, the quadratic expression X^tAX is transformed into $\sum_i \lambda_i y_i^2$. This fact can be seen in two different but equivalent ways. First, that by performing a change of coordinates $X = MY$, the quadric (which does not move) has a simpler expression. Or secondly, that the rigid motion given by $TX = Y = M^tX$ maps this quadric into another one with a simpler equation.

3.3.2 Let us consider now the curves in the plane given by quadratic equations, called *conics*. By the remark above we may assume that in certain cartesian coordinates the equation is

$$Ax^2 + By^2 + Cx + Dy + E = 0.$$

If $A \neq 0$, write $Ax^2 + Cx = A(x + \frac{C}{2A})^2 + E - \frac{C^2}{4A}$ and set $x' = x + \frac{C}{2A}$; analogously if $B \neq 0$. That is, we delete the linear term if the corresponding quadratic term is not zero. Doing this we obtain the following canonical forms:

$$Ax^2 + By^2 + E = 0, \quad Ax^2 + Dy + E = 0.$$

In the first case the conic is an ellipse if A, B have the same sign (in this case E must have the opposite sign, otherwise the conic is empty or a point), and a hyperbola otherwise. In the second case, we have a parabola if $D \neq 0$ or a couple of points otherwise.

The reduced form of an ellipse is usually written

$$\frac{x^2}{a^2} + \frac{y^2}{b^2} = 1,$$

with a, b the axis. Since this means that $(\frac{x}{a}, \frac{y}{b})$ is in the unit circle, we get the *standard parametrization*

$$x = a\cos t, \quad y = b\sin t, \quad 0 \leq t \leq 2\pi.$$

Exercise 3.1. Give a geometric interpretation of the parameter t in the parametrization of the ellipse.

Similarly, the reduced form of the hyperbola is written

$$\frac{x^2}{a^2} - \frac{y^2}{b^2} = 1,$$

so that its axis is the x axis. The hyperbola has two separated components, in the right- and left-hand plane, respectively. The standard parametrization of the branch in the right half plane is

$$x = a\cosh t, \quad y = b\sinh t, \quad t \in \mathbb{R}.$$

Finally, the obvious parametrization of the parabola is done using x as parameter.

3.3.3 Let us consider now the surfaces in space given by quadratic expressions, the *quadrics*, which we may assume in the form

$$Ax^2 + By^2 + Cz^2 + Dx + Ey + Fz + G = 0.$$

As before, we can delete the linear term in case the quadratic one is not zero, to reach the following *canonical forms*, modulo change of Cartesian coordinates:

(a) $Ax^2 + By^2 + Cz^2 + G = 0, ABC \neq 0.$
(b) $Ax^2 + By^2 + Fz + G = 0, AB \neq 0.$
(c) $Ax^2 + Ey + Fz + G = 0, A \neq 0.$

In the first case (a), if A, B, C have the same sign (and G the opposed one), they are *ellipsoids*, see Figure 3.6. We write them in the form

$$\frac{x^2}{a^2} + \frac{y^2}{b^2} + \frac{z^2}{c^2} = 1,$$

with a, b, c the half-axis. Since $(\frac{x}{a}, \frac{y}{b}, \frac{z}{c})$ is in the unit sphere, the standard parametrization is

$$x = a\sin\phi\cos\theta, \quad y = b\sin\phi\sin\theta, \quad z = c\cos\phi, \quad 0 \leq \phi \leq \pi, \ 0 \leq \theta \leq 2\pi.$$

Assume there are two coefficients with the same sign, say $A, B > 0, C < 0$; if $G < 0$, this is a *one-sheet hyperboloid* (Figure 3.7), usually expressed

$$\frac{x^2}{a^2} + \frac{y^2}{b^2} - \frac{z^2}{c^2} = 1.$$

Sections with $z = k$ are ellipses, while sections with $x = k$ or $y = k$ are hyperbolas. A global parametrization is given by

$$x = a\cos t\cosh s, \quad y = b\sin t\cosh s, \quad z = c\sinh s, \quad 0 \leq t \leq 2\pi, \ s \in \mathbb{R}.$$

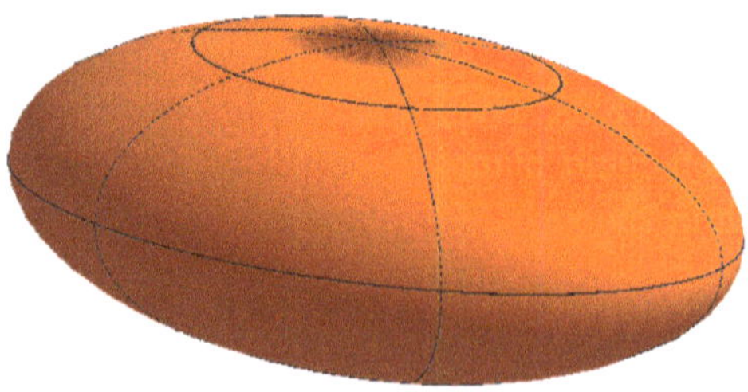

Figure 3.6. An ellipsoid.

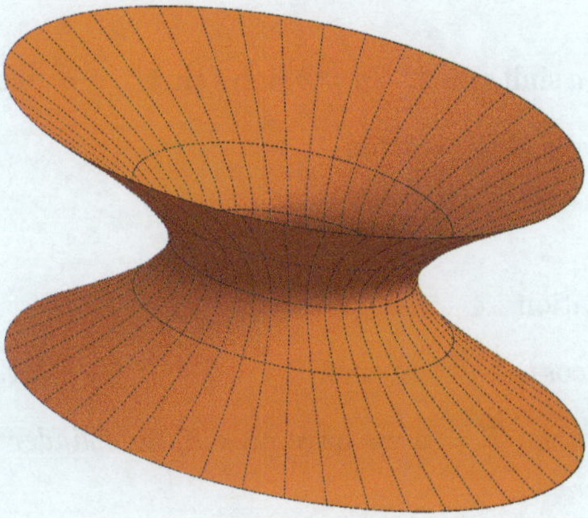

Figure 3.7. One-sheet hyperboloid.

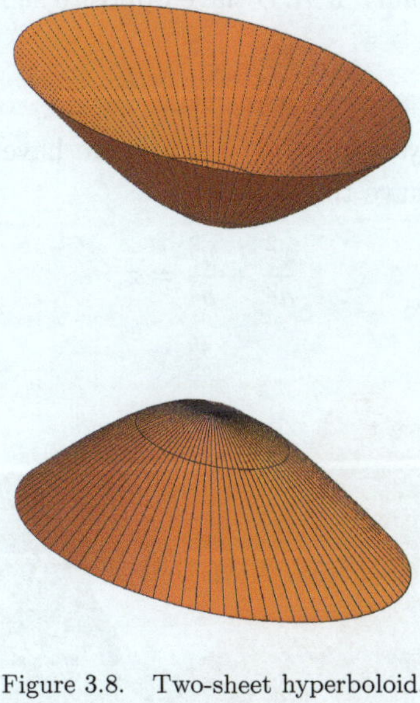

Figure 3.8. Two-sheet hyperboloid.

In case $G > 0$, this is a *two-sheet hyperboloid* (Figure 3.8), usually expressed

$$\frac{x^2}{a^2} + \frac{y^2}{b^2} + 1 = \frac{z^2}{c^2}.$$

The sheet with $z \geq c$ can be parametrized

$$x = a\sqrt{\frac{z^2}{c^2} - 1}\cos t, \quad y = b\sqrt{\frac{z^2}{c^2} - 1}\sin t, \quad z = z,\ z \geq c,$$

or else

$$x = a \sinh s \cos t, \quad y = b \sinh s \sin t, \quad z = c \cosh s.$$

If $G = 0$, we have a *cone*, written

$$\frac{x^2}{a^2} + \frac{y^2}{b^2} = z^2,$$

with parametrization

$$x = az \cos t, \quad y = bz \sin t, \quad z = z, \quad 0 \leq t \leq 2\pi, \ z \in \mathbb{R}.$$

In case (b), when $F = 0$ we have an *elliptic cylinder* if A, B have the same sign,

$$\frac{x^2}{a^2} + \frac{y^2}{b^2} = 1,$$

and an *hyperbolic cylinder* if A, B have opposite sign,

$$\frac{x^2}{a^2} - \frac{y^2}{b^2} = 1.$$

When $F \neq 0$, we may assume $G = 0$ and we have an *elliptic paraboloid* (Figure 3.10) if A, B have the same sign

$$\frac{x^2}{a^2} + \frac{y^2}{b^2} = z,$$

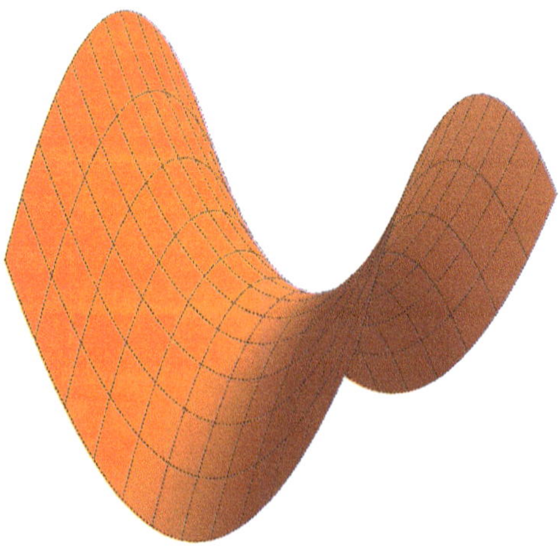

Figure 3.9. Hyperbolic paraboloid.

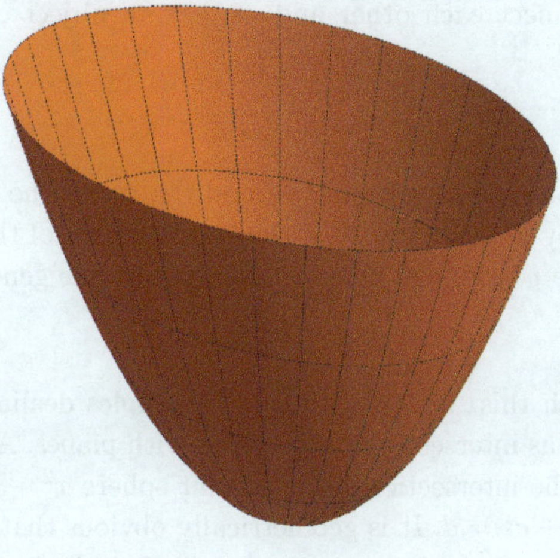

Figure 3.10. Elliptic paraboloid.

and an *hyperbolic paraboloid* if A, B have opposite sign (Figure 3.9),

$$\frac{x^2}{a^2} - \frac{y^2}{b^2} = z.$$

In case two, all parametrizations are obvious.

Finally, in case (c), when $E = F = 0$ we have two parallel planes $x = \pm\sqrt{-\frac{G}{A}}$. If E or F are not zero, say $E \neq 0$, then we may assume $F = G = 0$ and we have a *parabolic cylinder*

$$y = cx^2.$$

In conclusion, in dimension $n = 2, 3$, what we have seen is that there is an orthogonal matrix M and a translation vector p such that in the new coordinates Y defined by $X = MY + p$ the conic or quadric has an equation in a canonical form.

Let us explain the difference between the one-sheet hyperboloid and the two-sheet hyperboloid, assuming for simplicity that $A = B$. Note that

$$x^2 + y^2 = z^2 + 1$$

is the surface obtained by rotating around the z axis the curve whose equation in the plane zy is $y^2 = z^2 + 1$, a hyperbola having two disjoint branches, one in $y > 1$ and another one in $y < -1$; if these two branches rotate, they connect each other and we get an object of a single piece. Similarly,

$$x^2 + y^2 + 1 = z^2$$

is the surface obtained by rotating around the z axis the hyperbola in the zy plane given by $y^2 + 1 = z^2$. Now the two branches of this hyperbola are separated by the y axis, and when rotating, each one generates a different sheet.

3.3.4 We finish this section with some examples dealing with conics in space appearing as intersections of quadrics with planes. As a first example let us consider the intersection Γ of the unit sphere $x^2 + y^2 + z^2 = 1$ with a plane $ax + by + cz = d$. It is geometrically obvious that Γ, if not empty, is a circle; let us compute its center and radius and find a parametrization. We assume $a^2 + b^2 + c^2 = 1$ and say $c \neq 0$. We may solve for z in the plane, $z = \frac{1}{c}(d - ax - by)$ and replace it in the equation of the sphere to get

$$c^2(x^2 + y^2) + (d - ax - by)^2 = 1.$$

So, if $(x, y, z) \in \Gamma$, then (x, y) satisfies this equation, meaning that it defines the projection of Γ onto the xy-plane. Being bounded it must be an ellipse, which we might write in canonical form, parametrize $x = x(t), y = y(t)$ and then use $z = \frac{1}{c}(d - ax - by)$ to obtain $z = z(t)$. A better way to deal with this is using cartesian coordinates x', y', z' such that the plane becomes $z' = d$ and so parametrized by x', y'. So we choose the vector $n = (a, b, c)$, which is perpendicular to the plane, as unit vector in the z' axis and consider the two unit vectors,

$$w = \frac{1}{\Delta}(0, -c, b), \quad v = n \times w = \frac{1}{\Delta}(b^2 + c^2, -ab, -ac),$$

with $\Delta = \sqrt{b^2 + c^2}$. The new coordinates are thus

$$x' = \frac{1}{\Delta}(-cy + bz), \quad y' = \Delta x - \frac{a}{\Delta}(by + cz), \quad z' = ax + by + cz,$$

the inverse mapping given by the transpose matrix

$$x = \Delta y' + az', \quad y = -\frac{c}{\Delta}x' - \frac{ab}{\Delta}y' + bz', \quad z = \frac{b}{\Delta}x' - \frac{ac}{\Delta}y' + cz'.$$

In the new coordinates, the plane is given by $z' = d$ and the sphere by $x'^2 + y'^2 + z'^2 = 1$. Then Γ is the circle on the plane $z = d$ with equation

$x'^2 + y'^2 = 1 - d^2$ and parametrization

$$x' = \sqrt{1-d^2}\cos t, \quad y' = \sqrt{1-d^2}\sin t, \quad z' = d,$$

from which it follows that

$$(x, y, z) = (ad, bd, cd) + \sqrt{1-d^2}\,(\cos t\, w + \sin t\, v).$$

It is classical that the three types of conics (ellipses, parabolas and hyperbolas) arise as sections Γ of the cone $z^2 = x^2 + y^2$ with a plane. Let us check this analytically, and obtain at the same time a parametrization. Assume that the plane has equation

$$ax + by + cz = d,$$

where $\Delta^2 = a^2 + b^2 \neq 0$ (otherwise it is obvious that the section is a circle). As before we introduce the orthonormal basis, with $\alpha = \sqrt{a^2+b^2+c^2}$

$$n = \frac{1}{\alpha}(a, b, c), \quad w = \frac{1}{\Delta}(-b, a, 0), \quad v = n \times w = \frac{1}{\Delta\alpha}(-ca, cb, a^2 + b^2).$$

Let x', y', z' be the cartesian coordinates in these new axes,

$$x' = \frac{1}{\Delta}(-bx + ay), \quad y' = \frac{1}{\alpha\Delta}(-cax + cby + \Delta^2 z), \quad z' = \frac{1}{\alpha}(ax + by + cz),$$

with inverse transformation given by

$$x = -\frac{b}{\Delta}x' - \frac{ac}{\Delta\alpha}y' + \frac{a}{\alpha}z', \quad y = \frac{a}{\Delta}x' + \frac{cb}{\Delta\alpha}y' + \frac{b}{\alpha}z', \quad z = \frac{\Delta}{\alpha}y' + \frac{c}{\alpha}z'.$$

The plane has equation $z' = \frac{d}{\alpha}$, so that Γ is the conic lying in this plane with equation $2z^2 = x'^2 + y'^2 + z'^2$, or

$$x'^2 + \left(1 - 2\frac{\Delta^2}{\alpha^2}\right)y'^2 = 4\frac{cd\Delta}{\alpha^3}y' + \left(2\frac{c^2}{\alpha^2} - 1\right)z'^2,$$

or

$$\alpha^2 x'^2 + (c^2 - a^2 - b^2)y'^2 = 4\frac{cd\Delta}{\alpha}y' + (c^2 - a^2 - b^2)\frac{d_2}{\alpha^2}.$$

From this we see that Γ is an ellipse if $c^2 > a^2 + b^2$ (if not empty), a hyperbola if $c^2 < a^2 + b^2, d \neq 0$, a couple of lines if $c^2 < a^2 + b^2, d = 0$, a parabola if $c^2 = a^2 + b^2, d \neq 0$ and a line (a generatrix of the cone) if $c^2 = a^2 + b^2, d = 0$. In the first two cases, the center of Γ is $x' = 0$, $y' = 0, z' = \frac{d}{\alpha}$, or

$$\left(\frac{ad}{a^2+b^2+c^2}, \frac{bd}{a^2+b^2+c^2}, \frac{cd}{a^2+b^2+c^2}\right)$$

in the original coordinates.

3.4 Arcs in Euclidean Space

3.4.1 Here, we consider vector-valued functions of a single real variable; it will be convenient to adopt a cinematic point of view, so that we will think that this variable is time, denoted t, and use the notation γ instead of f. Thus, γ is a map

$$\gamma : (a, b) \longrightarrow \mathbb{R}^m,$$

with $m > 1$. We think of $\gamma(t)$ as the position at time t of a moving object. When $m = 1$, we are used to visualizing a real function $y = f(x)$ of a real variable x in terms of its graphic, the set of points in the plane of the form $(x, f(x))$. Here we will not consider any graphic in $\mathbb{R}^{m+1}$, but only the *range or trajectory* $\gamma^* = \gamma((a, b))$ in $\mathbb{R}^m$. So, the graph $y = f(x)$ is the range of $\gamma(x) = (x, f(x))$. We say that γ is a *parametrized arc*, or arc for short, and its range γ^* is a *path*. So, γ is a specific way of traveling along the path γ^*. Obviously, γ^* does not determine γ, we may travel through the same road at different speeds.

We will assume that γ is continuous at each point; this amounts to requiring that each component of γ, denoted $x_i(t), i = 1, \ldots, m$, is continuous. Intuitively, continuity means, say in $n = 2$, that we can draw γ^* on the blackboard without lifting up the chalk. In the plane and space we use the notations

$$\gamma(t) = (x(t), y(t)), \quad \gamma(t) = (x(t), y(t), z(t)),$$

respectively.

If $\phi : (c, d) \to (a, b), t = \phi(s)$ is a continuous bijection (hence, strictly monotone), and we compose γ with ϕ then $\hat{\gamma}(s) = \gamma(\phi(s))$ has the same range γ^*. We say that $\hat{\gamma}$ is a *reparametrization* of γ or that it is obtained by a *change of parameter*. In case ϕ is increasing, we say that $\hat{\gamma}$ *preserves orientation*, and *changes orientation* if ϕ is decreasing.

If γ is defined and continuous at the end points a, b (that is, γ has a well defined limit at a and b), and $p = \gamma(a), q = \gamma(b)$ we say that γ *goes from p to q*, or that p is the *initial point* and q the *final point*. If $p = q$, we say that the arc is *closed* (we use here the term "closed" in a different sense than in paragraph 2.1.4).

We think of γ^* as a thread in $\mathbb{R}^m$, but in fact it can be a rather general set. For instance, there exist (continuous) arcs γ, called generically *Peano arcs*, for which γ^* is the open unit square $(0, 1) \times (0, 1)$. By the theorem of

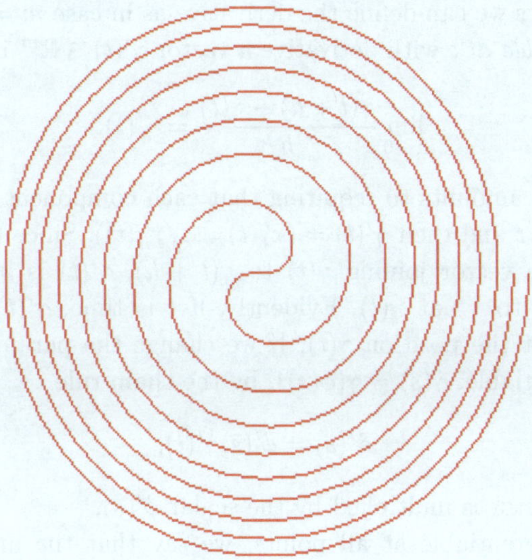

Figure 3.11. Fermat's double spiral.

invariance of dimension quoted in paragraph 2.4.8, though, such γ cannot be a homeomorphism.

One may define arcs using a general coordinate system; if $(u_1, \ldots, u_n)$ are coordinates in U, we can define an arc in U by setting all coordinates in terms of a parameter $t \in (a, b)$

$$u_1(t), u_2(t), \ldots, u_n(t), \quad a < t < b,$$

or may choose one of the coordinates as a parameter. For instance, in the plane with polar coordinates,

$$x(t) = t \cos t, \quad y(t) = t \sin t, \quad t > 0,$$

defines a spiral. Figure 3.11 is Fermat's double spiral defined by $r^2 = a^2\theta$ in polar coordinates.

Example 3.5. Consider, for instance, in the plane the path given by

$$\gamma(t) = (|\sin kt| \cos t, |\sin kt| \sin t), \quad t \in \mathbb{R}.$$

The shape of γ^* is a flower with $2k$ petals, auto-intersecting at the origin. For the origin there is no ball $B(0, r)$ for which $B(0, r) \cap \gamma^*$ is homeomorphic to an open interval. Obviously, $\gamma^* \setminus (0, 0)$ is a curve in the sense of Definition 3.2.

3.4.2 For arcs we can define the derivative as in case $m = 1$: we say that γ is *differentiable at* t with derivative a vector $\gamma'(t) \in \mathbb{R}^m$ if

$$\lim_{h\to 0} \frac{\gamma(t+h) - \gamma(t)}{h} = \gamma'(t).$$

Of course, this amounts to requiring that each component $x_i(t)$ of γ have a derivative at t and then $\gamma'(t) = (x_1'(t), \dots, x_n'(t))$. Since the incremental quotient is the vector joining $\gamma(t)$ to $\gamma(t+h)$, $\gamma'(t)$ is interpreted as a *tangent vector* to γ^* at $\gamma(t)$. Evidently, if t is time, $\gamma'(t)$ is the instant speed vector at the position $\gamma(t)$. If we change the parameter, $t = \phi(s)$, with ϕ differentiable, $\hat{\gamma}(s) = \gamma(\phi(s))$, by the chain rule

$$\hat{\gamma}'(s) = \phi'(s)\gamma'(t),$$

the tangent vector is multiplied by the scalar $\phi'(s)$.

If γ is differentiable at all points, we say that the arc is *regular*. If moreover $\gamma'(t)$ is continuous in t (that is, the tangent changes continuously with t), we say that γ is an arc *of class* C^1.

3.4.3 Tangent cone. If $M \subset \mathbb{R}^n$ is a general set and $p \in M$, we consider all differentiable arcs γ defined on $(-\varepsilon, +\varepsilon)$ for some ε, on M (that is $\gamma(t) \in M$ for all t), and passing through p, say $\gamma(0) = p$, and set

$$T_p(M) = \{v \in \mathbb{R}^n : v = \gamma'(0)\}.$$

Thus, $T_p(M)$ consists of all tangent vectors at p to the arcs on M through p. Obviously, $0 \in T_p(M)$; if $v \in T_p(M)$, say $v = \gamma'(0)$, and $\gamma_\lambda(t) = \gamma(\lambda t)$, then $\gamma_\lambda'(0) = \lambda v$, and so $\lambda v \in T_p(M)$. We call $T_p(M)$ the *tangent cone to* M *at* p.

It might be that $T_p(M)$ is reduced to zero; for instance in the plane, the set $M = \{(x, y) : y = |x|\}$ has no non-zero tangent vectors at $(0, 0)$, This is because if $x(t) = 0$ and both $x(t)$ and $y(t) = |x(t)|$ are differentiable at zero, then $x'(0) = 0, y'(0) = 0$. The sets M having non-trivial tangent cone at every point are the interesting ones from the point of view of differential calculus.

3.4.4 We introduce now a concept related to connectedness that will turn out to be equivalent. We say that an open set U in $\mathbb{R}^n$ is *arc-connected* if given two points $x, y \in U$ there exists an arc $\gamma : [a, b] \to U$ such that $x = \gamma(a), y = \gamma(b)$. Recall that U is said to be convex if given two points

$x, y \in U$ the segment joining x, y is contained in U. Obviously, a convex set is arc-connected but not conversely.

Theorem 3.1. *An open set U is arc-connected if and only if it is connected.*

Proof. Assume U is arc-connected, $U = U_1 \cup U_2, U_1 \cap U_2 = \emptyset$ with U_1, U_2 open and, to reach a contradiction, that none is empty. Let $x \in U_1, y \in U_2$ and consider an arc $\gamma : [a, b] \to U, \gamma(a) = x, \gamma(b) = y$. We define $A = \{t \in [a, b] : \gamma(t) \in U_1\}$. Since U_1 is open, contains a and γ is continuous, A contains an interval $[a, a+\delta)$. Analogously, there exists an interval $(b-\tau, b]$ such that $\gamma(t) \in U_2, b-\tau < t \le b$, whence $(b-\tau, b]$ does not meet A. As a consequence, $\alpha = \sup A$ satisfies $a < \alpha < b$. Now, either $\gamma(\alpha) \in U_1$ or $\gamma(\alpha) \in U_2$. In the former case, by continuity of γ there exists an interval $I = [\alpha - \varepsilon, \alpha + \varepsilon] \subset (a, b)$ such that $\gamma(t) \in U_1, t \in I$, implying $\alpha + \varepsilon \in A$ in contradiction with α being the supremum of A. In the later case, there exists an interval $(\alpha - \varepsilon, \alpha + \varepsilon)$ not meeting A, again in contradiction with the definition of α.

Conversely, assume that U is connected. Fix a point $x \in U$ and define V as the set of points $y \in U$ that can be joined to x by an arc. Then V is open, because if $y \in V$ and $y' \in B(y, r) \subset U$, joining an arc from x to y with the segment from y to y' we have an arc from x to y', whence $B(y, r) \subset V$. But $U \setminus V$ is also open, by the same argument; as $x \in V$, V is non-empty and $U \setminus V$ must be empty, that is $U = V$. □

Open connected sets are called *domains.*

Corollary 3.1. *In the real line, the domains are the open intervals.*

Proof. Assume U is open and arc-connected and define

$$a = \inf U, \quad b = \sup U,$$

where $a = -\infty$ if U is not bounded below and $b = +\infty$ if U is not bounded above. Then there exists $a' > a$ arbitrarily close to $a, a' \in U$, and there exists $b' < b$ arbitrarily close to $b, b' \in U$. Since U is connected, $[a', b'] \subset U$ and, this being true for all $a' > a$ arbitrarily close to p and all $b' < b$ arbitrarily close to b, it follows that $(a, b) = U$. □

In the next proposition, we show that in domains the arc joining two points can be chosen of a very particular nature, namely γ^* consisting of segments in the direction of the coordinate axes. We call them *step-paths.*

Proposition 3.1. *For an open set U, the following properties are equivalent:*

(a) *U is arc-connected.*
(b) *Two points in U can be joined by a step-path.*
(c) *Two points in U can be joined by an arc of class C^1.*

Proof. Assume that U is arc-connected, p, $q \in U$ and let $\gamma : [a,b] \to U$ be such that $p = \gamma(a), q = \gamma(b)$. The range γ^* is a compact set; let r be the distance from γ^* to U^c so that for each $x \in \gamma^*$, one has $Q(x, \frac{r}{\sqrt{d}}) \subset U$. By compactness of γ^*, a finite number of these cubes cover γ^*. If V is their union, V is an open set containing γ^* and it is evident that within V we can draw a step-arc joining p, q. To check that (b) implies (c), just note that the corners of a step-arc can be smoothened to produce an arc of class C^1. □

If U is a domain, using this proposition we can define the *Euclidean distance in U* between two points $x, y \in U$ as

$$d_U(x,y) = \inf L(\gamma), \tag{3.2}$$

where the infimum is taken for all arcs of class C^1 joining x, y within U. This infimum does not change if we use instead all piece-wise C^1-arcs, or simply polygonals. It satisfies the distance properties,

(a) $d_U(x,y) = d_U(y,x) \geq 0$.
(b) $d_U(x,y) = 0$ if and only if $x = y$.
(c) $d_U(x,y) \leq d_U(x,z) + d_U(z,y), x, y, z \in U$.

Obviously, $d_U(x,y) \geq |x-y|$, and $d_U(x,y) = |x-y|$ if U is convex.

Exercise 3.2. Prove that unless the segment joining x, y is included in U, the infimum in the definition of $d_U(x,y)$ is not attained. In fact, given an arbitrary rectifiable curve joining x, y within U, different from the segment joining x, y, there is a strictly shorter polygonal joining x, y in U. Therefore, $d = d_U$ if and only if U is convex.

In a general open set U, declaring that $x \equiv y$ if x, y can be joined by an arc within U establishes an equivalence relation. The same arguments as before show that each equivalence class is open, and arc-connected by definition. It follows that U is the union of domains U_i, called the *connected components* of U. By Theorem 2.5, the number of components is finite or at most countable. In the real line, every open set is a finite or countable union of open intervals.

3.4.5 Jordan arcs. If γ is one-to-one in (a,b), we say that the arc is *simple* or that it is a *Jordan arc.* Closed ($\gamma(a) = \gamma(b)$) and simple (γ injective on (a,b)) arcs can be thought as parametrized by the unit circle $\mathbb{T}$ in the plane and are called too *closed Jordan arcs.* Their range is called a *Jordan path.*

The *Jordan–Schoenflies* theorem states that if

$$\gamma : \mathbb{T} \to \mathbb{R}^2,$$

is a closed Jordan arc in the plane, and $\Gamma = \gamma(\mathbb{T})$, then there is an homeomorphism $\Phi : \mathbb{R}^2 \to \mathbb{R}^2$ such that $\Phi(\mathbb{T}) = \Gamma$. As a consequence, $\mathbb{R}^2 \setminus \Gamma$ has two connected components and Γ is a curve in the sense of Definition 3.2 [18]. However, in dimension 3 and higher this does not hold. The so-called *wild knots* are examples.

Thus a Jordan path is not always a curve. On the other hand, a curve is a Jordan path, in the sense that a global parametrization exists:

Exercise 3.3. Prove that a curve Γ, if connected, is either homeomorphic to an open interval (a,b) and then there exists a global parametrization $\Phi : (a,b) \to \Gamma$ or else is homeomorphic to a circle and there is a global parametrization $\Phi : [a,b] \to \Gamma$, with $\Phi(a) = \Phi(b)$.

Next theorem, which is a special case of the implicit function Theorem 6.3, provides a sufficient condition valid in general dimension. It is included here as it can already be proved with the concepts so far introduced.

Proposition 3.2. *Let $\gamma : [a,b] \to \mathbb{R}^n$ be a C^1 simple arc with $\gamma'(t) \neq 0$ for all t. Then $\gamma(a,b) = \Gamma$ is a curve. In fact, Γ is locally a graph parametrized by one of the coordinates. More precisely, for each $p = (p_1, \ldots, p_n) = (p_1, p') \in \Gamma$ there is a coordinate, say x_1, and an open interval $R = I \times R'$, where I is an open interval centered at p_1 and R' an open interval in $\mathbb{R}^{n-1}$ centered at p', and a C^1 map $\Phi : I \to \mathbb{R}^n$ such that $\Phi(p_1) = p'$ and $\Gamma \cap R = \{(x, x') \in R : x' = \Phi(x_1)\}$.*

Proof. We assume for simplicity that $n = 2$, $\gamma(t) = (x(t), y(t))$. Assume $\gamma(0) = p$. Assuming without loss of generality that $x'(0) > 0$, by continuity of x' one has $x'(t) > 0$ for $t \in I$ small enough, then $x(t)$ is strictly increasing, and so γ is one-to-one and γ^* can be parametrized by x. Thus, there is an open interval such that $\gamma(I)$ is a graph of a C^1 function $y = y(x)$. Now we must see that Γ does not reenter $B(p, r)$ if r is small enough. Let

$$r(t) = |\gamma(t) - \gamma(0)|^2 = (x(t) - x(0))^2 + (y(t) - y(0))^2;$$

then

$$\begin{aligned}\frac{1}{2}r'(t) &= (x'(t)(x(t) - x(0)) + y'(t)(y(t) - y(0))\\ &= x'(0)^2 t + o(t) + y'(t)(y(t) - y(0)).\end{aligned}$$

If $y'(0) \neq 0$, then $y'(t) \neq 0$ for t small enough and $y'(t)(y(t) - y(0)) \geq 0$; if $y'(0) = 0$, then $y'(t)(y(t) - y(0)) = o(t)$. In all cases, $r'(t) \geq x'(0)^2 t + o(t)$ for t small enough, and so $r(t)$ is strictly increasing. By shrinking I we may assume that $r(t)$ strictly increases for $t \in I$. Now let

$$\delta = \min\{|\gamma(t) - p| : t \notin I\}.$$

Since γ is simple and $\{\gamma(t) : t \notin I\}$ is compact, $\delta > 0$. Then, $\Gamma \cap B(p, \delta) \subset \gamma(I)$ is parametrized by x. □

If $\gamma : (a, b) \to \mathbb{R}^n$ is simple and C^1, then $\gamma(c, d)$ is a curve if $[c, d] \subset (a, b)$ but not necessarily $\gamma(a, b)$, because $\lim_{t\to a,b} \gamma(t)$ might be a point in $\gamma(a, b)$. For instance, the flower with two petals $r = \sin 2\theta$ can be parametrized as $\gamma(t), t \in (-\infty, +\infty)$, with $\gamma(0) = (0, 0), \lim_{t\to\pm\infty} \gamma(t) = (0, 0)$.

Chapter 4

Differentiation

In this chapter, the basic notions of multivariable calculus are introduced: partial derivatives, the differential, the gradient and its meaning, the chain rule, etc., considering both the analytic and geometric points of view. Dealing with just first-order derivatives is sufficient in order to solve certain optimization problems. The theoretical basis of the gradient descent method is carefully explained. Together with partial differentiation the student learns in the last section partial anti-differentiation, that is, solving elementary partial differential equations.

4.1 The Differential of a Function

4.1.1 Let $f : U \longrightarrow \mathbb{R}^m$ be a function defined on an open set $U \subset \mathbb{R}^n, n \geq 1$. Given $p \in U$, we would like to generalize the definition of the derivative $f'(p)$ for $n = 1$. We can look at $f'(p)$ in two (closely related) ways:

(a) The number $f'(p)$, by definition,

$$f'(p) = \lim_{x \to p} \frac{f(x) - f(p)}{x - p} = \lim_{h \to 0} \frac{f(p+h) - f(p)}{h},$$

measures *the rate of change* of f at p. If $f'(p) \neq 0$, the infinitesimal quantity $f(p+h) - f(p)$ is equivalent to $f'(p)h$.

(b) The affine function $f(p) + f'(p)h$ is the unique affine map $A + Bh$ which approximates $f(p+h)$ at first order near p, meaning

$$f(p+h) - (A + Bh) = o(h), h \to 0.$$

Recall that the notation $A(h) = o(B(h))$ as $h \to 0$ means

$$\lim_{h\to 0} \frac{|A(h)|}{|B(h)|} = 0.$$

Note that continuity of f at p means that the constant function $A = f(p)$ is the only one approximating f at order zero: $f(p+h) - A = o(1)$.

Let us start along the lines of the first point of view in case $n > 1$. The first obvious point to remark is that we cannot consider the quotient of increments, because $h = x - p$ is a vector and we cannot divide by a vector. Incidentally, the case $n = 2$ is an exception because identifying $\mathbb{R}^2$ with the field of complex numbers we *can* copy this definition; doing this would lead us to another area, *complex analysis and holomorphic functions*, see paragraph 9.5.2.

Instead, what we can do is to imitate the notion of directional limit and introduce *directional derivatives*: given a non-zero vector v, we consider the line through p with direction vector v, in parametric form $p + tv, t \in \mathbb{R}$ and see how this function changes at $t = 0$:

$$\lim_{t\to 0} \frac{f(p+tv) - f(p)}{t} = D_v f(p).$$

If this limit exists, we call it the *derivative of f at p in the direction v*. If $f = (f_1, \dots, f_m)$, it is evident that this limit exists if and only if it exists for each component f_j, and in this case $D_v f(p)$ is the vector in $\mathbb{R}^m$ with components $D_v f_j(p)$.

The intuitive interpretation of $D_v f(p)$ depends on whether $m = 1$ or $m > 1$. If $m = 1$, the number $D_v f(p)$ measures the rate of change of f at p in the direction v. If $D_v f(p)$ is a big positive number, then $f(p+tv)$ increases fast at $t = 0$, etc. If $m > 1$ and $\gamma(t) = f(p + tv)$, then $D_v f(p) = \gamma'(0)$ is the tangent vector of the arc $\gamma(t)$ at the point $\gamma(0) = f(p)$. For this reason, when $m > 1$, it is more convenient to have a geometric view of f, as a *transformation* moving points, rather than m scalar functions considered independently.

When v is the vector e_i of the canonical basis, we use the notation $D_i f(p)$, and also $\frac{\partial f}{\partial x_i}(p)$, $\frac{\partial f}{\partial x}(p)$, $\frac{\partial f}{\partial y}(p)$, $\frac{\partial f}{\partial z}(p)$ when $n = 2, 3$. We will use too the notations $f_x(p), f_y(p), f_z(p)$. If the directional derivative exists at all points of U, we have a new function $D_v f(x)$, or $\frac{\partial f}{\partial x}$, etc. again functions of n variables, called the *partial derivatives*.

By definition, $\frac{\partial f}{\partial x_i}(p), p = (p_1, \dots, p_n)$, is obtained keeping $x_j = p_j$ for $j \neq i$ and differentiating with respect to x_i at p_i. So, $\frac{\partial f}{\partial x_i}(x)$ is obtained differentiating with respect to x_i thinking that the other variables $x_j, j \neq i$ remain constant. In doing so, we can apply the familiar rules of the differential calculus in one variable. In particular, if f is given by a formula in the variables x_i, $\frac{\partial f}{\partial x_i}$ exists at all points where the formula makes sense. For instance, the x, y derivatives of $f(x,y) = \sin(xy^2)$ are

$$f_x = y^2 \cos(xy^2), \quad f_y = 2xy\cos(xy^2).$$

It is straightforward to see that if $D_v f(p)$ exists and $\lambda \in \mathbb{R}$, then $D_{\lambda v} f(p)$ also exist and equals $\lambda D_v f(p)$. However, for linearly independent directions v, w, $D_v f(p)$ may exist and $D_w f(p)$ not. For instance, $f(x,y) = x^2 y \sin\frac{1}{y}, y \neq 0, f(x,0) = 0$, has derivative $\frac{\partial f}{\partial x} = 0$ at $(0,0)$, but $\frac{\partial f}{\partial y}$ does not exist at $(0,0)$. More than that, as it might be expected, the existence of all directional derivatives does not even imply the continuity at p. Consider again the function, in $n = 2$,

$$f(x,y) = \frac{x^2 y}{x^4 + y^2}, \quad f(0,0) = 0,$$

that was shown to be discontinuous at $(0,0)$ in paragraph 2.4.1. This function has zero limit along every line through the origin, is zero on the x, y axis and for a direction $v = (1, \alpha)$, then

$$D_v f(0,0) = \lim_{t \to 0} \frac{f(t, \alpha t)}{t} = \lim_{t \to 0} \frac{\alpha}{t^2 + \alpha^2} = \frac{1}{\alpha}$$

exists.

4.1.2 Thus, the existence of the directional derivatives is not the right definition if $n > 1$. The correct definition arises if we adopt the second point of view, in (b) above, the *existence of a linear approximation of the increment* $f(p+h) - f(p)$.

Definition 4.1. We say that $f : U \longrightarrow \mathbb{R}^m$ is *differentiable at a point* $p \in U$ if there exists a linear map from $\mathbb{R}^n$ to $\mathbb{R}^m$, called *the differential of* f *at* p and denoted $df(p)$, such that

$$f(p+h) = f(p) + df(p)(h) + R(h), \quad |R(h)| = o(|h|), h \to 0,$$

that is,

$$\lim_{h \to 0} \frac{|f(p+h) - f(p) - df(p)(h)|}{|h|} = 0,$$

where $df(p)(h)$ stands for the action of $df(p)$ on the vector h.

Defining

$$\tau(\delta) = \sup_{|h| \le \delta} \frac{|R(h)|}{|h|},$$

then τ decreases to zero as $\delta \to 0$ and

$$|R(h)| \le \tau(|h|)|h|.$$

We will often use this expression of the rest.

With the notation $x = p + h$, this means that $f(p) + df(p)(x - p)$ is the *affine approximation* of $f(x)$ near p,

$$f(x) = f(p) + df(p)(x - p) + o(|x - p|).$$

Obviously, then f is continuous at p, because $df(p)(h) \to 0$ as $h \to 0$. Again, if $f = (f_1, \dots, f_m)$, f is differentiable at p if and only if every component is, and in this case $df(p)(h) = (df_1(p)(h), \dots, df_m(p)(h))$.

In dimension $n = 1$, every linear map $L : \mathbb{R} \to \mathbb{R}^m$ is of the form $L(h) = hv$ for some vector $v = L(1) \in \mathbb{R}^m$, so that f is differentiable at p if and only if the derivative

$$f'(p) = \lim_{h \to 0} \frac{f(p + h) - f(p)}{h}$$

exists, and then $df(p)(h) = hf'(p)$.

When $m = 1$, the geometrical interpretation is the same as in the case $n = 1, m = 1$: f being differentiable at p means that the affine sub-manifold in $\mathbb{R}^{n+1}$ with parametric equation $h \to f(p) + df(p)(h)$ is tangent at $(p, f(p))$ to the graphic $y = f(x)$. In $n = 2$, with the notations (x, y) instead of x and (a, b) instead of p, the plane of equation

$$z - f(a, b) = \frac{\partial f}{\partial x}(a, b)(x - a) + \frac{\partial f}{\partial y}(a, b)(y - b)$$

is tangent to the graphic $z = f(x, y)$ at $(a, b, f(a, b))$. See Figure 4.1.

4.1.3 If f is differentiable at p and taking $h = tv$, where v is a direction vector as above, and since $df(p)(tv) = tdf(p)(v)$, we find that

$$f(p + tv) = f(p) + tdf(p)(v) + o(t).$$

Hence, $D_v f(p)$ exists and equals $df(p)(v)$. In particular, this shows that the differential is unique and consists in assigning to each direction v

Figure 4.1. A tangent plane to a graphic.

the directional derivative $D_v f(p)$, the tangent vector at $f(p)$ to the arc $f(p + tv)$.

Thus, the existence of $df(p)$ not only implies that f is continuous and that all directional derivatives exist, but also that $v \to D_v f(p)$ is linear. As a consequence, if we know $D_{v_j} f(p)$ for n linearly independent vectors v_j, then $df(p)$ is known, and so is $D_v f(p)$ for all vectors v.

Example 4.1. Assume that $f(x, y)$ is differentiable at p and that $D_{v_1} f(p) = 1$ for $v_1 = (1, 2)$ and $D_{v_2} f(p) = -2$ for $v_2 = (-1, 3)$. To find $D_v f(p)$, we first compute the coefficients a, b in $v = av_1 + bv_2$ and then $D_v f(p) = a - 2b$.

The function $f(x, y) = 0, y \neq 0, f(x, 0) = x$ is continuous at $(0, 0)$ and all directional derivatives at $(0, 0)$ are zero, except with respect to x, which equals 1; thus $v \to D_v f(0, 0)$ is not linear and f is nor differentiable at $(0, 0)$.

In conclusion, differentiability at p amounts to:

(a) Continuity at p, that is, existence of $f(p) = \lim_{x \to p} f(x)$.

(b) Existence of the n partial derivatives

$$A_i = \frac{\partial f}{\partial x_i}(p) = \lim_{t \to 0} \frac{f(p + te_i) - f(p)}{t}.$$

(c) The linear map $h = (h_1, \ldots, h_n) \to L(h) = A_1 h_1 + \cdots + A_n h_n$ must satisfy

$$|f(p+h) - f(p) - L(h)| = o(h).$$

Example 4.2. Consider in $n = 2$ the continuous function $f(x, y) = x^2 + 2xy^2$ and $p = (1, 1)$. The derivatives with respect to x, y are respectively $2x + 2y^2, 4xy$, which equal $4, 4$ at p. The affine map

$$L(x, y) = 3 + 4(x-1) + 4(y-1),$$

satisfies $f(x, y) - L(x, y) = o(\sqrt{(x-1)^2 + (y-1)^2})$. To check this, with $x - 1 = x', y - 1 = y'$,

$$\begin{aligned} f(x, y) &= (1 + x')^2 + 2(1 + x')(1 + y')^2 = 1 + 2x' + x'^2 + 2(1 + x') \\ &\quad \times (1 + y'^2 + 2y') \\ &= 3 + 4x' + 4y' + R = L(x, y) + R, \end{aligned}$$

where R denotes the sum of all terms in $x'^2, y'^2, x'y'$. All quadratic terms are $o(\sqrt{x'^2 + y'^2})$.

In general, since

$$\frac{1}{n} \leq \frac{|h|}{|h_1| + \cdots + |h_n|} \leq 1,$$

all quadratic expressions in the h_i are of course $o(|h|)$.

Continuity at p and linearity of $v \to D_v f(p)$ do not together imply the existence of $df(p)$, as shown by the example

$$f(x, y) = \frac{x^{1+\varepsilon} y^2}{x^2 + y^4}, \quad 0 < \varepsilon \leq \frac{1}{2}.$$

Since $|f(x, y)| \leq x^\varepsilon$, f is continuous at $(0, 0)$; it is zero on both axis and on $y = \alpha x$,

$$f(x, \alpha x) = \frac{\alpha^2 x^{1+\varepsilon}}{1 + \alpha^4 x^2}$$

is $o(t)$, so all directional derivatives are zero. However, $f(x, y)$ is not $o(\sqrt{x^2 + y^2})$ because on $x = cy^2, f(cy^2, y) = ky^{2\varepsilon}$ while $\sqrt{x^2 + y^2}$ is like y.

4.1.4 The following is a sufficient condition for differentiability in terms of directional derivatives:

Proposition 4.1. *If $D_i f(x), i = 1, \ldots, n$ exist for $x \in B(p, \varepsilon)$ and are continuous at p, f is differentiable at p.*

Proof. We prove it for $n = 2$ with notation $f(x, y), p = (a, b)$, the proof for $n > 2$ being analogous. We decompose

$$f(x,y) - f(a,b) = f(x,y) - f(x,b) + f(x,b) - f(a,b).$$

By the mean-value theorem applied to the one-variable functions $y \to f(x, y)$, $x \to f(x, b)$ there exist ξ between y, b, and η_y between a, x such that

$$f(x,y) - f(a,b) - \frac{\partial f}{\partial x}(a,b)(x-a) - \frac{\partial f}{\partial y}(a,b)(y-b)$$

$$= \left(\frac{\partial f}{\partial x}(\eta, b) - \frac{\partial f}{\partial x}(a,b)\right)(x-a) + \left(\frac{\partial f}{\partial y}(x,\xi) - \frac{\partial f}{\partial y}(a,b)\right)(y-b).$$

The continuity of the partial derivatives at (a, b) implies that the two last terms are indeed $o(|h|)$. □

Functions f in U such that $D_i f$ exists and is continuous in U are called of *class* C^1, denoted $f \in C^1(U)$.

The main practical consequence of the proposition is that a function expressed by a formula is differentiable at all points where the formula makes sense.

All natural statements regarding operations with differentiable functions hold:

Exercise 4.1. If $f, g : U \longrightarrow \mathbb{R}^m$ are differentiable at p, then:

(a) The sum $f + g$ is differentiable with differential $df(p) + dg(p)$.
(b) The scalar product $E(x) = \langle f(x), g(x) \rangle$ is differentiable with differential

$$dE(p)(h) = \langle f(p), dg(p)(h) \rangle + \langle df(p)(h), g(p) \rangle.$$

In particular, for scalar functions, $d(fg)(p) = f(p)dg(p) + g(p)df(p)$.
(c) If $m = 3$ and $V(x) = f(x) \times g(x)$ is the cross product, V is differentiable at p with differential

$$dV(p)(h) = df(p)(h) \times g(p) + f(p) \times dg(p)(h).$$

4.1.5 We finish this section with a remark on notation. Classically, the increments $h_1, \ldots, h_n$, when thought to be infinitesimal, are denoted $dx_1, \ldots, dx_n$; the corresponding infinitesimal increment of f is denoted df and we write

$$df = \sum_i \frac{\partial f}{\partial x_i}\, dx_i.$$

Another (formal) way of looking at this is to interpret $dx_1, \ldots, dx_n$ as the dual basis of the canonical basis, the differentials of the coordinate functions, meaning that the action of dx_i on a vector h is $dx_i(h) = h_i$, its i-th component, and that the above is an equality between linear maps. The action of df on h is then

$$\sum_i \frac{\partial f}{\partial x_i} dx_i(h) = \sum_i \frac{\partial f}{\partial x_i} h_i.$$

4.2 The Jacobian Matrix: The Chain Rule

4.2.1 If f is differentiable at p and we write the increment h in the canonical basis, $h = (h_1, \ldots, h_n) = \sum_i h_i e_i$, by linearity

$$df(p)(h) = \sum_i h_i df(p)(e_i) = \sum_i h_i D_i f(p).$$

If $f = (f_1, \ldots, f_m)$, $D_i f(p)$ is the vector in $\mathbb{R}^m$ with components

$$(D_i f_1(p), \ldots, D_i f_m(p)).$$

Hence, the matrix of $df(p)$ in the canonical basis of $\mathbb{R}^n, \mathbb{R}^m$ is

$$\begin{pmatrix} D_1 f_1(p) & D_2 f_1(p) & \ldots & D_n f_1(p) \\ D_1 f_2(p) & D_2 f_2(p) & \ldots & D_n f_2(p) \\ \vdots & \vdots & \vdots & \vdots \\ D_1 f_m(p) & D_2 f_m(p) & \ldots & D_n f_m(p) \end{pmatrix}.$$

We use as well the notation $df(p)$ for this matrix, called the *Jacobian matrix* of f at p. Its ith column is $D_i f(p), 1 \leq i \leq n$ and its jth row is $df_j(p)$, $1 \leq j \leq m$.

The chain rule for real functions of one real variable states that $g(f(x))$ has derivative $g'(f(x))f'(x)$. This is because $f'(x)$ is the factor to multiply an infinitesimal increment dx to get the infinitesimal increment dy of

$y = f(x)$, and in turn $g'(y)$ is the one to multiply dy to get the infinitesimal increment dz of $z = g(y)$. Now we analyze the general version:

Proposition 4.2. *If $f : U \to \mathbb{R}^m$ is differentiable at p and $g : V \to \mathbb{R}^k$ is differentiable at $q = f(p)$, the map $F(x) = g(f(x))$ is differentiable at p with differential $dF(p)(h) = dg(q)(df(p)(h))$, the composition of the differentials.*

Proof. We denote by h the small increments of x and by r the small increments of $y = f(x)$. The hypothesis means

$$f(p+h) = f(p) + df(p)(h) + R(h), \quad g(q+r) = g(q) + dg(q)(r) + S(r),$$

with $R(h) = o(h), S(r) = o(r)$ as $h, r \to 0$, respectively. We may write $f(p+h) = q + r$ with $r = df(p)(h) + R(h)$ (that indeed goes to zero as $h \to 0$). Therefore, by linearity of $dg(q)$

$$g(f(p+h)) = g(q) + dg(q)(df(p)(h)) + dg(q)(R(h)) + S(r).$$

We must check that the last two terms are $o(h)$. But since

$$|r| \leq |df(p)(h)| + |R(h)| \leq |df(p)||h| + O(h) = O(h),$$

all quantities $o(r)$, like $S(r)$, are as well $o(h)$. Now it remains to show that $dg(q)(R(h)) = o(h)$. But this follows from $|dg(q)(R(h))| \leq |dg(q)||R(h)| = o(h)$. □

In terms of the matrices $df(p) = \left(\frac{\partial f_j}{\partial x_i}(p)\right)_{i,j}$, $dg(q) = \left(\frac{\partial g_l}{\partial y_j}(q)\right)_{l,j}$, the matrix $dF(p)$ is the product matrix $dg(q)df(p)$; the entry $\frac{\partial F_l}{\partial x_i}$ of row $l = 1, \ldots, k$ and column $i = 1, \ldots, n$ is

$$\frac{\partial F_l}{\partial x_i}(p) = \sum_{j=1}^{m} \frac{\partial g_l}{\partial y_j}(q)\frac{\partial f_j}{\partial x_i}(p).$$

Or in vector terms

$$\frac{\partial F}{\partial x_i}(p) = \sum_{j=1}^{m} \frac{\partial g}{\partial y_j}(q)\frac{\partial f_j}{\partial x_i}(p).$$

The intuition of this rule is as follows. We must find the derivative with respect to x_i of $F_l(x) = g_l(f(x)) = g_l(f_1(x), \ldots, f_m(x))$. When x_i suffers an increment dx_i, all $y_j = f_j(x)$ get incremented by $dy_j = \frac{\partial f_j}{\partial x_i}dx_i$; the increment of y_j leads to the increment $\frac{\partial g_l}{\partial y_j}dy_j$, and adding on j we get the total increment. The rule is thus that to differentiate $g(f(x))$ with respect to x_i, we differentiate g with respect to each of its variables, multiply by the derivative of the variable with respect to x_i, and add.

4.2.2 A restatement of the chain rule in terms of directional derivatives is as follows: if v is a direction in $\mathbb{R}^n$, and $w = df(p)(v) = D_v f(p)$, then

$$\begin{aligned} D_v(F)(p) &= d(F)(p)(v) \\ &= dg(f(p))(df(p)(v)) = dg(f(p))(w) = D_w g(f(p)). \quad (4.1) \end{aligned}$$

This leads to the following geometric interpretation of the differential. Given a direction v, $D_v f(p)$ is by definition the tangent vector at $f(p)$ to the arc $f(p + tv)$. Now consider instead of the segment $p + tv$, a regular arc $\gamma(t)$ through $p, \gamma(0) = p$ with tangent $v, \gamma'(0) = v$, and look as well to the composition map $\hat{\gamma}(t) = f(\gamma(t)) = f(\gamma_1(t)), \ldots, \gamma_n(t))$, through $f(p)$. Its tangent at $f(p)$ is by the chain rule

$$\hat{\gamma}'(0) = \sum_{i=1}^{n} \frac{\partial f}{\partial x_i}(p)x_i'(0) = \sum_{i=1}^{n} \frac{\partial f}{\partial x_i}(p)v_i,$$

that equals $D_v f(p)$. So this is how the differential works: given v, we consider any regular arc γ through p with tangent vector v, transport γ with f, and consider the tangent at $f(p)$ of this transported arc. In particular, this shows that $D_v f(p)$ just depends on the values of f on a path through p with tangent v.

With this interpretation in mind, the chain rule itself is geometrically evident: if v is a direction vector, $D_v F(p) = dF(p)(v)$ is the tangent vector to the arc $g(f(p + tv))$; but this is the image by g of the arc $f(p + tv)$; by what has been said, $dF(p)(v)$ equals $dg(f(p))(w) = D_w g(f(p))$ where w is the tangent vector of $f(p + tv)$ at $f(p)$, namely $w = df(p)(v) = D_v f(p)$.

4.2.3 A *diffeomorphism* between two domains $U \subset \mathbb{R}^n, V \subset \mathbb{R}^m$ is a one-to-one differentiable map f from U onto V such that the inverse f^{-1} is also differentiable. By the chain rule, $df(x)$ is a linear invertible map with inverse $df^{-1}(f(x))$, in particular $n = m$. Domains of different dimensions cannot be diffeomorphic. Later on we will prove that, conversely, a one-to-one map of class C^1 from a domain U to $\mathbb{R}^n$ with $df(x)$ invertible, that is $\det df(x) = 0$, is a diffeomorphism.

4.3 The Gradient of a Scalar Function

4.3.1 In this section, we assume that f is a scalar function of n variables, differentiable at all points of an open set U. For a vector $v = (v_1, \ldots, v_n)$, we have seen

$$D_v f(p) = df(p)(v) = \sum_{i=1}^{n} v_i D_i f(p).$$

In view of this, it is natural to introduce the vector

$$\nabla f(p) = (D_1 f(p), \ldots, D_n f(p)),$$

called the *gradient of f at p*, so that

$$\langle \nabla f(p), v\rangle = D_v f(p). \tag{4.2}$$

Note that the chain rule

$$\frac{\partial F}{\partial x_i}(p) = \sum_{j=1}^{m} \frac{\partial g}{\partial y_j}(q)\frac{\partial f_j}{\partial x_i}(p),$$

is written

$$\nabla F(p) = \sum_{j=1}^{n} \frac{\partial g}{\partial y_j}(q)\nabla f_j(p),$$

and also

$$\nabla F(p) = (df(p)^t)(\nabla g(q)),$$

where $(df(p))^t$ denotes the transpose matrix. This is another way of writing (4.1) because if $w = D_v f(p) = df(p)(v)$,

$$\begin{aligned} D_v F(p) &= \langle \nabla F(p), v\rangle = \langle df(p)^t(\nabla g(q)), v\rangle = \langle \nabla g(q), df(p)(v)\rangle \\ &= \langle \nabla g(q), w\rangle = D_w g(q). \end{aligned}$$

The gradient is a concept related to the scalar product, by (4.2). As a consequence, if $v_1, \ldots, v_n$ is an orthonormal basis of $\mathbb{R}^n$, its expression in this basis is also

$$\nabla f = \sum_i D_{v_i} f\, v_i.$$

The gradient has different interpretations that we explain below.

4.3.2 First, assume $v, |v| = 1$, and $\nabla f(p) \neq 0$. By the Cauchy–Schwarz inequality, $|\langle \nabla f(p), v\rangle| \leq |\nabla f(p)|$, that is,

$$-|\nabla f(p)| \leq D_v f(p)| \leq \nabla f(p)|.$$

For $v = \frac{\nabla f(p)}{|\nabla f(p)|}$, the directional derivative equals $|\nabla f(p)|$, while for $v = -\frac{\nabla f(p)}{|\nabla f(p)|}$, it equals $-|\nabla f(p)|$. This means that the direction v along which $D_v f(p)$ is maximum is that of the gradient, and equal to the length of the gradient, and minimum in the opposite direction. Since $D_v f(p)$ measures the rate of change in the direction v, it turns out that if $\nabla f(p) \neq 0$, its direction is the one along which f increases faster, and $|\nabla f(p)|$ is the highest rate of increase.

Thus, for each $p \in U$ one has a vector $\nabla f(p)$ that is convenient to view as a vector with origin at p. This is an application

$$\nabla f : U \longrightarrow \mathbb{R}^n.$$

Example 4.3. Assume in $n = 2$ that $f(x, y)$ represents temperature at point (x, y), for example

$$f(x, y) = 50 - 8x^8 - 81y^6,$$

the origin being at highest temperature. Assume someone is at point $(1, \frac{2}{3})$, feels too cold, and wishes to move to a warmer region. The person decides to move in the direction of maximum growth of the temperature, at every time, and reach $(0, 0)$. We ask ourselves what path $\gamma(t)$ will follow and its length. Moving in the direction of maximum increase means that the speed vector must have the direction of the gradient, that is, $\gamma'(t) = \lambda(t)\nabla f(\gamma(t))$, for some positive function $\lambda(t)$; if $\gamma(t) = (x(t), y(t))$, this means that

$$x'(t) = \lambda(t)\frac{\partial f}{\partial x}(x(t), y(t)) = -64\lambda(t)x(t)^7,$$

$$y'(t) = \lambda(t)\frac{\partial f}{\partial y}(x(t), y(t)) = -6 \cdot 81\,\lambda(t)y(t)^5.$$

We want as well that $\gamma(0) = (1, \frac{2}{3})$. The function $\lambda(t)$ will depend on how fast the traveller moves along γ^*; since we are interested only in the latter, we eliminate $\lambda(t)$ to get

$$\frac{x'(t)}{64x(t)^7} = \frac{y'(t)}{6 \cdot 81\,y(t)^5},$$

showing that $81x(t)^{-6} - 16y(t)^{-4}$ is a constant C; imposing that $x(0) = 1$, $y(0) = \frac{2}{3}$ we find $C = 0$, so that the path to reach zero is given by

$y = \frac{2}{3}x^{\frac{3}{2}}, 0 \leq x \leq 1$. Its length is

$$\int_0^1 \sqrt{1+x}\,dx = \frac{2}{3}(2^{\frac{3}{2}} - 1).$$

In general, a map $X : U \longrightarrow \mathbb{R}^n$, interpreted in this way — $X(p)$ a vector with origin at p — is said to be a *vector field*. Examples of vector fields are force fields created by a mass distribution, for instance the Newtonian field created by a point mass at a point p

$$X(q) = c\frac{q-p}{|q-p|^3}, \quad q \neq p.$$

Another example is the velocity field of a fluid in stationary motion, $X(p)$ being the velocity of the fluid at point p.

4.3.3 A second interpretation of the gradient, when non-zero, is in terms of the level sets $L_c = \{x \in U : f(x) = c\}$. Recall that a point $p \in U$ is in $L_c : c = f(p)$ and U is the disjoint union of all level sets. We think of this set as something of dimension $n-1$, although, as pointed out in paragraph 3.4.1, this is not always the case. In the next proposition, the tangent cone has been introduced in paragraph 3.4.3.

Proposition 4.3. *$\nabla f(p)$ is perpendicular to $T_p(L_c), c = f(p)$.*

Proof. Assume that $\gamma(t)$ is a parametrized regular arc lying on L_c, $\gamma(0) = p$. Differentiating $f(\gamma(t)) = f(x_1(t), \ldots, x_n(t)) = c$ at zero using the chain rule we obtain

$$\langle \nabla f(p), \gamma'(0) \rangle = \sum_{i=1} D_i f(p) x_i'(0) = 0.$$

This shows that $\nabla f(p)$ is perpendicular at p to all tangent vectors at p of regular curves lying on L_c. □

Thus, $\nabla f(p)$ is orthogonal to the tangent cone of the level set through p, and informally we say that it is orthogonal to L_c. Later on, in Theorem 7.2, we will see that for C^1 maps, conversely, given a non-zero vector v orthogonal to $\nabla f(p)$, there exists a parametrized curve γ lying on L_c with tangent $\gamma'(0) = v$, so that $T_p(L_c)$ is exactly the orthogonal of the gradient, in particular a linear subspace, of dimension $n-1$, that will be called the *tangent space* of L_c at p. Its equation is $\langle \nabla f(p), x-p\rangle = 0$, or in developed form, if $p=(p_1,\ldots,p_n)$,

$$\sum_{i=1}^{n} D_i f(p)(x_i - p_i) = 0.$$

In case M is an intersection of level sets of m functions,

$$M = \{x \in U : f_i(x) = c_i\}, \quad i = 1,\ldots,m,$$

and $p \in M$, the m gradients $\nabla f_j(p)$ are orthogonal to $T_p(M)$.

Example 4.4. The ellipsoid

$$f = \frac{x^2}{4} + y^2 + \frac{z^2}{9} = 3,$$

has a tangent plane at $p = (2,1,3)$. Its normal vector is $\nabla f(p) = (1,2,\frac{2}{3})$, therefore the plane has the equation

$$(x-2) + 2(y-1) + \frac{2}{3}(z-3) = 0.$$

The intersection with the plane $x+y+z=6$ is a curve Γ trough p whose tangent at p is orthogonal to $\nabla f(p)$ and $n=(1,1,1)$. Therefore, $\nabla f(p) \times n = \frac{1}{3}(4,-1,-3)$ is a tangent vector to Γ at p.

4.3.4 A third use of the gradient is regarding *local extrema.* A point $p \in U$ is called a *local minimum* of f if there exists a ball $B(p,r) \subset U$ such that $f(p) \le f(x), x \in B(p,r)$ and a *local maximum* in case $f(p) \ge f(x), x \in B(p,r)$. Then, if f is differentiable at p and v is any direction vector, the function $t \to f(p+tv)$ has a local extrema at $t=0$; therefore its derivative $D_v f(p)$ must be zero; since this holds for all v, it follows that $df(p) = 0, \nabla f(p) = 0$.

The points with $\nabla f(p) = 0$ are called *critical points* of f, so all extrema are critical points. Obviously, not all critical points are extrema; for instance $f(x,y) = x^2 - y^2$ has a critical point at $(0,0)$ which is not a local maximum neither a local minimum. In the direction of the x axis it has a minimum and in the direction of the y axis it has a maximum; in this situation when there are linearly independent directions along which the function has the two types of extrema, we say that p is a *saddle point.* The function $f(x,y) = x^3 + y^3$ also has a critical point at $(0,0)$, and is not an extrema nor a saddle point. Later on we will learn how to classify these points.

4.3.5 Functional dependence. In the situation of the chain rule,

$$F(x) = g(f_1(x), \dots, f_m(x)), \quad x \in U, \tag{4.3}$$

with g scalar-valued, one says that F *depends functionally* on $f_1, f_2, \dots, f_m$ in U. This amounts to saying that the value $F(x)$ just depends on the values $f_i(x)$, for all $x \in U$ in a differentiable way. More generally, a system of functions $f_1, f_2, \dots, f_m$ defined in U is called *functionally dependent in U* if some of the functions depend functionally on the remaining ones in U. If it is not the case, we call them *a system of functionally independent functions in U*. If they are functionally independent on all sub-domains $V \subset U$, we call them *strongly functionally independent in U*.

If (4.3) holds, the chain rule states

$$dF = \sum_j \frac{\partial g}{\partial y_j} df_j, \quad \nabla F = \sum_j \frac{\partial g}{\partial y_j} \nabla f_j.$$

If $m = n - 1$, the matrix with column vectors $\nabla F, \nabla f_j$ has zero determinant, and we find an expression

$$\sum_{i=1}^{n} A_i(x) D_i F(x) = 0. \tag{4.4}$$

This is called a *linear first-order partial differential equation.*

So we can state:

Theorem 4.1.

(a) C^1 *functions* F *depending functionally on* $n-1$ *functions* $f_1, \dots, f_{n-1}$ *satisfy a linear first-order equation* (4.4).

(b) *If* $f_1, f_2, \dots, f_m$ *are functionally dependent in* V, *the gradients*

$$\nabla f_1(x), \dots, \nabla f_m(x)$$

are linearly dependent for $x \in V$.

(c) *The system* $f_1, f_2, \dots, f_m$ *is strongly functionally independent in* U *if the gradients* $\nabla f_1(x), \dots, \nabla f_m(x)$ *are linearly independent for all* $x \in U$.

Thus, for $m \le n$ differentiable functions $f_1, f_2, \dots, f_m$, the rank of the system ∇f_i may indicate the existence of some functional dependence among the f_i.

Example 4.5. The functions

$$f = x^2 + y - z, \quad F = x^4 + y^2 + z^2 + 2x^2y - 2x^2z - 2yz - x^2 - y + z,$$

satisfy

$$\nabla F = (2x^2 + 2y - 2z - 1)\nabla f.$$

This indicates that $F = g(f)$ might hold with $g'(f) = (2x^2 + 2y - 2z - 1)$. By inspection, we see that $g'(t) = 2t - 1$, so $g(t) = t^2 - t + c$, and check that indeed $F = f^2 - f$.

In full generality, if $\nabla F(x)$ is a linear combination of $\nabla f_j(x), x \in U$ (a local condition) F is not necessarily functionally dependent on $f_1, \ldots, f_m$ in U (a global condition). For instance, paragraph 4.7.1 shows an example of a function F such that $\frac{\partial F}{\partial x} = 0$ (that is ∇F is a scalar multiple of ∇y), but F is not a function of y. Local results will be explained in paragraph 6.1.5.

4.4 Finding Extreme Values

Together with local extrema we must consider *absolute extrema* of a function f on a set A. A point $p \in A$ is called an *absolute maximum* (resp., *absolute minimum*) of f on A if $f(x) \leq f(p) = M$ ($f(x) \geq f(p) = m$) for $x \in A$. Then M, m are the extreme values of f on A. Of course, they do not always exist. Clearly, if $p \in \mathring{A}$, then p is a local extrema and so a critical point if f is differentiable. This simple observation allows to compute the extreme values in some cases. For instance, assume that f is differentiable in $\mathbb{R}^n$ and $\lim_{|x|\to\infty} f(x) = +\infty$. Then f has an absolute minimum p which must be a critical point.

4.4.1 An important example is Gauss' *least squares method.* The general framework is that of a system of equations

$$f_j(x_1, \ldots, x_n) = c_j, \quad j = 1, \ldots, m,$$

with presumably no exact solutions, for instance if m is much greater than n. One then looks for approximate solutions, attributing

$$f(x) = \sum_j (f_j(x_1, \ldots, x_n) - c_j)^2,$$

as error to the approximate solution $x = (x_1, \ldots, x_n)$. It is then natural to single out the approximate solution minimizing the error. When the f_j

are linear, then f has infinite limit at infinity and therefore an absolute minimum exists (see paragraph 4.5.2).

Here we will look at a particular case in the context of statistics, the *regression line.* Suppose we are analyzing two random variables X, Y associated to an experiment. The best predictor of Y linear in $X, aX + b$, is defined as the one for which the expected value of $[Y - (aX + b)]^2$ is minimal. Given a sample

$$(x_1, y_1), \ldots, (x_m, y_m),$$

an estimate of the parameters a, b is obtained by minimizing

$$f(a, b) = \sum_{j=1}^{m} (y_j - ax_j - b)^2.$$

Geometrically, the line $y = ax + b$ is the one best fitting (vertically) the cloud of points $(x_j.y_j)$. Note then the deviation of the point (x_j, y_j) from the line $y = ax + b$ is measured here by $|y_j - ax_j - b|$ and not by the Euclidean distance $(1 + a^2)^{-\frac{1}{2}} |y_j - ax_j - b|$. This is because we want to predict Y in terms of X, the variables do not play the same role. Working with the Euclidean distance would lead to something symmetric in X, Y.

For the computation, we consider

$$\overline{x} = \frac{1}{m} \sum_j x_j, \quad \overline{y} = \frac{1}{m} \sum_j y_j,$$

and

$$\sigma_x^2 = \frac{1}{m} \left(\sum_j x_j^2 \right) - \overline{x}^2, \quad \sigma_y^2 = \frac{1}{m} \left(\sum_j y_j^2 \right) - \overline{y}^2,$$

$$\mathrm{cov}_{x,y} = \frac{1}{m} \left(\sum_j x_j y_j \right) - \overline{xy}.$$

Exercise 4.2. Check that $f(a, b)$ has a unique critical point at

$$\hat{a} = \frac{\mathrm{cov}_{x,y}}{\sigma_x^2}, \quad \hat{b} = \overline{y} - \hat{a}\overline{x}.$$

4.4.2 By Weierstrass' Theorem 2.9, every continuous function f attains an absolute maximum and an absolute minimum on a compact set K. Finding these values in specific cases is a task that can be accomplished for differentiable functions and compacts with a particular structure. With the tools at our disposal at this stage — just the notion of critical point and one variable differential calculus — we can already deal with the compact sets K in the plane whose boundary bK is, with the exception of a finite set V of vertices, the union of a finite number of open regular paths γ_i^*.

Assume that f is differentiable and attains its maximum on K at $p \in K$. If $p \in \overset{\circ}{K}$, there is a ball $B(p,r) \subset K$, so that p is a local maximum, whence is critical, $\nabla f(p) = 0$. If $p \in bK$ and is not a vertex, say $p = \gamma(0), \gamma : (a,b) \to bK$, then $\phi(t) = f(\gamma(t))$ has a local maximum at $t = 0$, therefore

$$0 = \phi'(0) = \langle \nabla f(p), \gamma'(0) \rangle.$$

That is, the tangential derivative must vanish. Thus, to find out the extreme values, we consider the set E of candidate points consisting of the following:

(a) The critical points in $\overset{\circ}{K}$. Typically, the system of critical points $f_x = 0$, $f_y = 0$ will have a finite number of solutions. Of those we select the ones in $\overset{\circ}{K}$.
(b) For each γ_i, the critical points of $f(\gamma_i(t))$. Again, typically we will have a finite number of points.
(c) The vertices.

Then we simply check the values of f at all points of E.

In the above situation, absolute extrema on bK need not be critical, just the tangential derivative is zero. For a better understanding, consider the following example.

Example 4.6. Let

$$K = \{(r\cos\theta, r\sin\theta), 0 \le r \le 1, 0 \le \theta \le \theta_0\},$$

and suppose that f is differentiable and has an absolute extrema at $(0,0)$. Then $(0,0)$ is critical if $\theta_0 > \pi$; this is because along the x axis and also along the line with direction $(\cos\theta_0, \sin\theta_0)$ f has a local extrema, therefore both $\frac{\partial f}{\partial x}(0,0)$ and $D_v f(0,0)$ are zero, and being linearly independent directions, this implies $\nabla f(0,0) = 0$.

Example 4.7. We compute the absolute maximum and minimum of $f(x,y) = x^2y(4-x-y)$ on the triangle K limited by the lines $x = 0$, $y = 0$, $x + y = 6$. First

$$f_x = 2xy(4-x-y) - x^2y = 0, \quad f_y = x^2(4-x-y) - x^2y = 0,$$

has solutions $(x,0), (0,y), (2,1)$, of which $(2,1)$ is in $\overset{\circ}{K}$. For $y = 0, 0 < x < 6$ or $x = 0, 0 < y < 6, f(x,0) = 0 = f(0,y)$. On $x + y = 6, 0 <$

$x < 6$ we parametrize by x; $f(x, y)$ equals $-2x^2(6 - x) = 2x^3 - 12x^2$ with derivative $6x^2 - 24x$ vanishing at $x = 4$. Hence, there are five candidates: $(2, 1), (4, 2)$ and the three vertices $(0, 0), (0, 6), (6, 0)$. The values are, respectively, $4, -64, 0, 0, 0$, so 4 is the absolute maximum and -64 the absolute minimum.

4.4.3 Another special case is that of convex functions, in arbitrary dimension. Recall that a function ϕ defined on an open interval $I \subset \mathbb{R}$ is called *convex* if

$$\phi(tx + (1 - t)y) \leq t\phi(x) + (1 - t)\phi(y), \quad x, y \in I.$$

Geometrically, the graph of ϕ between two points is under the secant. It is easily seen that this is equivalent to

$$\frac{\phi(y) - \phi(x)}{y - x} \leq \frac{\phi(z) - \phi(y)}{z - y}, \quad x < y < z,$$

and implies that ϕ is continuous on I. If the inequalities are strict for $0 < t < 1$, the function is called *strictly convex*. If ϕ is differentiable on I, it is convex if and only if ϕ' is nondecreasing; if ϕ' is strictly increasing, ϕ is strictly convex. Whence a twice differentiable function ϕ is convex if and only if $\phi'' \geq 0$, and it is strictly convex if $\phi'' > 0$.

We say that a continuous function f in $\mathbb{R}^n$ is *convex* (resp., *strictly convex*) if its restriction $\phi(t) = f(p + tv)$ to every line is convex (resp., strictly convex). Then iterating one has

$$f\left(\sum_i \lambda_i x_i\right) \leq \sum_i \lambda_i f(x_i), \quad \sum \lambda_i = 1, \ \lambda_i \geq 0. \tag{4.5}$$

A differentiable function f is convex if and only if every directional derivative $D_v f(p + tv)$ is nondecreasing in t. A function f is *concave* if $-f$ is convex.

In the next statement, we need the concept of *convex envelope* $\widehat{A}$ of a given set A. This is the intersection of all convex sets containing A, or equivalently

$$\widehat{A} = \left\{ x = \sum_i \lambda_i x_i, x_i \in A, \lambda_i \geq 0, \sum_i \lambda_i = 1 \right\}.$$

Exercise 4.3. Prove that $\widehat{K}$ is compact if K is compact.

Theorem 4.2. *A convex continuous function attains its maximum value on a compact set at the boundary, that is*

$$M = \max_K f = \max_{bK} f.$$

Moreover, $\max_K f = \max_{\widehat{K}} f$.

Proof. The first statement follows from the fact that a convex continuous function in a closed interval attains its maximum at the end-points. Given $p \in K$ and a direction v, the set $\{t : p + tv \in \overset{\circ}{K}\}$ is open. If (a, b) is the connected component containing 0, then both $p + av, p + bv \in bK$ and

$$f(p) \leq \max(f(p + av), f(\alpha + bv)).$$

The second statement follows from (4.5). □

A *polytope* in $\mathbb{R}^n$ (polygon in $n = 2$, polyhedron in $n = 3$) is a compact set K such that $K = \overline{\overset{\circ}{K}}$, $\overset{\circ}{K}$ is connected and bK is covered by a finite number of hyperplanes. Then, there are a finite number of points in bK, the vertices, that determine edges, the faces, etc. and so on. In this situation, the reasoning just used can be done on bK; assuming for simplicity that $n = 3$, if f is convex, the maximum on the faces of bK will be attained on the edges, and in turn the maximum on the edges will be attained at the vertices. In conclusion, we can state:

Theorem 4.3. *A continuous convex (resp., concave) function attains its maximum (resp., minimum) value on a polytope at one of the vertices.*

This result of course applies to linear functions. The algorithm to find the extreme values is called *the simplex algorithm.*

4.4.4 We point out that these results can be understood and solved geometrically in terms of the level sets of f, in case f does not have critical points. It is geometrically clear that if bK has a tangent at p, the level set $\{f = f(p)\}$ must be tangent to bK at p. In many situations this is sufficient in order to find the extreme values.

Example 4.8. Consider

$$f(x, y) = y - x^2, \; K = \{(x, y) : 2x^2 + 4y^2 \leq 1\}.$$

The parabola $y = x^2 + c$, coming from $c = -\infty$, goes up as c increases. The minimum of f on K is the value of c for which the parabola and the ellipse first meet, which is $c = -\frac{5}{8}$. If the parabola comes from $c = +\infty$, it goes down as c decreases, and the maximum of f on K will occur when the vertex of the parabola $(0, c)$ equals $(0, \frac{1}{2})$, that is, $c = \frac{1}{2}$. See Figure 4.2.

Similarly, consider the problem of optimizing a linear function f on a compact polyhedron K. As the hyperplane $f = c$ moves, it is clear that it will first touch K at the boundary; either along a full face, or else along a full edge, or else just at one vertex. See Figure 4.3.

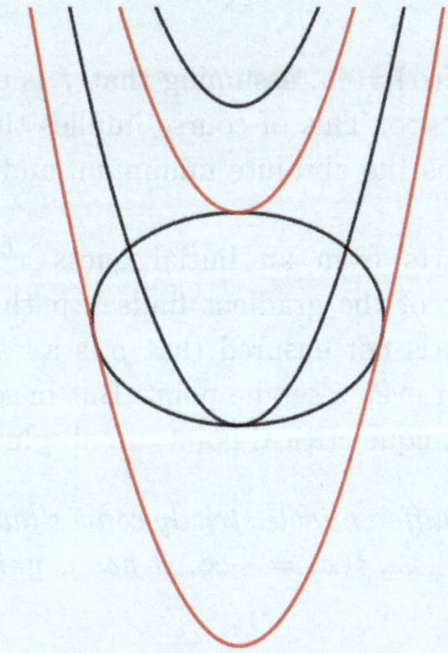

Figure 4.2. Finding extreme values with level curves.

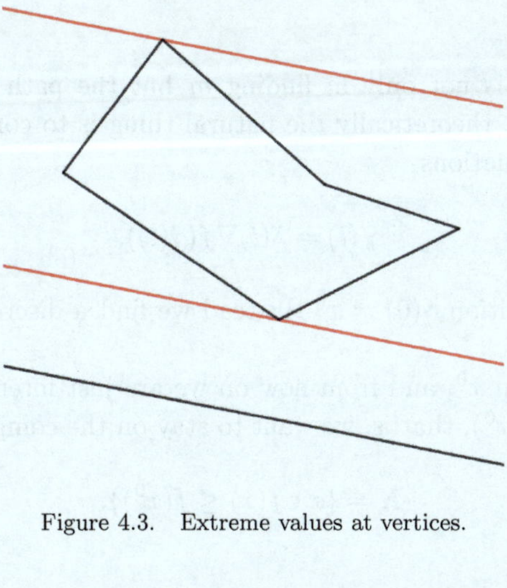

Figure 4.3. Extreme values at vertices.

4.5 The Gradient Descent Method

4.5.1 The simple fact that $\nabla f(x)$, if not zero, points in the direction of maximum increase of f is the basis for a very intuitive method to find extreme values, the so-called *gradient descent method* or *steepest descent algorithm.* We would like to find

$$m = \min_{x \in \mathbb{R}^n} f(x),$$

and a point α where $f(\alpha) = m$, assuming that f is of class C^1 in $\mathbb{R}^n$ and that $\lim_{|x|\to\infty} f(x) = +\infty$. This of course implies that there is at least a point α where f attains the absolute minimum, and it is a critical point, $\nabla f(\alpha) = 0$.

The algorithm starts from an initial guess x^0 and exploiting the characteristic property of the gradient finds a path leading to a critical point $p, \nabla f(p) = 0$. It is not ensured that p is an absolute minimum, it might be a local one, or even a saddle point. But in some cases, we know *a priori* that there is a unique critical point, an absolute minimum:

Proposition 4.4. *A differentiable strictly convex function has at most one critical point. If* $\lim_{|x|\to+\infty} f(x) = +\infty$, *it has a unique critical point, an absolute minimum.*

Proof. First note that if a differentiable convex function f has two critical points p, q and $v = p - q$, then $D_v f$ must be constant in the segment joining p, q and so it is linear there. This proves the first statement, and the second is obvious. □

The interest is not only in finding m but the path leading to p. As explained before, theoretically the natural thing is to consider the system of differential equations

$$\gamma'(t) = \lambda(t)\nabla f(\gamma(t)),$$

with initial condition $\gamma(0) = x^0$. Instead we find a discrete version of the path γ as follows.

We start from x^0, and from now on we are just interested in points x where $f(x) \le f(x^0)$, that is, we want to stay on the compact

$$K = \{x : f(x) \le f(x^0)\}.$$

If $\nabla f(x^0) = 0$, then we are done and set $p = x^0$. Otherwise, we look at f along the half-line through x^0 in the direction $-\nabla f(x^0)$ of fastest decrease:

$$\phi(t) = f(x^0 - t\nabla f(x^0)), \quad t > 0.$$

Then $\phi'(0) = -|\nabla f(x^0)|^2$ and so f decreases along this line. Next we minimize ϕ: if

$$\phi(t_1) = \min\{\phi(t), t > 0\},$$

which is strictly less than $f(x^1)$, we set

$$x^1 = x^0 - t_1\nabla f(x^0).$$

If $\nabla f(x^1) = 0$, we are done and set $p = x^1$. Otherwise, we look at f along $x^1 - t\nabla f(x^1)$, minimize in t, etc.

Note that if $\nabla f(x^1) \neq 0$, then it is orthogonal to $\nabla f(x^0)$, because the derivative in the direction $\nabla f(x^0)$ is zero at x^1. Thus, the line $x^0 - t_1\nabla f(x^0)$ is tangent to the level set of f through x^1 and the path changes to an orthogonal direction.

In this way, either in a finite number of steps we reach a critical point or else we produce a sequence (x^k) of points in K, with $x^{k+1} = x^k - t_k\nabla f(x^k)$, $f(x^{k+1}) < f(x^k)$. The path consists in a number of segments meeting orthogonally. Since K is compact, there is a subsequence (x^{k_m}) with a limit point $p \in K$. We claim that $\nabla f(p) = 0$. Indeed, if $\nabla f(p) \neq 0$, one would have

$$f(p - t\nabla f(p)) < f(p) \leq f(x^0)$$

for all $t > 0$ small enough. In particular, $p - t\nabla f(p)$ lies in the interior of K. Since $x^{k_m} - t\nabla f(x^{k_m}) \to p - t\nabla f(p)$, $f(x^{k_m} - t\nabla f(x_{k_m}))$ would be for m big enough strictly less than $f(p)$. But by construction, for all k

$$f(p) < f(x^k - t_k\nabla f(x^k)) \leq f(x^k - t\nabla f(x^k)),$$

the last inequality being the definition of t_k.

If f is strictly convex, one can prove in fact that the whole sequence x^k is convergent to the unique critical point and absolute minimum p. Indeed, the same argument proves then that whenever a subsequence of x^k is convergent, its limit is p. This implies that $x^k \to p$. Otherwise, one would have a subsequence x^{k_m} with $|x^{k_m} - p| > \varepsilon$; then x^{k_m} would have a convergent subsequence, whence with limit p, which is a contradiction.

There are plenty of computer programs with variants of this algorithm; with a prescribed tolerance level, the algorithms should find the path, which can be plotted, too. The speed of convergence is of course an issue; a well-known example of slow convergence is the *Rosenbrock function*

$$f(x, y) = (1 - x)^2 + 100(y - x^2)^2.$$

4.5.2 We finish this section with a particular case, dealing with the *least squares method.* With matrix notation, one wants to minimize

$$F(X) = |AX - B|^2 = X^t A^t AX - 2B^t AX + BB^t,$$

where A is a given $m \times n$ matrix that we look as a linear map from $\mathbb{R}^n$ to $\mathbb{R}^m$, X, B column vectors in $\mathbb{R}^n, \mathbb{R}^m$, respectively, B given. We first discuss the solution of this problem.

Let us denote by N, R the kernel and range of A, and let P be the orthogonal projection on R. If $X = X_1 + X_2$ is the orthogonal decomposition, $X_1 \in N, X_2 \in N^\perp$, then

$$F(X) = F(X_2) = |B - PB|^2 + |AX_2 - PB|^2.$$

Since the restriction of A to $N^\perp$ is an isomorphism onto R, there is a unique $X_0 \in N^\perp$ such that $AX_0 = PB$, the unique absolute minimum in $N^\perp$, with minimal value $|B - PB|^2$. Setting $X_0 = A^+ B$, the map A^+, which is linear, is called the *Moore–Penrose pseudo-inverse* of A. Obviously, the linear sub-manifold $X_0 + N$ consists of absolute minimums. One has

$$F(X) = |B - PB|^2 + |AX_2 - AX_0|^2 = |B - PB|^2 + |A(X_2 - X_0)|^2.$$

On $N^\perp$, F behaves like $|X_2 - X_0|^2$, so we see that X_0 is also the unique critical point in $N^\perp$. This also follows from

$$\nabla F(X) = 2A^t A(X - X_0),$$

and taking into account that $A^t A$ and A have the same kernel, and $A^t A$ is one-to-one on $N^\perp$. Therefore, we may consider X_0 as the unique solution in $N^\perp$ of $AX = PB$ or as the unique solution in $N^\perp$ of $A^t AX = A^t PB$. Since $A^t PB = A^t B$ (because $B - PB$ is orthogonal to R, whence in the kernel of A^t), one has $A^t AX = A^t B$. This latter formulation is more clear because it involves only the initial values. Altogether, $A^t AA^+ = A^t$ and we see that when A has rank n, $A^t A$ is invertible and $A^+ = (A^t A)^{-1} A^t$ is the linear solution operator.

Instead of inverting a matrix, let us check the descent method in this case. We simplify notations, delete constant terms and assume

$$F(X) = \frac{1}{2}X^tMX + C^tX,$$

with $M = A^tA$ a $n \times n$ symmetric matrix, positive definite. F is strictly convex with

$$\nabla F(X) = MX + C,$$

so the unique critical point and absolute minimum is $X_0 = -M^{-1}C$, with value $-\frac{1}{2}C^tM^{-1}C$. Let us denote by X one of the approximations obtained by the gradient method and see which is next. We must follow the direction $V = -\nabla F(X) = -MX - C$ and minimize in t

$$\begin{aligned} F(X + tV) &= \frac{1}{2}(X^t + tV^t)M(X + tV) + C^t(X + tV) \\ &= F(X) + \frac{1}{2}t^2V^tMV - tV^tV. \end{aligned}$$

The minimal value is at

$$t = \frac{V^tV}{V^tMV},$$

and so the next point is

$$X' = X + \frac{V^tV}{V^tMV}V,$$

where

$$F(X') = F(X) - \frac{1}{2}\frac{(V^tV)^2}{V^tMV}.$$

We next compare the error $F(X) - F(X_0)$ with the next one $F(X') - F(X_0)$. One has

$$\begin{aligned} \frac{F(X') - F(X_0)}{F(X) - F(X_0)} &= 1 + \frac{F(X') - F(X)}{F(X) - F(X_0)} \\ &= 1 - \frac{\frac{1}{2}\frac{(V^tV)^2}{V^tMV}}{\frac{1}{2}X^tMX + C^tX + \frac{1}{2}C^tM^{-1}C} \\ &= 1 - \frac{\frac{1}{2}\frac{(V^tV)^2}{V^tMV}}{\frac{1}{2}(MX + C)^tM^{-1}(MX + C)} \\ &= 1 - \frac{(V^tV)^2}{(V^tMV)(V^tM^{-1}V)}. \end{aligned}$$

To understand the quantity

$$\beta = \frac{(V^tMV)(V^tM^{-1}V)}{(V^tV)^2},$$

we consider the eigenvalues $0 < \lambda_1 \leq \cdots \leq \lambda_n$ of M. Since M becomes diagonal on a certain orthonormal basis (Theorem 1.4), one has

$$\beta = (\sum_i \lambda_i y_i^2)(\sum_i \frac{1}{\lambda_i} y_i^2),$$

for some vector $y = (y_1, \ldots, y_n)$, $|y| = 1$.

The so-called *Kantorovich inequality* states that

$$1 \leq \beta \leq \frac{(\lambda_1 + \lambda_n)^2}{4\lambda_1\lambda_n}. \tag{4.6}$$

This can be proved using Lagrange's multipliers, see paragraph 7.4.1. This implies

$$\frac{F(X') - F(X_0)}{F(X) - F(X_0)} \leq 1 - \frac{4\lambda_1\lambda_n}{(\lambda_1 + \lambda_n)^2} = \frac{(\lambda_n - \lambda_1)^2}{(\lambda_n + \lambda_1)} = \left(\frac{(\frac{\lambda_n}{\lambda_1}) - 1}{(\frac{\lambda_n}{\lambda_1}) + 1}\right)^2 = \delta < 1.$$

Iterating we see that in k steps the error is multiplied by δ^k. The quotient $\frac{\lambda_n}{\lambda_1}$ is called the *condition number* of M. The closest to 1, the faster the convergence to the optimal point.

There are many variants of this algorithm in which the choice of t at each step is done in an adaptative way. For practical and industrial applications, one does not have an analytic expression of f, just the values on a cloud of points, and one needs to consider a discrete version of the gradient.

4.6 Mean-Value Theorems

4.6.1 For continuous $f : [a, b] \to \mathbb{R}$ with derivative in (a, b), the mean-value theorem states that

$$f(b) - f(a) = f'(c)(b - a), \quad c \in (a, b).$$

The analogous of this result in a general situation, for a differentiable $f : U \to \mathbb{R}^m$ and $p, q \in U$ would state that

$$f(q) - f(p) = df(c)(q - p)$$

holds for some point c. For vector-valued functions, this cannot hold, as shown by the following examples. Consider $g(t) = (\cos t, \sin t), 0 \leq t \leq 2\pi$

or more generally a regular closed Jordan arc with non-zero tangent. One has $g(0) = g(2\pi)$, but g' never vanishes. As a second example, the function $f(x, y) = (e^x \cos y, e^x \sin y)$ (in complex coordinates, $f(z) = e^z$), satisfies $f(0) = f(2\pi)$, yet the differential $df(x, y)$ has determinant $e^{2x} \neq 0$ and so has zero kernel. For scalar-valued functions, we can state:

Proposition 4.5. *Let $f : U \to \mathbb{R}$ be differentiable in a domain $U, p, q \in U$ and let γ be a regular arc joining p, q in $U, \gamma(0) = p, \gamma(1) = q$. Then there is a point $c = \gamma(s)$ such that*

$$f(q) - f(p) = \langle \nabla f(c), \gamma'(s) \rangle.$$

In particular, if the segment joining p, q is contained in U, it contains c such that $f(q) - f(p) = \langle \nabla f(c), q - p \rangle$.

Proof. It is sufficient to apply the one-variable result to $f(\gamma(t))$. □

4.6.2 In the general case $m \geq 1$, inequalities hold; in the next theorem $d_U(p, q)$ has been defined in (3.2).

Theorem 4.4. *Assume $f : U \to \mathbb{R}^m$ is differentiable, $|df(x)| \leq M$, $p, q \in U$. Then*

$$|f(q) - f(p)| \leq M d_U(p, q).$$

In particular, if U contains the segment S joining p, q,

$$|f(q) - f(p)| \leq M|q - p|.$$

If $df = 0$ in U, f is constant.

Proof. If f is scalar-valued, this follows from the previous proposition. Assume $m > 1$, $S \subset U$ and set $h(t) = f((1-t)p + tq)$; then $h'(t) = df((1-t)p + tq)(q - p)$ has length

$$|h'(t)| \leq |df((1-t)p + tq)||q - p| \leq M|q - p|.$$

Then

$$|f(q) - f(p)| = |h(1) - h(0)| = \max_{|v|=1} \langle h(1) - h(0), v \rangle.$$

Now, $\langle h(t), v \rangle$ is scalar-valued, with derivative $\langle h'(t), v \rangle$, and therefore

$$\langle h(1) - h(0), v \rangle = \langle h'(c_v), v \rangle \leq |h'(c_v)| \leq M|q - p|.$$

For general points $p, q \in U$, and a polygonal $P : p = p_0, p_1, \ldots, p_N = q$ joining p, q inside U, we get

$$|f(q) - f(p)| \leq \sum_{i=0}^{N-1} |f(p_i) - f(p_{i+1})| \leq M \sum_{i=0}^{N-1} |p_i - p_{i+1}| = ML(P),$$

and the result follows by minimizing in P. □

4.7 The Concept of Partial Differential Equation

4.7.1 If f is differentiable in a domain U and does not depend on a variable x_i, that is f is a function of the remaining coordinates, of course $D_i f = 0$ in U. If the intersection of U with every line parallel to the x_i axis is connected, that is, an open interval in that line, then the converse is also true. For, assuming $n = 2, (0, y) \in U$ and say $D_1 f = 0$,

$$f(x, y) = f(0, y) + \int_0^x D_1 f(t, y)\, dt = f(0, y).$$

If U meets a line parallel to the x axis in two or more disjoint intervals, f will be constant on each, but possibly with a different constant. Consider for instance the U-shaped domain $U = U_1 \cup U_2 \cup U_3$ with

$$U_1 = [-2, -1] \times [-1, 1], U_2 = [-1, 1] \times [-1, 0], U_3 = [1, 2] \times [-1, 1],$$

and let f be defined by $f(x, y) = y^2$ when $y < 0$ or $x < 0$ and $f(x, y) = y^3$ when $x, y > 0$. Then f is differentiable in U, $\frac{\partial f}{\partial x} = 0$, but $f(x, y) \neq f(-x, y)$ if $y > 0$.

Domains U meeting every coordinate line along an interval will be called *p-domains*. For instance, products of open intervals, their images by a linear invertible transformation, are p-domains. Note that a p-domain might not be convex, for instance an L-shaped domain.

More generally, in a p-domain we can find the general solution of a *partial differential equation* like

$$\frac{\partial f}{\partial x_i} = g(x),$$

with g being a given continuous function on U. This is the simplest partial differential equation. A particular solution is obtained anti-differentiating in

x_i, $\int g(x)\,dx_i$, meaning that the other variables are thought as parameters. Since its difference with f has zero derivative, it turns out that

$$f(x) = \int g(x)\,dx_i + \Phi(\hat{x}_i),$$

is the general solution, $\Phi(\hat{x}_i)$ denoting an arbitrary function of the remaining variables. Note in particular that in general the solution need not be differentiable, for instance $\frac{\partial f}{\partial x_j}$ may fail to exist for some $j \neq i$.

Example 4.9. The general solution of

$$\frac{\partial f}{\partial y}(x, y) = xy,$$

is $f(x, y) = \frac{1}{2}xy^2 + C(x)$, with C arbitrary. The general solution $f(x, y)$ of

$$\frac{\partial f}{\partial x} = x \cos xy,$$

is

$$f(x, y) = f(0, y) + \int_0^x t \cos ty\,dt.$$

This equals $f(0, 0) + \frac{x^2}{2}$ if $y = 0$ and otherwise, integrating by parts,

$$f(x, y) = \Phi(y) + \frac{xy \sin xy + \cos xy - 1}{y^2}, \quad y \neq 0.$$

Note that the right-hand side has indeed limit $\frac{x^2}{2}$ as $y \to 0$. The same computation in $y \neq 0$ can be presented as follows: anti-differentiating in x we get

$$f(x, y) = \int x \cos xy\,dx = \frac{xy \sin xy + \cos xy}{y^2} + \Psi(y).$$

Now the left-hand side must have a finite limit as $y \to 0$; as the limit equals

$$x + \lim_{y \to 0} \frac{\cos xy - 1}{y^2} + \lim_{y \to 0} \left(\frac{1}{y^2} + \Psi(y)\right) = \frac{x^2}{2} + \lim_{y \to 0} \left(\frac{1}{y^2} + \Psi(y)\right),$$

$\Psi(y)$ must have the form $-\frac{1}{y^2} + \Phi(y)$, with Φ with a finite limit $\Phi(0)$ at zero.

Iterating we see that in a p-domain a function depends just on the variables $x_i, i \in A$ where A is a certain set of indexes, $A \subset \{1, 2, \ldots, n\}$ if and only if $D_i f = 0, i \notin A$. For instance, $f(x, y, z)$ depends just on x if and only if $\frac{\partial f}{\partial y}, \frac{\partial f}{\partial z}$ are both zero. In particular, if all derivatives are zero, then f is constant.

4.7.2 We state and prove now some results regarding differentiation of functions defined by one variable integrals,

$$f(x) = \int_a^b F(x,t)\,dt.$$

We assume that F is continuous on $U \times [a,b]$, U a domain in $\mathbb{R}^n$, so that f is well defined. We claim first that f is continuous on U. Indeed, given $x_0 \in U$, let $\overline{B}(x_0, r) \subset U$. Since $\overline{B}(x_0, r) \times [a,b]$ is compact, f is uniformly continuous there: for all $\varepsilon > 0$ there exists $\delta > 0$ such that $|t - t'| < \delta, |x - x'| < \delta, x, x' \in \overline{B}(x_0, r)$, implies $|F(x,t) - F(x',t')| < \varepsilon$. In particular, $|F(x,t) - F(x_0,t)| < \varepsilon$ for all t if $|x - x_0| < \delta < r$, and then

$$|f(x) - f(x_0)| \leq \int_a^b |F(x,t) - F(x_0,t)|\,dt \leq \varepsilon(b-a).$$

Proposition 4.6. *Assume that* $\frac{\partial F}{\partial x_i}(t,x)$ *exists and is continuous on* $U \times [a,b]$. *Then* $\frac{\partial f}{\partial x_i}(x)$ *exists and*

$$\frac{\partial f}{\partial x_i}(x) = \int_a^b \frac{\partial F}{\partial x_i}(x,t)\,dt.$$

Proof. To prove this we may assume that $n = 1$ and that U is an open interval. Given $c \in U$, consider a closed interval $[c-r, c+r] \subset U$. We must show that

$$\int_a^b \left| \frac{F(c+h,t) - F(c,t)}{h} - \frac{\partial F}{\partial x}(c,t) \right| dt \to 0, h \to 0.$$

By the mean-value theorem,

$$\frac{F(c+h,t) - F(c,t)}{h} - \frac{\partial F}{\partial x}(c,t) = \frac{\partial F}{\partial x}(c+\theta h, t) - \frac{\partial F}{\partial x}(c,t), 0 < \theta < 1.$$

By the uniform continuity of $\frac{\partial F}{\partial x}$ on $[c-r, c+r] \times [a,b]$ given $\varepsilon > 0$, there exists $\delta > 0$ such that the last expression is less than ε for all t and θ if $|h| < \delta$. □

More generally, assume we want to differentiate

$$g(x) = \int_{a(x)}^{b(x)} F(x,t)\,dt, \quad x \in U,$$

where a, b are real-valued and differentiable in U and F is as before. We introduce

$$f(x,y) = \int_a^y F(x,t)\,dt,$$

so that $g(x) = f(x, b(x)) - f(x, a(x))$. Then, by the proposition,

$$\frac{\partial f}{\partial x_i} = \int_a^y \frac{\partial F}{\partial x_i}(x,t)\,dt,$$

while by the fundamental theorem of calculus

$$\frac{\partial f}{\partial y} = F(x,y).$$

By the chain rule, we find

$$\begin{aligned} D_i g(x) = \frac{\partial f}{\partial x_i}(x, b(x)) + \frac{\partial f}{\partial y}(x, b(x)) D_i b(x) - \frac{\partial f}{\partial x_i}(x, a(x)) \\ - \frac{\partial f}{\partial y}(x, a(x)) D_i a(x) = \int_{a(x)}^{b(x)} \frac{\partial F}{\partial x_i}(x,t)\,dt + F(x, b(x)) D_i b(x) \\ - F(x, a(x)) D_i a(x). \end{aligned}$$

4.7.3 Functions with given gradient. Suppose now that all partial derivatives $D_i f$ are given in a p-domain,

$$D_i f(x) = g_i(x),$$

with g_i continuous. This amounts, of course, to giving $df(x)$ or $\nabla f(x)$ at each $x \in U$. Does this system of partial differential equations have a solution? In dimension $n = 1$, we know that this is indeed the case, the general solution being the indefinite integral of g, plus a constant term. Recall that a different issue is whether the indefinite integral can be expressed in terms of elementary functions; this is not always the case, for instance for the gaussian $f = e^{-x^2}$.

We will see that in order that a solution exists, the data g_i must satisfy a compatibility condition. Assume for simplicity that $n = 2$ and we want to solve in the plane

$$\frac{\partial f}{\partial x}(x,y) = g_1(x,y), \quad \frac{\partial f}{\partial y}(x,y) = g_2(x,y),$$

with g_1, g_2 of class C^1. The first one is equivalent to

$$f(x,y) = \int_0^x g_1(t,y)\,dt + A(y),$$

with A arbitrary. Now we would like to choose A so that the second equation holds, too. At this point, we use proposition 4.6 to obtain

$$\frac{\partial f}{\partial y}(x,y) = \int_0^x \frac{\partial g_1}{\partial y}\,dt + A'(y).$$

If we want this to equal $g_2(x,y)$, the difference

$$\int_0^x \frac{\partial g_1}{\partial y}(t,y)\,dt - g_2(x,y)$$

should depend only on y, that is, the x derivative should be zero, and we find that

$$\frac{\partial g_1}{\partial y} = \frac{\partial g_2}{\partial x} \tag{4.7}$$

is a necessary compatibility condition. The argument is reversible, for if this condition is satisfied, then

$$\frac{\partial f}{\partial y}(x,y) = \int_0^x \frac{\partial g_2}{\partial t}(t,y)\,dt + A'(y) = g_2(x,y) - g_2(0,y) + A'(y),$$

so now it is enough to solve $A'(y) = g_2(0,y)$.

Another way of presenting the same computation is writing

$$f(x,y) = f(x,0) + \int_0^y g_2(x,s)\,ds = f(0,0) + \int_0^x g_1(t,0)\,dt + \int_0^y g_2(x,s)\,ds.$$

Then $f_y = g_2$ and

$$f_x(x,y) = g_1(x,0) + \int_0^y \frac{\partial g_2}{\partial x}(x,s)\,ds,$$

so $f_x = g_1$ is equivalent to

$$g_1(x,y) - g_1(x,0) = \int_0^y \frac{\partial g_2}{\partial x}(x,s)\,ds,$$

or

$$\int_0^y \frac{\partial g_1}{\partial y}(x,s)\,ds = \int_0^y \frac{\partial g_2}{\partial x}(x,s)\,ds,$$

that is in turn equivalent to (4.7). So we have proved:

Theorem 4.5. *For g_i of class C^1 in a p-domain, the system*

$$D_i f(x) = g_i(x), \quad i = 1, \ldots, n,$$

has a solution if and only if (4.7) *holds.*

The function f is called a *potential function.* When dealing with vector analysis and conservative fields in Chapter 18, we will come back to this issue and see that potential functions exist in a larger class of domains.

To find potential functions, in practice we proceed as in the next example:

Example 4.10. We solve

$$f_x = y^2z^3 + y^2 + 2zx, f_y = 2xyz^3 + 2xy + z^2, f_z = 3xy^2z^2 + 2zy + x^2 + 2.$$

We can check that the compatibility conditions are satisfied. From the first one,

$$f = xy^2z^3 + xy^2 + zx^2 + A(y, z),$$

which replaced in the second one gives

$$2xyz^3 + 2xy + A_y = 2xyz^3 + 2xy + z^2, A_y = z^2.$$

Then $A = yz^2 + B(z), f = xy^2z^3 + xy^2 + zx^2 + yz^2 + B(z)$; replacing in the last one,

$$3xy^2z^2 + x^2 + 2yz + B' = 3xy^2z^2 + 2zy + x^2 + 2,$$

so $B' = 2, f = xy^2z^3 + xy^2 + zx^2 + yz^2 + 2z + c.$

In dimension $n > 2$, the same argument shows that if $I \subset \{1, 2, \ldots, n\}$ is a set of indexes,

$$\frac{\partial f}{\partial x_i} = h_i(x), \quad i \in I,$$

with h_i of class C^1, has a solution if and only if for each pair $i, j \in I$

$$\frac{\partial h_i}{\partial x_j} = \frac{\partial h_j}{\partial x_i}.$$

In this case, the solution is obtained as before, by partial anti-differentiation, step by step.

4.7.4 We have considered an equation like

$$D_i f(x) = g(x),$$

somehow the simplest partial differential equation, whose general solution is obtained by anti-differentiation in x_i,

$$f(x) = \int g(x)\, dx_i + \Phi,$$

with Φ is a function of all variables but x_i.

We consider now an equation involving more than one variable; assume for instance that $f(x, y)$ is a function of two variables satisfying

$$f_x = f_y.$$

It is obvious that $f(x, y) = A(x + y)$, with A a function of one variable, satisfies this equation, because $f_x = f_y = A'(x + y)$. Conversely, every solution is of this form; indeed, the equation means that f has zero derivative in the direction $(1, -1)$ at all points, whence it must be constant on every line $x + y = c$.

More generally, we consider a *linear partial differential equation with constant coefficients* in the whole of $\mathbb{R}^n$

$$\sum_{i=1}^{n} A_i \frac{\partial f}{\partial x_i} = B(x),$$

with $A_i \in \mathbb{R}$, not all zero, and the function B is given, say a continuous function. To find the general solution, we note that the left-hand side equals $D_v f(x)$ with $v = (A_1, \ldots, A_n)$. It is therefore natural to consider the lines with direction v. If $v_1, \ldots, v_{n-1}$ are linearly independent vectors orthogonal to v, the equations

$$\langle v_i, x \rangle = c_i, \quad i = 1, \ldots, n-1$$

define the general line with direction v, that is, the family of these lines is parametrized by $c_1, \ldots, c_{n-1}$. Let

$$x_i(t) = a_i + tA_i, \quad i = 1, \ldots, n,$$

a parametrization; if $h(t) = f(x_1(t), \ldots, x_n(t))$, then the equation becomes

$$h'(t) = B(x_1(t), \ldots, x_n(t)),$$

with general solution

$$h(t) = \int_0^t B(x_1(s), \ldots, x_n(s))\, ds + c,$$

with c constant. Now, this constant may vary with each line, that is

$$c = \Phi(c_1, \ldots, c_n) = \Phi(\langle v_1, x\rangle, \ldots, \langle v_{n-1}, x\rangle).$$

The integral above is a particular solution, while $\Phi(\langle v_1, x\rangle, \ldots, \langle v_{n-1}, x\rangle)$, with Φ an arbitrary differentiable function of $n-1$ variables, is the general solution of the homogeneous equation

$$\sum_{i=1}^{n} A_i \frac{\partial f}{\partial x_i} = 0.$$

If the value $f(p)$ is known for a point p on each line with direction vector v, for instance if f is known on the orthogonal $V^t X = 0$, then f is unique.

Example 4.11. Let us illustrate the method with the equation

$$\frac{\partial f}{\partial x} - 2\frac{\partial f}{\partial y} + 3\frac{\partial f}{\partial z} = xy.$$

Consider the line

$$2x + y = c_1, \quad 3x - z = c_2,$$

with direction $(1, -2, 3)$ and parametrization $x = t, y = c_1 - 2t, z = 3t - c_2$. If $h(t) = f(t, c_1 - 2t, 3t - c_2)$, then

$$h'(t) = xy = t(c_1 - 2t).$$

Therefore,

$$h(t) = \frac{c_1}{2}t^2 - \frac{2}{3}t^3 = \frac{2x+y}{2}x^2 - \frac{2}{3}x^3 = \frac{1}{3}x^3 + \frac{1}{2}x^2 y,$$

is a particular solution and

$$f(x, y, z) = \frac{1}{3}x^3 + \frac{1}{2}x^2 y + \Phi(2x + y, 3x - z)$$

is the general solution. If f satisfies say $f(x, 0, z) = h(x, z)$ with h given, then

$$\Phi(2x, 3x - z) + \frac{1}{3}x^3 = f(x, 0, z) = h(x, z),$$

so $\Phi(u, v) = h(\frac{u}{2}, \frac{3}{2}u - v) - \frac{1}{24}u^3$, and the unique solution is

$$\frac{1}{3}x^3 + \frac{1}{2}x^2 y - \frac{1}{24}(2x + y)^3 + h(x + \frac{y}{2}, \frac{3}{2}y + z).$$

Chapter 5

Higher-Order Derivatives

The differential leads to the best linear approximation of a function $f(x)$ around a fixed point p. In order to produce better approximations by higher-order polynomials in $x-p$, higher-order derivatives are introduced leading to Taylor's development. We next explain the standard applications of Taylor's formula to local extrema. The next section deals with infinitely differentiable functions, showing their abundance in different contexts. In the final section, we study Taylor's formula at infinite order, that is, real-analytic functions and their properties.

5.1 Schwarz's Rule

5.1.1 The main point of the differential is that $df(p)(x-p)$ is a linear approximation of $f(x)-f(p)$, with error $o(|x-p|)$. To get better approximations, one must consider higher-order derivatives. In what follows we assume f scalar-valued, $m=1$.

If f is differentiable at all points in some ball B centered at p, we may consider the map

$$df : B \to L(\mathbb{R}^n, \mathbb{R}).$$

Here $L(\mathbb{R}^n, \mathbb{R})$ stands for the space of linear maps from $\mathbb{R}^n$ to $\mathbb{R}$. Identifying df with ∇f, this is the map with components $D_i f$. If it is differentiable at p, we say that *f is twice differentiable at p*. This means that there exists a linear map

$$d_2f(p) : \mathbb{R}^n \to L(\mathbb{R}^n, \mathbb{R}),$$

such that

$$|df(p+h) - df(p) - d_2f(p)(h)| = o(|h|).$$

This amounts to saying that each component D_jf of ∇f is differentiable at p. The second differential $d_2f(p)$ may be thought of as a bilinear map with matrix

$$(D_iD_jf(p)), \quad i,j = 1,\dots,n.$$

Its action on (u,v) is

$$d_2f(p)(u,v) = \sum_{i,j} D_iD_jf(p)u_iv_j = D_u(D_vf)(p),$$

a second-order directional derivative, also denoted $D_{uv}f(p)$. When u, v are vectors in the canonical basis, we use the notation D_{ij} o $\frac{\partial^2 f}{\partial x_j \partial x_i}$. In two or three variables we use, too, f_{xx}, f_{xy}, etc.

By Proposition 4.1, if all second-order derivatives $D_{ij}f(x)$ exist for x close to p and are continuous at p, f is twice differentiable at p. This is the case for all functions given by analytic expressions. If we take such expressions and differentiate them with respect to two different variables, we realize that the result does not depend on the order of differentiation. That is a general fact that can be easily proved, the *Schwarz's rule*:

Theorem 5.1. *If f is twice differentiable at p, $d_2f(p)$ is symmetric: $D_{uv}f(p) = D_{vu}f(p)$ for all directions u, v.*

Proof. It is sufficient to prove it for $n = 2$ and the directions of the canonical basis: $D_{12}f(0,0) = D_{21}f(0,0)$. Using the notation $f(x,y)$, the left-hand side is formally

$$\begin{aligned}&\lim_{h\to 0}\frac{1}{h}(D_2f(h,0) - D_2f(0,0))\\&\quad = \lim_{h\to 0}\frac{1}{h}\left(\lim_{h\to 0}\frac{1}{h}(f(h,h) - f(h,0)) - \lim_{h\to 0}\frac{1}{h}(f(0,h) - f(0,0))\right),\end{aligned}$$

so it is natural to consider for small h

$$Q(h) = f(h,h) - f(h,0) - (f(0,h) - f(0,0)).$$

Setting $A(t) = f(t,h) - f(t,0)$, one has $Q(h) = A(h) - A(0)$, which by the mean-value theorem equals

$$A'(\theta h)h = h(D_1 f(\theta h, h) - D_1 f(\theta h, 0)).$$

Now, differentiability of $D_1 f$ means

$$D_1 f(x,y) = D_1 f(0,0) + x D_{11} f(0,0) + y D_{21} f(0,0) + o(\sqrt{x^2+y^2}),$$

so

$$D_1 f(\theta h, h) - D_1 f(\theta h, 0)) = h D_{21} f(0,0) + o(h),$$

and it follows that $Q(h) = h^2 D_{21} f(0,0) + o(h^2)$ meaning that

$$\lim_{h\to 0} \frac{Q(h)}{h^2} = D_{21} f(0,0).$$

If we write Q under the form $Q(h) = f(h,h) - f(0,h) - (f(h,0) - f(0,0))$ and argue in the same way, we find that the same limit equals $D_{12} f(0,0)$. □

The matrix of $d_2 f(p)$ in the canonical basis is called *the Hessian* $Hf(p)$

$$Hf(p) = (D_{ij} f(p)).$$

We recall at this point some standard notions from linear and tensor algebra on how to manipulate bilinear and n-linear maps. If E is a linear space, to deal with bilinear maps on E one chooses a basis $e_1, \ldots, e_n$ and its dual basis $\omega_1, \ldots, \omega_n$ so that $u = \sum_i \omega_i(u) e_i$. The general bilinear form is then given by

$$\Phi(u,v) = \sum_{i,j} a_{ij} \omega_i(u) \omega_j(v), \quad u, v \in E,$$

where (a_{ij}) is called the matrix of Φ in the given basis, usually written $\Phi = \sum_{ij} a_{ij} \omega_i \otimes \omega_j$. Thus,

$$d_2 f(p) = \sum_{ij} D_{ij} f(p)\, dx_i \otimes dx_j.$$

5.1.2 Higher-order derivatives are defined analogously. If f is twice differentiable at all points in a ball B centered at p, we say that f is *three times differentiable at* p if $d_2 f$, a map from B to the space of bilinear maps from $\mathbb{R}^n \times \mathbb{R}^n$ to $\mathbb{R}$, is differentiable at p. This amounts to saying that each second-order derivative D_{uv} is differentiable at p. Then, $d_3 f(p)$ is a 3-linear map acting on three directions u, v, w,

$$d_3 f(p)(u, v, w) = D_u(D^2_{v,w} f)(p),$$

that is symmetric. We use the notation $D_{uvw} f(p)$ or $D_{ikl} f(p)$ for directions in the canonical basis.

In general, $d_r f(p)$ is an r-linear symmetric map, that is invariant by permutations, assigning to each r-tuple of vectors $v^1, \dots, v^r$ the derivative of order r obtained by taking successive directional derivatives in the directions $v^1, \dots, v^r$. In canonical coordinates, the action of $d_r f(p)$ on the r-tuple $(v^1, \dots, v^r)$ is

$$\sum_{i_1=1}^{n} \sum_{i_2=1}^{n} \cdots \sum_{i_r=1}^{n} D_{i_1 i_2 \dots i_r} f(p) v^1_{i_1} v^2_{i_2} \cdots v^r_{i_r}.$$

In condensed notation,

$$d_r f(p) = \sum_{i_1=1}^{n} \sum_{i_2=1}^{n} \cdots \sum_{i_r=1}^{n} D_{i_1 i_2 \dots i_r} f(p)\, dx_{i_1} \otimes \cdots \otimes dx_{i_r}.$$

Every symmetric r-linear map is determined by its restriction to the diagonal, an homogeneous polynomial of degree r

$$d_r f(p)(v, v, \dots, v) = \sum_{i_1=1}^{n} \sum_{i_2=1}^{n} \cdots \sum_{i_r=1}^{n} D_{i_1 i_2 \dots i_r} f(p) v_{i_1} v_{i_2} \cdots v_{i_r}.$$

By symmetry, many of the n^r terms in this sum are equal, so it is convenient to introduce a notation that does not take into account the order. A *multi-index* α is an ordered n-tuple $\alpha = (\alpha_1, \alpha_2, \dots, \alpha_n)$, with $\alpha_i \in \mathbb{N}$. We set $|\alpha| = \alpha_1 + \alpha_2 + \cdots + \alpha_n$, $\alpha! = \alpha_1! \alpha_2! \cdots \alpha_n!$ and denote by

$$D^\alpha f(p) = \frac{\partial^{|\alpha|}}{\partial x_1^{\alpha_1} \partial x_2^{\alpha_2} \cdots \partial x_n^{\alpha_n}} f(p),$$

the derivative of order $|\alpha|$ including α_1 derivatives with respect to x_1, α_2 with respect $x_2, \dots, \alpha_n$ with respect to x_n. For $x \in \mathbb{R}^n$ and a multi-index

α, we set

$$(x-p)^\alpha = (x_1-p_1)^{\alpha_1}(x_2-p_2)^{\alpha_2}\cdots(x_n-p_n)^{\alpha_n}.$$

In the above expression of $d_r f(p)$ the number of r-ples $(i_1, i_2, \ldots, i_r)$ including, regardless of order, α_1 ones, α_2 two's, etc, α_n n's are exactly

$$\frac{r!}{\alpha_1!\alpha_2!\cdots\alpha_n!}.$$

Indeed, there are

$$\frac{r!}{\alpha_1!(r-\alpha_1)!} = \binom{r}{\alpha_1},$$

different ways of choosing the α_1 indexes among $i_1, i_2, \ldots, i_r$ to be equal to 1. Next, there are $\binom{r-\alpha_1}{\alpha_2}$ different ways of choosing the α_2 indexes among the remaining $r-\alpha_1$ ones to be equal to 2, etc. Therefore, we may write

$$d_r f(p)(v, v, \ldots, v) = \sum_{|\alpha|=r} \frac{r!}{\alpha!} D^\alpha f(p) v^\alpha.$$

For instance, when $r=2$, $n=2$, $a=(0,0)$ and using (x,y) instead of (u_1, u_2),

$$D_{20}f(0)x^2 + 2D_{11}f(0)xy + D_{02}f(0)y^2.$$

When $r=2, n=3$ with coordinates (x,y,z),

$$D_{200}f(0)x^2 + D_{020}f(0)y^2 + D_{002}f(0)z^2 + 2D_{110}f(0)xy + 2D_{101}f(0)xz$$
$$+2D_{011}f(0)yz.$$

If $r=3, n=2$,

$$D_{30}f(0)x^3 + D_{03}f(0)y^3 + 3D_{21}f(0)x^2y + 3D_{12}f(0)xy^2,$$

while if $r=3, n=3$,

$$D_{300}f(0)x^3 + D_{030}f(0)y^3 + D_{003}f(0)z^3 + 3D_{210}f(0)x^2y + 3D_{201}f(0)x^2z$$
$$+3D_{120}f(0)xy^2 + 3D_{021}f(0)y^2z + 3D_{102}f(0)xz^2 + 3D_{012}f(0)yz^2$$
$$+6D_{111}f(0)xyz.$$

In case f has continuous partial derivatives up to order m, we say that *f is of class C^m*, and if it has continuous derivatives of all orders, one says that f is of class C^∞ or *smooth*, for instance all functions given by formulas.

5.2 Taylor's Formula

5.2.1 Assume that $f : U \to \mathbb{R}$ is N times differentiable at $p \in U$ and $d_r f(p) = 0, r = 0, 1, \ldots, N$, that is, f and all its derivatives up to order N are zero at p. We say in this case that f *vanishes to order N at p*. We claim that the decay of f at p is then

$$|f(x)| = o(|x - p|^N).$$

Indeed, for $N = 1$, this is just the definition of the differential, because we are assuming $f(p) = 0, df(p) = 0$. In general we proceed by induction: assume this has been proved for $N - 1$ and that f vanishes to order N at p. We must show that given $\varepsilon > 0$ there exists $\delta > 0$ such that $|f(x)| \leq \varepsilon|x - p|^N$ if $|x - p| \leq \delta$. Since df (that is, the first-order derivatives) vanish to order $N - 1$, by the induction hypothesis, there exists $\delta > 0$ such that $|df(x)| \leq \varepsilon|x - p|^{N-1}$ if $|x - p| \leq \delta$. If $|x - p| \leq \delta$, then for all y in the ball of center p and radius $|x - p|$

$$|df(y)| \leq \varepsilon|y - p|^{N-1} \leq \varepsilon|x - p|^{N-1}.$$

The mean-value Theorem 4.4 applied to the ball of center p and radius $|x - p|$ with $M = \varepsilon|x - p|^{N-1}$ implies

$$|f(x)| = |f(x) - f(p)| \leq M|x - p| = \varepsilon|x - p|^N,$$

as desired.

Definition 5.1. Two functions f, g are said *to be equal to order N at one point p* if $d_r f(p) = d_r g(p), r = 0, 1, \ldots, N$, their difference vanishes to order N at p. Then

$$|f(x) - g(x)| = o(|x - p|^N).$$

The setting of Taylor's formula is the local approximation, around a point p, by polynomials. We start noticing that polynomials in $x_1, \ldots, x_n$ are the same as polynomials in $x_1 - p_1, \ldots, x_n - p_n$, being more convenient to work with than the latter. The multi-index notation is useful too for polynomials; altogether, we write a generic polynomial of degree at most N in the form

$$P(x) = \sum_{|\alpha| \leq N} c_\alpha (x - p)^\alpha.$$

Note that the advantage of using $x_i - p_i$ as generators is

$$c_\alpha = \frac{1}{\alpha!} D^\alpha P(p). \tag{5.1}$$

Theorem 5.2. *Assume f is defined around a point p. Given N, there is at most one polynomial P of degree at most N such that P equals f at order N at p. If f is N times differentiable at p, this polynomial exists and is given by*

$$P(x) = \sum_{|\alpha| \le N} \frac{1}{\alpha!} D^\alpha f(p)(x-p)^\alpha.$$

Proof. If P, Q are of degree at most N and both equal f to order N at p, then $P(x) - Q(x) = o(|x-p|^N$, which obviously implies $P = Q$. On the other hand, it is clear by (5.1) that P defined above has the same derivatives at p of order at most N as f. □

This polynomial, depending on f, p, N, is called *the Taylor polynomial of f at p of order N*; the expansion

$$f(x) = \sum_{|\alpha| \le N} \frac{1}{\alpha!} D^\alpha f(p)(x-p)^\alpha + R_N(x), \quad R_N(x) = o(|x-p|^N),$$

with R_N being called the *error or remainder of order N*, is called the *Taylor expansion.* An alternative expression for the Taylor polynomial is in terms of the higher-order differentials,

$$\begin{aligned} P(x) &= \sum_{r=0}^{N} \frac{1}{r!} d^r f(p)(x-p, \ldots, x-p) \\ &= \sum_{r=0}^{N} \frac{1}{r!} \sum_{i_1=1}^{n} \sum_{i_2=1}^{n} \cdots \sum_{i_r=1}^{n} D_{i_1 i_2 \ldots i_r} f(p)(x_{i_1} - p_{i_1})(x_{i_2} - p_{i_2}) \cdots (x_{i_r} - p_{i_r}). \end{aligned}$$

As in one variable, if f is $N+1$ times differentiable in a neighborhood of p, it is possible to obtain an explicit closed expression for the error term. Indeed, given p, x set $h(t) = f(p + t(x-p)), 0 \le t \le 1$; then

$$h'(t) = \sum_{i_1=1}^{n} D_{i_1} f(p + t(x-p))(x_{i_1} - p_{i_1}) = df(p + t(x-p))(x-p),$$

$$\begin{aligned} h''(t) &= \sum_{i_1=1}^{n}\sum_{i_2=1}^{n} D_{i_1 i_2} f(p+t(x-p))(x_{i_1}-p_{i_1})(x_{i_2}-p_{i_2}) \\ &= d_2 f(p+t(x-p))(x-p, x-p), \end{aligned}$$

and in general $h^{(r)}(t) = d^r f(p+t(x-p))(x-p,\ldots,x-p)$. Then the expansion

$$h(1) = h(0) + h'(0) + \cdots + \frac{1}{N!}h^{(N)}(0) + \frac{1}{(N+1)!}h^{(N+1)}(\theta), 0 < \theta < 1$$

becomes

$$f(x) = P(x) + \frac{1}{(N+1)!}d^{N+1} f(\xi)(x-p,\ldots,x-p),$$

where $\xi = a + \theta(x-p)$ is a point in the segment joining a, x. Another way of writing this rest is

$$\sum_{|\alpha|=N+1} \frac{1}{\alpha_1!\alpha_2!\cdots\alpha_n!} D^\alpha f(\xi)(x-p)^\alpha.$$

5.2.2 *Newton's method* to find roots of a single equation $f(x) = 0$ in an unknown $x \in \mathbb{R}$ consists in starting from an initial guess x_0, replacing $f(x)$ by its affine approximation at x_0,

$$f(x_0) + f'(x_0)(x - x_0) = 0,$$

to produce the next iteration

$$x_1 = x_0 - \frac{f(x_0)}{f'(x_0)}.$$

The same method can be applied for a system of n equations in n unknowns

$$f_1(x_1,\ldots,x_n) = 0, \ldots, f_n(x_1,\ldots,x_n) = 0,$$

in short $f(x) = 0$. We start from an initial guess q^0 and replace $f(x) = 0$ by its approximation $f(q^0) + df(q^0)(x - q^0) = 0$ to produce

$$q^1 = q^0 - [df(q^0)]^{-1}(f(q^0)).$$

In general,

$$q^{k+1} = q^k - [df(q^k)]^{-1}(f(q^k)).$$

In practice, in numerical computations one does not compute the inverse matrix but solves the system $df(q^k)(y) = -f(q^k)$ in some way and sets

$q^{k+1} = q^k + y$. Let us analyze this iteration process in the simplest situation: we assume that f is a diffeomorphism of class C^2 from $B(p,r)$ to some open set $V \subset \mathbb{R}^n$, $f(p) = 0$ and we start from an initial guess $q^0 \in B(p,r)$. We can write Taylor's development centered at $x \in B(p,r)$ evaluated at p:

$$0 = f(p) = f(x) + df(x)(p - x) + R,$$

with $R = (R_1, \ldots, R_d), R_i = \frac{1}{2} d_2 f_i(\xi_i)(p - x, p - x)$ (for vector-valued functions we cannot take in general the same ξ_i). By hypothesis, $df(x)$ is invertible; assume $|df(x)^{-1}| \leq k$ for all x; we assume too that $|d_2 f_i(x)| \leq K$, then $|R| \leq K'|p - x|^2$. With $x = q^0$, we have

$$0 = f(q^0) + df(q^0)(p - q^0) + R, |R| \leq K'|q^0 - p|^2,$$

and therefore $q^1 = q^0 + p - q^0 + [df(q^0)]^{-1}(R)$, implying

$$|q^1 - p| \leq kK'|q^0 - p|^2.$$

Shrinking r if necessary, we may assume that $kK'r < 1$. Then all iterates q^k remain in $B(p,r)$ and

$$|q^{k+1} - p| \leq kK'|q^k - p|^2 \leq kK'r|q^k - p|.$$

Iterating this implies

$$|q^k - p| \leq (kK'r)^k |q^0 - p|,$$

so that indeed $q^k \to p$. Moreover, $|q^{k+1} - p| \leq kK'|qk - p|^2$ means that the convergence is quadratic.

A particular case is when g is scalar and one wants to find its critical points, that is $f = \nabla g$. Then we need $d_2 g(x)$ to be bounded below, for instance g strictly convex. In this case, we know there is just a critical point p of g, and the above shows that if one starts sufficiently close to p, Newton's algorithm converges quadratically to p.

5.3 Second-Order Criteria for Local Extrema

5.3.1 In this section, we use Taylor's formula to analyze a function at a singular point, $\nabla f(p) = 0$. Assuming f is twice differentiable, Taylor's development holds,

$$f(x) - f(p) = \frac{1}{2} d_2 f(p)(x - p, x - p) + o(|x - p|^2).$$

Theorem 1.4 proves that a symmetric matrix A can be diagonalized in an orthonormal basis. The matrix A and the associated quadratic form are called *non-negative* (*resp., non-positive*) *definite*, if $A(v, v) \geq 0$ (resp., $A(v, v) \leq 0$); in case the inequality is strict for $v \neq 0$, A is called *positive* (*rest., negative*) *definite*. We use the notations $A \geq 0, A \leq 0, A > 0, A < 0$. In the coordinates $y_1, \ldots, y_n$ in which A diagonalizes,

$$A(v, v) = \sum_i \lambda_i y_i^2,$$

where the λ_i are the eigenvalues. Thus, $A \geq 0$ (resp., $A \leq 0$) if and only $\lambda_i \geq 0$ (resp., $\lambda_i \leq 0$) for all i, $A > 0$ (resp., $A < 0$) iff $\lambda_i > 0$ (resp., $\lambda_i < 0$). Since $\det A$ is the product of the eigenvalues, $A > 0$ (resp., $A < 0$) iff $A \geq 0$ (resp., $A \leq 0$) and $\det A \neq 0$. If $\det A \neq 0$ and neither $A > 0, A < 0$, there are at least two eigenvalues with different signs.

Note that if A is positive, then it is automatically bounded below,

$$A(v, v) \geq m|v|^2,$$

with m being the smallest eigenvalue. Similarly, if A is negative, then

$$A(v, v) \leq -m|v|^2.$$

Proposition 5.1. *If f is twice differentiable at p and has a local minimum (resp., maximum), then $d_2 f(p) \geq 0$ (resp., $d_2 f(p) \leq 0$). In the opposite direction, if p is a critical point and $d_2 f(p) > 0$ (resp., $d_2 f(p) < 0$), then f has a local minimum (resp., maximum) at p. In fact, the extreme is* strict, *meaning that there exists a ball $B(p, \delta)$ and $c > 0$ such that*

$$f(x) \geq f(p) + c|x - p|^2, \quad |x - p| < \delta$$

in the first case and

$$f(x) \leq f(p) - c|x - p|^2, \quad |x - p| < \delta$$

in the second case.

Proof. Assume f has a local minimum at p; for every direction v, with $x = p + tv$, one has for t small

$$\frac{1}{2} t^2 d_2 f(p)(v, v) + o(t^2) = f(p + tv) - f(p) \geq 0,$$

which implies that the Hessian is non-negative definite

$$d_2 f(p)(v, v) \geq 0.$$

In the same way, we see that the Hessian is non-positive definite if f has a local maximum at p.

Assume now that $df(p)$ is positive definite; then, by the observation preceding the statement,

$$f(x) - f(p) \geq m|x-p|^2 + R(x), \qquad \frac{f(x)-f(p)}{|x-p|^2} \geq m + \frac{R(x)}{|x-p|^2},$$

with $R(x) = o(|x-p|^2)$. Choosing δ such that the size of the last term is smaller than $\frac{m}{2}$ for $|x-p| < \delta$, we are done. □

In the hypothesis of the theorem something more can be said. The hessian $d_2f(p)$ is the differential of the gradient. If p is critical and $\det d_2f(p) \neq 0$, by the inverse function Theorem 6.2, the map $x \to \nabla f(x)$ is locally one-to-one, so p is an isolated critical point of f.

In case $d_2f(p)$ has two eigenvalues of different signs, there are two unit orthogonal vectors u, v such that

$$d_2f(p)(tu + sv) = at^2 - bs^2,$$

with $a, b > 0$. Then $f(a+tu) - f(p) \geq a't^2, f(a+sv) - f(p) \leq b's^2$ for s, t small enough, and p is called a *saddle point.*

A twice differentiable function $g(t)$ in an interval is convex if and only if $g''(t) \geq 0$, strictly convex if $g'(t) > 0$. Therefore,

Proposition 5.2. *A twice differentiable function f in a domain U is convex if and only if $d_2f(p) \geq 0, p \in U$. If $d_2f(p) > 0, p \in U$, f is strictly convex.*

5.3.2 It is possible to decide the character of a symmetric matrix A without computing the eigenvalues, using the next theorem by Silvester. In the statement, Φ denotes the bilinear form associated to A: $\Phi(x,y) = \langle Ax, y\rangle$ or X^tAY in matrix notation.

Theorem 5.3. *For a symmetric bilinear form $\Phi(x,y)$, with matrix $A = (a_{ij})$ in the canonical basis $e_1, \ldots, e_n$, $\Phi > 0$ if and only if $\Delta_k = \det(a_{ij})_{i,j=1,\ldots,k} > 0, k = 1, \ldots, n$. If $\Phi \geq 0$, then $\Delta_k \geq 0$.*

Proof. It is enough to consider the positive case. If $\Phi \geq 0$, its restriction to $\mathbb{R}^k = \{x_{k+1} = \cdots = x_n = 0\}$ is also positive and so $\Delta_k \geq 0$ ($\Delta_k > 0$ if $\Phi > 0$). We prove the converse by induction; if $\Delta_k > 0$ for all k, by the induction hypothesis $\Phi > 0$ on $\mathbb{R}^{n-1}$ and we may consider an orthonormal basis $v_1, \ldots, v_{n-1}$ of $\mathbb{R}^{n-1}$ in which Φ diagonalizes,

$\Phi(v_i, v_j) = \lambda_i \delta_{ij}, i, j = 1, \dots, n-1, \lambda_i > 0$. Then $v_1, v_2, \dots, v_{n-1}, e_n$ is a basis in which Φ has matrix B

$$B = \begin{pmatrix} \lambda_1 & 0 & \cdots & 0 & \Phi(v_1, e_n) \\ 0 & \lambda_2 & \cdots & 0 & \Phi(v_2, e_n) \\ \vdots & \ddots & \ddots & 0 & \vdots \\ 0 & \cdots & 0 & \lambda_{n-1} & \Phi(v_{n-1}, e_n) \\ \Phi(e_n, v_1) & \cdots & \cdots & \Phi(e_n, v_{n-1}) & \Phi(e_n, e_n) \end{pmatrix}.$$

We obtain another basis replacing e_n by

$$v_n = e_n - \sum_{j=1}^{n-1} \frac{\Phi(e_n, v_j)}{\lambda_j} v_j.$$

One has

$$\Phi(v_n, v_i) = \Phi(e_n, v_i) - \sum_{j=1}^{n-1} \Phi(e_n, v_j)\Phi(v_j, v_i) = \Phi(e_n, v_i) - \Phi(e_n, v_i) = 0.$$

So in the basis $v_1, v_2, \dots, v_n$, Φ has a diagonal matrix with entries $\lambda_1, \dots, \lambda_{n-1}$ and the last one

$$\Phi(v_n, v_n) = \Phi(v_n, e_n) = \Phi(e_n, e_n) - \sum_{j=1}^{n-1} \frac{\Phi(e_n, v_j)^2}{\lambda_j}.$$

If we subtract from the row F_n of B the row F_j multiplied by $\frac{\Phi(e_n, v_j)}{\lambda_j}, j = 1, \dots, n-1$, and compute the determinant, we obtain

$$\det B = \lambda_1 \cdots \lambda_{n-1} \Phi(v_n, v_n).$$

Since $\det B$ has the same sign as $\det A = \det \Delta_n$, it follows that $\Phi(v_n, v_n) > 0$ and so $\Phi > 0$. ☐

Of course, applied to $-\Phi$ we get that $\Phi < 0$ if and only if Δ_k has sign $(-1)^{k+1}$. The case $n = 2$ is particularly simple to state:

Proposition 5.3. *If f is twice differentiable at a critical point p, $\det Hf(p) = c \neq 0$, then:*

(a) *If $c < 0$, p is a saddle point.*
(b) *If $c > 0$ and $f_{xx}(p) > 0$, p is a local minimum.*
(c) *If $c > 0$ and $f_{xx}(p) < 0$, p is a local maximum.*

In general dimension, a saddle point occurs if $\det Hf(p) \neq 0$ and both the conditions for $Hf(p) > 0, Hf(p) < 0$ fail.

Another result relevant in this context is *Descartes' theorem*, which states that if a polynomial $P(\lambda)$ has only real roots (in our case the characteristic polynomial), then the number of positive roots is equal to the number of sign changes of its coefficients. With this criteria, without finding the roots, we may know the multiplicity of zero, if any, the number of positive roots and the number of negative ones.

Example 5.1. We analyze the critical points of

$$f(x, y, z) = x^4 + y^4 + z^4 - (x + y + z)^2.$$

These are $p = (0, 0, 0), q = (\pm\sqrt{\frac{3}{2}}, \pm\sqrt{\frac{3}{2}}, \pm\sqrt{\frac{3}{2}})$. The Hessian at p has eigenvalues $0, 0, -6$, so the criteria does not apply. However, along the plane $x + y + z = 0$ the function is positive, while on the line (t, t, t) it equals $3t^4 - 9t^2$, negative close to p, and so p is not an extremum. The Hessian at q is positive definite and so f has a local minimum. Since f has limit $+\infty$ at infinity, it has a global minimum which must be q.

5.3.3 It is a natural question to ask for higher-order sufficient criteria for local extrema. In dimension $n = 1$ and for a critical point p, if the Taylor development starts

$$f(x) - f(p) = \frac{1}{k!} f^{(k)}(p)(x - p)^k + \cdots,$$

it is straightforward to see that if k is even, then $f(x) - f(p)$ will have the same sign as $f^{(k)}(p)$ for $|x-p|$ small. The same occurs in several variables. If

$$f(x) - f(p) = \frac{1}{k!} d^k f(p)(x - p, \ldots, x - p) + \cdots,$$

the sign of $f(x) - f(p)$ will be that of the kth homogeneous polynomial. But which sign is this? The author does not know, for even k, algebraic criteria similar to the ones for $k = 2$, for a homogeneous polynomial $P_k(x)$ to be positive definite, that is,

$$P_k(x) \geq c|x|^k, \quad x \neq 0,$$

or for P to be a sum of squares of $k/2$ homogeneous polynomials.

5.4 Smooth Functions with Compact Support

5.4.1 Recall that a function f defined in a domain U is said to be of class C^∞, or simply *smooth*, if it has derivatives of all orders. The *closed support* of f is defined as the closure of $\{f \neq 0\}$, the complement of the largest open set $V \subset U$ where f vanishes. The existence of smooth functions with compact support is not easy to see, as no closed formula in terms of elementary functions will define such a function. In this section, we show the existence of plenty of smooth functions with compact support, even adapted to particular requirements, such as *partitions of unity.*

First, we define such functions in one variable. Let h denote the function

$$h(x) = 0, \;\; x \leq 0, \quad h(x) = e^{-\frac{1}{x^2}}, \;\; x > 0.$$

Since $h^{(k)}(x) = o(x^N)$ as $x \to 0$ for all N, h is a smooth function flat at zero. Now let

$$g(x) = h(x)h(1-x).$$

Then g is a smooth non-negative function supported in $[0, 1]$. Next,

$$\int_0^x g(t)\, dt,$$

is also a smooth function, zero for $x \leq 0$ and constant equal to $m = \int_0^1 g$ for $x \geq 1$. Thus,

$$S(x) = \frac{1}{m} \int_0^x g(t)\, dt,$$

smoothly jumps from 0 to 1 as $0 \leq x \leq 1$.

Rescaling S we can easily show that given two nested intervals $[a, b] \subset (c, d)$ there is a smooth non-negative function ϕ supported in (c, d) such that $\phi(x) = 1$ on $[a, b]$, namely

$$\phi(x) = S\left(\frac{x-c}{a-c}\right) S\left(\frac{x-d}{b-d}\right).$$

For the proof of the nth dimensional version that follows, we use the bell-type even function ψ corresponding to $[-\frac{1}{2}, +\frac{1}{2}] \subset (-1, 1)$,

$$\psi(x) = S(2(x+1))S(2(x-1)).$$

Theorem 5.4. *If U is open and $K \subset U$ is compact, there is a smooth function ϕ with compact support in U, such that $0 \leq \phi \leq 1, \phi(x) = 1, x \in K$.*

Proof. For each $p \in K$, let $B(p,r)$ be an open ball included in U and consider

$$\phi_p(x) = \psi\left(\frac{|x-p|}{r}\right),$$

a smooth non-negative function supported in $B(p,r)$ equal to 1 on $B(p,\frac{r}{2})$. Since a finite number of $B(p,\frac{r}{2})$ cover K, adding the corresponding ϕ_p we get a smooth non-negative function H supported in U and $H(x) \geq 1, x \in K$. Then

$$\phi(x) = S(H(x))$$

satisfies all requirements. □

5.4.2 At some points in this text, the following theorem will be quite useful to localize proof arguments. The collection of functions $\psi_1, \ldots, \psi_m$ is called a *partition of unity* subordinate to the covering U_i.

Theorem 5.5. *If $U_1, \ldots, U_m$ are open sets covering a compact set K, there are smooth non-negative functions $\psi_1, \ldots, \psi_m$, ψ_i supported in U_i, such that $\sum \psi_i(x) = 1, x \in K$.*

Proof. Break K into m compact pieces K_i, $i = 1, \ldots, m$, $K = \cup_i K_i$, $K_i \subset U_i$; for instance, if $U = \cup_i U_i$,

$$K_i = \{x \in K : d(x, U_i^c) \geq \varepsilon\}$$

with $\varepsilon < d(K,U)$, works. Let ϕ_i be as in the previous theorem relative to K_i, U_i. Then $\sum_k \phi_k(x) \geq 1, x \in K$, and therefore we may choose

$$\psi_i = \frac{\phi_i}{\sum_k \phi_k}.$$ □

5.4.3 Borel's theorem. Another use of the bell and jump functions is to construct infinitely differentiable functions with prescribed derivatives. Our aim is to prove Borel's theorem:

Theorem 5.6. *Given arbitrary constants $c_\alpha, \alpha \in \mathbb{N}^n$, there is an infinitely differentiable function f in $\mathbb{R}^n$ such that $D^\alpha f(0) = c_\alpha$ for all α.*

Of course, for a finite number of c_α we can simply take as f the polynomial

$$P(x) = \sum \frac{c_\alpha}{\alpha!} x^\alpha.$$

For an infinite number, the above is the natural choice if the series converges, see next section. In general, this series might be divergent, even for all $x \neq 0$, and some modification is needed. To construct smooth functions, we need the differentiable version of Weierstrass' criteria for uniform convergence, a criteria used for example in the construction of continuous nowhere differentiable functions.

Proposition 5.4. *Assume that f_k is a sequence of C^1 functions in a domain U, and $f, g_i, i = 1, \ldots, n$ are such that:*

(a) *$f_k \to f$ uniformly on compacts.*
(b) *For each $i = 1, \ldots, n$, $D_i f_k \to g_i$ uniformly on compacts.*

Then f is of class C^1 in U and $D_i f = g_i$.

Proof. First, note that f, g_i are continuous. Now, for all p and directions v,

$$f_k(p + tv) = f_k(p) + \int_0^t \sum_i v_i D_i f_k(p + sv)\, ds,$$

whence

$$f(p + tv) = f(p) + \int_0^t \sum_i v_i g_i(p + sv)\, ds.$$

This proves that $D_i f = g_i$ and by Proposition 4.1, f is of class C^1. □

Of course, the proposition can be stated for a series instead. Iterating, we see that if all series $\sum_k D^\beta f_k$ are uniformly convergent on compacts, then $\sum_k f_k$ is smooth and

$$D^\beta \sum_k f_k = \sum_k D^\beta f_k.$$

The standard way of checking uniform convergence on compacts $K \subset U$ of the series is Weierstrass' M-test:

$$|D^\beta f_k(x)| \leq M_k(K, \beta), \quad x \in K, \quad \sum_k M_k(K, \beta) < +\infty. \tag{5.2}$$

We can now prove Borel's theorem. The idea is to introduce convergence factors in the series

$$f = \sum_\alpha \frac{c_\alpha}{\alpha!} x^\alpha \Phi_\alpha(x),$$

so that (5.2) is satisfied (note that the index k is replaced by the index α). Moreover, we want that for fixed β

$$D^\beta f(0) = \sum_\alpha \frac{c_\alpha}{\alpha!} D^\beta [x^\alpha \Phi_\alpha(x)]_{x=0} = c_\beta.$$

This is accomplished if $\Phi_\alpha(x) = 1$ around the origin. We thus try

$$\Phi_\alpha(x) = \psi\left(\frac{|x|}{\varepsilon_\alpha}\right),$$

where ψ is a bell function equal to one for $|t| < \frac{1}{2}$ and zero for $|t| > 1$, and the $\varepsilon_\alpha \to 0$ are to be chosen. Note that with this choice, f is a smooth function in $x \neq 0$ because only a finite number of terms are non-zero around a fixed x. For fixed β, it is enough to bound

$$D^\beta \left[x^\alpha \psi\left(\frac{|x|}{\varepsilon_\alpha}\right)\right],$$

for $|\alpha|$ big enough, say $|\alpha| > 2|\beta|$. This derivative is a sum

$$\sum_{\gamma \leq \beta} \binom{\beta}{\gamma} D^\gamma [x^\alpha] D^{\beta-\gamma} \left[\psi\left(\frac{|x|}{\varepsilon_\alpha}\right)\right],$$

and therefore is bounded by

$$\sum_{\gamma \leq \beta} \frac{\beta!}{\gamma!(\beta-\gamma)!} \frac{\alpha!}{(\alpha-\gamma)!} |x|^{|\alpha|-|\gamma|} (\varepsilon_\alpha)^{|\gamma|-|\beta|} M_{|\beta|},$$

where $M_{|\beta|}$ is a bound for the derivatives of ψ of order at most $|\beta|$. Inserting $\frac{c_\alpha}{\alpha!}$ and taking into account that $|x| \leq \varepsilon_\alpha$ we get a bound

$$C_\beta \sum_{|\alpha|>2|\beta|} |c_\alpha| (\varepsilon_\alpha)^{|\alpha|-|\beta|} \leq \sum_{|\alpha|>2|\beta|} |c_\alpha| (\varepsilon_\alpha)^{|\alpha|/2},$$

and so it is clear that taking ε_α small enough the series converges.

5.4.4 Still another result showing the abundance of smooth functions is the following one:

Theorem 5.7. *If F is a closed set in $\mathbb{R}^n$, there exists a smooth function $f, f(x) \geq 0$ such that the zero set of f is exactly F.*

Proof. In the continuous category one would simply choose $f(x) = d(x, F)$. To obtain a smooth function we simply observe that the existence of compactly supported smooth functions implies that f exists locally, so we need to just add up these. Formally, the open set $\mathbb{R}^n \setminus F$ is a union $\cup_k B_k$ of open balls, with $\overline{B_k} \cap F = \emptyset$. Let f_k be a smooth function positive in B_k and zero outside B_k. We just need to choose positive constants λ_k such that the series $f = \sum_k \lambda_k f_k$ satisfies (5.2). If

$$C_k = \sup_{|\alpha| \leq k} \sup_{x \in B_k} |D^\alpha f_k(x)|,$$

it is enough to set $\lambda_k = \frac{1}{2^k C_k}$. □

Since

$$D^\alpha f = \sum \lambda_k D^\alpha f_k,$$

it follows that $D^\alpha f = 0$ on F for all α. Using particular coverings of $\mathbb{R}^n \setminus F$, in particular the so-called *Whitney covering*, it is possible to obtain other functions with additional control on the derivatives of F. In Theorem 16.1, we will see that for some closed sets, the sub-manifolds of dimension $n-1$, it is possible to choose f such that $\nabla f \neq 0$ on F.

5.5 Real Analytic Functions

5.5.1 Multi-power series. It is tempting, as for one-variable functions and for an infinitely differentiable function f, to let $N \to +\infty$ in the Taylor development of f at a fixed point p. This leads to the Taylor power series

$$\sum_\alpha \frac{D^\alpha f(p)}{\alpha!}(x-p)^\alpha.$$

In general, a *multi-power series*, in multi-index notation, is an expression

$$\sum_{\alpha \in \mathbb{N}^n} c_\alpha (x-p)^\alpha, \quad c_\alpha \in \mathbb{R}. \tag{5.3}$$

The first question is about the domain of convergence of such multi-series. The natural notion of convergence to consider for a series $\sum_{\alpha \in \mathbb{N}^n} a_\alpha$ is the so-called *inconditional summability*, meaning that for every bijection $\phi : \mathbb{N} \to \mathbb{N}^n$ the series $\sum_n a_{\phi(n)}$ is convergent. It then follows that the sum is independent of ϕ and unconditional summability is equivalent to *absolute summability*

$$\sum_{\alpha\in\mathbb{N}^n} |a_\alpha| < +\infty.$$

Alternatively, $\sum_{\alpha\in\mathbb{N}^n} a_\alpha$ is absolutely summable with sum S if and only if for each $\varepsilon > 0$ there exists a finite set $F \subset \mathbb{N}^n$ such that $\sum_{\alpha\in G} |a_\alpha| \leq \varepsilon$ for all finite sets $G, G \cap F = \emptyset$, and

$$|S - \sum_{\alpha\in F} a_\alpha| \leq \varepsilon.$$

Thus, we must consider the set A in $[0, +\infty)^n$ of $r = (r_1, \ldots, r_n), r_i \geq 0$ such that

$$\sum_\alpha |c_\alpha| r^\alpha < +\infty.$$

Together with A one considers the set B of r for which there exists $r' \in A, r_i < r'_i$.

Exercise 5.1. Show that $r \in B$ if and only if there exists $r', r_i < r'_i$, such that $|c_\alpha|(r')^\alpha \leq C$.

The *domain of convergence* D of (5.3) is the largest open set containing p where the series is absolutely summable, namely $(|x_i - p_i|) \in B$. It is non-empty if and only if $\sum_\alpha |c_\alpha| r^\alpha < +\infty$ for some $r_i > 0$. Note that it is a p-domain and always star-shaped with respect to p, that is, if $q \in D$, then the whole segment from p to q lies in D.

In dimension $n = 1$, by the ratio test, D is the open interval centered at p of *radius of convergence* ρ given by

$$\frac{1}{\rho} = \limsup_{k\to\infty} |c_k|^{\frac{1}{k}}.$$

For $n > 1$, the shape of D can be quite general. For instance, given a one-variable power series $\sum_k c_k t^k$ with $0 < \rho$, the double series $\sum_k c_k (x + y)^k$ has domain of convergence given by $|x| + |y| < \rho$ while $\sum_k c_k (x + y^2)^k$ has domain of convergence given by $|x| + |y|^2 < \rho$, $\sum_k c_k x^k y^k$ has domain $|x||y| < \rho$, etc.

If $r \in B, r'$ is as in the exercise above, and $|x_i - p_i| \leq r_i$, from

$$\sum_\alpha |c_\alpha||x_i - p_i|^{\alpha_i} \leq C \sum_\alpha \left(\frac{r'_i}{r_i}\right)^{\alpha_i} < +\infty,$$

and from Weierstrass' test it follows that the series is uniformly convergent in compacts of D and thus defines a continuous function f in D,

$$f(x) = \sum_\alpha c_\alpha (x-p)^\alpha.$$

Theorem 5.8. *The function f is infinitely differentiable in D and for every multi-index β, one has*

$$D^\beta f(x) = \sum_{\alpha \geq \beta} c_\alpha \frac{\alpha!}{(\alpha-\beta)!}(x-p)^{\alpha-\beta},$$

where the right-hand side multi-power series has the same domain of convergence. In particular,

$$c_\alpha = \frac{D^\alpha f(p)}{\alpha!}.$$

As a consequence, two multi-power series with the same sum have the same coefficients.

Proof. The statement about the domain of convergence follows form the exercise. It is sufficient to prove the other statement for $n = 2, p = (0,0)$ and a first-order derivative, say D_x. Assume that

$$f(x,y) = \sum_{k,l} a_{k,l} x^k y^l,$$

is absolutely summable in $D : |x| < r_1, |y| < r_2$. Then the triple series

$$\sum_{k,l} a_{k,l}(x+h)^k y^l,$$

is absolutely summable for $|h|$ small enough. A basic property of absolutely summable series is that the sum can be done in arbitrary packages, whence we may write

$$f(x+h, y) = \sum_p f_p(x,y) h^p.$$

For each p, f_p is absolutely summable in D and the power series in h has a positive radius of convergence. Since

$$\frac{f(x+h,y)-f(x,y)}{h}=f_1(x,y)+O(h),$$

it follows that $D_x f(x,y)$ exists and

$$D_x f(x,y)=f_1(x,y)=\sum_{k,l} k a_{k,l} x^{k-1} y^l.$$

□

Thus, every multi-power series with non-trivial domain of convergence is a Taylor series. In fact, every power series is a Taylor series, by Theorem 5.6.

5.5.2 Real-analytic functions. Given an infinitely differentiable function in a domain U, the Taylor series at $p \in U$ may have a non-trivial domain of convergence or not. For instance, by Borel's theorem just quoted, there is f with $f^{(k)}(0) = k!k^k$, so that the Taylor series at zero has zero radius of convergence. On the opposite side, the function

$$h(t)=e^{-\frac{1}{t^2}},\quad t\neq 0$$

is infinitely differentiable in $t \neq 0$ and $h^{(k)}(t) = o(|t|^p)$ for all k, p. By the mean-value theorem, this implies that h is smooth in the whole line and $h^{(k)}(0) = 0$ for all k, this function is *flat* at zero. Then the Taylor series of h at zero is identically zero, while $h(t) \neq 0$. If f is defined by a power series with non-trivial interval of convergence, the Taylor series for $f + h$ sums $f \neq f + h$.

For the Taylor series at p to have a non-trivial domain of convergence with sum f, it is necessary and sufficient that

$$\sum_\alpha \frac{|D^\alpha f(p)|}{\alpha!} r^\alpha < +\infty,$$

for some $r = (r_i), r_i > 0$ and that the rest

$$\sum_{|\alpha|=N+1} \frac{1}{\alpha!} D^\alpha f(\xi)(x-p)^\alpha$$

has limit zero as $N \to \infty$. Recall from Calculus I that this is the case for the usual functions $e^x, \sin x, \cos x$, etc.

If the Taylor series of f at each point $p \in U$ has a non-trivial domain of convergence, say it converges in the n-interval $|x_i - p_i| < r_i$, and its sum is

$f(x)$ we say that f is *real-analytic* in U. Informally, a real-analytic function is given locally by a polynomial in $x-p$ of infinite order. By the theorem, f is smooth, all derivatives are also real-analytic and they can be obtained using term by term differentiation.

Exercise 5.2. Assume that f is smooth in a domain U. Prove that f is real-analytic in U if for every compact $K \subset U$ there are positive constants C, r such that

$$|D^\alpha f(x)| \leq C r^{|\alpha|} \alpha!, \quad x \in K.$$

The converse holds too, see Exercise 17.2, together with Theorem 9.4.

A multi-power series $\sum_{\alpha \in \mathbb{N}^n} c_\alpha x^\alpha$ with non-trivial domain of convergence is real-analytic. Indeed, we already know that

$$D^\beta f(x) = \sum_{\alpha \geq \beta} c_\alpha \frac{\alpha!}{(\alpha-\beta)!} x^{\alpha-\beta}.$$

Then, using summation by packages and Newton's binomial,

$$\begin{aligned}\sum_\beta \frac{D^\beta f(x)}{\beta!}(y-x)^\beta &= \sum_\beta \sum_{\alpha \geq \beta} c_\alpha \frac{\alpha!}{\beta!(\alpha-\beta)!} x^{\alpha-\beta}(y-x)^\beta \\ &= \sum_\alpha c_\alpha \sum_{\beta \leq \alpha} \frac{\alpha!}{(\alpha-\beta)!} x^{\alpha-\beta}(y-x)^\beta = \sum_\alpha c_\alpha y^\alpha = f(y).\end{aligned}$$

Exercise 5.3. Check that compositions of real-analytic functions are real-analytic. In particular, every closed formula in terms of usual functions is a real-analytic function in its domain of definition.

Real-analytic functions enjoy a very important property, the *principle of analytic continuation.* This principle states that real-analytic functions, like polynomials, are rigid objects, unless the functions constructed in the previous section.

Theorem 5.9. *Assume f, g are real-analytic in a domain U, and satisfy the (equivalent) properties:*

(a) *they are equal to infinite order at some point $p \in U$, that is, $D^\alpha f(p) = D^\alpha g(p)$ for all multi-indexes.*
(b) *$f = g$ in some (tiny) open set $U' \subset U$.*

Then $f = g$ in U.

Proof. It is enough to prove it for $g = 0$. As in similar proofs, we consider the set V of points $p \in U$ satisfying the hypothesis. It is open and non-empty. But it is closed as well, because if $p \in U, p = \lim p_k, p_k \in V$, then $D^\alpha f(p) = \lim_k D^\alpha f(p_k) = 0$. □

In case $n = 1$, the zeroes of non-trivial real-analytic functions are isolated. This is because if $f(p) = 0$, the order of the zero is finite, so

$$f(x) = \frac{f^{(k)}(p)}{k!}(x-p)^k + \cdots, \quad f^{(k)}(p) \neq 0,$$

implying that $f(x) = (x-p)^k g(x)$ with $g(p) \neq 0$, whence $f(x) \neq 0$ for x close enough to p. As a consequence, two real-analytic functions f, g in an interval I are equal if $f(p) = g(p)$ for all points p in a set with an accumulation point in I (for instance, a sequence with limit point in I), a rigid property that of course does not hold if $n > 1$.

Chapter 6

The Inverse and Implicit Function Theorems

In this chapter, we study first the differentiable changes of coordinates and their associate vector fields. In later chapters, the use of changes of coordinates will become clear, to solve differential equations or to compute integrals. The *inverse function theorem*, which is an existence result, can be looked at as a local criteria to get coordinates. But, above all, is a paradigmatic exponent of a basic idea in differentiation, that the objects — functions in this case — inherit locally the properties of their linear approximations. The *implicit function theorem* is also an existence result. In this chapter, we present its analytic viewpoint, in terms of *implicit functions.*

6.1 Differentiable Changes of Coordinates

6.1.1 Recall that a *vector field* in U is simply a map $X : U \to \mathbb{R}^n$ that we view as assigning to each $x \in U$ the vector $X(x)$ with origin at x. A vector field X acts on differentiable functions f to produce a function Xf defined by

$$(Xf)(x) = D_v f(x) = \langle X(x), \nabla f(x)\rangle, \quad v = X(x).$$

In coordinates, if $X = (A_1, \dots, A_n)$,

$$Xf(x) = \sum_i A_i(x) D_i f(x).$$

The main example we have seen is the gradient of a function. Now we will introduce a different type of vector field.

We come back to the concept of change of coordinates in a domain $U \subset \mathbb{R}^n$. Recall that by this we mean a one-to-one map $\Phi : U \to \mathbb{R}^n$, that is bijective from U to $\Phi(U)$, and continuous in both senses, also called a *homeomorphism* between U and $\Phi(U)$. In fact, by Brouwer's theorem, if Φ is continuous and one-to-one, then $V = \Phi(U)$ is open and Φ^{-1} is continuous, so we might take this as a definition.

If Φ has components $(u_1(x), \dots, u_n(x))$, we think of $u_1(x), \dots, u_n(x)$ as the new coordinates of the point whose canonical coordinates are $x = (x_1, \dots, x_n)$. So we do not view Φ as effectively moving points, Φ expresses the new coordinates in terms of the old ones. This new system of coordinates has new coordinate axis through each point; each point $p \in U$ is exactly the intersection of n new axis, the level sets

$$\{x : u_i(x) = u_i(p), i \neq j\},$$

that we think as being nice curves. See Figure 3.3.

Definition 6.1. A *differentiable change of coordinates* is one for which Φ and its inverse Φ^{-1} are differentiable. We use as well the term *diffeomorphism*. We denote by $x_i(u), u = (u_1, \dots, u_n)$ the components of Φ^{-1}.

All partial derivatives $\frac{\partial u_j}{\partial x_i}, \frac{\partial x_i}{\partial u_j}$ exist. Since the composition of Φ and Φ^{-1} is the identity, the chain rule implies that

$$d(\Phi^{-1}) = (d\Phi)^{-1}.$$

Thus, the differential $d\Phi(x)$ is an invertible linear map in $\mathbb{R}^n$, that is, $\det d\Phi(x) \neq 0$. Later on we will see as a consequence of the inverse function theorem that if $\Phi : U \to \mathbb{R}^n$ is injective and $d\Phi(x)$ is invertible for all x, then in fact $\Phi(U) = V$ is open and Φ^{-1} is differentiable.

The new coordinate axes are the paths parametrized by u_i, keeping the $u_j, j \neq i$ fixed

$$u_i \to \Phi^{-1}(u) = (x_1(u), \dots, x_n(u)).$$

The tangent vector to this path, the ith column of $d\Phi^{-1}(u)$, is not zero at each point. For the time being let us denote by X_i this vector field, so $X_i(x)$ is the tangent vector at x to the ith new axis. In coordinates,

$$X_i(x) = \left(\frac{\partial x_1}{\partial u_i}(u), \dots, \frac{\partial x_n}{\partial u_i}(u) \right), \quad u = \Phi(x).$$

When the change of coordinates is affine, that is, $X = P + MV$, with M an invertible matrix with linearly independent column vectors $v_1, \ldots, v_n$, all level sets are lines and $X_i(x)$ is a constant field, equal to v_i. In general, the vector $X_i(x)$ may change direction, for instance this is the case for the polar coordinates.

Assume now that $f(x)$ is a scalar function of $x \in U$. Composing with $x = \Phi^{-1}(u)$ we have a function defined on V,

$$g(u) = f(\Phi^{-1}(u)), \quad \text{or} \quad g(\Phi(x)) = f(x).$$

The function g is nothing but f expressed in the new coordinates u, but by now we denote it differently, g. As Φ and its inverse are differentiable, f is differentiable if and only if g is. If we apply the chain rule, since by construction $X_i(x)$ is mapped by $d\Phi(x)$ to e_i, by (4.1) one has

$$(X_i f)(x) = \frac{\partial g}{\partial u_i}(u), \quad u = \Phi(x).$$

In a developed form, the left-hand side is

$$\sum_{j=1}^{n} \frac{\partial x_j}{\partial u_i}(\Phi(x)) \frac{\partial f}{\partial x_j}(x).$$

Said otherwise, $(X_i f)(x)$ is $\frac{\partial g}{\partial u_i}(u)$ expressed in the coordinates x. This justifies changing the notation and replacing X_i by the notation $\frac{\partial}{\partial u_i}$. We also use the notation ∂_i and call them *coordinate vector fields.* Thus, the partial derivative with respect to u_i, which is a constant vector field acting on functions g defined on $\Phi(U) = V$, is transported with the same notation to a vector field acting on f, in accordance with the idea that points have not moved but have changed coordinates. Then

$$\frac{\partial}{\partial u_i} = \left(\frac{\partial x_1}{\partial u_i}(u), \ldots, \frac{\partial x_n}{\partial u_i}(u) \right), \quad u = \Phi(x), \; i = 1, \ldots, n,$$

is a basis of vector fields. With this notation, $\frac{\partial}{\partial x_i} = e_i$. See Figure 6.1.

Thus, we do not distinguish between f, g and use the notation $f(u_1, \ldots, u_n)$ instead of $g(u_1, \ldots, u_n)$. Thus,

$$\frac{\partial f}{\partial u_i} = \sum_{j=1}^{n} \frac{\partial f}{\partial x_j} \frac{\partial x_j}{\partial u_i}.$$

Obviously,

$$\frac{\partial u_i}{\partial u_j} = \delta_{i,j},$$

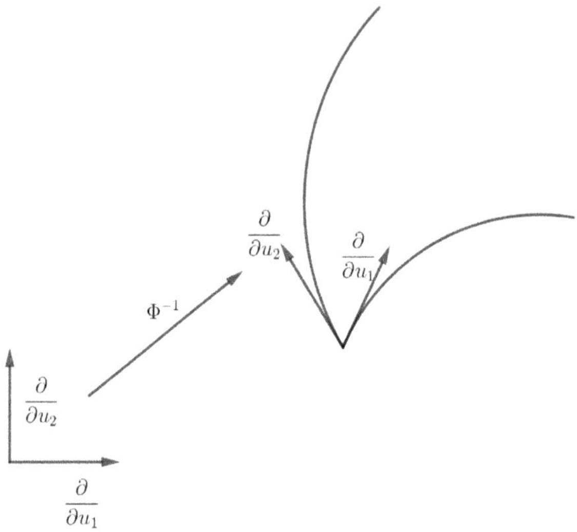

Figure 6.1. Coordinate vector fields.

that is also the statement that $d\Phi$ and $d\Phi^{-1}$ are inverse matrices. Since the left-hand term is the directional derivative of u_i in the direction $\frac{\partial}{\partial u_j}$, so equal to $du_i(\frac{\partial}{\partial u_j})$, it turns out that

$$\frac{\partial}{\partial u_1}, \ldots, \frac{\partial}{\partial u_n},$$

a basis of vectors, and $du_1, \ldots, du_n$, a basis of linear forms, are dual basis, that in general change with x. So we may write, exactly as with canonical coordinates,

$$df = \sum_i df\left(\frac{\partial}{\partial u_i}\right) du_i = \sum_i \frac{\partial f}{\partial u_i} du_i. \tag{6.1}$$

In terms of the gradients ∇u_i,

$$\left\langle \nabla u_i, \frac{\partial}{\partial u_j} \right\rangle = \delta_{ij}.$$

This means that ∇u_i, $\frac{\partial}{\partial u_j}$ are dual basis in $\mathbb{R}^n$ at every point, with respect to the usual Euclidean product. Also,

$$\nabla f = \sum_i \frac{\partial f}{\partial u_i} \nabla u_i.$$

We point out a remark regarding the notation $\frac{\partial}{\partial u_i}$. If the same function, say h, appears in two different systems of coordinates

$$(v_1, v_2, \ldots, v_n),\ (w_1, \ldots, w_n),\ v_1 = w_1 = h,$$

then $\frac{\partial}{\partial v_1}$ might be different than $\frac{\partial}{\partial w_1}$. Consider, for example, together with canonical coordinates (x, y) in the plane the coordinates $u = x, v = x + y$; then

$$\frac{\partial}{\partial u} = \frac{\partial}{\partial x} - \frac{\partial}{\partial y}.$$

When there is no ambiguity in the coordinate system, we use the shorter notation ∂_i.

6.1.2 Let us consider in full detail the polar and spherical coordinates. For the polar coordinates r, θ,

$$\partial_r = (\cos\theta, \sin\theta), \quad \partial_\theta = (-r\sin\theta, r\cos\theta),$$

are the radial and tangential fields to circles. For spherical coordinates (ρ, ϕ, θ),

$$x = \rho\sin\phi\cos\theta, \quad y = \rho\sin\phi\sin\theta, \quad z = \rho\cos\phi,$$

∂_ρ has the radial direction at every point, ∂_ϕ is the tangent vector to meridians and ∂_θ is the tangent field to parallels. In coordinates,

$$\begin{aligned}
\partial_\rho &= (\sin\phi\cos\theta, \sin\phi\sin\theta, r\cos\phi),\\
\partial_\phi &= (\rho\cos\phi\cos\theta, \rho\cos\phi\sin\theta, -\rho\sin\phi),\\
\partial_\theta &= (-\rho\sin\phi\sin\theta, \rho\sin\phi\cos\theta, 0).
\end{aligned}$$

These are the expressions in terms of the coordinates (ρ, ϕ, θ). In terms of x, y, z, they become

$$\begin{aligned}
\partial_\rho &= \frac{1}{\sqrt{x^2+y^2+z^2}}(x, y, z),\\
\partial_\phi &= \frac{1}{\sqrt{x^2+y^2}}(zx, zy, -x^2+y^2),\\
\partial_\theta &= (-y, x, 0).
\end{aligned}$$

Strictly speaking, they are a coordinate system only in the complement of a half-plane $\theta = c$ containing the z-axis.

The *spherical coordinates* in dimension n consist in ρ and $n-1$ angles $\phi_i, i = 1, 2, \ldots, n-1$, where

$$x_n = \rho \cos\phi_1, x_{n-1} = \rho \sin\phi_1 \cos\phi_2, x_{n-2} = \rho \sin\phi_1 \sin\phi_2 \cos\phi_3, \ldots,$$

$$x_2 = \rho \sin\phi_1 \cdots \sin\phi_{n-2} \cos\phi_{n-1}, x_1 = \rho \sin\phi_1 \cdots \sin\phi_{n-2} \sin\phi_{n-1},$$

with

$$\rho \geq 0, 0 < \phi_1 < \pi, \ldots, 0 < \phi_{n-2} < \pi, 0 < \phi_{n-1} < 2\pi.$$

6.1.3 Assume that a differentiable arc γ is given in U using a coordinate system $u_1, \ldots, u_n$

$$u_i = u_i(t), \quad i = 1, \ldots, n, \;\; a \leq t \leq b.$$

In terms of the canonical coordinates, the arc is given by

$$x_j = x_j(u_1(t), \ldots, u_n(t)), \quad j = 1, \ldots, n.$$

The tangent T has components

$$x_j'(t) = \sum_{i=1}^{n} \frac{\partial x_j}{\partial u_i} u_i'(t),$$

that is, we have

$$T = \sum_i u_i'(t)\partial_i.$$

Then

$$|T|^2 = \langle T, T\rangle = \sum_{i,j} \langle \partial_i, \partial_j\rangle u_i' u_j'.$$

The matrix with entries

$$g_{ij} = \langle \partial_i, \partial_j\rangle,$$

is the Gram matrix of the basis ∂_i. The above expression is written

$$ds^2 = \sum_{ij} g_{ij}\, du_i \otimes du_j,$$

and expresses how to compute lengths in the u coordinates. One says that this is the expression of the Euclidean metric in these coordinates. For polar coordinates, this coincides with the formulas found in paragraph 14.1.4.

6.1.4 A change of coordinates may be useful to simplify a specific problem. In the same way that to discuss a linear system of equations it is simpler to write it say in a triangular form, a problem in differential calculus can become simpler expressed in a convenient coordinate system. To illustrate this we reinterpret the method explained in paragraph 4.7.4 to solve a linear partial differential equation

$$\sum_{i=1}^{n} A_i \frac{\partial f}{\partial x_i} = B(x),$$

where $v = (A_1, \ldots, A_n) \neq 0$, in terms of changes of coordinates. If $u_1, \ldots, u_n$ are linear coordinates so that the lines with direction vector v are given by $u_i = c_i, i = 2, \ldots, n$, the equation will take the form

$$\frac{\partial f}{\partial u_1} = \hat{B},$$

and therefore can be solved by one anti-differentiation in u_1. Consider again Example 4.11; if x', y', z' are new linear coordinates with

$$x' = 2x + y, \quad y' = 3x - z,$$

and $z' = \alpha x + \beta y + \gamma z, \Delta = -2\beta + 3\gamma + \alpha \neq 0$, then

$$\frac{\partial}{\partial z'} = \frac{1}{\Delta}\left(\frac{\partial}{\partial x} - 2\frac{\partial}{\partial y} + 3\frac{\partial}{\partial z}\right).$$

Choosing $z' = x$, the equation becomes

$$\frac{\partial f}{\partial z'} = xy = z'(x' - 2z'),$$

with solution

$$f(x', y', z') = \frac{1}{2}x'(z')^2 - \frac{2}{3}(z')^3 + \Phi(x', y') = \frac{1}{3}x^3 + \frac{1}{2}yx^2 + \Phi(2x+y, 3x-z).$$

The same considerations apply to a system of equations

$$\sum_{i=1}^{n} A_{ij} \frac{\partial f}{\partial x_i} = B_j(x), \quad j \in J,$$

where $J \subset \{1, \ldots, n\}$. If k is the cardinal of J and $u_1, \ldots, u_n$ are linear coordinates so that lines with direction vector $v_j = (A_{1j}, \ldots, A_{nj})$ are the first u_j axis, $j = 1, \ldots, k$, the system will take the form

$$\frac{\partial f}{\partial u_j} = \hat{B}_j, \quad j = 1, \ldots, k.$$

As indicated before, if the compatibility conditions

$$\frac{\partial \hat{B}_j}{\partial u_l} = \frac{\partial \hat{B}_l}{\partial u_j},$$

hold, which amount to $D_{v_j} B_l = D_{v_l} B_j, i, j \in J$, then the general solution can be obtained by successive anti-differentiation, one for each variable $u_j, j \in J$. The general solution will involve a general function $\Phi(u_{k+1}, \ldots, u_n)$ of $n - k$ variables. Therefore, if f is prefixed on a set A meeting every linear sub-manifold spanned by the $v_j, j \in J$ in just one point, then f is unique.

Example 6.1. Let us discuss the system

$$2f_x + f_y - f_z = x + y + cz, \quad f_x - f_y + f_z = 2x + y,$$

prefixing the directional derivatives along $v_1 = (2, 1, -1), v_2 = (1, -1, 1)$. We consider linear coordinates x', y', z' defined by

$$(x, y, z) = x'(2, 1, -1) + y'(1, -1, 1) + z'(0, 0, 1),$$

the choice of the last vector being irrelevant as soon as it is linearly independent with v_1, v_2. Then the equations become

$$f_{x'} = (3 - c)x' + cy' + cz', \quad f_{y'} = 5x' + y'.$$

The compatibility condition is $c = 5$ (which is $D_{v_1}(2x+y) = D_{v_2}(x+y+cz)$); in this case the general solution of the first equation is

$$-(x')^2 + 5x'y' + 5x'z' + \Phi(y', z'),$$

which replaced in the second gives $(\Phi)_{y'} = y', \Phi(y', z') = \frac{1}{2}(y')^2 + h(z')$, so the general solution is

$$f(x, y, z) = -\frac{1}{9}(x + y)^2 + \frac{5}{9}(x + y)(x - 2y) + \frac{5}{3}(x + y)(z + y)$$
$$+ \frac{1}{18}(x - 2y)^2 + \Phi(z + y),$$

with $\Phi(u)$ arbitrary. If say f is known along the y-axis, $f(0,t,0) = h(t)$, then

$$-\frac{1}{9}t^2 - \frac{10}{9}t^2 + \frac{5}{3}t^2 - \frac{4}{18}t^2 + \Phi(t) = h(t),$$

so $\Phi(t) = h(t) + \frac{1}{3}t^2$ and the unique solution is

$$f(x,y,z) = -\frac{1}{9}(x+y)^2 + \frac{5}{9}(x+y)(x-2y) + \frac{5}{3}(x+y)(z+y) + \frac{1}{18}(x-2y)^2 + \frac{1}{3}(z+y)^2 + h(z+y).$$

6.1.5 We consider again the concept of functional dependence in the context of coordinate systems:

Theorem 6.1. *Assume $u_1,\ldots,u_m$ form part of a coordinate system in which U is a p-domain. Then F depends functionally on $u_1,\ldots,u_m$ in U if and only if $\nabla F(x)$ is a linear combination of $\nabla u_1(x),\ldots,\nabla u_m(x), x \in U$.*

Proof. Let $u_1,\ldots,u_m,\ldots,u_n$ be a coordinate system in U. By (6.1), $\nabla F(x)$ is a linear combination of $\nabla u_1(x),\ldots,\nabla u_m(x), x \in U$ if and only if $\frac{\partial F}{\partial u_j} = 0, j = m+1,\ldots,n$, that is, F depends just on $u_1 = u_1,\ldots,u_m$. □

When $m = n-1$, we already noticed that the condition is simply that the determinant of the matrix with column vectors $\nabla u_j(x), j = 1,\ldots,n-1$ and $\nabla F(x)$ is zero at every point. This gives a *linear partial differential equation* of type

$$\sum_{i=1} A_i(x)\frac{\partial F}{\partial x_i} = 0.$$

In general, one has $n-m$ such equations. In Proposition 7.1, we further comment on Theorem 6.1.

Consider, for instance, the polar coordinates (r,θ). A function $f(x,y)$ is radial, that is of the form $f(x,y) = h(\sqrt{x^2+y^2})$ if and only if

$$\partial_\theta f = f_x(-r\sin\theta) + f_y r\cos\theta = xf_y - yf_x = 0,$$

and constant on rays, that is a function of θ if and only if

$$\partial_r f = f_x\cos\theta + f_y\sin\theta = xf_x + yf_y = 0.$$

In space, $f(x, y, z)$ depends functionally on $r, \phi = \arccos \frac{z}{r}$ if and only if

$$-yf_x + xf_y = 0.$$

6.2 The Inverse Function Theorem

6.2.1 The inverse function theorem is an important result aligned with the main spirit of differential calculus, the fact that a differentiable function inherits locally, that is around p, the properties of $df(p)$. It is also the basis leading to other results, such as the implicit function theorem and the constant rank theorem, important both from the analytic and geometric point of view.

The local property involved with these theorems relates to existence and uniqueness of solutions of a system of equations and can be explained as follows. Assume we want to discuss a system of k equations in n unknowns $x = (x_1, \dots, x_n)$

$$f_1(x_1, \dots, x_n) = y_1, \quad f_2(x_1, \dots, x_n) = y_2, \dots, \quad f_k(x_1, \dots, x_n) = y_k,$$

in short $f(x) = y$, with $y = (y_1, \dots, y_k)$ given. The emphasis is in the case the f_j are not linear, otherwise we know very well from basic linear algebra the structure of the solutions. Assume that for some value q close to y we know a solution p of $f(x) = q$, so it is to be expected that $f(x) = y$ will have a solution x close to p. Since $f(x)$ is well approximated by $f(p)+df(p)(x-p)$ for x close to p, it is tempting to replace the nonlinear system $f(x) = y$, for y close to q, by the linear system

$$f(p) + df(p)(x - p) = y, df(p)(x - p) = y - q,$$

and expect that the structure of solutions of both systems, linear and nonlinear, is similar. We know from linear algebra that the structure of solutions of the linear system depends on three parameters: n, k and the rank of $df(p)$.

The inverse function theorem confirms that this intuitive expectation is indeed right, the nonlinear system behaves like the linear one for y close to q. In the inverse function theorem, the assumption is that existence and uniqueness holds for the linear system, in particular $k = n$, and the conclusion is that the nonlinear system has too a unique solution x close to p for y close to q. Recall that a diffeomorphism f between two open sets W, V is a one-to-one map from W onto V, both f and its inverse being differentiable.

Theorem 6.2. *Assume $f : U \to \mathbb{R}^n$ is of class C^1 in a domain U in $\mathbb{R}^n$, $p \in U, q = f(p)$ and $df(p)$ is invertible, that is, $\det df(p) \neq 0$. Then there is an open set $W \subset U, p \in W$, such that f is a C^1-diffeomorphism from W to an open set $V = f(W), q \in V$. In short, we say that f is a* local diffeomorphism. *If f is of class C^N, so is the inverse map.*

The theorem can be restated by saying that the components of f constitute a system of coordinates around p.

If $g : V \to W$ is the inverse map, by the chain rule $dg(q) = (df(p))^{-1}$, therefore, for $y \in V$ the solution $g(y)$ of $f(x) = y$ is approximated by $x = g(q) + dg(q)(y - q) = p + dg(q)(y - q)$, which is the solution of the approximate system $df(p)(x - p) = y - q$.

Example 6.2. Consider for instance the nonlinear system in three unknowns x, y, z

$$x^3y^2z + x^2y^3z^2 + x = 3.1; \; x^2y^3 + y^3z^2 + z = 2.9; \; y^3z^3 + xz + x = 3.2,$$

in short $f(x, y, z) = (3.1, 2.9, 3.2)$. We notice that with $3.1, 2.9, 3.2$ replaced by $3, 2, 3$, the system has the solution $a = (1, 1, 1)$ and that

$$df(p) = \begin{pmatrix} 6 & 5 & 3 \\ 2 & 6 & 1 \\ 2 & 3 & 4 \end{pmatrix}$$

is invertible. Solving $df(p)(h) = (0.1, -0.1, 0.2)$ gives $h = (-0.44, 0.03, 0.96)$, so $(0.56, 1.03, 1.96)$ is the approximate solution.

Before providing the proof of the theorem, we point out three remarks. First, the theorem is trivial if $n = 1$, so its interest is in case $n > 1$. Assume that f is of class C^1 in an open interval containing p. If say $f'(p) > 0$, by continuity $f'(x) > 0$ in an open interval $W, p \in W$, so f is strictly increasing in W, $f(W)$ is an open interval V containing $f(p)$ and the inverse function $g = f^{-1}$ is also strictly increasing. If $y \in V, f(x) = y$ and $k \to 0$, then $h = g(y + k) - g(y) \to 0$ (because g is continuous) and

$$\frac{g(y + k) - g(y)}{k} = \frac{h}{f(x + h) - f(x)},$$

shows that g is diffferentiable at y with derivative $g'(y) = 1/f'(x)$.

The second remark is that the theorem does not provide any information about the sizes of W, V. The third remark is that there exist homeomorphisms of class C^1 which are not diffeomorphisms, already in dimension one, e.g., $f(x) = x^3$.

6.2.2 The first step of the proof deals precisely with this situation:

Proposition 6.1.

(a) *Suppose f is a homeomorphism between two open sets W, V in $\mathbb{R}^n$ with inverse g, and that f is differentiable at $p \in U$. Then g is differentiable at $q = f(p)$ if and only if $df(p)$ is invertible as a linear map, that is* $\det df(p) \neq 0$ *and in this case $dg(q) = (df(p))^{-1}$.*
(b) *Let f be a homeomorphism of class C^1 between two open sets W, V in $\mathbb{R}^n$. Then f is a C^1 diffeomorphism if and only if $df(x)$ is invertible for all $x \in U$, that is,* $\det df(x) \neq 0$.
(c) *If f is of class C^N, so is g.*

Proof. We already know, by the chain rule, that $\det df(p) \neq 0$ is necessary. To show it is sufficient, set $x = p + h$, $y = f(p+h) = q + r$; since f is a homeomorphism, h, r determine each other and the statements $h \to 0, r \to 0$ are equivalent. By hypothesis,

$$r = f(p+h) - f(p) = df(p)(h) + \epsilon(h)|h|,$$

with $\epsilon(h) \to 0$ as $h \to 0$, in particular $r = O(h)$. Applying the inverse map $L = df(p)^{-1}$, we obtain

$$h = L(r) - |h|L(\epsilon(h)),$$

or

$$g(q+r) - g(q) = L(r) - |h|L(\epsilon(h)).$$

Therefore, we must show that the last term is $O(r)$. Being $O(h)$, it is enough to prove that $h = O(r)$. But

$$|h| \leq |L(r)| + |h||L(\epsilon(h))| \leq |L||r| + |h||L||\epsilon(h)|,$$

where $|L|$ stands for the norm of the linear map L. Since $\epsilon(h) \to 0$ as $h \to 0$, there exists $\delta > 0$ such that $|\epsilon(h)||L| < \frac{1}{2}$ if $|h| < \delta$. Then, if $|h| < \delta$,

$$|h| \leq |L||r| + \frac{|h|}{2},$$

so $|h| \leq 2|L||r|$ and $h = O(r)$.

The second and third statements are trivial, because the partial derivatives of g, the entries of the inverse matrix of df, are continuous expressions in terms of the partial derivatives of f and $\det df(x)$. □

The second step is to quantify the assumption on $df(p)$: if L is the inverse map, one has

$$|L(w)| \leq |L||w|, \quad w \in \mathbb{R}^n,$$

which setting $v = Lw, w = df(p)(v), m = |L|^{-1}$ amounts to

$$m|v| \leq |df(p)(v)|, \quad v \in \mathbb{R}^n.$$

The linear map $df(p)$ being invertible means that $df(p)(v) = 0$ implies $v = 0$, now we have quantified this fact.

In a third step, we claim that there exists a closed ball $\overline{B} = \overline{B}(p, r) \subset U$ such that for $x \in \overline{B}$,

(a) $\det df(x) \neq 0$,
(b) $|df(x) - df(p)| \leq \frac{m}{2}$, the left-hand side being the norm of the linear map $df(x) - df(p)$.

It is clear that the first statement holds for r small enough, by continuity of $\det df(x)$. The second follows as well from the C^1 hypothesis because all possible norms being equivalent in a finite dimensional space (in this case the space of matrices) one has

$$|df(x) - df(p)| \leq C \sum_{i,j} \left| \frac{\partial f_i}{\partial x_j}(x) - \frac{\partial f_i}{\partial x_j}(p) \right|,$$

for some constant C. Next, we claim that these imply

$$|f(x) - f(y)| \geq \frac{m}{2}|x - y|, \quad x, y \in \overline{B}. \tag{6.2}$$

Consider $\hat{f}(x) = f(x) - df(p)(x)$, with differential $d\hat{f}(x) = df(x) - df(p)$; hence, $|d\hat{f}(x)| \leq \frac{m}{2}$ for $x \in \overline{B}$, and so by the mean-value theorem

$$|f(x) - f(y) - df(p)(x - y)| = |\hat{f}(x) - \hat{f}(y)| \leq \frac{m}{2}|x - y|, \quad x, y \in \overline{B}.$$

But $|df(p)(x - y)| \geq m|x - y|$, so

$$\begin{aligned} \big||f(x) - f(y)| - |df(p)(x - y)|\big| &\leq |f(x) - f(y) - df(p)(x - y)| \\ &\leq \frac{m}{2}|x - y| \leq \frac{1}{2}|df(p)(x - y)|, \end{aligned}$$

meaning that

$$\frac{1}{2}|df(p)(x - y)| \leq |f(x) - f(y)| \leq \frac{3}{2}|df(p)(x - y)|.$$

Therefore, with $C = |df(p)|$,

$$\frac{m}{2}|x-y| \le |f(x)-f(y)| \le \frac{3C}{2}|x-y|, \quad x,y \in \overline{B}.$$

We next consider the boundary S of $\overline{B}$, which is compact; then $f(S)$ is also compact, and $f(p) \notin f(S)$, by (6.2). With $d = d(f(p), f(S)) > 0$, we define $V = B(f(p), \frac{d}{2})$, so that $|y - f(p)| < |y - f(x)|$ for all $x \in S$. Next, we define the open set

$$W = \{x \in B : f(x) \in V\}.$$

By (6.2), f is one-to-one in B, so f is one-to-one from W into V. We will prove now that for $y \in V$ there exists $x \in B$ such that $f(x) = y$. Consider

$$h(x) = |y - f(x)|^2 = \sum_{i=1}^{n}(y_i - f_i(x))^2.$$

The function h has an absolute minimum in $\overline{B}$, and since $h(p) < h(x), x \in S$, it is attained at some $x \in B$. Then x is also a local minimum, therefore $D_j h(x) = 0$, that is

$$\sum_i D_j f_i(x)(y_i - f_i(x)) = 0, \; j = 1, \ldots, n.$$

But $df(x)$ is invertible for $x \in B$, and therefore $y_i = f_i(x), y = f(x)$. Thus, we have proved that f is one-to-one from W onto V; its inverse is also continuous, again by (6.2), and the proposition finishes the proof.

6.2.3 If $\det df(x) \neq 0$ for all $x \in U$, it follows that f is locally one-to-one and that $f(U)$ is open, so we can state:

Corollary 6.1. *A map $f : U \to \mathbb{R}$ of class C^1 is a diffeomorphism onto its image if and only if it is one-to-one and $\det df(x) \neq 0$ for all $x \in U$.*

Except if $n = 1$, the assumption $\det df(x) \neq 0$ does not imply that f is globally one-to-one. For instance, the map in the plane given by $f(x, y) = (x^2 - y^2, 2xy)$ (which is $f(z) = z^2$ in complex notation $z = x + iy$) satisfies $\det df(x, y) \neq 0$ for $(x, y) \neq (0, 0)$, is locally one-to-one but not globally one-to-one.

6.2.4 From the point of view of coordinate systems, the inverse function theorem ensures that if $u_1, \ldots, u_n$ are of class C^1 with linearly independent gradients $\nabla u_j(p)$, then they constitute a coordinate system around p. We call them *a local system of coordinates* at p.

Proposition 6.2. *A system of k functions $u_1, \ldots, u_k$ of class C^1 can be enlarged to a local system of coordinates at p if and only if they have linearly independent gradients (or differentials) at p.*

Proof. The matrix with rows $\nabla u_j(p)$ has a non-vanishing minor of order k. Assume it is the one corresponding to the first k columns, $\det(D_i u_j(p)) \neq 0, i, j = 1, \ldots, k$; then the system $(u_1, \ldots, u_k, x_{k+1}, \ldots, x_n)$ has linearly independent gradients at p and therefore is a local coordinate system. □

The next corollary follows from the proposition and Theorem 6.1.

Corollary 6.2. *If $f_1, \ldots, f_m$ are of class C^1 with linearly independent gradients at p, a function F of class C^1 defined around p depends functionally on $f_1, \ldots, f_m$ around p if and only if $\nabla F(x)$ is a linear combination of $\nabla f_1(x), \ldots, \nabla f_m(x)$ for x close to p.*

6.3 The Implicit Function Theorem, Analytic Version

6.3.1 The implicit function theorem can be viewed in two equivalent ways, the analytic and the geometric one. The first one deals with the concept of *implicit function.* To explain this concept, assume for simplicity that $n = 2$ and $f(x, y)$ is defined in a plane domain U. We say that the equation

$$f(x, y) = 0,$$

defines implicitly $y = h(x)$ in U if $(x, y) \in U, f(x, y) = 0$ is equivalent to $y = h(x)$. Geometrically, it means that the set $M = \{(x, y) \in U : f(x, y) = 0\}$ is the graphic of some function h, that is, M meets every vertical line at most once. This fact can be checked graphically in many circumstances, and also analytically. For example,

$$f(x, y) = y + x^2 y + e^x y^3 + x^4 y^5 + \sin x = 0,$$

defines $y = h(x)$ for $x \in \mathbb{R}$, because given $x \in \mathbb{R}$ the above is a polynomial of odd degree in y with partial derivative

$$f_y(x, y) = 1 + x^2 + 3e^x y^2 + 5x^4 y^4 > 0,$$

so there is a unique value $y = h(x)$ solving the equation. We know this value $y = h(x)$ exists, yet we are not able to solve for y in terms of elementary functions, that's why we use the term *implicit* in contraposition to *explicit*, when we are able to solve for y in some way.

The implicit function theorem provides a local criteria for the existence of the implicit function. More significantly, for instance in cases as the above where the existence is obvious, it ensures that the implicit function has the same degree of smoothness as the given equation; this makes it possible to write arbitrarily long Taylor expansions of the implicit function, see paragraph 6.3.2.

More generally, instead of a single equation we consider a system of m-equations

$$f_j(x_1, \ldots, x_n) = y_j, \quad j = 1, \ldots, m,$$

in short $f(x) = y$, with $m < n$, the case $m = n$ being covered by the inverse function theorem.

For linear systems $AX = Y$ of m equations with $n > m$ unknowns, we know that the rank r of A governs everything. Suppose $r = m$, so there is minor of order $m \times m$ of A with non-zero determinant; assume without loss of generality that it is the one consisting of the first m columns, and write $\mathbb{R}^n = \mathbb{R}^m \times \mathbb{R}^k, m = n - k, x = (x', x''), A = A'|A''$. Then the m equations are linearly independent; the general solution of the linear system is obtained by expressing the system as

$$A'X' = Y - A''X'',$$

and solving for X', that is, the variables X'' are thought as parameters and X' is a linear *explicit* function of X'' and Y. So, the system has an affine sub-manifold of dimension k of solutions for all Y.

We assume that for $y = q$ we know a solution $p, q = f(p)$; as in the inverse function theorem, we expect that the qualitative properties of the system for y close to q are the same as the approximate linear system

$$f(p) + df(p)(x - p) = y, \quad df(p)(x - p) = y - q,$$

so $X = x - p, Y = y - q, A = df(p)$. The implicit function theorem confirms that this expectation is true.

Theorem 6.3. *Let $f_1, f_2, \ldots, f_m, m < n$ be functions of class C^1 defined around $p \in \mathbb{R}^n$ with linearly independent gradients at p and write $f = (f_1, \ldots, f_m), q = f(p)$. Assume without loss of generality that*

$$\det(D_i f_j(p))_{i,j=1,\ldots,m} \neq 0.$$

Then there is an open set $B \subset \mathbb{R}^m, q \in B$, an open set $A = A' \times A'', A' \subset \mathbb{R}^m, A'' \subset \mathbb{R}^{n-m}, p = (p', p'') \in A$ and a C^1 function $h : A'' \times B \to A'$, with $h(p'', q) = p'$, such that for $y \in B$ the statement $x = (x', x'') \in A, f(x) = y$ is equivalent to $x' = h(x'', y)$. If the f_j are of class C^N, so is h.

Note that the assumption implies, by Theorem 4.1, that none of the functions f_j depend functionally on the others, meaning that no equation is consequence of the remaining ones, as in the linear case. The statement means that $f(x', x'') = y$, for $x'' \in A'', y \in B$, has a unique solution $x' = h(x'', y) \in A'$. In general, the variables appearing in the $m \times m$ minor of $df(p)$ with non-zero determinant are the ones that the system $f(x) = y$ defines as functions of y and the remaining ones.

Proof. To prove it, as in Proposition 6.2, we consider an open set W containing p such that $(f_1, \ldots, f_m, x_{m+1}, \ldots, x_n)$ is a C^1-diffeomorphism from W to an open set $V \subset \mathbb{R}^n, (q, p'') \in V$. Let Ψ be the inverse map, that as Φ will keep fixed the last k coordinates,

$$\Psi(y, x'') = (h_1(y, x''), \ldots, h_m(y, x''), x''),$$

with h_j of class C^1 in V. We choose now A'', B such that $(q, p'') \in B \times A'' \subset V$, define $U = \Psi(B \times A'')$, A' the projection of U on the first m coordinates, and define

$$h(x'', y) = (h_1(y, x''), \ldots, h_m(y, x'')). \qquad \square$$

We point out that the implicit function h is unique if $h(p'', q) = p'$. But for a given p'' there might exist different p' with $f(p', p'') = q$, each giving rise to a different implicit function defined around p''. For instance, for $x^2 + y^2 = 1$ there exist two implicit functions $y = y(x)$ defined around 0: $y = \sqrt{1 - x^2}$, corresponding to $(0, 1)$ and $y = -\sqrt{1 - x^2}$ corresponding to $(0, -1)$.

We have obtained the implicit function theorem from the inverse function theorem. In fact, they are equivalent statements, for if $f = (f_1, \ldots, f_n)$ satisfies $f(p) = q, \det df(p) \neq 0$, then $F(x, y) = f(x) - y$ defines x as a function of y near (p, q).

The existence of a continuous implicit function can be proved under a more general assumption. We state it for simplicity for a single equation:

Exercise 6.1.

(a) Prove that if $F(x,y)$ is continuous in an interval $R : |x| < a, |y| < b$, $F(0,0) = 0$ and is contractive in y uniformly in x,

$$|F(x,y_1) - F(x,y_2)| \leq C|y_1 - y_2|, \quad k < 1,$$

then there is $\alpha < a$ and a unique continuous function $h(x), |x| < \alpha, h(0) = 0$, such that $F(x,h(x)) = h(x)$. *Hint*: consider $h_0(x) = 0, h_{n+1}(x) = F(x,h_n(x))$.

(b) Prove that if $f(x,y), D_2f(x,y)$ are continuous in $R, f(0,0) = 0, D_2f(0,0) \neq 0$, then there is a unique continuous function $h(x), |x| < \alpha, h(0) = 0$, such that $f(x,h(x)) = 0$. *Hint*: consider $F(x,y) = \frac{f(x,y)}{D_2f(0,0)} - y$.

6.3.2 We study now Taylor's development of implicit functions. We assume that f is of class C^N, $D_nf(p) \neq 0$, and let $x_n = h(x')$ be the implicit function defined by $f = 0$ near p. We start by computing the first-order derivatives of h by differentiating the identity $f(x',h(x')) = 0$:

$$D_if(x',h(x')) + D_nf(x',h(x'))D_ih(x') = 0, \; i = 1,\ldots,n-1, \tag{6.3}$$

from which

$$D_ih(x') = -\frac{D_if(x',h(x'))}{D_nf(x',h(x'))}.$$

Using this expression we see again that if f is of class C^2, then h is also C^2.

As we know, the value of h at $p' = (p_1,\ldots,p_{n-1}), h(p') = p_n$, this formula allows us to compute the first derivatives of h at p'. We can in fact compute recursively derivatives of h at p' of order less or equal than N. For the second-order derivatives $D_{ij}h$ we could differentiate again the previous formula, but it is often easier to differentiate again in (6.3):

$$D_{ij}f + D_{in}fD_jh + (D_{jn} + D_{nn}D_jh)D_ih + D_nfD_{ij}h = 0,$$

from which we can derive the value of $D_{ij}h(p')$. Analogously, with higher-order derivatives.

Example 6.3. Let us consider

$$f(x,y,z) = xy^2z^3 + yz^2x^3 + zx^2y^3 - 3 = 0,$$

defining $z = z(x,y)$ around $p = (1,1,1)$. One has

$$0 = f_x + f_z z_x = y^2z^3 + 3yz^2x^2 + 2zxy^3 + (3xy^2z^2 + 2yzx^3 + x^2y^3)z_x = 0,$$

which at $(1,1,1)$ gives $0 = 6 + 6z_x, z_x = -1$, and analogously $z_y = -1$. Differentiating again in x,

$$3y^2z^2z_x + 6yz^2x + 6yx^2zz_x + 2zy^3 + 2xy^3z_x + (3y^2z^2 + 6xy^2zz_x + 6yzx^2$$
$$+2yx^3z_x + 2xy^3)z_x + (3xy^2z^2 + 2yzx^3 + x^2y^3)z_{xx} = 0,$$

from which $3z_x+6+6z_x+2+2z_x+(3+6z_x+6+2z_x+2)z_x+6z_{xx} = 0, z_{xx} = 1$; differentiating in y

$$2yz^3 + 3y^2z^2z_y + 3z^2x^2 + 6yx^2zz_y + 6zxy^2 + 2xy^3z_y + (6xyz^2 + 6xy^2zz_y$$
$$+2zx^3 + 2yz_yx^3 + 3x^2y^2)z_x + (3xy^2z^2 + 2yzx^3 + x^2y^3)z_{xy} = 0,$$

giving $2+3z_y+3+6z_y+6+2z_y+(6+6z_y+2+2z_y+3)z_x+6z_{xy} = 0, z_{xy} = -\frac{1}{2}$. By symmetry, $z_{yy} = 1$. It follows that the second-order expansion of the implicit function is

$$z = 1 - (x-1) - (y-1) + \frac{1}{2}(x-1)^2 + \frac{1}{2}(y-1)^2 - (x-1)(y-1)$$
$$= 3 - x - y + \frac{1}{2}(x^2 + y^2).$$

An analogous procedure can be followed in the general case of m equations $f_j(x) = 0$. Let us illustrate it for two equations

$$f(x,y,z) = 0, \quad g(x,y,z) = 0, \; f(a,b,c) = g(a,b,c) = 0.$$

Assuming the minor

$$M = \begin{pmatrix} f_y(a,b,c) & f_z(a,b,c) \\ g_y(a,b,c) & g_z(a,b,c) \end{pmatrix},$$

has non-zero determinant, we have two implicit functions $y = h(x), z = k(x)$ with $h(a) = b, k(a) = c$ such that

$$f(x,h(x),k(x)) = 0, \quad g(x,h(x),k(x)) = 0.$$

Differentiating in x and evaluating at a, we reach the linear system

$$f_x(a,b,c) + f_y(a,b,c)h'(a) + f_z(a,b,c)k'(a) = 0,$$
$$g_x(a,b,c) + g_y(a,b,c)h'(a) + g_z(a,b,c)k'(a) = 0,$$

in the unknowns $h'(a), k'(a)$. Since the matrix of this system is precisely the non-zero minor, we may solve for $h'(a), k'(a)$. Further differentiations always lead to linear system in the unknowns $h^{(j)}(a), k^{(j)}(a)$ with matrix M.

Example 6.4. We consider the system

$$x + y + z = 3, \quad y^2 z + z^2 x + 2x^2 y = 4,$$

which around $(1,1,1)$ defines $y = y(x), z = z(x)$. Differentiating we get

$$1 + y' + z' = 0, \quad 2yy'z + y^2 z' + 2zz'x + z^2 + 4xy + 2x^2 y' = 0,$$

which at point $(1,1,1)$ gives

$$1 + y' + z' = 0, \quad 4y' + 2z' + 5 = 0,$$

and so $y' = -\frac{3}{2}$, $z' = \frac{1}{2}$.

6.3.3 The same considerations apply to the inverse function theorem; if possibly non-explicit, we can obtain arbitrarily long Taylor's expansion of the inverse function if f is smooth.

Example 6.5. Consider the nonlinear system

$$xy^2 + yx^2 = u, \quad xy^3 - yx^3 = v,$$

which for (u,v) close to $(2,0)$ has a unique solution (x,y) close to $(1,1)$. Differentiating in u,v we get

$$x_u y^2 + +2yxy_u + y_u x^2 + 2xyx_u = 1, \; x_u y^3 + 3y^2 y_u x - y_u x^3 - 3yx^2 x_u = 0,$$
$$x_v y^2 + +2yxy_v + y_v x^2 + 2xyx_v = 0, \; x_v y^3 + 3y^2 y_v x - y_v x^3 - 3yx^2 x_v = 1,$$

giving at $(1,1)$ that $x_u = y_u = \frac{1}{6}, x_v = -\frac{1}{4}, y_v = \frac{1}{4}$. Differentiating in u,v again and evaluating at $(1,1)$ leads to the equations

$$y_{uu} - x_{uu} = 0, \quad y_{uu} + x_{uu} = -\frac{1}{9}, \quad x_{uv} + y_{uv} = 0,$$
$$y_{uv} - x_{uv} = \frac{1}{16}, \quad x_{vv} + y_{vv} = \frac{1}{6}, \quad -x_{vv} + y_{vv} = \frac{3}{8},$$

from which

$$x_{uu} = y_{uu} = -\frac{1}{18}, \quad x_{uv} = -y_{uv} = -\frac{1}{32}, \quad y_{vv} = \frac{13}{48}, \quad x_{vv} = -\frac{5}{48}.$$

The approximate inverse of second order is thus

$$x(u,v) = 1 + \frac{1}{6}(u-2) - \frac{1}{4}v - \frac{1}{36}(u-2)^2 - \frac{1}{16}v(u-2) - \frac{5}{96}v^2,$$

$$y(u,v) = 1 + \frac{1}{6}(u-2) + \frac{1}{4}v - \frac{1}{36}(u-2)^2 + \frac{1}{16}v(u-2) + \frac{13}{96}v^2.$$

Chapter 7

Regular Sub-Manifolds

In this chapter, we explain the geometric point of view of the implicit function theorem, in terms of *regular sub-manifolds.* The main new concept is that of *tangent space*, a linear approximation of the sub-manifold. Then functional dependence, introduced in paragraph 4.3.5, is interpreted geometrically. The *constant rank* theorem is also presented from a geometric viewpoint. In the last sections, the tangent space is used to discuss existence of tubular neighborhoods and to give criteria for constrained optimization.

7.1 Geometric Implicit Function Theorem

7.1.1 Recall from Definition 3.2 the concept of topological sub-manifold of dimension k in $\mathbb{R}^n$, a set having locally k degrees of freedom, dimension k. We specialize this concept to the differentiable setting.

Definition 7.1. A subset $M \subset \mathbb{R}^n$ is said to be a *regular sub-manifold of dimension k and of class C^1* if for all $p \in M$ there exists a C^1-map $\Phi : U \longrightarrow \mathbb{R}^n$, where U is an open set in $\mathbb{R}^k$, with $0 \in U, \Phi(0) = p$, and a ball $B(p, r)$ such that Φ is a homeomorphism between U and $M \cap B(p, r)$ and $d\Phi(u)$ has rank k for $u \in U$.

We denote the n components of Φ by

$$x_i(u_1, \ldots, u_k), \quad i = 1, \ldots, n.$$

Φ being a homeomorphism means that $u = (u_1, \ldots, u_k)$ are coordinates of $\Phi(u) \in M$ around p. We call Φ a *local parametrization* of M, or a *local chart.* Given all values of u_j except one, u_i, that varies, $\Phi(u)$ is an arc on M, the u_i-axis, with non-zero tangent, the ith column of $d\Phi(u)$, that we denote by ∂_i, as in Section 6.1.

When $k = 2$, we use the term *regular surface* and when $k = 1$, *regular curve*. When $k = n-1$, also the term *hypersurface* is used. A particular case is that of the graphics, for which k of the variables, say $x_1, \ldots, x_k$ are the parameters, and the others are given by $x_i = h_i(x_1, \ldots, x_k), i = k+1, \ldots, n$. That is, Φ has the form, up to permutations,

$$\Phi(x_1, \ldots, x_k) = (x_1, \ldots, x_k, \phi_{k+1}(x_1, \ldots, x_k), \ldots, \phi_n(x_1, \ldots, x_k)).$$

A *regular atlas* is a collection of local charts (U_i, Φ_i) such that the $\Phi_i(U_i)$ cover M.

We point out that in the definition it is enough to assume that $d\Phi(0)$ has rank k; this means that some minor of order k of $d\Phi(0)$ is not zero, and by continuity it will be non-zero in a ball. Also, as shown in what follows, the fact that $d\Phi(0)$ has maximal rank implies that Φ is one-to-one in some ball. This has been proved for $k = 1$ in Proposition 3.2.

Of course, the simplest examples are the affine sub-manifolds of dimension k, $M = p + V$ where V is a linear subspace of dimension k. If $v_1, \ldots, v_k$ is a basis of V, M has the global parametrization

$$x = p + \sum_{i=1}^{k} u_i v_i.$$

The ellipsoid

$$\frac{x^2}{a^2} + \frac{y^2}{b^2} + \frac{z^2}{c^2} = 1$$

has the parametrization $(x, y, z) = \Phi(\phi, \theta)$

$$x = a \sin\phi \cos\theta, \quad y = b \sin\phi \sin\theta, \quad z = c \cos\phi, \quad 0 \le \phi \le \pi, \quad 0 \le \theta \le 2\pi,$$

where strictly speaking we should consider just $0 < \phi < \pi, 0 < \theta < 2\pi$, to meet the definition above. When the parametrization omits, as in this case, a set of lower dimension, we call it an *essential global parametrization*. It is easily checked that the parametrization of conics and quadrics described in paragraphs 3.3.2 and 3.3.3 are all of this type.

The $(n-1)$-dimensional sphere $|x| = 1$ in $\mathbb{R}^n$ has an essential global parametrization obtained by setting $\rho = 1$ in the spherical coordinates

$$x_n = \cos\phi_1, \quad x_{n-1} = \sin\phi_1 \cos\phi_2, \quad x_{n-2} = \sin\phi_1 \sin\phi_2 \cos\phi_3, \ldots,$$

$$x_2 = \sin\phi_1 \cdots \sin\phi_{n-2} \cos\phi_{n-1}, \quad x_1 = \sin\phi_1 \cdots \sin\phi_{n-2} \sin\phi_{n-1}.$$

7.1.2 In all the examples of conics and quadrics, we realize that M is an intersection of level sets

$$M = \{x : f_1(x) = c_1, \ldots, f_m(x) = c_m\}, \quad m = n - k.$$

Intuitively, if the f_i are functionally independent, meaning that no equation is consequence of the others, each time we impose an equation the number of degrees of freedom goes down by one, so M should be an object of dimension $k = n - m$. The implicit function theorem tells us that these points of view are equivalent:

Theorem 7.1. *For $M \subset \mathbb{R}^n$ and $k < n, m = n - k$, the following are equivalent:*

(a) *M is a regular sub-manifold of dimension k.*

(b) *M is locally defined by m strongly functionally independent functions: for each $p \in M$ there is a ball $B(p,r)$ and functions $f = (f_1, \ldots, f_m)$ of class C^1 in $B(p,r)$ such that $\nabla f_1(x), \ldots, \nabla f_m(x)$ are linearly independent for $x \in B(p,r)$ and*

$$M \cap B(p,r) = \{x \in B(p,r) : f_1(x) = 0, \ldots, f_m(x) = 0\}.$$

(c) *M is locally a graph of a C^1- function: for each $p \in M$ there is a decomposition of variables $\mathbb{R}^n = \mathbb{R}^m \times \mathbb{R}^k$, an open set $A = A' \times A'', p \in A$, and a C^1 function $h : A'' \to A'$ such that $M \cap A$ is the graph of h,*

$$M \cap A = \{x = (x', x'') : x' = h(x'')\}.$$

(d) *For each $p \in M$ there is a local coordinate system $u_1, \ldots, u_n$ around p such that M is given locally by $u_{k+1} = \cdots = u_n = 0$.*

Proof. By Proposition 6.2, (b) and (d) are equivalent. The implicit function theorem is the statement that (b) implies (c). It is obvious that (c) implies (a), because a graph is just a parametrization where the parameters are some of the variables, or in other words,

$$\Phi(x'') = (h(x''), x''),$$

satisfies (a). It is also straightforward that (c) implies (b), because if M is locally a graph $x' = h(x'')$, then it is locally defined by $F(x', x'') = x' - h(x'') = 0$. Now we prove that (a) implies (c). Let (U, Φ) be a local chart around p; in fact the proof will show that it is enough to assume that Φ is of class C^1 and that $d\Phi(0)$ has rank k. If $\Phi = (\phi_1, \ldots, \phi_n)$, since

$d\Phi(0)$ has rank k there exists a minor of order k with non-zero determinant. We assume without loss of generality that it is the one consisting in the last k components $\Phi'' = (\phi_{m+1}, \ldots, \phi_n)$, and write again $x = (x', x'')$, etc. Then $\det \Phi''(0) \neq 0$, and by the inverse function theorem Φ'' is a C^1-diffeomorphism between an open set in $\mathbb{R}^k$ that we continue to denote by U and an open set V in $\mathbb{R}^k$ containing p''. This implies that Φ is a homeomorphism between U and the image $\Phi(U)$ and that $d\Phi(u)$ has rank k for all $u \in U$. Let $G : V \to U$ be the inverse map of Φ''; by hypothesis, the points of M around p are

$$(\Phi'(u), \Phi''(u)), u \in U.$$

Setting $x'' = \Phi''(u)$, $u \in U$ amounts to $x'' \in V$, so if we choose x'' as a parameter, we see that M consists of points

$$(\Phi'(G(x'')), x''), x'' \in V,$$

and therefore it is the graph of $h(x'') = \Phi'(G(x''))$. □

The meaning of part (d) is that in suitable coordinates M is like $\mathbb{R}^k$ within $\mathbb{R}^n$.

In most examples, M is given globally: if $f_j, j = 1, \ldots, m$ are of class C^1 in a domain U and $\nabla f_j(x), j = 1, \ldots, m$ are linearly independent for all $x \in U$, then

$$M = \{x \in U : f_j(x) = c_j, j = 1, 2, \ldots, m\},$$

if not empty, is a regular sub-manifold of dimension $k = n - m$.

7.1.3 Tangent space to a sub-manifold. When comparing topological sub-manifolds with regular sub-manifolds, the difference is that in the regular case there are plenty of regular arcs on M through p, and $T_p(M)$ is a linear space of dimension k, the *tangent space to* M:

Theorem 7.2. *If M is a regular k-sub-manifold, the tangent cone $T_p(M)$*

$$T_p(M) = \{\gamma'(0) : \gamma(t) \in M, \gamma(0) = p\},$$

is a linear space of dimension k. In terms of a local chart (U, Φ) around p, one has

$$T_p(M) = d\Phi(0)(\mathbb{R}^k).$$

In terms of local defining functions $f_1, \ldots, f_m$,

$$T_p(M) = \{v \in \mathbb{R}^n : \langle \nabla f_j(p), v\rangle = 0, j = 1, \ldots, m\}.$$

Proof. If $d\Phi(0)(w) = v$, then $\gamma(t) = \Phi(tw)$ is a curve on M with tangent v. On the other hand, if $\gamma(t)$ is a curve on M through p, then $f_i(\gamma(t))$ is constant, and by the chain rule

$$\langle \nabla f_i(p), \gamma'(0)\rangle = 0.$$

This means that

$$d\Phi(0)(\mathbb{R}^k) \subset T_p(M) \subset E,$$

where E is the orthogonal complement of the linear space spanned by $\nabla f_j(p)$, $j = 1, \ldots, m$. These vectors being linearly independent, E has dimension k, same as $d\Phi(0)$, and the result follows. □

Recall that the column vectors of $d\Phi(0)$, $D_i\Phi(0), i = 1, \ldots, k$, constitute a basis of $T_p(M)$; these are the tangent vectors to the curves on M obtained by keeping all parameters fixed but one, the "axis" u_i and written

$$\partial_i = D_i\Phi(0) = \left(\frac{\partial x_1}{\partial u_i}, \ldots, \frac{\partial x_n}{\partial u_i}\right).$$

The affine sub-manifold $p + T_p$ is the one we visualize as being tangent to M at p.

In case $k = n - 1$, the orthogonal $N_p(M)$ of $T_p(M)$ is spanned by ∇f if f is a defining function. In terms of a local chart, with the notation of paragraph 1.3.2 by

$$\partial u_1 \times \cdots \times \partial u_{n-1}.$$

The line $p + N_p(M)$ is called the *normal line.*

The case $k = n$ is not excluded, the n-dimensional sub-manifolds of $\mathbb{R}^n$ being the open sets U. Of course, $T_p(U) = \mathbb{R}^n$; it is useful to think of $v \in R^n = T_p(U)$ as a free vector v with origin at p.

Certain familiar objects, like the cone C defined by the equation $z^2 = x^2 + y^2$ have isolated singular points, in this case the origin 0, where C has no tangent. Strictly speaking C is not a surface but $C \setminus 0$ is. We call these objects *sub-manifolds with singularities.* The singularity can be of various types, for instance $S : z^4 = x^2 + y^2$ has a cusp at the origin. This is the surface of revolution spanned by the curve $z^2 = y$ on the zy-plane when rotating around the z-axis.

7.1.4 Note that if $v \in T_p(M)$ and g is defined around p, $D_v g(p)$ depends only on the values of g on M: if $g_1 = g_2$ on M around p, then $D_v g_1(p) = D_v g_2(p)$. Said otherwise, $D_v g(p)$ makes sense for g defined just on M. This leads to the notion of *tangential differential* $d_t g(p)$ for g defined just on M at p as the linear map

$$v \in T_p(M) \to d_t g(p)(v) = D_v g(p) = \frac{d}{dt}_{t=0}(g(\gamma(t)),$$

with γ on M, $\gamma(0) = p$. We can consider as well the *tangential gradient* $\nabla_t g(p)$, the unique vector $u \in T_p(M)$ such that $D_v g(p) = \langle u, v\rangle$. If g is defined in a ball $B(p,r)$, $d_t g(p)$ is just the restriction of $dg(p)$ to $T_p(M)$, and $\nabla_t g(p)$ is the orthogonal projection of $\nabla g(p)$ on $T_p(M)$.

Example 7.1. Consider $S = \{f = z^2 - xy = 0\}$ around $p = (1,1,1)$ and assume that g is differentiable and equals xyz on S. If we know a normal derivative of g at p, say $D_n g(p) = 1$ with $n = \nabla g(p) = (-1,-1,2)$, then the full gradient $\nabla g(p)$ is known. We first compute the tangential derivatives $D_v g(p), D_w g(p)$ with two linearly independent vectors $w, v \in T_p(S)$ (that is, orthogonal to n), say $w = (1,-1,0), v = (1,1,1)$; since g equals $h = xyz$ on S, we find

$$D_w g(p) = D_w h(p) = h_x(p) - h_y(p) = 0,$$

$$D_v g(p) = D_v h(p) = h_x(p) + h_y(p) + h_z(p) = 3.$$

Then $\nabla g(p) = (\alpha, \beta, \gamma)$ is the solution of

$$1 = D_n g(p) = -\alpha - \beta - 2\gamma, 0 = D_w g(p) = \alpha - \beta = 0, 3 = D_v g(p) = \alpha + \beta + \gamma,$$

which is $(\frac{7}{2}, \frac{7}{2}, -4)$.

7.1.5 The inverse function theorem can now be understood in a geometric way. Let us assume that $n = 2, f(x,y) = (u(x,y), v(x,y)), f(p) = (0,0)$ and that $df(p)$ is invertible. The linear independence of the gradients $\nabla u(p), \nabla v(p)$ means that the level sets

$$u(x,y) = 0, \quad v(x,y) = 0$$

are regular curves through p intersecting transversally. The theorem states that for small values of c_1, c_2 the level sets $u = c_1, v = c_2$ meet transversally at only one point close to p, a fact that is intuitively clear. This is a clear example of how an obvious intuitive fact needs a non-trivial analytical proof.

7.1.6 When $k = 1$, we make a distinction between *regular arcs*, which are simply differentiable maps $\gamma(t)$, and *regular curves*, that are sets M locally parametrized by maps $\gamma(t)$ with derivative $\gamma'(t) \neq 0$ (and so locally one-to-one). In Theorem 3.2, we have seen that a regular (connected) curve admits a global parametrization by arc-length. If $\gamma(t)$ is a regular arc and γ' vanishes at some point, the path γ^* is not necessarily a regular curve. For instance, $\gamma(t) = (t^2, t^3), -1 < t < 1$ has the range

$$\gamma^* = \{(x, y) : x = y^{\frac{2}{3}}\},$$

which is not a regular curve at $(0, 0)$, where no tangent exists; equivalently, if both $x(t), y(t)$ are differentiable at 0 and $x(t)^3 = y(t)^2$, then $x'(0) = y'(0) = 0$. Still, γ' may vanish and γ^* be a regular curve, for instance $\gamma(t) = (t^3, t^6)$ whose range $y = x^2$ is a regular curve.

Exercise 7.1. Assume $\Phi : U \to V$ is a C^1-diffeomorphism and $M \subset U$ a regular sub-manifold in U. Prove that $\Phi(M)$ is a regular sub-manifold in V and that $T_{\Phi(p)}(\Phi(M)) = d\Phi(p)(T_p(M))$. Check how local charts and local defining functions are transported from M to $\Phi(M)$.

7.2 Functional Dependence: The Constant Rank Theorem

7.2.1 We review here the concept of functional dependence using the inverse and implicit function theorems. Recall from paragraph 4.3.5 that a system of functions $f_1, f_2, \ldots, f_m$ defined in U is called *functionally dependent in U* if some of the functions depend functionally on the remaining ones on U. When this is the case around every point of U, we say that they are *locally functionally dependent in U*. Assuming all functions of class C^1, by the implicit function theorem this means that a functional relation

$$g(f_1, \ldots, f_m) = 0,$$

with g of class C^1 and $\nabla g \neq 0$, holds locally in U. Then the rank of the system ∇f_i is less than m. Geometrically it means that the map $f = (f_1, \ldots, f_m)$ takes values in the hyper-surface $g = 0$.

Example 7.2. For the functions $f_1(x, y) = 2xy + 2x + 1, f_2(x, y) = x^2y^2 + 2x^2y + x^2 - 1$, a computation shows that the Jacobian of the transformation $f = (f_1, f_2)$ is identically zero, indicating that some dependence might occur. Indeed, $4f_2 = f_1^2 - 2f_1 - 3$, so f takes values in the graph $u = \frac{1}{4}(v^2 - 2v - 3)$.

If $\nabla f_1(x), \ldots, \nabla f_m(x)$ are linearly independent for all $x \in U$, no functional relation can occur. Then, we saw in Corollary 6.2 that F locally depends functionally on $f_1, \ldots, f_m$ if and only if $\nabla F(x)$ is a linear combination of the $\nabla f_i(x)$ for all $x \in U$. The next result can be seen as a global version. We have seen that in this situation the level sets

$$M = \{x \in U : f_1(x) = c_1, \ldots, f_m(x) = c_m\}$$

are regular sub-manifolds.

Proposition 7.1. *If M is arc-connected for all values of $c_1, \ldots, c_m$, a differentiable function F depends functionally on $f_1, \ldots, f_m$ in U if and only if $\nabla F(x)$ is a linear combination of $\nabla f_1(x), \ldots, \nabla f_m(x)$ for all $x \in U$.*

Proof. The necessity of the condition is a general fact, so we prove only the sufficiency. For $c_1, \ldots, c_m$ fixed, since $\nabla f_1(x), \ldots, \nabla f_m(x)$ span the orthogonal to $T_x(M)$, one has $D_v F(x) = \langle \nabla F(x), v \rangle = 0$ for all $v \in T_x(M)$. Therefore, F is constant along all arcs on M, and M being arc-connected, F is constant on M. Thus, F is constant whenever the f_j are constant, which amounts to saying that $F(x) = g(f_1(x), \ldots, f_m(x))$ for some function g. It remains to prove that g is differentiable at $(c_1, \ldots, c_m)$. If $p \in M$, in some ball $B(p, r)$, $f_1, \ldots, f_m$ form part of a coordinate system $u_1, \ldots, u_n$. In this system, $F = g(u_1, \ldots, u_m)$; since F is differentiable, g is differentiable. □

7.2.2 Up to now we have considered a system of equations

$$f_j(x_1, \ldots, x_n) = y_j, \quad j = 1, \ldots, m,$$

in short $f(x) = y$, and studied it around a solution $(p, q), f(p) = q$. The inverse function theorem ($m = n$) is an existence and uniqueness result, solutions are points, while in the implicit function theorem (where $m < n$) solutions are regular sub-manifolds. In both cases, the local behavior of the system is the same as the linear approximating one, $f(p) + df(p)(x - p) = y$, and in both cases $df(p)$ has maximum rank m.

In this paragraph, we consider the general case and assume $df(x)$ has constant rank $k \leq \min(m, n)$ in a domain. For instance, if k is the maximum of the rank of $df(x)$ in U, and $df(p)$ has rank k, then $df(x)$ has rank k in a neighborhood of p.

When $L : \mathbb{R}^n \to \mathbb{R}^m$ is linear with rank k, the range $L(\mathbb{R}^n)$ is a linear subspace of dimension k and each fiber $L^{-1}(y)$ is an affine sub-manifold of dimension $n-k$. The constant rank theorem states the analogous result for differentiable maps:

Theorem 7.3. *Assume $f : U \to \mathbb{R}^m, f = (f_1, \ldots, f_m)$ is of class C^1 in a domain $U \subset \mathbb{R}^n$ and that the rank of $df(x)$ is k for all $x \in U$. Then,*

(a) *For each $p \in U$ there is a ball $B(p, \varepsilon)$ such that k among the functions $f_1, \ldots, f_m$ are strongly functionally independent and the remaining ones depend functionally on them.*
(b) *For all $p \in U$, there is a ball $B(p,r)$ such that $f(B(p,r))$ is a regular sub-manifold M of dimension k of R^m.*
(c) *For $y \in f(U)$, the set $f^{-1}(y)$ is a regular sub-manifold of dimension $n-k$ of U (thus a discrete at most countable set if $k = n$).*
(d) *If $k = m, f(U)$ is open in $\mathbb{R}^m$. If $k < m$ and f is one-to-one, $f(U)$ is a regular sub-manifold of dimension k.*

As a restatement, we may say that the system

$$f_j(x_1, \ldots, x_n) = y_j, \quad j = 1, \ldots, m$$

behaves like the linear one: in order that some solutions exist, $y = (y_1, \ldots, y_m)$ must satisfy a compatibility condition, $y \in M$, and in this case the solutions form a regular sub-manifold of dimension $n-k$.

Proof. Fix $p \in U$; there exists k among the f_j, say $f_1, \ldots, f_k$ such that $\nabla f_j(p)$ are linearly independent and so in some ball $B(p,r)$ there is a coordinate system $u_1, \ldots, u_n$ with $f_j = u_j, j = 1, \ldots, k$. Since the rank is constantly k, $\nabla f_{k+1}, \ldots, \nabla f_m$ are linear combinations of $\nabla u_j, j = 1, \ldots, k$, therefore by Corollary 6.2 $f_{k+1}, \ldots, f_m$ depend functionally on $f_1, \ldots, f_k$,

$$f_j = \phi_j(f_1, \ldots, f_k), \quad j = k+1, \ldots, m,$$

with ϕ_j a C^1 function of k variables. This proves (a) and shows that $f(B(p,r))$, around $f(p)$ is parametrized by $u_1, \ldots, u_k$ and the map

$$G(u_1, \ldots, u_k) = (u_1, \ldots, u_k, \phi_{k+1}(u_1, \ldots, u_k), \ldots, \phi_m(u_1, \ldots, u_k)).$$

If y is fixed and $f(p) = y$, this shows as well that the set $f^{-1}(y)$ is described in $B(p,r)$ by $u_i = y_i, i = 1, \ldots, k$. If $k = n$, this means that $f^{-1}(y) \cap B(p,r) = p$. If $k < n$, this is a regular sub-manifold parametrized by $u_{k+1}, \ldots, u_n$. □

Setting $v_i = y_i, i = 1, \ldots, k, v_j = y_j - \phi_j(y_1, \ldots, y_k), j = k+1, \ldots, m$, which is a coordinate system in $\mathbb{R}^m$, we may rephrase the statement by saying that in the coordinate systems $u_1, \ldots, u_n, v_1, \ldots, v_m$ the map f is a projection

$$(u_1, \ldots, u_n) \to (u_1, \ldots, u_k, 0, \ldots, 0).$$

If $k < m$ and f is not one-to-one, $f(U)$ is not necessarily a sub-manifold of dimension k. If say $q = f(p_1) = f(p_2)$ and $B(p_1, r_1), B(p_2, r_2)$ are as in (b), $f(B(p_1, r_1)), f(B(p_2, r_2))$ might be transverse at q. Think for instance of a path $\gamma(t), \gamma'(t) \neq 0$ such that γ^* auto-intersects.

Example 7.3. Consider in $\mathbb{R}^3$ with coordinates x, y, z to $\mathbb{R}^2$ with coordinates u, v, $f = (f_1, f_2)$ with

$$f_1(x,y,z) = x^2y^2 + x^2z^2 + 2x^2yz + 1, f_2(x,y,z) = x^3y^3 + 3x^3y^2z + 3x^3yz^2 + x^3z^3.$$

It is easily checked that

$$\nabla f_1 = 2x(y+z)(y+z, x, x), \quad \nabla f_2 = 3x^2(y+z)^2(y+z, x, x),$$

so the rank is one in the complement U of $x = y+z = 0$. So $f(U)$ is a regular curve Γ in the plane. To identify it, we must find a functional relation among f_1, f_2. With $g = x(y+z)$ we realize that $\nabla f_1 = 2g\nabla g, \nabla f_2 = 3g^2\nabla g$, from which it follows that $f_1 = g^2 + a, f_2 = g^3 + b$. We check by inspection that $a = 1, b = 0$. To find an equation for Γ, we must eliminate t from the equations

$$P(t) = t^2 + 1, \quad Q = t^3.$$

This is always possible for arbitrary polynomials P, Q. In our case,

$$(P-1)^3 = Q^2.$$

So the curve in the uv plane has the equation

$$(u-1)^3 = v^2,$$

which is indeed a regular curve outside the origin. The fibers are the surfaces $g = x(y+z) = t$.

Exercise 7.2. Prove that $x+y+z, x^2+y^2+z^2-xy-yz-xz, x^3+y^3+z^3-3xyz$ are functionally dependent. Describe the sub-manifold $f(\mathbb{R}^3)$ and the fibers $f^{-1}(y)$.

7.3 Tubular Neighborhoods

7.3.1 In the following paragraphs, we deal with the existence of *tubular neighborhoods* of regular surfaces S in $\mathbb{R}^3$. First, note that if $x \notin S$, and $d(x,S) > 0$ (S is not necessarily closed), then $d(x,S) = d(x,p)$ is attained at some point $p \in S$. If $\Phi(s,t)$ is a local parametrization of S around $p, \Phi(0,0) = p$, the function $\phi(s,t) = |x - \Phi(s,t)|^2$ has a local minimum at $(0,0)$; since

$$\phi_s(0,0) = 2\langle x - p, \Phi_s(0,0)\rangle,$$

and similarly for ϕ_t, we see that $x - p$ is orthogonal to the tangent space $p + T_p(S)$.

Let $p \in S$ and $\Phi(s,t)$ be a local parametrization around p. In some ball $B(p,r)$ the range of Φ is $B(p,r) \cap S, \Phi(0,0) = p$, and Φ_s, Φ_t are tangent vectors to S so that $\Phi_s \times \Phi_t$ is a normal vector. We normalize it, that is we consider the unit normal vector

$$N(s,t) = \frac{\Phi_s \times \Phi_t}{|\Phi_s \times \Phi_t|},$$

and define

$$\Psi(s,t,\lambda) = \Phi(s,t) + \lambda N(s,t).$$

In order to apply the inverse function theorem to Ψ, we require Φ to be of class C^2; the differential $d\Psi(0,0,0)$ has column vectors $\Phi_s(0,0), \Phi_t(0,0), \Phi_s \times \Phi_t$, and so it is invertible. We conclude that Ψ is a local diffeomorphism at $(0,0,0)$, whence there exists $\varepsilon > 0$ such that the map

$$F(q,\lambda) = q + \lambda N(q), \quad q \in B(p,\varepsilon) \cap S, \quad |\lambda| < \varepsilon,$$

is one-to-one onto some neighborhood U of p. By shrinking it if necessary, we may assume that $U \subset B(p,r)$, and so $F(q,\lambda) \in S$ iff $\lambda = 0$. For $\lambda \neq 0$, the distance from $x = F(q,\lambda)$ to S might be smaller than λ and attained outside $B(p,\varepsilon) \cap S$; but in case it is attained at $y \in B(p,\varepsilon) \cap S$, and since $x - y$ is orthogonal to $T_y(S)$, necessarily $y = q$ and $d(x,S) = |\lambda|$. We call (F,U) a *local tube* around p.

Now let $K \subset S$ be a compact set. A finite number of local tubes $U_1, \ldots, U_N$ cover K, and by Theorem 2.8, there is $\tau > 0$ such that every closed ball $\overline{B}(q,\tau), q \in K$, is included in some U_i. With $\delta = \frac{\tau}{2}$, we claim that F is one-to-one from $K \times [-\delta, +\delta]$ onto the compact set

$$K_\delta = \{x : d(x,K) = d(x,S) \le \delta\}.$$

Indeed, assume $x \notin S$ and $d(x,S) = d(x,K) = d \le \delta$ is attained at $q, q' \in K$; then x, q, q' all lie in a single local tube U, and by the remark above, necessarily $q = q'$ and $x = F(q,\lambda)$ with $\lambda = \pm d$. The same argument shows that for $x = F(q,\lambda), q \in K, |\lambda| \le \delta$ one has $d(x,S) = d(x,K)$.

So we have proved the case $n = 3, k = 2$ case of the following theorem:

Theorem 7.4. *Let M be a k-dimensional regular sub-manifold in $\mathbb{R}^n$ of class C^2. For $p \in M$, let $N_p(\delta)$ be a ball centered at p of radius δ in the orthogonal $p + (T_p(M))^\perp$. Then for each compact set $K \subset M$ there is $\delta > 0$ such that all $N_p(\delta)$ are pairwise disjoint and fill up the compact set*

$$K_\delta = \{x : d(x,K) = d(x,M) \le \delta\}.$$

7.3.2 The theorem does not hold for C^1 sub-manifolds. Indeed, consider for $n = 2$ the curve $y = \frac{1}{\alpha}|x|^\alpha$ for $\alpha > 1$. The map Φ is

$$\Phi(x,t) = (x,y) + \frac{t}{\sqrt{1 + |x|^{2\alpha-2}}}(-|x|^{\alpha-1}, 1).$$

Let us check for solutions $x, t \ge 0$ of $\Phi(x,t) = (0,a)$ with small $a > 0$; this means

$$x - \frac{t}{\sqrt{1 + x^{2\alpha-2}}}x^{\alpha-1} = 0, \quad \frac{1}{\alpha}x^\alpha + \frac{t}{\sqrt{1 + x^{2\alpha-2}}} = a.$$

Besides the trivial solution $x = 0, t = a$, another solution is

$$t = \sqrt{1 + x^{2\alpha-2}}x^{2-\alpha},$$

with $x^\alpha + x^{2-\alpha} = a$. Clearly, if $1 < \alpha < 2$, x is of the order of $a^\beta, \beta(2-\alpha) = 1$ and t is of the order of a. This shows that there is no disc centered at $(0,0)$ in which Φ is one-to-one. Note that when $\alpha = 2$, the argument breaks down.

7.4 Constrained Optimization

7.4.1 In this section, we deal with optimization with constraints, that is to say, finding extreme values of scalar functions among points satisfying a number of constraints $f_j(x) = c_j$. To be precise, assume that $f_j, j = 1, \ldots, m$ are given continuous functions (constraints) defined on a domain $U \subset \mathbb{R}^n$,

$$M = \{x \in U : f_j(x) = c_j, j = 1, \ldots, m\},$$

is the set of points satisfying the constraints, and we are interested in finding the maximum and minimum value of g on M, if any.

We say that $p \in M$ is a *local constrained maximum* (*resp., minimum*) if there is a ball $B(p, r)$ such that $g(x) \leq g(p)$ (resp., $g(x) \geq g(p)$) for $x \in M \cap B(p, r)$.

Proposition 7.2. *Assume g, f_j of class $C^1, m < n$. If $\nabla f_j(p), j = 1, \ldots, m$ are linearly independent (that is, around p, M is a regular sub-manifold of dimension $k = n - m$) and p is a local constrained extremum, then $\nabla g(p)$ is a linear combination of $\nabla f_j(p), j = 1, \ldots, m$.*

Proof. We provide two proofs. Consider $v \in T_p(M), v = \gamma'(0), \gamma(0) = p$ as in Theorem 7.2. Then $g(\gamma(t))$ has a local extremum at $t = 0$, and hence $D_v g(p) = \langle \nabla g(p), v \rangle = 0$, proving $\nabla g(p) \in T_p(M)^{\perp}$. But $T_p(M)$ is the orthogonal complement of the linear space spanned by $\nabla f_j(p), j = 1, \ldots, m$, and so the result follows.

From a geometric point of view, the conclusion amounts to saying that if $\nabla g(p) \neq 0$, the hypersurface $S = \{g = g(p)\}$ and M are tangent at p. This can be seen geometrically as follows. If it were not the case and S, M meet transversally, then $\{g = g(p) \pm \varepsilon\}$ would meet M near p in every neighborhood of p, and p could not be an extremum. This is intuitively clear, as illustrated by Figure 7.1; analytically this is a consequence of the implicit function theorem. If $\nabla g(p), \nabla f_j(p)$ are linearly independent, then the system

$$g(x) = g(p) \pm \varepsilon, f_j(x) = 0, \quad j = 1, \ldots, m,$$

has indeed solutions in every small ball around p. In Figure 7.2, the level curves are tangent and only one of the curves $\{g = g(p) \pm \varepsilon\}$ meets the level curve of f. □

While in unconstrained optimization one has $\nabla g(p) = 0$, here $\nabla g(p)$ is a linear combination of $\nabla f_j(p), j = 1, \ldots, m$, maybe not zero. In both cases this is a necessary condition, not always a sufficient one: in paragraph 7.4.3 we will see a sufficient condition expressed in terms of higher-order derivatives.

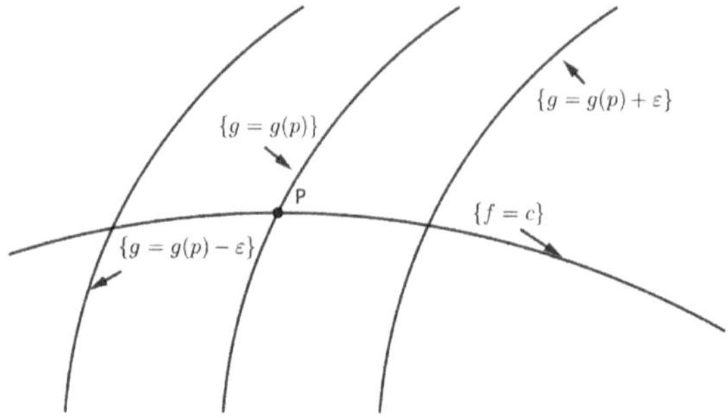

Figure 7.1. Transverse intersection, not a constrained extrema.

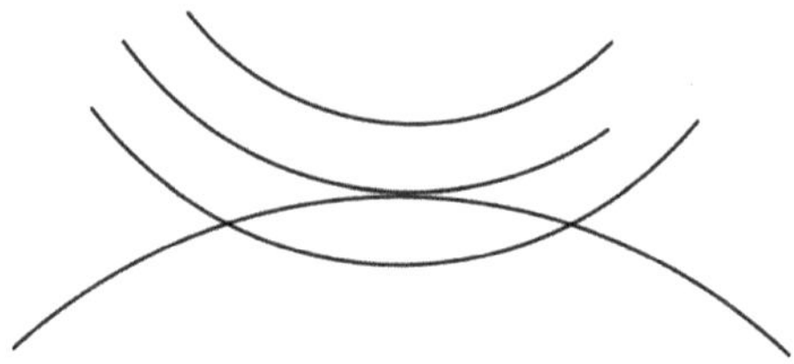

Figure 7.2. A constrained local extrema.

The proposition leads to the following strategy to solve the constrained optimization problem of finding the *absolute maximum and minimum* of g on M, in case they exist, for instance when M is compact or $|g|$ has infinite limit at infinity: the points $p \in M$ where the absolute extreme values occur are either points $x \in M$ where $\nabla f_j(x)$ are linearly dependent (when $m = 1$ this means $\nabla f_1(x) = 0$) or else points $x \in M$ such that

$$\nabla g(x) = \lambda_1 \nabla f_1(x) + \cdots + \lambda_m \nabla f_m(x),$$

for some coefficients $\lambda_1, \ldots, \lambda_m$ called the *Lagrange multipliers*. Introducing the so-called *Lagrangian*

$$\mathcal{L}(x_1, \ldots, x_n, \lambda_1, \ldots, \lambda_m) = g(x) - \sum_j \lambda_j (f_j(x) - c_j),$$

the latter are the critical points of $\mathcal{L}$, the solutions of the system of $n + m$ equations in $n + m$ unknowns

$$\frac{\partial \mathcal{L}}{\partial x_i} = \frac{\partial g}{\partial x_i} - \sum_j \lambda_j \frac{\partial f_j}{\partial x_i} = 0, \quad i = 1, \ldots, n,$$

$$\frac{\partial \mathcal{L}}{\partial \lambda_j} = (f_j - c_j) = 0, \quad j = 1, \ldots, m. \tag{7.1}$$

Thus, a set of candidate points is identified, typically finite, and it is enough to compute the value of f at each point to decide.

Example 7.4. Let us find the point in $f = y^5 - x^2 = 0$ closest to $(\alpha, 0), \alpha > 0$. Here $g(x,y) = (x-\alpha)^2 + y^2$ and the Lagrange system, for candidates $(x,y) \neq (0,0)$ (where $\nabla f = 0$), is

$$x(1+\lambda) = \alpha, \quad 2y = 5\lambda y^4, \quad y^5 = x^2.$$

The solution is $x = \frac{\alpha}{1+\lambda}, y^3 = \frac{2}{5\lambda}$ with $\lambda > 0$ satisfying

$$(1+\lambda)\lambda^{-\frac{5}{6}} = \left(\frac{5}{2}\right)^{\frac{5}{6}} \alpha.$$

The function $\phi(\lambda)$ on the left decreases in $(0,5)$ from $+\infty$ to a certain $\alpha_1 = \phi(5)$ and increases for $\lambda > 5$ to $+\infty$ again. So for $\alpha < \alpha_1$, the Lagrange system has no solution and the singular point $(0,0)$ is the closest one. For $\alpha = \alpha_1$, a solution appears, two for $\alpha > \alpha_1$, corresponding to two values of λ. Now, a computation shows that at these points g is smaller than α^2 exactly when $\lambda < 2$. If $\alpha_2 = \phi(2)$, it follows that for $\alpha_1 \leq \alpha < \alpha_2$ the solutions of the Lagrange system are not the closest points. For $\alpha > \alpha_2$, one of the two solutions is the closest point.

7.4.2 Assume now that U is a bounded domain defined by a finite number of inequalities

$$f_j(x_1, \ldots, x_n) < c_j, \quad j = 1, \ldots, m,$$

with $f_j \in C^1$, and we wish to find the absolute maximum and minimum of g on $\overline{U}$. If $p \in \overline{U}$ is an extremum, either $p \in U$, and then p must be a critical point of g, or else $p \in bU$. Now, bU is the union of the sets A_J, where J is a subset of $\{1, 2, \ldots, m\}$, with

$$A_J = \{x : f_j(x) = c_j, j \in J, f_j(x) < c_j, j \notin J\}.$$

A terminology at use is that the constraints $f_j = c_j, j \in J$ are the *active ones.* If $p \in A_J$ and $\nabla f_j(p), j \in J$ are linearly independent, then p must be among the solutions of the Lagrange system

$$\nabla g(x) = \sum_{j \in J} \lambda_j \nabla f_j(x), \quad f_j(x) = c_j, \quad j \in J, \tag{7.2}$$

satisfying $f_j(x) < c_j, j \notin J$. Thus, we consider the set E of candidate points:

(a) The singular points of g in U.
(b) For each J, the points $p \in A_J$ where $\nabla f_j(p), j \in J$ are not linearly independent.
(c) For each J, the points $p \in A$ which are solutions of the Lagrange system (7.1).

Typically, E will consist in a finite union of points, so it is sufficient to check the value of g at these points.

Exercise 7.3. Find the extreme values of

$$g = z^2 - 2x^2 - y^2 - 4xy - 2xz - z + x$$

on

$$K = \{f_1 = 6x^2+y^2+z^2+4xy-2zx \le 1, f_2 = z^2-(4x^2+y^2+4xy+2xz) \ge 0\}.$$

Note that K is the intersection of an ellipsoid and a cone, a compact set. Check the following:

(a) The function g has only one critical point, $(0, 0, \frac{1}{2})$; since it satisfies the strict inequalities, it is a candidate point.
(b) There are two points in bK with $f_1 = 0, f_2 > 0$ satisfying the corresponding Lagrange system.
(c) There is a point, the vertex $(0, 0, 0)$ of the cone, where ∇f_2 vanishes and $f_1 < 0$.
(d) There are four points in $f_2 = 0$ satisfying the Lagrange system, but only two of them with $f_1 < 0$.
(e) The curve $f_1 = 0, f_2 = 0$ has no singular points and there are eight points solving the corresponding Lagrange system.

Altogether there are 14 candidate points.

If p is an absolute maximum of g on $\overline{U}$ and just one of the constraints is active, say

$$f_1(p) = 0, \quad f_2(p) < c_2, \dots, \quad f_m(p) < c_m,$$

then the Lagrange multiplier λ_1 for which $\nabla g(p) = \lambda_1 \nabla f_1(p)$ must be non-negative; indeed, if $\lambda_1 < 0$, then along the direction of $\nabla g(p)$, g strictly increases and f_1 strictly decreases so $q = p + \varepsilon \lambda g(p)$, for ε small enough, would be a point in U with $g(q) > g(p)$. Similarly, $\lambda_1 \le 0$ if p is an absolute minimum.

7.4.3 Next we analyze second-order conditions for constrained local extrema. We consider again the situation in paragraph 7.4.1 and ask whether it is possible to formulate a sufficient condition in terms of second-order derivatives, in similar terms as for unconstrained extrema.

We start pointing out a procedure in terms of a local chart. If $p \in M$ satisfies the Lagrange system, let $\Phi(u_1, u_2, \ldots, u_k), \Phi(0) = p$, with $d\Phi(0)$ of rank $k = n - m$ a local chart around p. Obviously, g has a constrained local extremum at p if and only if $h(u) = g(\Phi(u))$ has an unconstrained extremum at $u = 0$. Imposing $\nabla h(0) = 0$ is equivalent to the Lagrange system. One could then compute the hessian $d_2h(0)$ and apply the above criteria. What we will show is that these criteria can be reformulated in terms of the Lagrangian

$$\mathcal{L}(x) = h(x) - \sum_j \lambda_j f_j(x),$$

without making use of a local chart.

One might think, by analogy with the unconstrained case, that a sufficient condition might be that the Hessian d_2g be positively or negatively defined on the tangent space $T_p(M)$. However, if this were the case, a strictly convex function, with positive hessian at all points, for instance $x^2 + 2y^2$, would have only local unconstrained minima, something obviously absurd. This means that the criteria we are searching for cannot depend only on the hessian of g and $T_p(M)$, but also on the second-order derivatives of the constraints. This is clear geometrically, looking for example at the extrema of $x^2 + 2y^2$ on $x^2 + y^2 = 1$, Figure 7.3. The maxima, which are absolute, are $(0,1), (0,-1)$ and the minima are $(1,0), (-1,0)$. At points $(0,1), (0,-1)$ the level curve $x^2 + y^2 = 1$ is within the level curve $x^2 + 2y^2 = 2$, while

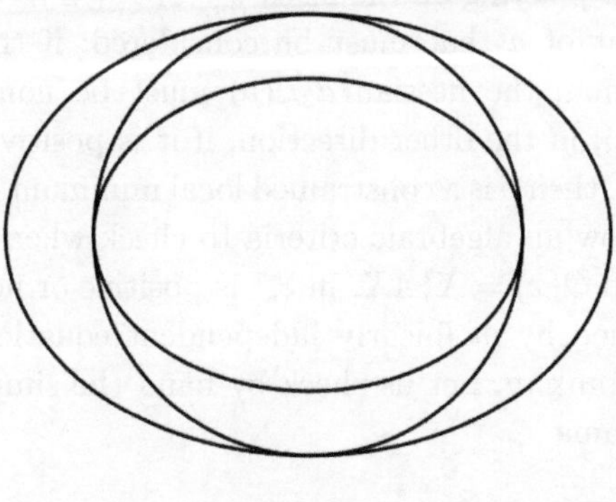

Figure 7.3. Relative position of level curves.

at points $(1,0),(-1,0)$ it is the other way around. The relative position of these curves depends on second-order derivatives, not only first-order derivatives.

To guess which is the right sufficient condition, we consider as above the local chart $\Phi(u)$, $h(u)=g(H(u))$, and compute a second-order derivative $\frac{\partial^2 h}{\partial u_i \partial u_j}(0)$. If $x_l = x_l(u)$ are the components of Φ, one has

$$\frac{\partial h}{\partial u_j} = \sum_{l=1}^{n} \frac{\partial g}{\partial x_l}(\Phi(u)) \frac{\partial x_l}{\partial u_j},$$

and differentiating again

$$\frac{\partial^2 h}{\partial u_i \partial u_j} = \sum_{l,n=1}^{n} \frac{\partial^2 g}{\partial x_l \partial x_n} \frac{\partial x_n}{\partial u_i} \frac{\partial x_l}{\partial u_j} + \sum_{l=1}^{n} \frac{\partial g}{\partial x_l}(\Phi(u)) \frac{\partial^2 x_l}{\partial u_j \partial u_i}.$$

Since $f_j(\Phi(u)) = c_j$, the same expression with f_j replacing g is zero; on the other hand, at $u=0$, we know that $\nabla g(p) = \sum_{j=1}^{m} \lambda_j \nabla f_j(p)$. Therefore, the last sum equals

$$\sum_{l=1}^{n} \sum_{j=1}^{m} \lambda_j \frac{\partial f_j}{\partial x_l}(p) \frac{\partial^2 x_l}{\partial u_j \partial u_i}(0) = -\sum_{l=1}^{n} \sum_{j=1}^{m} \lambda_j \frac{\partial^2 f_j}{\partial x_l \partial x_n}(p) \frac{\partial x_n}{\partial u_i}(0) \frac{\partial x_l}{\partial u_j}(0),$$

and finally

$$\frac{\partial^2 h}{\partial u_i \partial u_j}(0) = d_2\mathcal{L}(p)\left(\frac{\partial}{\partial u_i}, \frac{\partial}{\partial u_j}\right),$$

the hessian of $\mathcal{L}$ at p acting on the basis $\frac{\partial}{\partial u_i}$ of $T_p(M)$. So, it is the hessian of $\mathcal{L}$ and not that of g that must be considered: if there exists a local constrained extrema, the hessian $d_2\mathcal{L}(p)$ must be non-negative or non-positive on $T_p(M)$; in the other direction, if it is positive (resp., negative) definite on $T_p(M)$, then p is a constrained local minimum (resp., maximum).

So, we need now an algebraic criteria to check when a quadratic form, in matrix notation $Q(x) = X^t AX$, in $\mathbb{R}^n$ is positive or negative on a linear subspace M defined by m linearly independent equations $BX = 0$, B a $m \times n$ matrix of rank m. Let us check by hand the simplest possible case $n=2, m=1$: assume

$$Ax^2 + 2Bxy + Cy^2 > 0, \quad \text{if } Ex + Fy = 0,\ x, y \neq 0.$$

Setting $x = -\frac{F}{E}y$, one has

$$\left(\frac{AF^2}{E^2} - \frac{2BF}{E} + C\right) y^2 > 0,$$

so $AF^2 - 2BEF + CE^2 > 0$, that is the determinant of the matrix

$$\begin{pmatrix} 0 & E & F \\ E & A & B \\ F & B & C \end{pmatrix},$$

is positive. To deal with the general case, assume that the minor B' of order m of B consisting of the first m columns is not zero, and write correspondingly $x' = (x_1, \ldots, x_m), x'' = (x_{m+1}, \ldots, x_n)$. The constraint $BX = B'X' + B''X'' = 0$ means

$$X' = -(B')^{-1}B''X'' = CX'',$$

(that is to say, we are parametrizing M by X''). Consider now the same block partition for A

$$\begin{pmatrix} A_{11} & A_{12} \\ A_{12}^t & A_{22} \end{pmatrix},$$

where A_{11} is a $m \times m$ matrix, etc. We consider the so-called *bordered Hessian*, which is the $(n+m) \times (n+m)$ matrix defined by

$$A^* = \begin{pmatrix} 0 & B' & B'' \\ (B')^t & A_{11} & A_{12} \\ (B'')^t & A_{12}^t & A_{22} \end{pmatrix}.$$

The matrix of the restriction of Q to the subspace M is the $(n-m)\times(n-m)$ matrix

$$\begin{aligned} D &= \begin{pmatrix} C^t & I \end{pmatrix} \begin{pmatrix} A_{11} & A_{12} \\ A_{12}^t & A_{22} \end{pmatrix} \begin{pmatrix} C \\ I \end{pmatrix} \\ &= C^t A_{11} C + C^t A_{12} + A_{12}^t C + A_{22}. \end{aligned}$$

We have

$$\begin{pmatrix} I_k & 0 & 0 \\ 0 & I_k & 0 \\ 0 & C^t & I_{n-k} \end{pmatrix} \begin{pmatrix} 0 & B' & B'' \\ (B')^t & A_{11} & A_{12} \\ (B'')^t & A_{12}^t & A_{22} \end{pmatrix} \begin{pmatrix} I_k & 0 & 0 \\ 0 & I_k & C \\ 0 & 0 & I_{n-k} \end{pmatrix}$$

$$= \begin{pmatrix} 0 & B' & 0 \\ (B')^t & A_{11} & C_{12} \\ 0 & C_{12}^t & D \end{pmatrix}$$

for some matrix C_{12}. Let $D_l, 0 \le l \le n-m-1$, denote the principal minors of D obtained when deleting its last l rows and columns, whose signs we are interested in. Analogously, A_l^* denotes the minor of A^* obtained deleting its last l rows and columns. The last matrix identity implies

$$(-1)^m \det A_l^* = (\det B')^2 \det D_l.$$

It follows that the restriction of Q to M is positive defined if and only if

$$(-1)^m \det A_l^* > 0, \quad l = 0, \dots, n-m-1,$$

and it is negative defined if and only if these quantities alternate signs, starting in -1 for $l = n-m+1$, that is, $\det A_l^*$ has the sign $(-1)^{n-l}$.

This algebraic criteria, when applied to the constrained optimization, leads to the matrix

$$A^*(p,\lambda) = \begin{pmatrix} 0 & df(p) \\ df(p)^t & d_2\mathcal{L}(p) \end{pmatrix},$$

where $f = (f_1, \dots, f_m)$. We have shown that if (p, λ) is a critical point of the Lagrangian

$$\mathcal{L}(x,\lambda) = g(x) - \sum_j \lambda_j (f_j(x) - c_j),$$

with the c_j fixed, then a condition on A^* implies that p is a strict local extremum of g on M.

Example 7.5. Let us find the local extrema of $g = xyz$ on $x+y+2z = 6$. The Lagrangian $\mathcal{L} = xyz - \lambda(x+y+2z-6)$ has a unique critical point $p = (2,2,1)$, the bordered hessian being

$$\begin{pmatrix} 0 & 1 & 1 & 2 \\ 1 & 0 & 1 & 2 \\ 1 & 1 & 0 & 2 \\ 2 & 2 & 2 & 0 \end{pmatrix}.$$

The minors of interest have signs $-, +, -$ and thus this point is a local maximum, as can be seen geometrically, too.

7.4.4 Now let us look at level curves $f_j = c_j'$ for c_j' close to c_j, and the Lagrange system

$$f_j(x) - c_j' = 0, \quad j = 1, \dots, m,$$

$$\nabla g(x) + \sum_j \mu_j \nabla f_j(x) = 0.$$

We look at this as a system $F(x, \mu, c') = 0$ of $n+m$ equations in the $n+2m$ variables $x, \mu_1, \ldots, \mu_m, c'_1, \ldots, c'_m$. Now note that

$$\frac{\partial F}{\partial \mu \partial x},$$

evaluated at (p, λ, c), the minor of $dF(p, \lambda, c)$ corresponding to the μ and x' derivatives, is precisely A^*. Since the condition on A^* includes $\det A^* \neq 0$, by the implicit function theorem it turns out that the Lagrange system defines x, μ as functions of $c'_1, \ldots, c'_m$. In other words, for c'_j close to c_j the Lagrange system has solutions x, μ close to p, λ. Moreover, by continuity, the matrix A^* corresponding to this new point (x, μ) would satisfy the same hypothesis, whence x has the same character as p. If say p satisfying $f_j(p) = c_j$ is a local maximum there, there is a point $x = x(c')$ in $f_j = c'_j$ which is also a local maximum there.

For instance, if we know *a priori* that f has a unique global minimum $x = x(c)$ on each manifold $f_j = c_j$ and the hypothesis on A^* is fulfilled at p, we know now that x depends smoothly on c.

We can then look at the optimal value $g(x(c)) = g^*(c)$ as a function of c and its rate of change:

$$\frac{\partial g^*}{\partial c_j} = \sum_{i=1}^{n} \frac{\partial g}{\partial x_i} \frac{\partial x_i}{\partial c_j} = \sum_{i=1}^{n} \sum_{l=1}^{m} \lambda_l \frac{\partial f_l}{\partial x_i} \frac{\partial x_i}{\partial c_j}.$$

Differentiating $f_j(x(c)) = c_j$ with respect to c_l

$$\sum_{i=1}^{n} \frac{\partial f_j}{\partial x_i} \frac{\partial x_i}{\partial c_l} = \delta_{jl},$$

whence the last sum equals λ_j.

This result, $\frac{\partial g^*}{\partial c_j} = \lambda_j$, provides an interpretation of the Lagrange multiplier, particularly useful in applications to economic models.

Chapter 8

Ordinary Differential Equations

In previous chapters, some vector fields have appeared as tangent fields to given geometric objects, for instance, tangent fields to sub-manifolds and coordinate vector fields. In this and the next chapters, we study, conversely, how tangent fields define geometric objects. Here, we deal with the *existence and uniqueness theorem* for the Cauchy problem of an ordinary differential equation, a fundamental result in Analysis and already hinted at in Example 4.3. We first present the analytic viewpoint, where time-dependence or parametrizations matter, and then the geometric viewpoint in terms of integral curves or integral sub-manifolds.

8.1 Vector Fields and Differential Forms

8.1.1 We consider again vector fields X, always of class C^1 in a domain U. The most obvious ones are the constant ones, $X(x) = v$, a fixed vector. We have seen some non-constant vector fields: the gradient fields, the *coordinate vector fields* ∂_i of a coordinate system, and the fields X tangent to a regular sub-manifold M, that is, such that $X(p) \in T_p(M)$, for all $p \in M$. These are called *tangent vector fields.*

Incidentally, non vanishing continuous tangent vector fields to a sub-manifold may exist or not. In the unit circle S^1 in the plane, $X(x, y) = (-y, x)$ is such a field. However, a famous theorem by Poincaré (also known as the *hairy ball theorem*) states that every continuous vector field X on the unit sphere S^2 in space has a singularity, that is, there exists $p \in S$ such that $X(p) = 0$. This is another important result in Algebraic Topology.

Usually we identify a vector field $X = (A_1(x), \dots, A_n(x))$ with its directional derivative

$$X = \sum_i A_i(x) D_i,$$

acting on differentiable functions f by

$$Xf(x) = \sum_i A_i(x) D_i f(x).$$

8.1.2 The concept of a 1-*differential form* ω is dual to the one of vector field. This is a map assigning to each $x \in U$ a linear map $\omega(x)$ acting on $T_x(\mathbb{R}^n) = \mathbb{R}^n$. In canonical coordinates,

$$\omega = \sum_i B_i(x)\, dx_i,$$

usually assuming $B_i \in C^1(U)$. Then ω acts on vector fields to produce functions,

$$\omega(X) = \sum B_i(x) A_i(x).$$

Clearly, a form is defined by its action on fields. Of course, every linear map consists in scalar multiplication with a vector, so one could identify ω with the vector field $\sum_i B_i D_i$ but conceptually in some situations it is useful not to do so.

We have seen as only example up to now the differential df of a function, including the differential dx_i of cartesian coordinates, $df = \sum_i D_i f\, dx_i$. We saw in (6.1) that if $u_1, \dots, u_n$ is a coordinate system, then the basis $\partial_1, \dots, \partial_n$ of vector fields and the basis $du_1, \dots, du_n$ are dual, in particular $df = \sum_i \partial_i f\, du_i$. A general 1-form can be written, too, $\omega = \sum_i B_i(u)\, du_i$.

One motivation to consider such general forms is, for instance, if we think in the *incident space* of $\omega(x)$ at each $x \in U$, the linear space

$$L(x) = \{v \in \mathbb{R}^n : \omega(x)(v) = 0\}.$$

To give a distribution of hyperplanes, one through every $x \in U$ is the same as to give a non-vanishing 1-differential form, and this turns out to be a convenient language, as we will see in the next section.

If M is a regular sub-manifold, the *restriction* of ω to M is obtained by restricting the action of ω to $T_p(M)$, for $p \in M$. A statement like $\omega_1 = \omega_2$ on M means that they have the same restriction to M.

Not all 1-forms are of type df for some function f; in terms of vector fields, not every vector field is a gradient. As we saw in Theorem 5.1, a necessary condition is that $D_iA_j = D_jA_i$, and it is sufficient for p-domains, too. The forms $\omega = df$, corresponding to gradient fields ∇f, are called *exact.*

8.1.3 Assume f is twice differentiable in U, $X = \sum_i A_iD_i, Y = \sum_j B_jD_j$ two vector fields; we compute XYf:

$$XYf(x) = \sum_i A_i(x)D_i(Yf)(x) = \sum_{i,j} A_i(x)B_j(x)D_{ij}f(x)$$
$$+\sum_{i,j} A_i(x)D_iB_j(x)D_jf(x),$$

which we may write

$$XY(f)(x) = d_2f(x)(X(x), Y(x)) + df(x)(D_XY(x)),$$

where $d_2f(x)$ is the second differential of f and

$$D_XY = \sum_j (XB_j)D_j.$$

We may think that D_XY, which arises naturally in this computation, measures how Y changes along the direction of X.

In particular, if $(u_1, \ldots, u_n)$ is a general coordinate system,

$$\frac{\partial^2 f}{\partial u_i \partial u_k} = d_2f(\partial_i, \partial_k) + \sum_{j=1}^{n} \frac{\partial f}{\partial x_j}\frac{\partial^2 x_j}{\partial u_i \partial u_k}.$$

Thus, in general, the matrix of $d_2f(x)$ in this basis is not $(\frac{\partial^2 f}{\partial u_i \partial u_k})_{ik}$; it is so if the change of coordinates is linear, for then $\frac{\partial^2 x_j}{\partial u_i \partial u_k} = 0$.

Thus, to extend the action of d_2f on a couple (X, Y) of non-constant vector fields X, Y so that $d_2f(X, Y)(x)$ depends just on $X(x), Y(x)$ (this is called the *tensor property*), we must define

$$d_2f(X, Y) = XYf - df(D_XY) = X(df(Y)) - df(D_XY).$$

In this context, when acting on general vector fields, d_2f is called the *second covariant derivative* of f, sometimes denoted $(d^\nabla)^2 f$. Note that $XYf = X(df(Y))$ and $df(D_XY)$ do not have separately the tensor property, in the sense that their values at x not only depend on $X(x), Y(x)$, but their difference does.

Replacing *df* by a general 1-form ω, leads to the *covariant derivative* of a 1-form ω,

$$d^{\nabla}\omega(X,Y) = X(\omega(Y)) - \omega(D_X Y).$$

If $\omega = \sum_k C_k dx_k$, this equals

$$\sum_i A_i D_i \left(\sum_k C_k B_k\right) - \sum_k C_k \sum_i A_i (D_i B_k) = \sum_{i,k} (D_i C_k) A_i B_k,$$

which we may write as

$$d^{\nabla}\omega = \sum_{i,k} (D_i C_k) dx_i \otimes dx_k,$$

and is a 2-tensor.

8.1.4 Now, the statement of Schwarz's rule in this context is that the anti-symmetric part $d_2 f(X,Y) - d_2 f(Y,X)$ is zero. This leads to consider the anti-symmetric part of $d^{\nabla}\omega$,

$$d^{\nabla}\omega(X,Y) - d^{\nabla}\omega(Y,X) = X(\omega(Y)) - Y(\omega(X)) - \omega(D_X Y - D_Y X).$$

The vector field

$$[X,Y] = D_X Y - D_Y X,$$

whose action on f is thus

$$[X,Y]f = X(Yf) - Y(Xf)$$

is called the *commutator or Lie bracket* of X, Y. In coordinates,

$$X(Yf) - Y(Xf) = \sum_{i,j} (A_i D_i B_j - B_i D_i A_j) D_j f.$$

The above is the definition of the *exterior derivative* $d\omega$,

$$d\omega(X,Y) = X(\omega(Y)) - Y(\omega(X)) - \omega([X,Y]).$$

Thus, the statement of Schwarz rule in this context is $d(df) = 0$. Thus, an exact form ω must satisfy $d\omega = 0$; the latter are called *closed* forms.

8.1.5 Anti-symmetric bilinear maps on $\mathbb{R}^n$ are called 2-*forms*. We recall at this point some standard notions from linear and tensor algebra on how to manipulate such forms. The general bilinear form is given in a basis ω_i of 1-forms by

$$\Phi = \sum_{ij} a_{ij}\omega_i \otimes \omega_j,$$

meaning that

$$\Phi(u, v) = \sum_{ij} a_{ij}\omega_i(u)\omega_j(v).$$

It is anti-symmetric, $\Phi(u, v) = \Phi(v, u)$, iff $a_{ij} = -a_{ji}$, that is

$$\Phi(u, v) = \sum_{i<j} a_{ij}(\omega_i \otimes \omega_j - \omega_j \otimes \omega_i).$$

For a couple of 1-forms ω, η, their *wedge product* is defined by

$$\omega \wedge \eta = \omega \otimes \eta - \eta \otimes \omega, \quad (\omega \wedge \eta)(u, v) = \omega(u)\eta(v) - \omega(v)\eta(u).$$

Thus, the general 2-form is given in coordinates by

$$\Phi = \sum_{i<j} a_{ij}\omega_i \wedge \omega_j.$$

From this we see that the space of such forms have the dimension $n(n-1)/2$.

With this notation, let us compute $d\omega$ using a coordinate system in which $\omega = \sum_i B_i(u)\, du_i$. First, applying the definition, for a function λ and a 1-form η,

$$\begin{aligned} d(\lambda\eta)(X, Y) &= X(\lambda\omega(Y)) - Y(\lambda\omega(X)) - \lambda\omega([X, Y]) = \lambda d\omega(X, Y) \\ &\quad + X(\lambda)\omega(Y) - Y(\lambda)\omega(X), \end{aligned}$$

that is, $d(\lambda\omega) = \lambda d\omega + d\lambda \wedge \omega$. Therefore,

$$d\omega = \sum_{i<j} \left(\frac{\partial}{\partial u_i} B_j(u) - \frac{\partial}{\partial u_j} B_i(u) \right) du_i \wedge du_j = \sum_j dB_j \wedge du_j.$$

We will encounter again the concept of differential form in the context of vector analysis, in Section 18.5.

8.2 Vector Fields and Ordinary Differential Equations

8.2.1 Given a vector field X on U, it is natural to look at the parametrized arcs $\varphi(t)$ within U (that is, $\varphi^* \subset U$) such that $\varphi'(t) = X(\varphi(t))$ for all t. In coordinates, if the equations of φ are $x_i(t), i = 1, \ldots, n$, and $X = (A_1, \ldots, A_n)$, this amounts to n equations

$$x_i'(t) = A_i(x_1(t), x_2(t), \ldots, x_n(t)), \quad i = 1, \ldots, n. \tag{8.1}$$

This is called an *autonomous system of ordinary differential equations.* Thinking of t as time and $\varphi(t)$ as position at time t of a moving particle, this means that at every instant t the instant speed, when the particle is at $\varphi(t)$, is the one defined by the field. If X is a force or velocity field and at time $t = 0$ we place a particle at point p, the particle will follow one of these paths. On an intuitive basis, one would say that the arc $\varphi(t)$ is unique. This can be indeed proven if X satisfies certain conditions, and it is the contents of the *existence and uniqueness theorem* for the Cauchy problem for ordinary differential equations. This is a central result in analysis, which we describe in following paragraphs.

It will be convenient to adopt a slightly more general frame. An *ordinary differential equation* (ODE) is an expression of type

$$x' = f(t, x), \tag{8.2}$$

where $\Omega \subset \mathbb{R} \times \mathbb{R}^n$ is open and $f : \Omega \longrightarrow \mathbb{R}^n$ continuous. We say that t is the *time variable* and x the *space variable.* We say that a parametrized regular arc

$$\varphi : I \longrightarrow \mathbb{R}^n$$

is a solution if $(t, \varphi(t)) \in \Omega, t \in I$ and $\varphi'(t) = f(t, \varphi(t)), t \in I$. In case I contains an end-point the derivative, ϕ' is understood as a lateral one.

Example 8.1. The simplest situation is $f(t, x) = f(t)$. Solving the equation means finding a primitive of f. The general solution depends on an additive constant. Given (s, y),

$$\varphi(t) = y + \int_s^t f(\tau)\, d\tau,$$

is the unique solution such that $\varphi(s) = y$.

If f is independent of t we say that the equation is *autonomous*, like (8.1). For instance, $x' = x$ is autonomous, $\varphi(t) = ke^t$ is a solution for each k. The equation

$$(x', y') = (y, -x)$$

is also autonomous, and

$$\varphi(t) = (A\cos t + B\sin t, -A\sin t + B\cos t)$$

is a solution. Again, for each (s, p) there is a unique solution such that $\varphi(s) = p$. In fact, every ODE can be reduced to an autonomous ODE in one more variable, by treating t as a variable and setting $(x, t)' = (f(t, x), 1)$.

An equation of type

$$x^{(n)} = f(t, x, x^{(1)}, \ldots, x^{(n-1)}),$$

where $f : \Omega \longrightarrow \mathbb{R}, \Omega \subset \mathbb{R} \times \mathbb{R}^n$ is called a *differential equation of order n*. A solution is a $\varphi : I \longrightarrow \mathbb{R}$, n times differentiable, such that

$$\varphi^{(n)}(t) = f(t, \varphi(t), \varphi^{(1)}(t), \ldots, \varphi^{(n-1)}(t)).$$

These equations can be reduced to an ODE of order one setting $x_i = x^{(i-1)}$ $i = 1, \ldots, n$. We obtain the equation

$$(x_1', x_2', \ldots, x_n') = (x_2, x_3, \ldots, f(t, x_1, \ldots, x_n)).$$

In all the examples, we have seen that there is a unique solution satisfying the *initial condition* $\varphi(s) = y$. However, the next example shows that this is not always the case.

Example 8.2. We search for solutions of $x' = 3x^{2/3}$ such that $\varphi(0) = 0$. A computation shows that for all $c \geq 0$,

$$\varphi_c(t) = (t - c)^3, \quad t \geq c, \; \varphi_c(t) = 0, \; t < c,$$

is a solution.

The problem

$$x' = f(t, x), \quad x(s) = y,$$

is called *the Cauchy problem*. In the following examples, we solve the Cauchy problem in an elementary way.

Example 8.3. Let $\Omega = \mathbb{R} \times (a,b), f(t,x) = f(x)$, with f continuous in (a,b) and $f(x) \neq 0$. We write the equation in the form

$$\frac{\varphi'}{f(\varphi)} = 1.$$

We may consider

$$F(x) = \int_y^x \frac{d\tau}{f(\tau)},$$

so that $F' = \frac{1}{f}$, whence F is a bi-continuous map from (a,b) to some interval (c,d) where F^{-1} is defined and is continuous. The equation is written now as

$$F'(\varphi)\varphi' = 1, \quad (F \circ \varphi)'(t) = 1,$$

which amounts to

$$F(\varphi(t)) - F(\varphi(s)) = t - s, \quad F(\varphi(t)) = t - s,$$

and so the solution of the Cauchy problem is unique and given by

$$\varphi(t) = F^{-1}(t-s), \quad t \in (s+c, s+d).$$

Example 8.4. An ODE with *separate variables* is one of type

$$x' = g(t)f(x),$$

where g, f are continuous in $(t_1, t_2), (a,b)$, respectively, and $f(x) \neq 0$. Proceeding as in the previous example, we see that for $s \in (t_1, t_2), y \in (a,b)$, the Cauchy problem has the unique solution

$$\varphi(t) = F^{-1}(G(t)), \quad G(t) = \int_s^t g(\tau)d\tau,$$

with F as above.

In practice, one proceeds by *separating* the variables

$$\frac{dx}{f(x)} = g(t)dt$$

and integrating both terms, $F(x) = G(t) + c$. Then imposing $x(s) = y$ we find c. For instance, consider the *Verhulst or logistic* equation modeling population growth

$$N' = rN\left(1 - \frac{N}{M}\right).$$

Here, $N(t)$ denotes population at time t, r is interpreted as the growth rate of a population and M is a constant, interpreted as a critical value for N, while $N(t) < M$ the population grows, and decreases while $N(t) > M$. Given an initial population $N(0) = N_0$ we are interested in the evolution of $N(t)$. Clearly, $N(t) = M$ is a solution. To find other solutions we separate variables

$$M\frac{dN}{N(M-N)} = r\,dt,$$

and integrate

$$\log N - \log|M-N| = rt + c, \; \frac{|M-N|}{N} = Ke^{-rt}.$$

While $N < M, N(t) = \frac{M}{1+Ke^{-rt}}$ with $N_0 = \frac{M}{1+K}, K = \frac{M}{N_0} - 1$,

$$N(t) = \frac{MN_0}{N_0 + (M - N_0)e^{-rt}},$$

increases to M as $t \to +\infty$. While $N > M, N(t) = \frac{M}{1-Ke^{-rt}}$ decreases to M as $t \to +\infty$.

Example 8.5. We consider the *linear equation*

$$x' = a(t)x + b(t),$$

with $a(t), b(t)$ continuous in (t_1, t_2). If

$$A(t) = \int_s^t a(\tau)\,d\tau,$$

we perform the change of variable $x(t) = e^{A(t)}y(t)$. Then the equation becomes

$$y'e^A + yA'e^A = ax + b = aye^A + b, y' = be^{-A},$$

so $y(t) = c + \int_s^t b(\tau)e^{-A(\tau)}\,d\tau$, and

$$x(t) = ce^{A(t)} + e^{A(t)}\int_s^t b(\tau)e^{-A(\tau)}\,d\tau.$$

The first term is the general solution of the *homogeneous equation* $x' = a(t)x$ and the second term is a particular solution x_p of the equation. The structure of the solution is thus similar to that of a linear system.

8.2.2 There is a connection between the implicit function theorem and ordinary differential equations. To explain this we shift the notation to $y = y(x)$ instead of $x = x(t)$. Assume that

$$F(x, y) = 0,$$

is an equation in the conditions of Theorem 6.3, F is of class C^1, $F(a, b) = 0, D_2F(a, b) \neq 0$. If $y(x)$ is the implicit function, differentiating $F(x, y(x)) = 0$ we get

$$D_1F(x, y(x)) + y'(x)D_2F(x, y(x)) = 0.$$

Conversely, if $y(a) = b$ and $y(x)$ satisfies this equation, then $F(x, y(x)) = 0$. An ODE

$$A(x, y) + y'(x)B(x, y) = 0,$$

usually written $A\,dx + B\,dy = 0$, for which $A = D_1F, B = D_2F$ for some $F \in C^1$ is called *exact*, and F is called the *potential function* Thus the Cauchy problem for an exact equation is equivalent to the implicit function theorem, the solution of the Cauchy problem is the implicit function defined by $F(x, y) = c$.

Example 8.6. The equation

$$(3x^2 + 6xy^2)dx + (6x^2y + 4y^3)dy = 0,$$

is exact with

$$F(x, y) = \int (3x^2 + 6xy^2)dx = x^3 + 3x^2y^2 + G(y)$$

$$6x^2y + G'(y) = 6x^2y + 4y^3, G(y) = y^4.$$

Thus, the general solutions of the equation are the implicit functions $y = y(x)$ defined by

$$x^3 + 3x^2y^2 + y^4 = c.$$

When A, B are of class C^1 we know from paragraph 4.7.3 that in p-domains the necessary and sufficient condition for an equation to be exact is that $A_y = B_x$. Since the equation is equivalent to

$$\lambda(x, y)A(x, y) + y'(x)\lambda(x, y)B(x, y) = 0, \quad \lambda \neq 0,$$

it is natural to try to find λ such that $(\lambda A)_y = (\lambda B)_x$, that is

$$\lambda(A_y - B_x) = B\lambda_x - A\lambda_y.$$

Such λ is called an *integrating factor*. We will see later that integrating factors always exist for these equations, in fact this statement is equivalent to Picard's Theorem 8.2 that follows. For specific equations, an integrating factor can be found by inspection, whence the implicit function theorem solves Cauchy problem for those. For instance, this is the case if $\frac{A_y - B_x}{B}$ depends only on x.

Example 8.7. For

$$\left(2xy + x^2y + \frac{1}{3}y^3\right)dx + (x^2 + y^2)dy = 0,$$

one has $\frac{A_y - B_x}{B} = 1$, whence we may take $\lambda = e^x$ as integrating factor. Indeed, multiplying by λ the equation becomes exact with potential function $ye^x(x^2 + \frac{1}{3}y^2)$.

8.3 Existence and Uniqueness for the Cauchy Problem

8.3.1 In the previous examples, we have seen by direct methods that the Cauchy problem has a unique solution. In this paragraph, we will prove the basic existence and uniqueness result for the Cauchy problem for an ODE.

The first step is to reformulate the problem in terms of an *integral equation*. If φ is a solution of the Cauchy problem, then

$$\varphi(t) = y + \int_s^t f(\tau, \varphi(\tau))d\tau. \tag{8.3}$$

Conversely, if this holds, then $\varphi(s) = y$ and by the fundamental theorem of calculus, φ satisfies the e.d.o.

The second step is to rephrase the integral equation in terms of a fix-point of a certain transformation T. Clearly, we must consider

$$T\varphi(x) = y + \int_s^t f(\tau, \varphi(\tau))d\tau.$$

Thus, a continuous φ is a solution of the Cauchy problem if and only if $T\varphi = \varphi$. There is a problem with this latter statement, for we do not know the domain of definition of φ, only that it has to contain s. In some sense, finding the domain of definition is part of the problem.

The third step is having a criteria on a transformation ensuring the existence of a unique fixed point. In paragraph 2.1.3, we pointed out that the fixed point theorem holds in a general complete metric space:

Theorem 8.1. *If T is a transformation in a complete metric space and T^m is contractive for some m, T has a unique fixed point.*

So we need to choose a complete metric space (of functions) in which T is a contractive.

For $A \subset \mathbb{R}^n$ we denote by $C_b(A, \mathbb{R}^k)$ the class of continuous bounded functions $f : A \to \mathbb{R}^k$. This is a metric space with the distance

$$d(f, g) = \sup_{x \in A} |f(x) - g(x)|.$$

Convergence in this space, $d(f_m, f) \to 0$, means that f_m *converges to* f *uniformly on* A: given ε there is m_0 such that $|f_m(x) - f(x)| \leq \varepsilon, x \in A, m \geq m_0$.

This space is complete. For if (f_m) is a Cauchy sequence,

$$|f_l(x) - f_m(x)| \leq \varepsilon, \quad x \in A, l, \quad m \geq m_0(\varepsilon),$$

then $(f_m(x))$ is a Cauchy sequence for every $x \in A$, whence $f(x) = \lim_m f_m(x)$ exists, and passing to the limit as $l \to +\infty$ we find that f_m converges uniformly to f,

$$|f_m(x) - f(x)| \leq \varepsilon, \quad x \in A, \quad m \geq m_0(\varepsilon).$$

So f is bounded. Since the f_m are continuous, f is also continuous:

$$|f(x) - f(y)| \leq |f(x) - f_m(x)| + |f_m(x) - f_m(y)| + |f_m(y) - f(y)|.$$

Given $\varepsilon > 0$, the first and third terms are less than $\varepsilon > 0$ for all x, y, and the second too if y is close enough to x. Thus, $f \in C_b(A, \mathbb{R}^k)$ and $d(f_m, f) \to 0$.

If $F \subset \mathbb{R}^k$ is closed, the same proof shows that the space $C_b(A, F)$ of continuous bounded functions taking values on F is complete. Recall that if A is compact, continuity implies boundedness, by Weierstrass' Theorem 2.9.

Next, we specify the metric space X in which T acts. Let $\Omega \subset \mathbb{R} \times \mathbb{R}^n$, f be continuous in Ω and $(s, y) \in \Omega$. We denote by $I = I(s, a)$ the closed interval $[s - a, s + a]$ and $\overline{B} = \overline{B}(y, r)$. We want φ to have its graph within Ω so to start with we consider $X = C(I, \overline{B})$, continuous functions

$$\varphi : I(s, a) \to \overline{B}(y, r).$$

For such φ,

$$T\varphi(t) - y = \int_s^t f(\tau, \varphi(\tau))\, d\tau.$$

In order that $T\varphi \in X$ this should have size $\leq r$. If $|f| \leq M$, we see that $|T\varphi(t) - y| \leq M|t - s| \leq Ma$. Thus, we need $Ma \leq r$. So we take $\alpha = \min(a, \frac{r}{M})$; with this choice, T acts in the complete metric space $X = C(I(s, \alpha), \overline{B}(y, r))$.

Theorem 8.2. (Picard's theorem). *Assume $f : I(s, a) \times \overline{B}(y, r) \longrightarrow \mathbb{R}^n$ is continuous and Lipschitz in x,*

$$|f(t, x) - f(t, y)| \leq C|x - y|, \quad t \in I(s, a), \quad x, y \in \overline{B}(y, r).$$

Let $M = \max\{|f(t, x)| : (t, x) \in I(s, a) \times \overline{B}(y, r)\}$. Then there is a unique solution of the Cauchy problem

$$x' = f(t, x), \quad x(s) = y,$$

defined in $I(s, \alpha), \alpha = \min\{a, r/M\}$ and taking values in $\overline{B}(y, r)$.

Proof. We need to prove that some iterate of T is contractive. So we start with $d(T\varphi, T\psi) = \max |T\varphi(t) - T\psi(t)|$:

$$|T(\varphi)(t) - T(\psi)(t)| = |\int_s^t (f(\tau, \varphi(\tau)) - f(\tau, \psi(\tau)))d\tau|$$

$$\leq \int_s^t |f(\tau, \varphi(\tau)) - f(\tau, \psi(\tau))|\, d\tau \leq \int_s^t C|\varphi(\tau) - \psi(\tau)|\, d\tau \leq C|t - s|d(\varphi, \psi).$$

If $C\alpha < 1$, T itself is contractive. But we do not need to shrink α further, for if we iterate we get

$$|T^2(\varphi)(t) - T^2(\psi)(t)| \leq \int_s^t CC|\tau - s|d(\varphi, \psi)\, d\tau$$
$$= \frac{C^2(t - s)^2}{2} d(\varphi, \psi) \leq \frac{C^2\alpha^2}{2} d(\varphi, \psi).$$

By induction we see that

$$d(T^m\varphi, T^m\psi) \leq \frac{C^m}{m!}\alpha^m d(\varphi, \psi).$$

Since $\frac{(C\alpha)^m}{m!}$ has limit zero, the proof is finished. □

By the mean-value Theorem 4.4, a function of class C^1 satisfies the Lipschitz assumption.

Under the sole assumption of continuity of f, one can still prove the *existence* of a solution of the Cauchy problem in $I(s, \alpha)$. This is *Peano's theorem*, see [17]. Unicity can be proved too in some cases, for instance exact equations of type

$$D_1F(x, y(x)) + y'(x)D_2F(x, y(x)) = 0,$$

with F of class C^1 and not C^2, enjoying existence and uniqueness, as they are equivalent to $F(x, y(x)) = c$.

Note that the solution is defined in an interval depending on the initial condition. In some special cases, though, the solutions are global:

Theorem 8.3. *If f is defined in $[a, b] \times \mathbb{R}^n$ and is Lipschitz in x, for all (s, y) there is a unique solution of the Cauchy problem defined in $[a, b]$. The same statement holds replacing $[a, b]$ by a general interval I, bounded or unbounded.*

Proof. The same proof as in Picard's theorem works taking $X = C([a, b], \mathbb{R}^n)$. For the second statement, consider an increasing sequence $[a_n, b_n]$ exhausting I, $s \in [a_n, b_n]$ and the solution φ_n. By unicity, $\varphi_{n+1} = \varphi_n$ in $[a_n, b_n]$, so we get a solution defined in the whole I. □

This theorem applies to the linear equation in n unknowns

$$x' = f(t, x) = A(t)x + B(t), \quad t \in [a, b],$$

where $A(t)$ is an $n \times n$ matrix and $B(t)$ a column vector with continuous entries because

$$|f(t, x) - f(t, y)| = |A(t)(x - y)| \leq |A(t)||x - y| \leq C|x - y|,$$

$|A(t)|$ denoting the matrix norm.

Coming back to the general situation, it is clear that if f satisfies locally a Lipschitz condition in Ω, the unicity allows to talk about a unique maximal solution φ of the Cauchy problem. Formally, if $(I, \varphi), (J, \psi)$, $s \in I \cap J$ are solutions, then $\varphi = \psi$ on $I \cap J$, so we get a solution in $I \cup J$. This leads to a maximal interval $I(s, y)$ and a maximal solution φ; formally, $I(s, y)$ is the union of all intervals $I, s \in I$ for which some solution exists in I. It is an open interval, because if $b \in I(s, y)$ is an end-point, the Cauchy problem with initial data $(b, \varphi(b)) \in \Omega$ has a solution and φ can be continued beyond b.

It is natural to look at the behavior of the maximal solution at the end-points of $I(s, y)$.

Exercise 8.1. Prove that if b is a finite end-point of $I = I(s, y)$, then $\varphi(t)$ approaches $b\Omega$ as $t \to b$ meaning that for each compact $K \subset \Omega$ there is $\varepsilon > 0$ such that $\varphi(t) \notin K, t \in I, |t - b| < \varepsilon$.

In general, $\lim_{t\to b} \varphi(t)$ does not exist. It does if f is globally bounded and b is finite, for then $\varphi'(t)$ is bounded and $|\varphi(t) - \varphi(t')| = O(|t - t'|)$.

8.3.2 We study now the dependence of the solution of Cauchy problem on the initial conditions and possible parameters. The latter refers for instance to coefficients that f might have. From an intuitive point of view, we expect that if the initial conditions or the parameters suffer small changes, the solution will not be much different.

To be precise, assume $f : \Omega \longrightarrow \mathbb{R}^n$ is continuous in an open set Ω in $\mathbb{R} \times \mathbb{R}^n \times \mathbb{R}^k$. In $f(t, x, \lambda)$, λ is the multi-parameter. We assume that for each λ, $f(t, x)$ is locally Lipschitz in x, so that we can consider the unique maximal solution φ of the Cauchy problem

$$x' = f(t, x, \lambda), \quad x(s) = y, \quad (s, y, \lambda) \in \Omega.$$

The solution φ and its domain will depend on (s, y, λ) so we use the notation $\varphi(t, s, y, \lambda), t \in I(s, y, \lambda)$. The question is to study the dependence on the initial condition (s, y) and λ.

The following construction shows that parameters can be assimilated to the initial conditions. We simply consider λ as a variable, that is, we introduce the new variable $w = (x, \lambda)$, and the new k equations $\lambda' = 0$. Thus, our system of n equations with k parameters is equivalent to a system of $n + k$ equations with no parameters. So, from now on we concentrate on the dependence on the initial condition. Next result is known as Gromwall's lemma:

Proposition 8.1. *Let u, v be non-negative continuous functions on $[a, b]$ such that*

$$u(t) \leq \alpha + \int_a^t u(\tau)v(\tau)d\tau, \quad t \in [a, b].$$

Then

$$u(t) \leq \alpha \quad e^{\int_a^t v(\tau)d\tau}, \quad t \in [a, b].$$

Proof. Replacing α by $\alpha+\varepsilon$ and letting $\varepsilon \to 0$ we may assume $\alpha > 0$. Let $w(t) = \alpha + \int_a^t u(\tau)v(\tau)d\tau$. Then $w(t) > 0$ and

$$\frac{w'(t)}{w(t)} \le v(t),$$

which implies

$$w(t) \le w(a)e^{\int_a^t v(\tau)\,d\tau} = \alpha e^{\int_a^t v(\tau)\,d\tau}. \qquad \square$$

Theorem 8.4. *The set*

$$D = \{(t, s, y) : t \in I(s, y)\},$$

is an open set in $\mathbb{R}^{n+2}$, *and* $\varphi(t, s, y)$ *is locally Lipschitz in* D *in the three variables.*

Proof. First we show that if $K \subset \Omega$ is compact, there is $\alpha = \alpha(K)$ such that for $(s, y) \in K$ the solution of the Cauchy problem is defined for $|t-s| < \alpha$. Consider

$$d = d(K, \Omega^c) > 0, \quad L = \left\{p = (t, x) \in \Omega : d(p, K) \le d' = \frac{d}{2}\right\},$$

and let $M = \max_{p\in L}|f(p)|$. For $p = (s, y) \in K, B(p, d') \subset \Omega$, whence $I(s, \frac{d'}{\sqrt{2}}) \times B(y, \frac{d'}{\sqrt{2}}) \subset \Omega$, and $|f| \le M$ there. That is, a, b, M in Picard's theorem can be chosen uniformly in $(s, y) \in K$.

Assume $t_0 \in (a, b), [a, b] \subset I(s_0, y_0)$ and consider the graph

$$G = (t, \varphi(t, s_0, y_0)), \quad t \in [a, b].$$

Choosing $K = \{p = (s, y) \in \Omega, d(p, G) \le \tau\}$ for τ small enough, we see that there exists $\alpha > 0$ such that whenever $d((s, y), G) \le \tau$ the Cauchy problem with initial condition $x(s) = y$ is defined in the interval $J_s = (s-\alpha, s+\alpha)$ and with graph within a compact $L \subset \Omega$. Let $M = \max_{p\in L}|f(p)|$. By compactness, we can take the Lipschitz constant C uniformly in L:

$$|f(s, y) - f(s, y')| \le C|y - y'|, \quad (s, y), (s, y') \in L. \tag{8.4}$$

Now we claim that there is a constant C' such that whenever the solutions of Cauchy problem with initial data $(s, y), (s', y') \in K$ are defined in a common interval $I \subset [a, b]$ containing s, s', with graph within L, then

$$|\varphi(t, s, y) - \varphi(t, s', y')| \le C'(|s - s'| + |y - y'|), \quad t \in I.$$

To see this, we write

$$\begin{aligned}&|\varphi(t,s,y)-\varphi(t,s',y')|\\&\quad\le|\varphi(t,s,y)-\varphi(t,s,y')|+|\varphi(t,s,y')-\varphi(t,s',y')|=I+II.\end{aligned}\tag{8.5}$$

Let $u(t)=\varphi(t,s,y)-\varphi(t,s,y')$. Using (8.3) and (8.4)

$$\begin{aligned}|u(t)|&\le|y-y'|+\int_s^t|f(\tau,\varphi(\tau,s,y))-f(\tau,\varphi(\tau,s,y'))|\,d\tau\\&\le|y-y'|+C\int_s^t|u(\tau)|\,d\tau.\end{aligned}$$

By Gromwall's proposition above,

$$I=|u(t)|\le|y-y'|e^{C|t-s|}\le e^{C(b-a)}|y-y'|.$$

As for II, we may write $\varphi(t,s',y')=\varphi(t,s,\varphi(s,s',y'))$. Then II can be estimated as I obtaining

$$II\le C'|y'-\varphi(s,s',y')|.$$

But by (8.3) again, obviously $|y'-\varphi(s,s',y')|\le M|s-s'|$, and the claim is proved.

Now we show that if $|s-s_0|+|y-y_0|=r$ is small enough, the solution $\varphi(t,s,y)$ is defined for $t\in[a,b]$. Set $\beta=\alpha/2, s_m=s_0+m\beta$; if $r<\beta$, $\varphi(t,s,y)$ is defined at least in $[s_0,s_1]$. By the claim, with $I=[s_0,s_1]$,

$$|\varphi(s_1,s,y)-\varphi(s_1,s_0,y_0)|\le C''(|s-s_0|+|y-y_0|)\le C''r.$$

So, if $C''r<\tau,(s_1,\varphi(s_1,s,y))\in K$. Then $\varphi(t,s,y)=\varphi(t,s_1,\varphi(s_1,s,y))$ is defined in I_{s_1} and so can be continued to reach s_2. Using the claim for $I=[s_0,s_2]$ we see that the previous estimate holds with s_1 replaced by s_2 and so on. After a finite number of steps it is clear that $\varphi(t,s,y)$ is defined in the whole of $[a,b]$. This shows that D is open.

Finally, if $t,t'\in(a,b),(s,y),(s',y')\in K$,

$$\begin{aligned}&|\varphi(t,s,y)-\varphi(t',s',y')|\\&\quad\le|\varphi(t,s,y)-\varphi(t,s',y')|+|\varphi(t,s',y')-\varphi(t',s',y')|.\end{aligned}$$

We have seen already in the claim that the first term on the right is $O(|s-s'|+|y-y'|)$. Evidently, by (8.3) the second one is $O(|t-t'|)$. Altogether, locally

$$|\varphi(t,s,y)-\varphi(t',s',y')|=O(|t-t'|+|s-s'|+|y-y'|).\qquad\square$$

The map φ is called the *flux* of the ODE.

8.3.3 Next, we study the differentiability properties of φ. For convenience we reintroduce the parameters

$$x' = f(t, x, \lambda),$$

f being a continuous function in $\Omega \subset \mathbb{R} \times \mathbb{R}^n \times \mathbb{R}^k$. Assimilating parameters to initial conditions, we have seen that if f is locally Lipschitz in x, for each $(s, y, \lambda) \in \Omega$ there is a unique maximal solution $\varphi_{(s,y,\lambda)}$, of the Cauchy problem

$$x' = f(t, x, \lambda), \quad x(s) = y,$$

defined in an open interval $I(s, y, \lambda)$. Moreover,

$$D = \{(t, s, y, \lambda) : (s, y, \lambda) \in \Omega, t \in I(s, y, \lambda)\},$$

is open in $\mathbb{R}^{2+n+k}$ and the map $\varphi : D \longrightarrow \mathbb{R}^n$ defined by

$$\varphi(t, s, y, \lambda) = \varphi_{(s,y,\lambda)}(t)$$

is continuous in D.

Theorem 8.5. *If f is differentiable in x and $d_x f(t, x, \lambda)$ is continuous in Ω, then φ is of class C^1 in y. For $i = 1, \ldots, n$, $\frac{\partial \varphi}{\partial y_i}(t, s, y, \lambda)$ is the solution of the Cauchy problem for the linear system*

$$z' = (d_x f)(t, \varphi(t, s, y, \lambda), \lambda) z, \quad z(s) = e_i, \tag{8.6}$$

where e_i denotes the ith vector in the canonical basis.

So the equation is the same for each derivative, the initial condition changes with i. Setting

$$J(t) = (d_x f)(t, \varphi(t, s, y, \lambda), \lambda),$$

the n Cauchy problems together define a matrix-valued Cauchy problem

$$\Sigma'(t) = J(t)\Sigma(t), \quad \Sigma(s) = Id.$$

These equations are named the *variational equations* or *equations of first variation.*

Proof. We introduce the matrix-valued auxiliary function $F(t, x_1, x_2, \lambda)$

$$F(t, x_1, x_2, \lambda) = \int_0^1 (d_x f)(t, \tau x_2 + (1-\tau)x_1, \lambda) d\tau.$$

It is continuous and

$$F(t, x, x, \lambda) = d_x f(t, x, \lambda), f(t, x_2, \lambda) - f(t, x_1, \lambda) = F(t, x_1, x_2, \lambda)(x_2 - x_1).$$

Given $i \in \{1, \ldots, n\}$ we consider for small h

$$w_h(t) = \varphi(t, s, y + he_i, \lambda), x_h(t) = \frac{w_h(t) - w_0(t)}{h}.$$

We must prove that $\lim_{h \to 0} x_h(t)$ exists. First we show that $x_h(t)$ satisfies a linear ODE. Indeed,

$$\begin{aligned} x_h'(t) &= \frac{w_h'(t) - w_0'(t)}{h} = \frac{f(t, w_h(t), \lambda) - f(t, w_0(t), \lambda)}{h} \\ &= \frac{F(t, w_0(t), w_h(t), \lambda)(w_h(t) - w_0(t))}{h} = F(t, w_0(t), w_h(t), \lambda) x_h(t). \end{aligned}$$

On the other hand, $x_h(s) = \frac{w_h(s) - w_0(s)}{h} = e_i$. Thus, $x_h(t)$ is the solution of the Cauchy problem

$$z' = F(t, w_0(t), w_h(t), \lambda) z, \quad z(s) = e_i.$$

Note that this family of linear Cauchy problems is continuous in the parameter h. By Theorem 8.4, $\lim_{h \to 0} x_h(t)$ exists and equals the solution of the Cauchy problem obtained letting $h \to 0$, which is (8.6). By hypothesis, this family of linear Cauchy problems depends continuously on the paràmetres s, y, λ, whence Theorem 8.4 finishes the proof. □

Theorem 8.6. *Under the same hypothesis, if f is differentiable in λ and $d_\lambda f(t, x, \lambda)$ is continuous, then φ is of class C^1 in λ. For $j \in \{1, \ldots, k\}$,*

$$\frac{\partial \varphi}{\partial \lambda_j}(t, s, y, \lambda)$$

is the solution of the Cauchy problem

$$z' = (d_x f)(t, \varphi(t, s, y, \lambda), \lambda) z + d_\lambda f(t, \varphi(t, s, y, \lambda), \lambda) e_j, \quad z(s) = 0.$$

Proof. Set $w = (x, \lambda)$ and consider

$$F : \Omega \longrightarrow \mathbb{R}^n \times \mathbb{R}^k, \quad F(t, w) = (f(t, w), 0).$$

It is differentiable in the variable w, so by Theorem 8.5 the solutions of the Cauchy problem are of class C^1 in w, too. But its solutions are

$$\Psi(t, s, x, \lambda) = (\varphi(t, s, x, \lambda), \lambda).$$

Therefore, $\varphi(t, s, x, \lambda)$ is of class C^1 in λ. Also by Theorem 8.5, $\frac{\partial \varphi}{\partial \lambda_j}(t, s, x, \lambda)$ is a solution of the Cauchy problem

$$z' = d_{x,\lambda}F(t, \varphi(t, s, x, \lambda))z, \quad z(s) = (0, \ldots, 0, e_j).$$

Note that

$$d_{x,\lambda}F = \begin{pmatrix} d_x f & d_\lambda f \\ 0 & 0 \end{pmatrix}, \quad \frac{\partial \varphi}{\partial \lambda_j} = \begin{pmatrix} \frac{\partial \varphi}{\partial \lambda_j} \\ e_j \end{pmatrix},$$

so the result follows. □

Exercise 8.2. In the situation of Theorem 8.5, with f of class C^1, prove that φ is differentiable in s. Which equation and initial condition satisfies $\frac{\partial \varphi}{\partial s}$?

8.4 The Geometric Point of View of Autonomous ODEs

8.4.1 We come back to equation (8.1) for a C^1 field X on the domain $U \subset \mathbb{R}^n$. All results of the previous section apply. For $p \in U$, we denote by $\gamma_p(t)$ the solution of the Cauchy problem

$$\gamma'(t) = X(\gamma(t)), \quad \gamma(0) = p, \tag{8.7}$$

which we assume defined in its maximal interval. It is called the *integral arc of X through p*. It is customary to use the notation $\gamma_t(p) = \gamma(t, p)$, too. By Theorem 8.5, on the smooth dependence on the initial condition, γ_t is C^1 in p.

Note that fixed $s \in (\alpha, \beta)$, $\tau(t) = \gamma(s + t)$ is a solution of the system with $\tau(0) = \gamma_s(p)$, so by unicity $\gamma(t, \gamma_s(p)) = \gamma(t+s, p)$ for $t \in (\alpha - s, \beta - s)$. In other words, one has

$$\gamma_t \circ \gamma_s = \gamma_{t+s}, \quad \gamma_0 = Id,$$

and in particular γ_t, γ_{-t} are inverse maps. Thus, γ_t is a diffeomorphism in its domain of definition. The family γ_t is called accordingly a *local uniparametric group*, sometimes also denoted *exponential map* e^{tX} for obvious reasons.

We reformulate Theorem 8.2 in a more geometric language, where now we make the basic assumption that $X(x) \neq 0, x \in U$. Since $X(x) \neq 0$, the integral paths γ^* of X are 1-sub-manifolds, C^1 curves in our notation, and $X(x)$ is tangent at every point. Conversely, if Γ is a regular curve such that $X(x) \in T_x(\Gamma)$ for all $x \in \Gamma$, a parametrization $\gamma(t)$ of Γ will satisfy

$$\gamma'(t) = \lambda(t)X(\gamma(t))$$

for some $\lambda(t) \neq 0$; by a change of parametrization we can get $\lambda = 1$. Thus, we can omit the parametrization and phrase the situation as follows: we are given a line $L(x)$ through each $x \in U$ in a smooth way, the one spanned by $X(x)$; an *integral curve* Γ, is one satisfying $T_x(\Gamma) = L(x), x \in \Gamma$. In cartesian coordinates, $X = (A_1, \ldots, A_n), \gamma$ with components $x_i(t)$, one writes

$$\frac{dx_1}{A_1(x)} = \frac{dx_2}{A_1(x)} = \cdots = \frac{dx_n}{A_n(x)}, \tag{8.8}$$

where it is understood that one seeks for curves Γ for which these equalities among 1-forms hold when restricted to Γ, and $dx_i = 0$ when $A_i = 0$. From this point of view, we emphasize the path followed by a solution rather than the specific parametrization satisfying (8.7).

We will now combine Theorems 8.5 and 8.6 with the inverse and implicit function theorems to understand the structure of integral curves. To begin with, more generally, we may ask for higher-dimensional *integral sub-manifolds* M, meaning $L(x) \subset T_x(M), x \in M$; intuitively, a sub-manifold *generated* by integral curves will have this property.

Theorem 8.7. *If N is a regular sub-manifold of dimension $m \leq n-1$ transverse to X at every point, that is $X(x) \notin T_x(N), x \in N$, there is a unique integral sub-manifold M of dimension $m+1$ containing N.*

Proof. It is enough to prove the statement locally, around a fixed $p \in N$. Let Φ be a local chart at p. We consider the integral curves of X through points $q \in N$ close to p, that is the solution $\gamma(t, q)$ of

$$\gamma'(t) = X(\gamma(t)), \quad \gamma(0) = q.$$

If $q = \Phi(u)$, by Theorem 8.5, the function $\Psi(u, t) = \gamma(t, \Phi(u))$ is C^1, $\Psi(0, 0) = p$. Since $X(p)$ is transversal to $T_p(N)$, $d\Psi(0, 0)$ has rank $m+1$, and Ψ is a chart of the desired M. □

Example 8.8. With coordinates x, y, z, let $X = (4x - y + z, x + 2y - 5z, 3x - 3y)$ for $(x, y, z) \neq (0, 0, 0)$, and N the curve $z = x, 2y = x$, with parametrization $x = 2u, y = u, z = 2u$. The tangent to M is $(2, 1, 2)$ while $X(2u, u, 2u) = (9u, -6u, 0) = u(9, -6, 0)$, so the transversality condition is fulfilled. With $D = \frac{d}{dt}$, the autonomous system is written

$$(D - 4)x + y - z = 0, \quad -x + (D - 2)y + 5z = 0, \quad -3x + 3y + Dz = 0.$$

Triangulation of this system leads to the general solution

$$x = Ae^{6t} + Ce^{3t}, \quad y = -Ae^{6t} + Be^{-3t} + Ce^{3t}, \quad z = Ae^{6t} + Be^{-3t}, \quad (8.9)$$

equivalent to

$$3Ae^{6t} = x - y + z, \quad 3Be^{-3t} = 2z - x + y, \quad 3Ce^{3t} = 2x + y - z. \quad (8.10)$$

Now we must find the integral curve starting when $t = 0$ at $(x, y, z) = (2u, u, 2u)$. Solving for A, B, C gives $A = B = C = u$ and thus (8.9) with $A = B = C = u$ is the parametrization of M. Eliminating u, t in (8.10) we find the equation of M in continuous form

$$(2x + y - z)^3 - (2z - x + y)(x - y + z)^2 = 0.$$

8.4.2 In case $m = n - 1$, Φ is a local diffeomorphism, by the inverse function theorem. The inverse map defines local coordinates $u_1, \ldots, u_{n-1}, u_n$ around p such that the integral curves of X are given by $u_i = c_i, i = 1, \ldots, n - 1$ and are parametrized by u_n. In fact, by construction,

$$X = \partial_n.$$

Thus, the following theorem is proved:

Theorem 8.8. *If X is a non-vanishing C^1 vector field in U and $p \in U$, there are local coordinates $u_1, \ldots, u_n$ around p such that $X = \partial_n$ and the integral paths of X are the curves $u_1 = c_1, \ldots, u_{n-1} = c_{n-1}$.*

In Theorem 5.1, we saw that a gradient vector field X of class C^1 satisfies the condition $D_iA_j = D_jA_i$. Theorem 8.8 states that coordinate vector fields, unless gradient vector fields, have nothing special, since a general non-vanishing vector field is locally one of those.

8.4.3 We consider now *two* vector fields and ask whether they form part of the same coordinate system.

Theorem 8.9. *Given two non-vanishing vector fields X, Y, there exists a local system of coordinates $u_1, \ldots, u_n$ such that*

$$X = \partial_1, \quad Y = \partial_2,$$

if and only if X, Y commute: $X(Yf) = Y(Xf)$, that is, $[X, Y] = 0$.

Proof. The condition is obviously necessary by Schwarz's rule, Theorem 5.1. For the converse we may assume $n = 2$, and by Theorem 8.8 we may work with coordinates x, y centered at p such that $X = \frac{\partial}{\partial x}$. Then the assumption means that Y does not depend on x, that is, $Y = (A(y), B(y))$. We take as N in Theorem 8.8 the integral curve $\tau(s)$ of Y through $(0, 0)$, so $\gamma(t, s) = t + \tau(s)$ and indeed

$$\frac{\partial \gamma}{\partial s} = \tau'(s) = Y(\tau(s)) = Y(\gamma(t, s)).$$ □

The above proof shows too that the map $\gamma(t, s)$ above does not depend on the order in which we consider the integral curves of X, Y, that is, one may start with the integral curve $\gamma(t)$ of X through $(0, 0)$ and then the integral curve $\tau(s, t)$ of Y through $\gamma(t)$. As with one vector field, sometimes it is denoted e^{tX+sY}, the *exponential map* of X, Y.

In the same way, a family $X_1, X_2, \ldots, X_k$ of vector fields are coordinate fields of a single coordinate system if and only if $[X_i, X_j] = 0$. In this case, there is a well-defined exponential map $e^{\sum_i t_i X_i}$ parametrizing a k-dimensional sub-manifold whose tangent space is spanned by X_i. However, this latter fact, the existence of such sub-manifold, is a less restrictive condition, see Theorem 9.2.

Chapter 9

Linear Partial Differential Equations

Solving some elementary partial differential equations, as a converse operation of partial differentiation, has already been considered in Section 4.7. In this chapter, we provide a systematic treatment of some types of linear equations, always paying attention to their geometric interpretation. In particular, we carefully discuss Frobenius' theorem on the existence of sub-manifolds with given tangent spaces. Next, second-order constant coefficient equations and their classification are presented. The elliptic ones, particularly Laplace equation, is then used to open one of the many entrance doors to *complex analysis*. We relate complex analysis with *complex power series* and *complex-analytic functions* along the lines of Section 5.5.

9.1 First Integrals of Ordinary Differential Equations

9.1.1 In the following sections, we use the inverse and implicit function theorems, combined with Theorem 8.5, to study solutions of first-order partial differential equations in a domain U. All our considerations will be of local nature, although the examples considered will be global.

Definition 9.1. A *first integral* of a vector field X is a C^1 function F on U, which is constant on every integral curve of X.

So, F is a first integral iff the level hypersurfaces $F = c$ are tangent to X at every point; with every point p they contain the integral curve through p, we call them *integral hypersurfaces*. For exact equations in $n = 2$,

$$A(x, y)\, dx + B(x, y)\, dy = 0,$$

a first integral and a potential function are the same thing.

Since this simply means that $\langle \nabla F, \gamma'(t) \rangle = 0$ whenever $\gamma'(t)$ has the direction of X, F is a first integral if and only if, with $X = (A_1, \ldots, A_n)$,

$$XF = \langle \nabla F, X \rangle = \sum_{i=1}^{n} A_i(x) D_i F(x) = 0. \tag{9.1}$$

In this context, the integral curves of X are called *characteristic* curves of the equation (9.1).

Theorem 8.8 proves the local existence of $n-1$ functionally independent first integrals $u_1, \ldots, u_{n-1}$. By setting $u_i = c_i$ one has the general integral curve of X. If F is a first integral, then ∇F is a linear combination of the ∇u_i and by Corollary 6.2, F depends functionally on $u_1, \ldots, u_{n-1}$, $F = \Phi(u_1, \ldots, u_{n-1})$. This is therefore the general solution of (9.1).

Example 9.1. In Example 8.8, for a general integral curve as described in (8.9) and (8.10), we see that $(2x + y - z)(2z - x + y) = CB$ and $(2z - x + y)^2(x - y + z) = B^2 A$ are constants on each integral curve. These are two functionally independent first integrals.

9.1.2 We can as well solve the non-homogeneous linear equation

$$\sum_{i=1}^{n} A_i(x) D_i f(x) = B(x), \tag{9.2}$$

in short $Xf = B$, for B given in U. The main point is that if f is a solution of (9.2) and γ a characteristic, then $h(t) = f(\gamma(t))$ satisfies

$$h'(t) = B(\gamma(t)). \tag{9.3}$$

That is, a solution f must satisfy an ordinary differential equation along every characteristic, and so it is completely determined by its value at one point. If $u_1, \ldots, u_{n-1}$ are as before, fixed an integral curve, that is fixed $c_1, \ldots, c_{n-1}$, if $\gamma(t), \gamma(0) = p$ is a parametrization, a particular solution is

$$f_0(\gamma(t)) = h(t) = \int_0^t B(\gamma(s))\, ds.$$

Of course, the general solution is obtained by adding the general solution of the homogeneous equation,

$$f = f_0 + \Phi(u_1, u_2),$$

with Φ an arbitrary function.

Example 9.2. Let us find the general solution in $\mathbb{R}^3$ of

$$xf_x + f_y + 2yf_z = x,$$

and check that there is a unique solution f satisfying the additional condition $f(x, 0, z) = z$. The integral curves are the solutions of

$$x' = x, \quad y' = 1, \quad z' = 2y,$$

that is

$$x(t) = ae^t, \quad y = t + b, \quad z = (t + b)^2 + c.$$

Eliminating t, we see that

$$u(x, y, z) = xe^{-y} = c_1, \quad v(x, y, z) = z - y^2 = c_2,$$

is the general integral curve, parametrized $x = c_1e^t, y = t, z = t^2 + c_2$. If $h(t) = f(c_1e^t, t, t^2 + c_2)$, the equation reads

$$h'(t) = c_1e^t f_x + f_y + 2tf_z = xf_x + f_y + 2yf_z = x = c_1e^t,$$

implying that $h(t) = c_1e^t + \Phi(c_1, c_2)$ and therefore

$$f(x, y, z) = x + \Phi(xe^{-y}, z - y^2)$$

is the general solution. Imposing $f(x, 0, z) = z$ gives $x + \Phi(x, z) = z$, that is $\Phi(u, v) = v - u$, so that $x + z - y^2 - xe^y$ is the unique solution. To check unicity from another point of view, note that each integral curve meets $y = 0$ only at the point with $t = 0, (c_1, 0, c_2)$, therefore,

$$h(t) = h(0) + \int_0^t h'(s)\, ds = c_2 + \int_0^t c_1e^s\, ds = c_2 + c_1(e^t - 1)$$

$$= z - y^2 + x - xe^{-y}.$$

From this point of view, note that the plane $y = 0$ where the solution is prefixed may be replaced by an arbitrary surface meeting each integral curve at just one point.

An equivalent procedure is to work in the coordinate system $u_1, \ldots, u_n$.

Example 9.3. Consider again Example 9.2, for which the characteristics are given by

$$u(x, y, z) = xe^{-y} = c_1, \quad v(x, y, z) = z - y^2 = c_2.$$

Choose $w = w(x, y, z)$ such that u, v, w is a coordinate system, say $w = y$. Then $x = ue^w, y = w, z = v + w^2$

$$\frac{\partial}{\partial w} = ue^w \frac{\partial}{\partial x} + \frac{\partial}{\partial y} + 2w \frac{\partial}{\partial z},$$

and the equation becomes

$$\frac{\partial f}{\partial w} = x = ue^w,$$

with general solution

$$f = ue^w + \Phi(u, v) = x + \Phi(xe^{-y}, z - y^2).$$

9.1.3 In the previous example, we have seen how to get first integrals from the general solution of the autonomous system. In practice, one may proceed the other way around and try to find by inspection $n - 1$ functionally independent first integrals.

In case $n = 2$, where we write (8.8) in the form

$$\omega = M(x, y)\, dx + N(x, y)\, dy = 0,$$

there is a particular comment to be done. We know that there always exists a first integral F so that the level curves $F = c$ are the integral curves. The integral curve through p is $F = F(p)$. Implicit functions $y = y(x)$ defined by these equations are solutions of $\frac{dy}{dx} = -\frac{M}{N}$. Since ∇F is orthogonal to level curves, we must have $\nabla F = \lambda X$ for some function λ,

$$F_x = \lambda(x, y)M(x, y), \quad F_y = \lambda(x, y)N(x, y),$$

or $\lambda\omega = dF$, that is, λ is an integrating factor. Thus, in dimension two integrating factors always exist. A way to solve equations in dimension two is to find an integrating factor and then find the potential function (first integral) by partial antidifferentiation.

Example 9.4. The equation $y(e^{xy} + y)\, dx + (y + x(e^{xy} + 2y))\, dy = 0$ is exact. Solving $F_x = y(e^{xy} + y)$ gives $F = e^{xy} + xy^2 + C(y)$, with $y + x(e^{xy} + 2y) = F_y = xe^{xy} + 2xy + C'(y)$, so $C' = y$ and the general solution is $e^{xy} + (x + \frac{1}{2})y^2 = c$.

Example 9.5. The equation $y(2+xy)\,dx + x(1+xy)\,dy = 0$ is not exact. Seeking for an integrating factor of the form $\lambda = x^a y^b$ we find that $\lambda = 1/(xy)$ works. Now $\frac{1}{x}(2+xy)\,dx + \frac{1}{y}(1+xy)\,dy = 0$ is exact. From $F_x = \frac{2}{x} + y$ we get $F = 2\log x + xy + C(y)$, with $x + C'(y) = F_y = \frac{1}{y} + x$, so $\log(x^2 y) + xy = c$ is the general solution. Near points where $xy \neq -1$, there is a unique solution of the equation

$$\frac{dy}{dx} = -\frac{2y + xy^2}{x + x^2 y}.$$

9.1.4 In higher dimension, to find first integrals for (8.8) some comments are in order. First, if two of the variables are separated, that is, A_i, A_j depend solely on x_i, x_j, one may deal with these, find an integrating factor and a first integral depending only on x_i, x_j. A second method consists in identifying by inspection solutions M_i of $\sum_i M_i A_i = 0$, called classically *multipliers*, and F with $\nabla F = (M_1, \ldots, M_n)$.

A general fact is that once a first integral F has been found, it is possible to reduce the order of the system; indeed, if $u_1, \ldots, u_n$ is a coordinate system with $u_1 = F$ and

$$\frac{du_1}{B_1(u)} = \frac{du_2}{B_2(u)} = \cdots = \frac{du_n}{B_n(u)}$$

is the expression of the system in the new coordinates, since u_1 is constant on integral curves, necessarily $B_1 = 0$ and we work with $u_2, \ldots, u_n$ treating u_1 as constant.

Example 9.6. Consider the system in $\mathbb{R}^3$

$$\frac{dx}{y-z} = \frac{dy}{z-x} = \frac{dz}{x-y}.$$

We find by inspection that $(1,1,1)$ and (x,y,z) are multipliers with first integral $x+y+z, x^2+y^2+z^2$. These are two functionally independent first integrals outside the origin, the integral curves are circles $x+y+z = a, x^2+y^2+z^2 = r^2$. At this point, note that we are not establishing the parametrization $x(t), y(t), z(t)$ such that

$$x' = y - z, \quad y' = z - x, \quad z' = x - y.$$

If v_1, v_2 are unit vectors that together with $\frac{1}{\sqrt{3}}(1,1,1)$ form an orthonormal basis, the parametrization

$$(x(t), y(t), z(t)) = (a,a,a) + r\cos\tau t v_1 + r\sin\tau t v_2$$

works only if $\tau = \pm\sqrt{3}$.

Example 9.7. The system

$$\frac{dx}{-2z(x+y)} = \frac{dy}{2z(x+y)} = dz$$

admits the multipliers $(1,1,0)$ with first integral $u = x + y$. In the new variables $u = x + y, v = x, w = z$ the system becomes $du = 0, dv = -2wu\,dw$, whence $u = c_1, dv = -2c_1 w\,dw$. Thus, the general solution is given by $u = c_1, v + c_1 w^2 = c_2$, $x + y$ and $x + z^2(x + y)$ are two first integrals.

9.1.5 Assume N of dimension $n - 2$ in Theorem 8.7 is given by two equations $f(x) = 0, g(x) = 0$ and that $n - 1$ functionally independent first integrals $u_1, \ldots, u_{n-1}$ are known. To find the equation of the integral submanifold M through N we must obtain a relation among $c_1, \ldots, c_{n-1}$, say $\Phi(c_1, \ldots, c_{n-1}) = 0$, from the equations

$$f(x) = 0, \quad g(x) = 0, \quad u_1(x) = c_1, \ldots, \quad u_{n-1}(x) = c_{n-1},$$

and then M is given by $\Phi(u_1, \ldots, u_{n-1}) = 0$.

Example 9.8. To find the integral surface for $X = (y - z, z - x, x - y)$ of example 9.6 containing the line $z = 0, x + 2y = 1$, we start from

$$x + y + z = a, \quad x^2 + y^2 + z^2 = b, \quad z = 0, \quad x + y = 1,$$

and eliminate x, y, z. First $x + y = a, x^2 + y^2 = b, x + 2y = 1$, then $y = 1 - a, x = 2a - 1$, so $(2a - 1)^2 + (a - 1)^2 = b$, and

$$5(x + y + z)^2 - 6(x + y + z) + 2 = x^2 + y^2 + z^2,$$

is the required solution.

9.2 The Linear and Quasi-Linear First-Order Equation

9.2.1 Let us analyze the analytical meaning of (9.1) in terms of the implicit functions defined by F. It will be convenient to change a bit the notation, moving to $n+1$ variables $(x, z) \in \mathbb{R}^n \times \mathbb{R}, x = (x_1, \ldots, x_n)$, writing $X = (A_1, \ldots, A_n, B)$ and the equation in the form

$$\sum_{i=1}^{n} A_i(x, z)\frac{\partial F}{\partial x_i} + B(x, z)\frac{\partial F}{\partial z} = 0.$$

Assume that $F = 0$ is parametrized by x, that is, it defines an implicit function $z = f(x)$. We find that f satisfies the so-called *quasi-linear equation* (linear in the first-derivatives)

$$\sum_{i=1}^{n} A_i(x, f(x))D_i f(x) = B(x, f(x)). \tag{9.4}$$

Let N' be an $n-1$ sub-manifold in $\mathbb{R}^n$, say parametrized by $x = \Phi(u)$, and $\varphi(u)$ defined on N'. The *Cauchy problem* for (9.4) consists in finding a solution f of (9.4) taking prescribed values on N': $f(\Phi(u)) = \varphi(u)$.

In the following, we assume $n = 2$, N' is a curve in the plane, and use the notation $\Phi(u) = (x(u), y(u)), u \in \mathbb{R}$. Assume f solves this problem and set $p(u) = D_1 f(\Phi(u)), q(u) = D_2 f(\Phi(u)), a_i(u) = A_i(\Phi(u), \phi(u)), b(u) = B(\Phi(u))$. Functions x, y, a_i, b just depend on the data. From the equation, one has

$$a_1(u)p(u) + a_2(u)q(u) = b(u),$$

and differentiating $f(x(u), y(u)) = \varphi(u)$

$$x'(u)p(u) + y'(u)q(u) = \varphi'(u).$$

Thus,

$$(a_2 x' - a_1 y')p = a_2\varphi' - by', \quad (a_2 x' - a_1 y')q = bx' - a_1\varphi'.$$

To discuss the situation, it is useful to have in mind the graph N of φ over N', with parametrization $(x(u), y(u), \varphi(u))$. There are three cases to consider.

If $a_2 x' - a_1 y' \neq 0$, then X is transverse to N at every point, we can apply Theorem 8.7 and there is a unique integral surface containing N, with parametrization $x(u,t), y(u,t), z(u,t)$ describing for fixed u the characteristic through the point $(x(u), y(u), \varphi(u))$. Now, since

$$\left.\frac{\partial(x, y)}{\partial(u, t)}\right|_{t=0} = a_2 x' - a_1 y' \neq 0,$$

this integral surface is indeed a graphic $z = f(x, y)$ near N. Thus, we have the following uniqueness and existence result for the Cauchy problem, that we state for general dimension:

Theorem 9.1. *Let N' be an $n-1$-sub-manifold in $\mathbb{R}^n$ with parametrization $x = \Phi(u)$ and φ defined on N'. If the vector $v(x) = (A_1(x, \varphi(x)), \ldots, A_n(x, \varphi(x)))$ is transversal to $T_x(N')$ for $x \in N'$ (that is, the matrix with column vectors $v(\Phi(u)), D_i\Phi(u), i = 1, \ldots, n-1$ has non-zero determinant for all u) the Cauchy problem has a unique solution defined near N'.*

If $a_2x' - a_1y'$ is identically zero, for the Cauchy problem to have a solution it is necessary that

$$a_2\varphi' - by' = 0, \quad bx' - a_1\varphi' = 0.$$

These compatibility conditions are classically called *transport equations.* This means that N is characteristic. In this case, there are infinite integral surfaces through N and so the Cauchy problem has infinite solutions. This can happen only if N' is a *base-characteristic curve*, the projection of a characteristic,

A third case that might occur is that $a_2x'-a_1y', a_2\varphi'-by' = 0, bx'-a_1\varphi'$ are zero at some singular points, typically in finite number. Outside these points the Cauchy problem has a unique solution f, and in case f extends in a C^1 way to these points, f is the unique solution, too.

Example 9.9. Consider the equation

$$(y-z)p + (z-x)q = x - y,$$

for the unknown function $z = f(x,y), p = f_x, q = f_y$. The system of characteristics is the one in Example 9.6, and the general integral surface is given by $\Phi(x+y+z, x^2+y^2+z^2) = 0$.

Let N' be the hyperbola $xy = 1$, parametrized $x = u, y = \frac{1}{u}$, and consider the Cauchy problem with data $\varphi(u) = 0$. In the above notations, $a_1(u) = \frac{1}{u}, a_2(u) = -u$, so that

$$a_2x' - a_1y' = u^{-3} - u.$$

At $u = \pm 1$ there are singular points. However, at these points the transport equations are satisfied. This means that the compatibility conditions hold everywhere and there is uniqueness outside $(1,1), (-1,-1)$. To find this unique solution we must seek for the integral surface through $xy = 1, z = 0$, that is, eliminate x, y, z from

$$x + y + z = a, \quad x^2 + y^2 + z^2 = b, \quad xy = 1, \quad z = 0.$$

We find $a^2 = (x+y)^2 = x^2 + y^2 + 2xy = b + 2$, so the integral surface is $(x+y+z)^2 = x^2 + y^2 + z^2 + 2$, or $xy + yz + zx = 1$. This defines

$$h(x,y) = \frac{1 - xy}{x + y}.$$

Since this is regular near the hyperbola, this is the unique solution.

For the same equation, we consider now a base-characteristic curve N', say the projection of $x + y + z = 0, x^2 + y^2 + z^2 = 2$, given by $x^2 + y^2 + xy = 1$. If we prescribe the value $-x - y$ on N', so that N is characteristic, the solution is not unique, both $-x - y$ and $\sqrt{2 - x^2 - y^2}$ are solutions. But if we prescribe the zero value on N', the compatibility condition $a_2x' - a_1y' \neq 0$ holds, To obtain the unique solution, we must eliminate a, b from

$$x + y + z = a, \quad x^2 + y^2 + z^2 = b, \quad x^2 + y^2 + xy = 1, \quad z = 0.$$

One gets $a^2 = b + 2$, so $x^2 + y^2 + z^2 + xy + yx + xz = 1$ is the unique integral surface containing N and

$$h(x, y) = \frac{1}{2}(-x - y + \sqrt{4 - 2xy - 3x^3 - 3y^2})$$

is the unique solution of the Cauchy problem with zero value on N'.

9.2.2 Of course, it is possible to solve the Cauchy problem solving first the autonomous system, as shown by the following example.

Example 9.10. We consider the Cauchy problem for

$$(x - y)p + (y - x - z)q = z,$$

with value 1 on $x = y$; one has here $a_2x' - a_1y' = 1$, so there is a unique solution. The system governed by X is

$$x' = x - y, \quad y' = y - x - z, \quad z' = z,$$

whose general solution is $z = Ae^t, x = C - Ae^t + Be^{2t}, y = C - Be^{2t}$. Imposing that at $t = 0$ starts at $(u, u, 0)$ and we obtain M in parametrized form

$$x(u, t) = -e^t + \frac{1}{2}e^{2t} + u + \frac{1}{2}, \quad y(u, t) = -\frac{1}{2}e^{2t} + u + \frac{1}{2}, \quad z(u, t) = e^t.$$

Eliminating u, t we find M in closed form, $z^2 + y = x + z$, and so

$$z = h(x, y) = \frac{1}{2}(1 + \sqrt{1 - 4(y - x)^2}),$$

is the solution.

9.2.3 As noted before, the compatibility condition depends both on N' and the data φ on N'. For the same N' one might have existence and uniqueness for some date φ, and non-existence for other data φ. This is not so for the *semi-linear* equation

$$\sum_{i=1}^{n} A_i(x)\frac{\partial f}{\partial x_i}(x) = B(x, f(x)),$$

in which the A_i just depend on x. The transversality assumption in Theorem 9.1 depends only on N' and not on the data φ. For these N', for all data φ the Cauchy problem has a unique solution. The base-characteristic curves are the solutions of the system

$$\frac{dx_1}{A_1(x)} = \cdots = \frac{dx_n}{A_n(x)},$$

and the transversality condition is simply the statement that they meet N' transversally. An alternative way to look at this situation is explained in paragraph 4.7.4. If $\tau(t)$ is a base-characteristic curve and f is a solution of the equation, then $g(t) = f(\tau(t))$ satisfies an ordinary differential equation

$$g'(t) = \sum D_i f(\tau(t))A_i(\tau(t)) = B(\tau(t), g(t)).$$

As such, g is completely determined by its value at one point of N', and this explains unicity.

The solutions considered here are generally speaking of local character. When considering global solutions other compatibility conditions play a role. Assume $n = 2$ and assume that the base-characteristic curves are horizontal lines $y = c$. If the curve N' is transversal but meets each horizontal line at more than one point, the value of the data at these points cannot be arbitrary.

9.3 Pfaff Systems: Frobenius' Theorem

9.3.1 Theorem 8.2 can be rephrased as stating that given a line $L(x)$ through each $x \in U$ in a smooth way, there exists a unique *integral curve* Γ, that is $T_x(\Gamma) = L(x), x \in \Gamma$, through each point $p \in U$.

More generally, assume we are given a k-dimensional linear space $L(x)$ for each $x \in U$ in a smooth way. A k-dimensional regular sub-manifold M is called an *integral sub-manifold* if $T_x(M) = L(x)$. We ask about an existence and uniqueness result in this context, assuming of course $k \geq 2$.

One way to give the data $L(x), x \in U$ is by means of $n-k$ linearly independent 1-forms $\omega_i = \sum_j A_{ij}(x)\,dx_j$ with $A_{ij} \in C^1(U)$, and define $L(x)$ as the incident space

$$L(x) = \{v \in \mathbb{R}^n : \omega_i(x)(v) = 0, i = 1, \ldots, n-k\}.$$

The integral sub-manifolds are those for which $\omega_i = 0$ on M. Alternatively, we may think of the vector fields $X_i = (A_{i1}, \ldots, A_{in})$ and define $L(x)$ as the orthogonal of the linear span of $X_i(x)$. Still, the distribution $L(x)$ can be given by m linearly independent m vector fields spanning $L(x)$ at each point.

We will deal in detail with the co-dimension-one case, hypersurfaces. For notational convenience we will work in $\mathbb{R}^{n+1}$ with coordinates $x' = (x_1, \ldots, x_n, z)$ and $\omega = \sum_j A_j(x,z)\,dx_j + B(x,z)\,dz \neq 0$, to stress the analytic counterpart of this geometrical problem that arises when (without loss of generality) we consider hyper-surfaces defined by graphs $z = u(x)$ of a C^1-function, in which case $\omega = 0$ on M means

$$A_j(x, u(x)) + B(x, u(x))\frac{\partial u}{\partial x_j}(x) = 0.$$

Thus, assuming $B \neq 0$, with $F_j = -\frac{A_j}{B}$, an equivalent way of posing the problem is as follows: we are given C^1 functions $F_j(x,z)$ on U and seek for $u(x)$ of class C^1 such that

$$D_j u(x) = F_j(x, u(x)), \quad j = 1, \ldots, n, \tag{9.5}$$

or simply $du - \sum_j F_j(x,u)dx_j = 0$. This is a direct generalization of (8.2). We will deal simultaneously with the two points of view, geometric and analytic. For existence and uniqueness, we start with the latter.

If $n > 1$, the system may have no solutions at all, for instance $u_x = 0, u_y = x$ has no solutions. If u is a solution, $p \in U$ and v is a direction vector, then $h(t) = u(p+tv)$ satisfies an ordinary differential equation

$$h'(t) = D_v u(p+tv) = \sum_j v_j D_j u(p+tv) = \sum_j v_j F_j(p+tv, h(t)),$$

and as such is determined by $h(0) = f(p)$. Thus, we see that the system (9.5) has at most one solution u with $u(p) = u_0$ given; equivalently, there is at most one integral hyper-surface through each point (p, u_0). If there is a (unique) integral hyper-surface through each (p, u_0), the system (9.5) is called *completely integrable.*

Assume u is a solution. Then u is twice differentiable, therefore $D_{ij}u = D_{ji}u$, that is $D_i(F_j(x, u(x))) = D_j(F_i(x, u(x)))$, or using the chain rule

$$\frac{\partial F_j}{\partial x_i}(x, u(x)) + \frac{\partial F_j}{\partial u} F_i(x, u(x)) = \frac{\partial F_i}{\partial x_j}(x, u(x)) + \frac{\partial F_i}{\partial u} F_j(x, u(x)).$$

If the system is completely integrable, $(x, u(x))$ is an arbitrary point of U, whence

$$D_i F_j + F_i D_u F_j = D_j F_i + F_j D_u F_i \tag{9.6}$$

identically in U. The converse is the *Frobenius' theorem.*

Theorem 9.2. *The system* (9.5) *is completely integrable if and only if* (9.6) *holds identically in* U.

Proof. We assume (9.6) and will prove that for every $(p, z_0) \in U$ there is a unique solution of (9.5) such that $u(p) = z_0$, using induction on n. When $n = 1$, this is Theorem 8.2. Assuming that the result holds for $n - 1$, let $g(x_2, \ldots, x_n) = g(x')$ be the unique solution of

$$\frac{\partial g}{\partial x_j}(x') = F_j(p_1, x', g(x')), \quad j = 2, \ldots, n, \quad g(p') = z_0.$$

Consider now the ordinary differential equation

$$\frac{dh}{dx_1} = F_1(x_1, x', h),$$

in which $x' = (x_2, \ldots, x_n)$ are regarded as parameters; let $u(x_1, x')$ be the solution with initial condition $u(p_1, x') = g(x')$. Since F_1, g are C^1 in the parameters x', the solution is also C^1 in those. So, u is C^1 and by construction, $u(p) = g(p') = z_0$ and $D_1 u(x) = F_1(x, u(x))$. We can write the equation satisfied by u as a function of x_1 in the equivalent integral form

$$u(x_1, x') = g(x') + \int_{p_1}^{x_1} F_1(t, x', u(t, x'))\, dt,$$

from which it follows that for $j \geq 2$

$$D_j u(x) = D_j g(x') + \int_{p_1}^{x_1} D_j \left[F_1(t, x', u(t, x'))\right] dt = F_j(p_1, x')$$
$$+ \int_{p_1}^{x_1} \left[D_j F_1(t, x', u(t, x')) + D_u F_1(t, x', u(t, x')) \times D_j u(t, x')\right] dt.$$

So $D_1 D_j u(x)$ exists and

$$D_1 D_j u(x) = D_j F_1(x, u(x)) + D_u F_1(x, u(x))\, D_j u(x).$$

This implies, with $a_j(x) = D_j u(x) - F_j(x, u(x))$, and making use of (9.6),

$$\begin{aligned} D_1 a_j(x) &= D_j F_1 + (D_u F_1)(D_j u) - D_1 F_j - (D_u F_j) F_1 \\ &= D_j F_1 + (D_u F_1)(D_j u) - D_j F_1 - (D_u F_1) F_j = a_j D_u F_1. \end{aligned}$$

Therefore,

$$a_j(x) = a_j(p_1, x') \exp \int_{p_1}^{x_1} \Psi(t, x')\, dt,$$

for some function Ψ. But

$$\begin{aligned} a_j(p_1, x') &= D_j u(p_1, x') - F_j(p_1, x', u(p_1, x')) = D_j g(x') - F_j(p_1, x', g(x')) \\ &= 0, \end{aligned}$$

and so $D_j u = F_j(x, u(x))$. □

In terms of $\omega = \sum_j A_j(x, u)\, dx_j + B(x, u)\, du \neq 0$, (9.6) takes the form

$$A_i(D_j B - D_u A_j) + A_j(D_u A_i - D_i B) + B(D_i A_j - D_j A_i) = 0.$$

For the geometrical problem there is no point in distinguishing one of the variables. Shifting the notation back again to $\mathbb{R}^n$, we conclude that the distribution of hyperplanes given by $\omega = \sum_{j=1}^n A_j(x)\, dx_j$ is completely integrable, that is there exists an integral hyper-surface through every point if and only if

$$A_i(D_j A_k - D_k A_j) + A_j(D_k A_i - D_i A_k) + A_k(D_i A_j - D_j A_i) = 0, i, j, k. \quad (9.7)$$

In fact, as shown, if this condition holds for a fixed k_0 and all $i \neq j \neq k_0$, then it holds for all i, j, k. There are thus $\binom{n-1}{2}$ equations.

9.3.2 There is a somehow more intrinsic point of view to understand Theorem 9.2 and state the integrability condition in terms of the spanning vector fields $X_1, \ldots, X_{n-1}$ and the incident form ω. If X, Y are tangent fields to some sub-manifold M, their commutator $[X, Y]$ is tangent too: indeed, if M is locally given by $f_j = c_j$, tangent fields are those X such that $X f_j = 0$, so $[X, Y] f_j = X(Y f_j) - Y(X f_j) = 0$ if both X, Y are tangent. In view of the definition

$$d\omega(X, Y) = X(\omega(Y)) - Y(\omega(X)) - \omega([X, Y]),$$

we see that the two conditions

(a) Each commutator $[X_i, X_j]$ is a linear combination of $X_1, \ldots, X_{n-1}$,
(b) $d\omega = 0$ on $L(x)$,

are equivalent. Now, if $\omega = \sum_j A_j(x,u)\,dx_j + B(x,u)\,du$ and say $B \neq 0$, the fields

$$X_j = -BD_j + A_j D_u, \quad j = 1, \ldots, n,$$

span the tangent space. It is easily checked that (9.6) means exactly that $d\omega(X_j, X_i) = 0$. Note that, of course, the integrability condition is stable under scalar multiplication.

The exterior algebra is again a very convenient language to deal with this situation, since an alternate 2-form η vanishes on the incident of a 1-form ω if and only if $\eta \wedge \omega = 0$. The integrability condition may thus be written $d\omega \wedge \omega = 0$.

9.3.3 In dimension $n = 3$ with notation $\omega = Adx + Bdy + Cdz$ and associated field $X = (A, B, C)$, the integrability condition reads

$$\langle \nabla \times X, X \rangle = 0, \tag{9.8}$$

where $\nabla \times X$ denotes the field

$$\nabla \times X = (C_y - B_z, A_z - C_x, B_x - A_y).$$

The notation is due to the formal fact that it is computed like the cross product between the vector $\nabla = (\partial_x, \partial_y, \partial_z)$ and X; it is somehow misleading, because in general $\nabla \times X$ is not orthogonal to X, that is precisely the integrability condition. This vector field is called the *curl or rotational* of X. We will encounter it again later on and see its physical meaning. For the time being, we have seen that it arises as an obstruction to complete integrability. In terms of two fields Y, Z spanning the orthogonal complement of X, the two forms of the integrability condition

$$\langle [Y, Z], X \rangle = 0, \quad \langle \nabla \times X, X \rangle = 0,$$

can be seen to be equivalent too by means of the formula

$$\nabla \times (Y \times Z) + [Y, Z] = (\operatorname{div} Z)Y - (\operatorname{div} Y)Z,$$

where $\operatorname{div} X = D_1 X_1 + D_2 X_2 + D_3 X_3$ for a vector field X. This function is called the *divergence* of X and will be encountered and studied in the later chapters.

9.3.4 At this point, we can repeat the procedure of Theorem 8.7 exploiting the fact that the solution $u(x,z)$ of (9.5) with $u(p) = z$ is C^1 in z. Assuming $B \neq 0$ near (p, z_0), the map $\Phi(x,z) = (x, u(x,z))$ has invertible differential $d\Phi(p, z_0)$, so by the inverse function theorem it is a diffeomorphism. The inverse map defines local coordinates $u_1, \ldots, u_{n+1}$ for which the integral hypersurfaces are the level sets $u_{n+1} = c$, parametrized by $u_1, \ldots, u_n$ (in fact by $x_1, \ldots, x_n$).

A function F which is constant on every integral hypersurface is called a *first integral* or *potential function* of X. We have shown that if X is completely integrable, locally there always exists a first integral F. Moreover, the general form is $\Phi(F)$ with Φ arbitrary. Since dF and ω have the same incident space at the very point, it follows that

$$dF = \lambda\omega,$$

that is, $\lambda\omega$ is exact. As before, the function λ is called an *integrating factor.* Thus, while in $\mathbb{R}^2$ every 1-form admits integrating factors, this is not the case in higher dimension. A restatement of Theorem 9.2 is thus:

Theorem 9.3. *A differential* 1*-form* $\omega = \sum A_j(x)\,dx_j$ *admits locally an integrating factor if and only if* (9.7) *holds.*

9.3.5 An alternative proof of Theorem 9.2, by Natani, consists in proving directly the existence of a first integral, also by induction. Assume for simplicity that $n = 3$ with notation $\omega = A(x,y,u)dx + B(x,y,u)dy + C(x,y,u)du$ and consider the ordinary differential equation

$$B(x,y,u)dy + C(x,y,u)du = 0,$$

with x as parameter. Let $G(x,y,u)$ be a first integral, that is, $d_{y,u}G = \mu(Bdy + Cdu)$. Let $v = G(x,y,u)$ and use the variables x, y, v, assuming $G_u \neq 0$. Then the equation becomes $dv + (\mu A - G_x)\,dx = 0$. We claim that $\mu A - G_x$ depends functionally on x, G. This amounts to linear dependence of the differentials,

$$\frac{\partial(x, G, G_x - \mu A)}{\partial(x,y,u)} = \frac{\partial(G, G_x - \mu A)}{\partial(y,u)} = 0,$$

that is, $G_u(G_x - \mu A)_y = G_y(G_x - \mu A)_u$. Using that $G_y = \mu B, G_u = \mu C$ it is easily checked that this is precisely the integrability condition. Thus, $\mu A - G_x = \Phi(x, G)$, so in the new variables the equation is $dv + \Phi(x,v)dx = 0$, which has a first integral.

9.3.6 There are a number of methods to find the general solution of a completely integrable system. Under the analytic formulation, the most trivial case is in $n = 2$

$$u_x = A(x, y), \quad u_y = B(x, y),$$

with $A_y = B_x$. The general solution is

$$u(x, y) = \int_0^x A(t, y)\, dt + \int_0^y B(0, t)\, dt + c.$$

To solve the general

$$u_x = A(x, y, u), \quad u_y = B(x, y, u),$$

assuming (9.6), one may follow the proof of the theorem above.

Example 9.11. For

$$u_x = 1 + y, \quad u_y = \frac{u}{1 + y},$$

we solve first $u_x = 1 + y$ for a particular value of y, say $y = 0$, obtaining $u = x + c$. Next we solve $u_y = \frac{u}{1+y}$, $u = K(x)(1 + y)$ with initial value $x + c$ at $y = 0$, obtaining $u = (x + c)(1 + y)$.

9.3.7 One may try too to find an integrating factor, either by inspection or using Natani's method.

Example 9.12. The equation

$$\omega = (x^2 z - y^3)\, dx + 3xy^2\, dy + x^3\, dz = 0$$

is completely integrable. Trivially $x = 0$ is an integral surface. In $x \neq 0, \lambda = x^{-2}$ is an integrating factor,

$$x^{-2}\omega = d(xz) + d(y^3/x),$$

and so $x^2 z + y^3 = cx$ is the general integral surface.

Example 9.13. The system

$$u_x = -\frac{y + u}{x + y}, \quad u_y = -\frac{u + x}{x + y},$$

is completely integrable. We solve the first equation, $\log(y+u)+\log(x+y) = c, (y + u)(x + y) = k$, and consider $v = (y + u)(x + y)$; then the equation becomes $dv = 2ydy$, whose general solution is $v - y^2 = c$. Thus, the general integral surface is

$$(y + u)(x + y) - y^2 = c, \quad u = \frac{c - xy}{x + y}.$$

Example 9.14. The system in $\mathbb{R}^3$

$$yz(y+z)dx + xz(x+z)dy + xy(x+y)dz = 0,$$

is completely integrable. With z as parameter, $y(y+z)dx + x(x+z)dy = 0$, whose general solution is

$$G(x,y,z) = \frac{xy}{(x+z)(y+z)},$$

and use now the variables $x, v = G(x,y,z), z$. The equation becomes $zdv = 2(v-1)vdz$, whose general solution is $\frac{v-1}{v} = kz^2$, that is, $v(z) = \frac{1}{1+kz^2}$. The general integral surface is

$$\frac{(x+z)(y+z)}{xy} = 1 + kz^2, \quad x+y+z = kxyz.$$

9.3.8 There is a variant in Natani's method to find the solution $v = v(z)$ in the last step consisting in fixing another variable, say $x = \alpha$, and consider the ordinary differential equation.

$$B(\alpha,y,z)dy + C(\alpha,y,z)dz = 0.$$

Typically, we would consider a value of α for which this is as simple as possible. Assume we know a first integral $F(y,z)$, so $F(y,z) = c$ is the family of curves along which the integral surfaces meet $x = \alpha$. Then, $G(\alpha,y,z) = v(z)$ must define the same family of curves than $F(y,z) = c$ and this allows finding the general v.

Example 9.15. For

$$z(z+y^2)dx + z(z+x^2)dy - xy(x+y)dz = 0,$$

we first consider y as a parameter, $z(z+y^2)dx - xy(x+y)dz = 0$,

$$\left(\frac{1}{x} - \frac{1}{x+y}\right)dx + \left(\frac{1}{z+y^2} - \frac{1}{z}\right)dz = 0,$$

whose general solution is

$$G(x,y,z) = \frac{x(z+y^2)}{z(x+y)} = v(y).$$

Next we take say $z = 1$, $(1+y^2)dx + (1+x^2)dy = 0$, with general solution $\arctan x + \arctan y = c$ or $1 - xy = k(x+y)$. Then, $x(1+y^2) = v(y)(x+y)$ and $1 - xy = k(x+y)$ must be the same family of curves, from which it follows that $v(y) = 1 - ky$ and $x(z+y^2) = (1-cy)z(x+y)$ is the general solution.

9.3.9 Still another point of view is as follows. Let $Y = (A', B', C')$ a vector field orthogonal to $X = (A, B, C)$ at every point; for instance $Y = \nabla \times X$, in view of (9.8). Then the integral surfaces will contain the integral curves of Y. Let u, v be two functionally independent first integrals of the system

$$\frac{dx}{A'} = \frac{dy}{B'} = \frac{dz}{C'} = 0.$$

Since ω, du, dv are incident to X, one must have $\omega = Pdu + Qdv$. Now, working in a coordinate system u, v, w, complete integrability means $PQ_w = QP_w$, that is $(\frac{P}{Q})_w = 0$, so ω is a multiple of some form $H(u, v)du + dv$. In fact, this is a proof that ω has an integrating factor if it is completely integrable and can be applied in practice.

Example 9.16. Consider again $yz(y + z)dx + xz(x + z)dy + xy(x + y) dz = 0$. Here $Y = \nabla \times X = (x(y - z), y(z - x), z(x - y))$. By inspection we see that $u = x + y + z, v = xyz$ are first integrals of the system governed by Y. We set $\omega = Pdu + Qdv$ and check that $P = -xyz = -v$, $Q = x + y + z = u$ so the equation is $udv - vdu = 0$ and $u = kv$ is the general solution.

9.3.10 *Mayer's method* is also worth-mentioning, as it requires solving just one differential equation. We set $y = \lambda x$ in the equation $\omega = 0$; assume $G(x, z, \lambda)$ is a first integral of the resulting equation in x, z, that is,

$$G(x, \lambda, z) = c, \quad y = \lambda x,$$

are the curves in $y = \lambda x$ meeting the integral surfaces. Thus, for $p = (0, 0, c)$

$$G(x, \lambda, z) = G(0, \lambda, c), \quad y = \lambda x,$$

is the curve intersection of the integral surface through $(0, 0, c)$ with $y = \lambda x$ and eliminating λ,

$$G\left(x, \frac{y}{x}, z\right) = G\left(0, \frac{y}{x}, c\right)$$

is the integral surface through $(0, 0, c)$.

Example 9.17. Consider again $(y + z)dx + (z + x)dy + (x + y)dz = 0$. With $y = \lambda x$ we get $[(\lambda + 1)z + 2\lambda x]dx + x(\lambda + 1)dz = 0$, so $G(x, \lambda, z) = \lambda x^2 + (\lambda + 1)xz$. Then

$$G\left(x, \frac{y}{x}, z\right) = xy + zy + zx = G(0, 0, c),$$

is the integral surface through $(0, 0, c)$.

9.3.11 In case of higher co-dimension, the system $\omega_i = 0, i = 1, \ldots, n-k$ is completely integrable if and only if $d\omega_i$ is zero restricted to the common incident space $\omega_1 = \cdots = \omega_{n-k} = 0$, that is,

$$d\omega_i \wedge \omega_1 \wedge \cdots \wedge \omega_{n-k} = 0.$$

The analytic version consists in a system of equations for $n-k$ unknown functions $u_1, \ldots, u_{n-k}$ in k variables $x_1, \ldots, x_k$ of type

$$\frac{\partial u_i}{\partial x_j} = F_{ij}(x_1, \ldots, x_k, u_1, \ldots, u_{n-k}), \quad i = 1, \ldots, n-k, \; j = 1, \ldots, k.$$

The integrability condition reads

$$\frac{\partial F_{ij}}{\partial x_r} + \sum_{l=1}^{n-k} \frac{\partial F_{ij}}{\partial u_l} F_{lr} = \frac{\partial F_{ir}}{\partial x_j} + \sum_{l=1}^{n-k} \frac{\partial F_{ir}}{\partial u_l} F_{lj}.$$

9.4 Elementary Second-Order PDEs

9.4.1 Exactly as for first-order derivatives, if we know a partial derivative of order r of f, we can obtain f by repeated anti-differentiation. For instance, in the plane, if

$$\frac{\partial^2 f}{\partial x \partial y} = 0,$$

then f_y is just a function of y whence

$$f(x, y) = A(x) + B(y),$$

for some one-variable functions A, B. And conversely, every f of this form satisfies the equation, that is the *general solution.* If say

$$\frac{\partial^2 f}{\partial x^2} = xy,$$

then $f_x = \frac{1}{2}x^2 y + A(y)$ and so the general solution is $f(x, y) = \frac{1}{6}x^3 y + xA(y) + B(y)$. The function $\frac{1}{6}x^3 y$ is a particular solution, while $xA(y) + B(y)$ is the general solution of the homogeneous equation $f_{xx} = 0$. This structure of the general solution is of course due to the linearity of the equation.

As for first-order equations, a change of coordinates may be useful to solve an equation. Consider for example the *wave equation*

$$\frac{\partial^2 f}{\partial x^2} - \frac{\partial^2 f}{\partial y^2} = 0.$$

With the change of coordinates $u = x + y, v = x - y$ one has $f_x = f_u u_x + f_v v_x = f_u + f_v, f_y = f_u - f_v$, and so

$$f_{xx} = f_{uu}u_x + f_{uv}v_x + f_{vu}u_x + f_{vv}v_x = f_{uu} + 2f_{uv} + f_{vv},$$
$$f_{yy} = f_{uu} - 2f_{uv} + f_{vv},$$

so that the equation is written $f_{uv} = 0$, implying that $f = A(u) + B(v)$, that is

$$f(x, y) = A(x + y) + B(x - y),$$

with A, B general functions of one variable is the general solution of the wave equation. Thinking of y as time t, and since the graph of $x \to A(x+t)$ is the one of A shifted t to the left, one says that $A(x+t)$ is a wave moving left at speed one; then f is the superposition of two waves moving in opposite directions.

For later purposes it is worth presenting this computation in a different way: we may decompose the second-order operators as a product of two operators of order one

$$\frac{\partial^2}{\partial x^2} - \frac{\partial^2}{\partial y^2} = \left(\frac{\partial}{\partial x} - \frac{\partial}{\partial y}\right)\left(\frac{\partial}{\partial x} + \frac{\partial}{\partial y}\right).$$

Then

$$\frac{\partial f}{\partial x} + \frac{\partial f}{\partial y} = \phi(x + y),$$

which in the variables u, v can be written $2f_u = \phi(u)$ and we get again $f = A(u) + B(v)$.

9.4.2 Let us analyze, more generally, a general linear partial differential equation of second order with constant coefficients in the plane, of the form

$$af_{xx} + 2bf_{xy} + cf_{yy} = \Phi(x, y),$$

where a, b, c are constants and Φ is a given function. Formally, we can write the left-hand side as the action on f of the linear operator

$$L = aD_x^2 + 2bD_xD_y + cD_y^2.$$

In Theorem 1.4, we saw that a homogenous polynomial of degree two may be written as

$$aX^2 + 2bXY + cY^2 = \lambda(\alpha X + \beta Y)^2 + \mu(-\beta X + \alpha Y)^2.$$

We may look at this as a formal equality valid whenever X, Y commute, hence,

$$L = \lambda(\alpha D_x + \beta D_y)^2 + \mu(-\beta D_x + \alpha D_y)^2.$$

In case one of the eigenvalues is zero, that is $ac = b^2$, then we get after a linear change of variables the reduced form

$$f_{xx} = \Phi(x, y),$$

which is solved by two anti-differentiations in x. This means that the original equation is of the form $D_{vv}f = \Phi$ for some direction v. These equations are called of *parabolic type*. If λ, μ are of different signs, after another linear change of variables we get the reduced form

$$f_{xx} - f_{yy} = \Phi(x, y),$$

which we just learned to handle, reducing it in turn to the form $D_{uv} = \Phi$. These equations are called of *hyperbolic type*. In case λ, μ have the same sign, the equation is said to be of *elliptic type*, and we are led to the reduced form

$$f_{xx} + f_{yy} = \Phi(x, y).$$

The left-hand side operator is called the *Laplacian* $\Delta = D_x^2 + D_y^2$, and $\Delta f = 0$ is called the *Laplace equation*.

We leave for the next section the discussion of the elliptic type equation. Both for the parabolic and hyperbolic cases, we have seen how to find the solution by solving two first-order equations, and the general solution depends on two arbitrary functions of one variable. Let us analyze what other condition should be imposed so that the solution is unique. For the first-order linear equation $D_v f = \Phi$, we know that if f is prefixed on some set A meeting once and only once each line with direction v, then f is unique. Analogously, for the parabolic equation

$$D_{vv}f = \Phi(x, y),$$

the solution f is unique if at A both f and $D_v f$ are prefixed. Usually, A is a level curve $A = \{g = 0\}$ with $\nabla g \neq 0$ at A, with $\nabla g, v$ linearly independent at points in A. In this situation, prefixing $f, D_v f$ at points in A amounts to prefixing f and the normal derivative $D_n f$ at points of A, $n = \nabla g$, because then the tangential derivative of f along the curve is known, too. Altogether, the parabolic equation has a unique solution satisfying

$$f(x) = g(x), \quad D_n f(x) = h(x), \quad x \in A,$$

with h, g given functions on a curve A meeting transversally every line with direction v at just one point. These are called *the Cauchy conditions.*

Example 9.18. Let us solve

$$f_{xx} + 4f_{xy} + 4f_{yy} = xy,$$

and find the unique solution satisfying $f(x,0) = g(x), f_y(x,0) = h(x)$. The left hand side is $D_{vv}f$ for $v = (1,2)$. Instead of changing variables we work on a fixed line with direction v, say $(x,y) = (c,0) + t(1,2), x = c+t, y = 2t$; if $h(t) = f(c+t, 2t)$, the equation becomes $h''(t) = xy = 2t(c+t)$, so

$$h'(t) = ct^2 + \frac{2}{3}t^3 + A, \quad h(t) = \frac{c}{3}t^3 + \frac{1}{6}t^4 + At + B,$$

where A, B may depend on $c = x - t = x - \frac{y}{2}$ and we find the general solution

$$f(x,y) = \frac{1}{3}\left(x - \frac{y}{2}\right)\frac{1}{8}y^3 + \frac{1}{6}\frac{1}{16}y^4 + yA(2x-y) + B(2x-y),$$

with A, B arbitrary functions of one variable. The last two terms constitute the general solution of the homogeneous equation. Then

$$g(x) = f(x,0) = B(2x), \quad h(x) = f_y(x,0) = A(2x) - B'(2x),$$

so $B(x) = g(\frac{x}{2}), A(x) = h(\frac{x}{2}) - \frac{1}{2}g'(x)$. Note the structure of the solution: the solution of the homogeneous equation satisfies the Cauchy conditions, while the particular solution has zero Cauchy conditions.

The solution of the wave equation $f = A(x+y) + B(x-y)$ is unique too if $f(x,0) = g(x), f_y(x,0) = h(x)$. Indeed,

$$A(x) + B(x) = g(x), \quad A'(x) - B'(x) = h(x),$$

implies $2A' = g' + h, 2B' = g' - h$, so

$$2A(x) = g(x) + \int_0^x h(t)\,dt + c, 2B(x) = g(x) - \int_0^x h(t)\,dt + d,$$

with $c+d=0$ and so

$$f(x,y)=\frac{1}{2}(g(x+y)+g(x-y))+\frac{1}{2}\int_{x-y}^{x+y}h(t)\,dt.$$

9.5 A Hint to Complex Analysis

In relation to the Laplace equation, in this section, we open one of the many entrance doors of Complex Analysis. An excellent textbook is [1], see [3] for additional topics.

9.5.1 We are interested in finding real-valued solutions f of the Laplace equation

$$\Delta f=\frac{\partial^2 f}{\partial x^2}+\frac{\partial^2 f}{\partial y^2}=0.$$

These functions are called *harmonic.* Of course, linear functions are solutions; for a homogeneous polynomial of degree two

$$f(x,y)=ax^2+2bxy+cy^2,$$

the equation becomes $a=-c$, so all are linear combinations of x^2-y^2 and xy. Similar computations can be done for higher degree polynomials. But this is not the right approach.

It is tempting to decompose, similarly as was done in the previous section, with $i^2=-1$,

$$\frac{\partial^2}{\partial x^2}+\frac{\partial^2}{\partial y^2}=\left(\frac{\partial}{\partial x}+i\frac{\partial}{\partial y}\right)\left(\frac{\partial}{\partial x}-i\frac{\partial}{\partial y}\right),$$

so every twice differentiable solution of one of the first-order equations

$$\frac{\partial f}{\partial x}-i\frac{\partial f}{\partial y}=0,\quad \frac{\partial f}{\partial x}+i\frac{\partial f}{\partial y}=0,$$

would be a solution of the Laplace equation, too. But we at once realize that there are no real solutions of these equations other than constants, because if f is real, they both amount to $\nabla f=0$. But we may consider *complex-valued* solutions of these.

9.5.2 First, let us rewrite linear maps $A:\mathbb{R}^2\to\mathbb{R}^2$ when we identify $\mathbb{R}^2$ with $\mathbb{C}$ via $z=x+iy$. If

$$A(x,y)=(ax+by,cx+dy),$$

then

$$Az = \lambda z + \mu \overline{z},$$

with

$$\lambda = \alpha + i\beta, \quad \mu = \overline{\alpha} + i\overline{\beta}, \quad \alpha = \frac{1}{2}(a - bi), \quad \beta = \frac{1}{2}(c - di).$$

Of course, given $\lambda, \mu \in \mathbb{C}$, these can be solved in α, β, so $Az = \lambda z + \mu\overline{z}$ is the general $\mathbb{R}$-linear map in $\mathbb{C}$. It is the sum of a $\mathbb{C}$-linear map, $z \to \lambda z$, and a anti-linear map $z \to \mu\overline{z}$. The $\mathbb{C}$-linear maps have the matrix

$$\begin{pmatrix} a & -b \\ b & a \end{pmatrix}.$$

For further use, we compute the determinant of A in terms of $\lambda = \lambda_1 + \lambda_2 i, \mu = \mu_1 + \mu_2 i$; the matrix of A is

$$\begin{pmatrix} \lambda_1 + \mu_1 & \mu_2 - \lambda_2 \\ \lambda_2 + \mu_2 & \lambda_1 - \mu_1 \end{pmatrix},$$

whose determinant is $\det A = |\lambda|^2 - |\mu|^2$. The conformal maps are those with $\mu = 0$, that is $\mathbb{C}$-linear or those with $\lambda = 0$, that is anti-linear.

When applied to a differential $A = df, f = u + iv, a = u_x, b = u_y, c = v_x, d = v_y$, this gives

$$\lambda = \frac{1}{2}(u_x - u_y i) + i\frac{1}{2}(v_x - v_y i), \quad \mu = \frac{1}{2}(u_x + u_y i) + i\frac{1}{2}(v_x + v_y i).$$

That's why the following operators are introduced:

$$\frac{\partial}{\partial z} = \frac{1}{2}\left(\frac{\partial}{\partial x} - i\frac{\partial}{\partial y}\right), \quad \frac{\partial}{\partial \overline{z}} = \frac{1}{2}\left(\frac{\partial}{\partial x} + i\frac{\partial}{\partial y}\right),$$

so that $\lambda = \frac{\partial f}{\partial z}, \mu = \frac{\partial f}{\partial \overline{z}}$,

$$df = \frac{\partial f}{\partial z}\, dz + \frac{\partial f}{\partial \overline{z}}\, d\overline{z}.$$

Here, dz is the identity map in $\mathbb{C}$ and $d\,\overline{z}$, its conjugate. Also, the Jacobian

$$Jf = \left|\frac{\partial f}{\partial z}\right|^2 - \left|\frac{\partial f}{\partial \overline{z}}\right|^2.$$

Of course, we have

$$\frac{\partial}{\partial z} z = 1, \quad \frac{\partial}{\partial z}\overline{z} = 0, \quad \frac{\partial}{\partial \overline{z}} z = 0, \quad \frac{\partial}{\partial \overline{z}}\overline{z} = 1.$$

We use the notations $D_{\overline{z}}, D_z$, too; note that $\Delta = 4D_zD_{\overline{z}}$. So we are interested in complex-valued solutions of either $D_zf = 0, D_{\overline{z}}f = 0$. Since

$$\overline{\frac{\partial f}{\partial z}} = \frac{\partial \overline{f}}{\partial \overline{z}},$$

we deal just with $D_{\overline{z}}f = 0$. This is called the *Cauchy–Riemann equation.* In terms of u, v

$$u_x = v_y, \quad u_y = -v_x. \tag{9.9}$$

By what has been said, this means that *df* is not only $\mathbb{R}$-linear, but also $\mathbb{C}$-linear. Then in Definition 4.1

$$f(x, y) = f(a, b) + D_zf(a, b)((x - a) + i(y - b)) + o(\sqrt{(x - a)^2 + (y - b)^2}),$$

or in complex notation

$$f(z + h) = f(z) + D_zf(z)h + o(|h|),$$

meaning that

$$\lim_{w \to z} \frac{f(w) - f(z)}{w - z} = D_zf(z).$$

This shows that f satisfies the Cauchy–Riemann equations at one point z if and only if it *has a complex derivative*, in the sense that the above limit exists; the function f is called *holomorphic* at z. An alternative notation for D_zf is then $f'(z)$. Said otherwise, $df(z)$ being $\mathbb{C}$-linear, consists in multiplication by $f'(z)$, a dilation followed by a rotation. If f is holomorphic at all points $z \in U$, we call it *holomorphic in* U. Holomorphic functions in the whole complex plane are called *entire*. If $\overline{f}$ is holomorphic, that is $D_zf = 0$, we say that f is *anti-holomorphic*. Holomorphic and anti-holomorphic maps with non-zero derivative are *conformal*, that is, if two curves meet with angle of size α, their images meet with the same angle, holomorphic functions keeping the orientation and anti-holomorphic reversing it. Since compositions of $\mathbb{C}$-linear maps are $\mathbb{C}$-linear, the chain rule implies that composition of holomorphic functions is also holomorphic.

We will see in Section 17.3 that holomorphic functions are automatically infinitely differentiable. Then from the Cauchy–Riemann Equations (9.9), we see that

$$\Delta u = 0, \quad \Delta v = 0, \quad \Delta f = 0,$$

so real and imaginary parts of holomorphic functions are harmonic.

The usual rules for differentiation hold too for complex-valued functions and these operators; for instance, an arbitrary polynomial $P(x,y)$, with the substitution $x = \frac{1}{2}(z+\overline{z}), y = \frac{1}{2i}(z-\overline{z})$ may be written as a polynomial in $z, \overline{z}$

$$P(x,y) = Q(z,\overline{z}) = \sum c_{kl} z^k (\overline{z})^l.$$

Then

$$D_z Q = \sum k c_{kl} z^{k-1} (\overline{z})^l, \quad D_{\overline{z}} Q = \sum l c_{kl} z^k (\overline{z})^{l-1}.$$

In particular, the holomorphic polynomials are exactly the polynomials in z, and their real and imaginary parts are harmonic functions. For instance, the harmonic homogeneous polynomials of degree three are the linear combinations of $x^3 - 3xy^2, y^3 - 3yx^2$.

9.5.3 Complex power series and complex-analytic functions. In Section 5.5, we have introduced (real) power series in several variables that can be looked as infinite degree polynomials in the variables $x_1, \ldots, x_n$. Analogously, a *complex power series* centered at $a \in \mathbb{C}$ is an expression

$$\sum_{k=0}^{\infty} c_k (z-a)^k, \quad c_k \in \mathbb{C}.$$

Taking real and imaginary parts, this amounts to two real power series in two real variables $x, y, z = x + iy$, so the results in that section apply. But complex power series, as well as their real and imaginary parts, have additional interesting properties. The first one is regarding the domain of convergence which, if non-empty, is always an open disc. Indeed, by the ratio test, exactly as for one real variable series, the series

$$\sum_k |c_k||z|^k,$$

is convergent if $|z| < \rho$ and divergent if $|z| > \rho$, where

$$\frac{1}{\rho} = \limsup_k |c_k|^{\frac{1}{k}},$$

and therefore the domain of convergence is the open disc $B(a,\rho)$. Later we show that this is also the domain of convergence of both the real and the imaginary parts.

Secondly, let us look at the function defined in this disc,

$$f(z) = \sum_{k=0}^{+\infty} c_k(z-a)^k, \ |z-a| < \rho,$$

that might be called a polynomial in z of infinite degree. As expected, f is holomorphic. To see this, recall from Theorem 5.8 that power series can be differentiated term-wise. Therefore,

$$D_{\overline{z}}f = 0, \quad D_z f = f'(z) = \sum_k k c_k z^{k-1}.$$

Iterating, we see that f is in fact infinitely holomorphic,

$$f^{(m)}(z) = \sum_{k \geq m} k(k-1)\cdots(k-m+1)c_k(z-a)^{k-m},$$

and a fortiori $c_k = \frac{f^{(k)}(a)}{k!}$. Thus, two complex power series with the same sum have the same coefficients.

Analogously, as in the real case, a complex-valued function f in a domain $U \subset \mathbb{C}$ is called *complex-analytic* if around each $a \in U$, f is the sum of a complex power series centered at a. Then f is infinitely holomorphic and this series is unique,

$$f(z) = \sum_k \frac{f^{(k)}(a)}{k!}(z-a)^k.$$

Exactly as for the real case, a function defined by a complex power series is complex-analytic. Exactly as in one real variable and with the same proof, the zeros of such functions are isolated and so the principle of analytic continuation holds: if two such functions in a domain agree in a set with an accumulation point, they are identical.

In paragraph 17.3.3, it will be shown that conversely and in striking difference with the real case, a holomorphic function is complex-analytic.

The following result explains the importance of this principle and justifies the terminology:

Theorem 9.4. *A real-analytic function defined in an open interval I is the restriction to I of a complex-analytic function defined in a domain U symmetric with respect the real line.*

Proof. For each $p \in I$, we complexify the Taylor expansion

$$f(x) = \sum_k c_k (x-p)^k, \quad |x-p| < r = r(p),$$

replacing x by a complex variable z. Since the radius of convergence is unchanged, this defines a local extension in discs D_p around points $p \in I$:

$$f_p(z) = \sum_k c_k (z-p)^k, \quad |z-p| < r = r(p).$$

If two such discs D_p, D_q overlap, then f_p, f_q are equal on a segment, whence they are equal in $D_p \cap D_q$. In this way, an extension of f to $U = \cup_{p \in I} D_p$ is obtained. □

The theorem means that the natural domain of definition of complex-analytic functions is the complex plane.

In particular, real and imaginary parts of complex power series are harmonic functions. For instance,

$$e^z = \sum_k \frac{z^k}{k!}, \quad \sin z = \sum_k (-1)^k \frac{z^{2k+1}}{(2k+1)!}, \quad \cos z = \sum_k (-1)^k \frac{z^{2k}}{(2k)!},$$

are entire functions, and their real and imaginary parts

$$e^x \cos y, \quad e^x \sin y, \quad \sin x \sinh y, \quad \sin x \cosh y, \quad \cos x \sinh y, \quad \cos x \cosh y,$$

are harmonic functions.

9.5.4 We have seen how holomorphic functions produce harmonic functions. In fact, there is essentially a one-to-one correspondence:

Theorem 9.5. *In a disk $\mathbb{D}, 0 \in \mathbb{D}$ (or more generally in a p-domain), there is a one-to-one correspondence between holomorphic functions f with $f(0) \in \mathbb{R}$, and real harmonic functions u given by $u = \operatorname{Re} f$. More precisely, a holomorphic function f, $f(0) \in \mathbb{R}$, is completely determined by its real part. In particular, harmonic functions are smooth.*

Proof. We must prove that if u is harmonic in $\mathbb{D}$, there exists v, also harmonic, such that $f = u + iv$ is holomorphic, that is, such that (9.9) holds. But u being harmonic precisely means that the data $u_x, -u_y$ in the problem $v_x = -u_y, v_y = u_x$ satisfy the compatibility condition (4.7) in Theorem 4.5. □

The harmonic function v, defined by the condition $\nabla v = J\nabla u$, where J is multiplication by i, is called *the harmonic conjugate* of u. The notation $(\nabla u)^{\perp}$ is also used. The previous theorem holds more generally for *simply connected domains*, meaning that its complement is connected.

Using this result we prove now that the real and imaginary parts of a complex power series $\sum_k c_k z^k$ have the same domain of convergence as the whole series. Assume that the real part $\sum a_{ij}x^i y^j = \sum u_{ij}$ has domain of convergence D, that is the series is unconditionally convergent in D. By Theorem 5.8, the series of gradients is also unconditionally convergent, and uniformly on compacts, in D. Then $\sum(\nabla u_{ij})^{\perp} = \sum \nabla v_{ij}$ converges uniformly on compacts in D, v_{ij} being a normalized conjugate of u_{ij}. Since D is a p-domain, $\sum v_{ij}$, the imaginary part of $\sum_k c_k z^k$ converges unconditionally in D. We may reverse the roles of the u_{ij}, v_{ij}, whence the three series have the same domain of convergence.

9.5.5 It is clear that if $u = P(x, y)$ is a harmonic polynomial, the function f is a polynomial $Q(z)$. It can be found algebraically, with no integration. Indeed, from

$$2P(x, y) = Q(x + iy) + \overline{Q(x + iy)},$$

we see that $2P(\frac{z}{2}, \frac{z}{2i}) = Q(z) + Q(0), Q(0) = P(0, 0)$, and therefore

$$Q(z) = 2P\left(\frac{z}{2}, \frac{z}{2i}\right) - P(0, 0).$$

The formalism with $D_z, D_{\overline{z}}$ is useful too to solve inhomogeneous equations such as

$$f_{xx} + f_{yy} = P(x, y),$$

where P is a polynomial in x, y. Consider for example

$$f_{xx} + f_{yy} = 2xy.$$

We write $2xy = \operatorname{Im} z^2 = \frac{1}{i}(z^2 - (\overline{z})^2)$ and $\Delta = 4D_z D_{\overline{z}}$. Then we obtain a particular solution by two anti-differentiations in $z, \overline{z}$:

$$D_{\overline{z}} f = \frac{1}{4i}\left(\frac{1}{3}z^3 - \frac{1}{2}z(\overline{z})^2\right),$$

and

$$f = \frac{1}{24i}\overline{z}z^3 - \frac{1}{24i}z(\overline{z})^3) = \frac{1}{12}\operatorname{Im}\overline{z}z^3 = \frac{1}{12}\operatorname{Im}((x-iy)(x^3+3x^2iy$$
$$-3xy^2 - iy^3)) = \frac{1}{12}(x(3x^2y-y^3) - y(x^3-3xy^2)) = \frac{1}{6}xy(x^2+y^2).$$

The general solution is then $\frac{1}{6}xy(x^2+y^2)+H$, with H harmonic.

Chapter 10

Orthogonal Families of Curves and Surfaces

This chapter has a classical geometric character. Basically, we use existence results for differential equations to construct bi-orthogonal families of curves in the plane, bi-orthogonal families of curves and surfaces in space, or triply orthogonal families of curves or surfaces in space. In particular, this leads to *orthogonal coordinate systems*, among which the *conformal* ones are particularly important. The conformal transformations are shown to be abundant if $n = 2$ and very few if $n > 2$, this is *Liouville's theorem.* The last section is devoted to a somehow forgotten topic, the *Lamé surfaces*, whose characterization was an important problem among geometers in the 19th century. It is much based on the great monography [5].

10.1 Families of Plane Curves

10.1.1 As seen in paragraph 9.1.3, every ordinary differential equation in the plane

$$\omega = M(x, y)\, dx + N(x, y)\, dy = 0,$$

has an integrating factor λ so that $\lambda\omega = df$ is exact, and the general integral curve is given by $f(x, y) = c$. In accordance with Theorem 8.2, there is a unique integral curve through every point p, namely $f(x, y) = f(p)$. This family is *uni-parametric* depending on the parameter c.

Conversely, assume that we have a family of curves

$$\Gamma_\lambda : G(x, y, \lambda) = 0,$$

with G of class C^1, $\nabla_{x,y} G \neq 0$, with the *unicity property* in some domain U: for every point $p = (a, b) \in U$, there is a unique curve in the family containing p. For instance, if $G_\lambda \neq 0$, by the implicit function theorem $G = 0$ defines $\lambda = \lambda(x, y)$ locally. The unique curve in the family through p is given by the system

$$G(x, y, \lambda) = 0, \quad G(a, b, \lambda) = 0.$$

We may view the family as given by $\lambda(x, y) = c$, too. Every curve in this family satisfies the ordinary differential equation

$$G_x(x, y, \lambda)\, dx + G_y(x, y, \lambda)\, dy = 0,$$

with $G(x, y, \lambda) = 0$, so eliminating $\lambda = \lambda(x, y)$ we get a differential equation for all λ. Of course, the easiest situation is $y = f(x, \lambda)$ or $G(x, y, \lambda) = u(x, y) - \lambda$.

Thus, there is a one-to-one correspondence between ordinary differential equations, uniparametric families of curves with the unicity property, and vector fields. Moreover, there always exist local coordinates u, v where the family is given by $u = c$ and each curve is parametrized by v.

Example 10.1.

(a) The family of curves $y = \tan(x + \lambda)$ has the unicity property. Differentiating, $y' = 1/\cos^2(x + \lambda)$; eliminating $\lambda, 1 + y^2 = 1/\cos^2(x + \lambda)$, so $y' = 1 + y^2$ is the differential equation of the family.

(b) The family of curves $y = x^3 + \lambda x$ has the unicity property in the domain $x \neq 0$. One has $y' = 3x^2 + \lambda$, so $xy' = y + 2x^3$ is the differential equation of the family.

(c) For the family $y^2 + xy + ce^x = 0$, one has $(2y + x)\, dy + (y + ce^x)\, dx = 0$ and the equation is $(2y + x)\, dy + (y - y^2 - xy)\, dx = 0$.

10.1.2 In the following, for a vector field $X = (M, N)$ in the plane, we denote by JX the orthogonal field $JX = (-N, M)$. If $\mathcal{F}$ is a family of curves with the unicity property with associated differential equation

$$M(x, y)\, dx + N(x, y)\, dy = 0,$$

that is, curves orthogonal to (M, N), the family $\mathcal{G}$ given by

$$-N(x,y)\,dx + M(x,y)\,dy = 0$$

is *orthogonal* to $\mathcal{F}$, meaning that for each point p the curves of $\mathcal{F}, \mathcal{G}$ through p meet orthogonally. We call it a *bi-orthogonal system of curves* in the plane. Clearly, each uni-parametric family of curves with the unicity property can be completed to a bi-orthogonal system.

In other words, for every smooth function u with $\nabla u \neq 0$, the integral curves of

$$\gamma'(t) = (\nabla u)(\gamma(t))$$

constitute an orthogonal family to $u = c$. If v is a first integral of this system, that is

$$\langle \nabla u, \nabla v \rangle = 0,$$

the integral curves are $v = k$, so the two systems

$$u = c, \quad v = k,$$

are orthogonal. The condition on v amounts to $\nabla v = \lambda J(\nabla u)$

$$v_x = -\lambda u_y, \quad v_y = \lambda u_x,$$

for some integrating factor $\lambda \neq 0$. The equation for λ, imposing $(\lambda u_y)_y = -(\lambda u_x)_x$, is

$$\lambda \Delta u = -\lambda_y u_y - \lambda_x u_x, \tag{10.1}$$

or in terms of $\lambda = e^{\phi}$,

$$-\phi_y u_y - \phi_x u_x = \Delta u.$$

The transformation $(x, y) \to (u, v)$ has the differential

$$\begin{pmatrix} u_x & u_y \\ -\lambda u_y & \lambda u_x \end{pmatrix},$$

whose inverse is

$$\frac{1}{\lambda |\nabla u|^2} \begin{pmatrix} \lambda u_x & -u_y \\ \lambda u_y & u_x \end{pmatrix}.$$

Then u, v is a local coordinate system with

$$\partial_u = \frac{\nabla u}{|\nabla u|^2}, \quad \partial_v = \frac{J(\nabla u)}{\lambda|\nabla u|^2},$$

that is, the new axes meet orthogonally at each point and the metric is diagonal

$$ds^2 = \frac{1}{|\nabla u|^2}\left(du^2 + \frac{1}{\lambda^2}dv^2\right).$$

We call it an *orthogonal coordinate system.* Given u there are many choices of λ, v, since v can be replaced by $\phi(v)$.

Conversely, an orthogonal coordinate system u, v corresponds to a bi-orthogonal system of curves. The most basic examples are polar coordinates r, θ for which

$$ds^2 = dr^2 + r^2 d\theta^2,$$

with circles and lines through the origin as bi-orthogonal families, see Figure 10.1.

Example 10.2. Let us find the orthogonal family to the ellipses

$$\frac{x^2}{\cosh^2 \lambda} + \frac{y^2}{\sinh^2 \lambda} = 1,$$

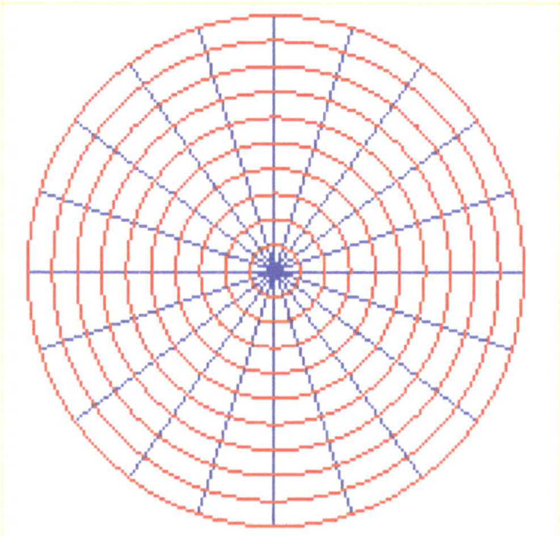

Figure 10.1. A bi-orthogonal family of lines and circles.

with focus at $1, -1$. The differential equation they satisfy is

$$\frac{x}{\cosh^2 \lambda} = -\frac{yy'}{\sinh^2 \lambda} = x + yy',$$

with x, y, λ related as above, the second equality being a consequence of the first. Denoting $\eta = x + yy'$, from the ellipse equation we get $\eta(x - \frac{y}{y'}) = 1$, that is, they satisfy

$$(x + yy')\left(x - \frac{y}{y'}\right) = 1.$$

Note this is quadratic in y' and invariant replacing y' by $\frac{1}{y'}$, so the orthogonal family must be the other solution of this equation. Guessing that it must be a one-parameter family of quadrics of the same form, we try

$$A(\lambda)x^2 + B(\lambda)y^2 = 1.$$

Then $y' = -\frac{Ax}{By}$, so

$$\left(x - \frac{Ax}{B}\right)\left(x + \frac{By^2}{Ax}\right) = 1, \quad \left(1 - \frac{A}{B}\right)\left(x^2 + \frac{B}{A}y^2\right) = 1.$$

Being quadratic in $\mu = \frac{A}{B}$, there are two families. For $0 < \mu < 1$, we have the family of ellipses, and for $\mu < 0$ we find the orthogonal family of hyperbolas, with the same focus, that can be written

$$\frac{x^2}{\cos^2 \theta} - \frac{y^2}{\sin^2 \theta} = 1.$$

See Figure 10.2.

Example 10.3. The family of circles centered at $(0, a)$ through $(-1, 0), (0, 1)$,

$$x^2 + y^2 - 2ay = 1,$$

has differential equation

$$2xy\,dx + (y^2 - x^2 + 1)\,dy = 0,$$

and

$$(x^2 - y^2 - 1)\,dx + 2xy\,dy = 0,$$

the one of the orthogonal family. By inspection, $\lambda = \frac{1}{x^2}$ is an integrating factor and we get another family of circles $x^2 + y^2 + kx + 1 = 0$ centered at the x axis as the orthogonal family. See Figure 10.3.

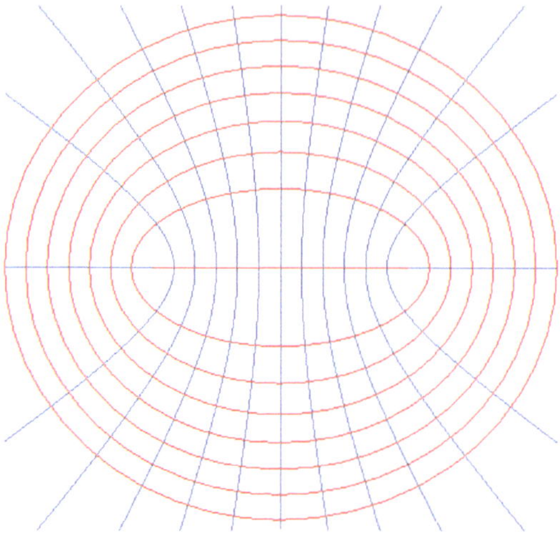

Figure 10.2. A bi-orthogonal family of ellipses and hyperbolas.

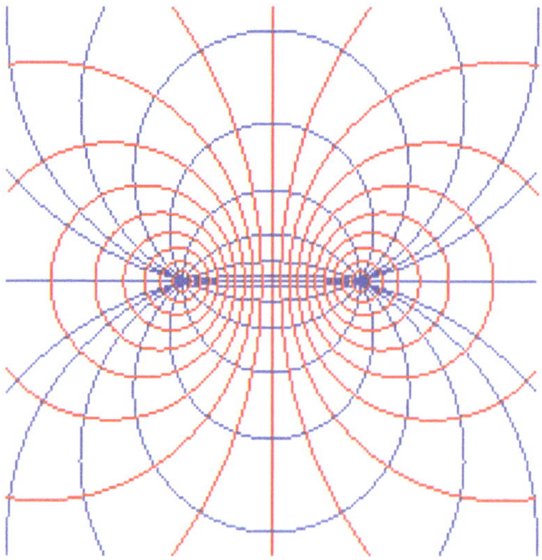

Figure 10.3. A bi-orthogonal family of circles.

Example 10.4.

(a) The family of curves $y = cx^\alpha$ in $x > 0$ has differential equation $x\,dy - \alpha y\,dx = 0$. The orthogonal family, solutions of $x\,dx + \alpha y\,dy = 0$, are the ellipses $x^2 + \alpha y^2 = k$ of fixed excentricity.

(b) The family $u = x^4 + y^4 = c$ has differential equation $x^3dx + y^3dy = 0$; the one of the orthogonal family is $y^{-3}dy = x^{-3}dx$ and the orthogonal family is $v = y^{-2} - x^{-2} = c$, that is, $x^2 - y^2 = cx^2y^2$.

10.1.3 Note that $\frac{\partial}{\partial v} = J(\frac{\partial}{\partial u})$ only when $\lambda = 1, \phi = 0$. Then by equation (10.1) $\Delta u = 0$, so u is harmonic and v is a harmonic conjugate of u considered in Theorem 9.5. We call it a *isothermal coordinate system.* In the coordinates u, v, the metric has the form

$$ds^2 = \mu(x, y)(du^2 + dv^2), \quad \mu = \frac{1}{|\nabla u|^2}.$$

The general isothermal coordinate system (u, v), that is

$$u_x^2 + u_y^2 = \mu = v_x^2 + v_y^2, \quad u_x v_x + u_y v_y = 0,$$

corresponds to either $\nabla v = \pm J(\nabla u)$.

Looking dynamically to the transformation $\Phi : (x, y) \to (u, v)$ this means that $d\Phi$ scales lengths in all directions by the factor μ, so Φ preserves angles. We say that Φ is *conformal.* Thus, conformal maps in the plane are given by a holomorphic or anti-holomorphic transformation, as already shown in paragraph 9.5.2.

We reach the same conclusion if we require that the vector fields $X = (M, N), JX = (-N, M)$ are coordinate vector fields. From Theorem 8.9, we know that the condition to impose is that $[X, JX] = 0$. This leads to the system

$$M(N_x + M_y) + N(N_y - M_x) = 0, \quad -N(N_x + M_y) + M(N_y - M_x) = 0,$$

implying, since $(M, N) \neq 0$,

$$N_x = -M_y, \quad N_y = M_x,$$

so M, N is a pair of conjugate harmonic functions.

Example 10.5.

(a) The functions $u = x^2 - y^2, v = 2xy$ constitute an isothermal coordinate system whose axes are the two families of hyperbolas

$$x^2 - y^2 = c, \quad xy = d,$$

corresponding to $w = \Phi(z) = z^2$, in complex coordinates $z = x + iy$. See Figure 10.4.

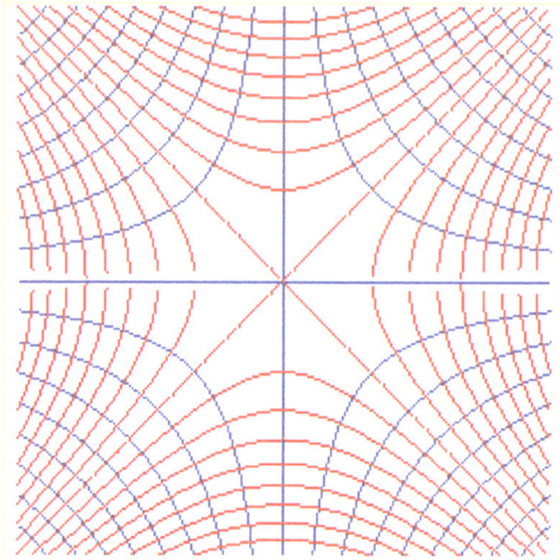

Figure 10.4. An isothermal coordinate system of hyperbolas.

(b) The *inversion map* $\Phi(z) = \frac{1}{\bar{z}} = \frac{1}{x^2+y^2}(x, y)$ gives isothermal coordinates with the orthogonal families

$$x = \lambda(x^2 + y^2), \quad y = \mu(x^2 + y^2),$$

of circles through the origin and centered at the axis. See Figure 10.5.

(c) Using $z = w^2$ instead, we get the two families of confocal parabolas

$$x + \sqrt{x^2 + y^2} = k_1, \quad \sqrt{x^2 + y^2} - x = k_2.$$

10.1.4 Obviously, a conformal map transforms a system of bi-orthogonal curves into another one. A computation shows that the confocal ellipses and hyperbolas of Example 10.2 are in fact the image of vertical and horizontal axes by the holomorphic map $\Phi(z) = \frac{1}{2}(z + \frac{1}{z})$.

For a conformal map, if we additionally require that X has length one, $M^2 + N^2 = 1$, then

$$MM_x - NM_y = 0, \quad NM_x + MM_y = 0,$$

so that M, N are constant. This means that the only orthogonal coordinate systems u, v with ∂_u, ∂_v unitary are cartesian coordinate systems obtained by rotating and translating the canonical one.

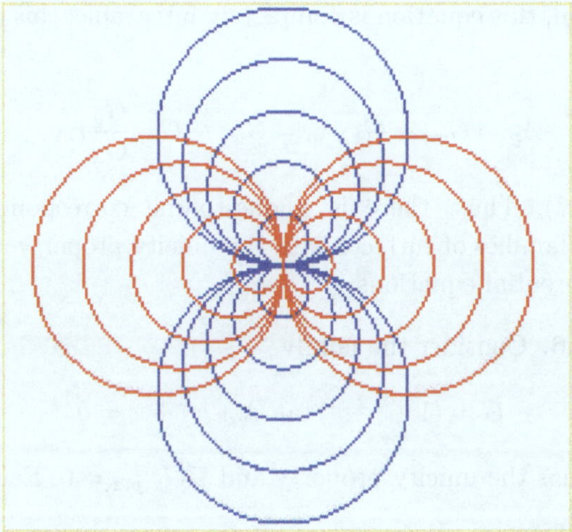

Figure 10.5. An isothermal coordinate system of circles.

10.2 Families of Curves and Surfaces in Space

10.2.1 As seen in paragraph 9.3.4, every completely integrable equation

$$A(x,y,z)\,dx + B(x,y,z)\,dy + C(x,y,z)\,dz = 0,$$

has locally an integrating factor and gives rise to a uniparametric family of surfaces $F(x,y,z) = c$, one through each point p.

Consider a uniparametric family S_λ of surfaces in $\mathbb{R}^3$

$$G(x,y,z,\lambda) = 0, \quad \nabla_{x,y,z} G \neq 0,$$

with the *unicity property*, meaning that for each p there is one and only one surface S_λ through p. For instance, by the implicit function theorem if $G_\lambda \neq 0$, $G = 0$ defines $\lambda = \lambda(x,y,z)$ locally. S_λ is a solution of the differential equation

$$G_x\,dx + G_y\,dy + G_z\,dz = 0,$$

where $G(x,y,z,\lambda) = 0$, so eliminating λ we get a differential equation of the form above with

$$\begin{aligned} &A = G_x(x,y,z,\lambda(x,y,z)), \quad B = G_y(x,y,z,\lambda(x,y,z)), \\ &C = G_z(x,y,z,\lambda(x,y,z)). \end{aligned}$$

By construction, this equation is completely integrable; this can be checked too using that

$$A_y = G_{xy} + G_{x\lambda}\lambda_y = G_{xy} - G_{x\lambda}\frac{G_y}{G_\lambda}, \ldots,$$

and using (9.7). Thus, there is a one-to-one correspondence between uniparametric families of surfaces with the unicity property and completely integrable differential equations.

Example 10.6. Consider the family

$$G = (1 + y^2z^2)\tan(\lambda + x) + x = 0,$$

which clearly has the unicity property and $\nabla G_{x,y,z} \neq 0$. Each satisfies

$$\left[(1 + y^2z^2)\frac{1}{\cos^2(\lambda + x)} + 1\right] dx + 2yz^2\tan(\lambda + x)\, dy + 2zy^2\tan(\lambda + x)\, dz = 0,$$

with $G(x, y, z, \lambda) = 0$, whence

$$[(1 + y^2z^2)^2 + x^2]\, dx - 2xyz^2\, dy - 2xzy^2\, dz = 0,$$

is the equation of this family.

10.2.2 Similar considerations apply to general bi-parametric families $\Gamma_{\lambda,\mu}$ of curves

$$G_1(x, y, z, \lambda, \mu) = 0, \quad G_2(x, y, z, \lambda, \mu) = 0,$$

with $\nabla_{xyz}G_i, i = 1, 2$, linearly independent, with again the unicity property, for each p there is a unique curve in the family through p. By the inverse function theorem, this is locally so if $\nabla_{\lambda,\mu}G_i, i = 1, 2$ are linearly independent.

These satisfy the equation

$$\frac{dx}{A_1(x, y, z, \lambda, \mu)} = \frac{dy}{A_2(x, y, z, \lambda, \mu)} = \frac{dz}{A_3(x, y, z, \lambda, \mu)},$$

with $X = (A_1, A_2, A_3) = \nabla_{xyz}G_1 \times \nabla_{xyz}G_2$. Upon elimination of λ, μ one gets the equation of the family

$$\frac{dx}{B_1(x, y, z)} = \frac{dy}{B_2(x, y, z)} = \frac{dz}{B_3(x, y, z)}.$$

10.2.3 A uni-parametric family of surfaces and a bi-parametric family of curves, both with the unicity property, are said to be *orthogonal* if for each point p the surface and curve through p meet orthogonally. This means that the associated equations have the form

$$A(x, y, z)\, dx + B(x, y, z)\, dy + C(x, y, z)\, dz = 0,$$

$$\frac{dx}{A(x, y, z)} = \frac{dy}{B(x, y, z)} = \frac{dx}{C(x, y, z)},$$

respectively. It follows from Theorem 8.5 that every family of surfaces has an orthogonal family of curves, but not the other way around, as the tangent field $X = (A, B, C)$ must satisfy the complete integrability condition (9.8).

Example 10.7.

(a) The family of ellipsoids

$$\frac{x^2}{a} + \frac{y^2}{b} + \frac{z^2}{c} = r^2,$$

with $a, b, c > 0$ fixed has the orthogonal family of curves given by

$$a\frac{dx}{x} = b\frac{dy}{y} = c\frac{dz}{z},$$

that is $x^a = y^b = z^c$.

(b) The family of surfaces $xy + xz + yz = c$ has the orthogonal family given by

$$\frac{dx}{y + z} = \frac{dy}{x + z} = \frac{dz}{x + y}.$$

With $u = x - y, v = y - z, w = z - y$, this amounts to

$$\frac{du}{u} = \frac{dv}{v} = \frac{dw}{w},$$

thus obtaining $(x - y)^2 = k_1(y - z)^2, (y - z)^2 = k_2(z - x)^2$.

The following is a well-known example of a bi-parametric family of lines with no orthogonal family of surfaces.

Example 10.8. The bi-parametric family consists of lines through points $(\lambda, \mu, 1)$ in $z = 1$ with direction vector $(\lambda + \mu, \mu - \lambda, 2)$, that is

$$(x, y, z) = (\lambda, \mu, 1) + t(\lambda + \mu, \mu - \lambda, 2).$$

Solving for t, λ, μ gives

$$\lambda = \frac{(1+t)x - ty}{t^2 + (1+t)^2}, \quad \mu = \frac{(1+t)y + tx}{t^2 + (1+t)^2}, \quad 2t = z - 1,$$

so that the direction vector $(\lambda+\mu, \mu-\lambda, 2)$ through (x, y, z) is proportional to

$$X = (zx + y, zy - x, 1 + z^2).$$

Now, a computation shows that the left-hand side of (9.8) is $-x^2 - y^2 - 2(1 + z^2)$, so no orthogonal surface exists at any point.

10.3 Triply Orthogonal Families of Curves and Surfaces

10.3.1 In this section, we deal with *triply orthogonal systems* of curves and surfaces in a domain U. In the first case, it is meant three systems Γ_i, $i = 1, 2, 3$ of curves such that for every point $p \in U$ there is a unique curve in each system through p meeting orthogonally. In the second case, we mean three systems S_i, $i = 1, 2, 3$ of surfaces such that for every $p \in U$ there is a unique surface in each system through p meeting orthogonally. Of course, the second situation implies the first one, with $\Gamma_1 = S_2 \cap S_3, \Gamma_2 = S_1 \cap S_3, \Gamma_3 = S_1 \cap S_2$.

Both situations are coded by three mutually orthogonal vector fields X_i, X_i being the tangent vector to Γ_i, normal to S_i, respectively. In the first case there is no other requirement on the X_i; however, in case of surfaces, by Theorem 9.2 they must satisfy the integrability condition (9.8)

$$\langle \nabla \times X_i, X_i \rangle = 0, \quad i = 1, 2, 3. \tag{10.2}$$

Being orthogonal, X_i, X_j span the tangent space to S_k, so in terms of commutators we must have

$$\langle [X_i, X_j], X_k \rangle = 0, \quad i \neq j \neq k. \tag{10.3}$$

10.3.2 We can move to $\mathbb{R}^n$ and consider n systems of mutually orthogonal hyper-surfaces S_i given (locally) by $u_i = c_i$, that is ∇u_i are orthogonal. Then $u_1, \ldots, u_n$ is a local coordinate system; we use the notations in paragraph 6.1.1 for the coordinate fields

$$\partial_i = \sum_j \frac{\partial x_j}{\partial u_i} D_j,$$

and define

$$\frac{1}{H_i} = |\nabla u_i|.$$

Then the matrix $(H_i D_j u_i)$ is orthogonal, whence its inverse matrix equals its transpose,

$$\partial_j x_i = H_j^2 D_i u_j, \tag{10.4}$$

or $\partial_j = H_j^2 \nabla u_j$. In the coordinates $u_1, \ldots, u_n$, the Euclidean metric diagonalizes, it is given by

$$ds^2 = \sum_j H_j^2 du_j^2.$$

We say that $u_1, \ldots, u_n$ are *orthogonal coordinates*, as before when $n = 2$.

Thus, all three settings — mutually orthogonal fields satisfying (10.2) or (10.3), mutually orthogonal systems of hyper-surfaces, orthogonal coordinates — are equivalent, each X_i being proportional to ∇u_i. Mutually orthogonal fields satisfying the stronger condition $[X_i, X_j] = 0$ means that the orthogonal coordinates can be chosen so that $X_i = \partial_i$.

10.3.3 As before, when all $H_j = H$ are equal, we say that the coordinates are *conformal*. In these coordinates, the metric takes the form

$$ds^2 = H^2 \sum_j du_j^2,$$

H being called the *dilation factor*. Looking at $\Phi : (x_1, \ldots, x_n) \to (u_1, \ldots, u_n)$ dynamically, this means that the differential $d\Phi$ is a linear map

$$d\Phi(x) = \frac{1}{H(x)} M(x),$$

with $M(x)$ an orthogonal matrix, that is, for all directions v, w

$$\langle D_v\Phi, D_w\Phi\rangle = \frac{1}{H(x)} \langle v, w\rangle,$$

meaning that Φ preserves angles, and we call Φ a *conformal mapping*, see paragraph 1.2.4. In this case, the column vectors $D_i\Phi$ are also orthogonal, meaning that the images of coordinate axes and hypersurfaces are mutually orthogonal, too. Thus, with every conformal map one has two systems of mutually orthogonal hypersurfaces, images and pre-images of cartesian hypersurfaces.

Besides the trivial cases — linear orthonormal transformations and dilations — an important example is the *inversion map* with pole at the origin

$$\rho(x) = \frac{x}{|x|^2}.$$

Indeed,

$$\nabla \rho_i = \nabla \frac{x_i}{|x|^2} = \frac{1}{|x|^2} e_i - 2\frac{x_i}{|x|^4} x,$$

so

$$\langle \nabla \rho_i, \nabla \rho_k \rangle = |x|^{-4} \delta_{ik},$$

with dilation factor $|x|^2$. The inversion with pole at p is obtained by replacing x by $x - p$.

In dimension $n = 2$, we have seen before that there are plenty of conformal maps, all holomorphic or anti-holomorphic transformations are. For $n \geq 3$, however, we will see in the next section that a sufficiently smooth conformal map is a composition of linear maps and inversions.

10.3.4 In the following, we provide some examples of triply orthogonal systems of surfaces. In $\mathbb{R}^3$, the easiest genuine example of orthogonal coordinates are the spherical ones ρ, ϕ, θ for which

$$ds^2 = d\rho^2 + \rho^2 \sin^2 \phi \; d\theta^2 + \rho^2 d\phi^2,$$

with spheres, cones with axis z and planes containing this axis as orthogonal surfaces. See Figure 10.6

To produce other examples, we point out some general facts.

(a) First, starting from such a situation in dimension n and adding $u_{n+1} = x_{n+1}$ one gets a system in $\mathbb{R}^{n+1}$. This is the case of cylindrical coordinates in $\mathbb{R}^3$. See Figure 10.7.
(b) Secondly, applying a conformal mapping to a system we obtain another system. Similarly as in dimension two, the images of cartesian coordinate planes by the inversion ρ lead to the system of spheres through the origin centered at the coordinate axis, that is,

$$x^2 + y^2 + z^2 = ax, \quad x^2 + y^2 + z^2 = by, \quad x^2 + y^2 + z^2 = cz.$$

See Figure 10.8.

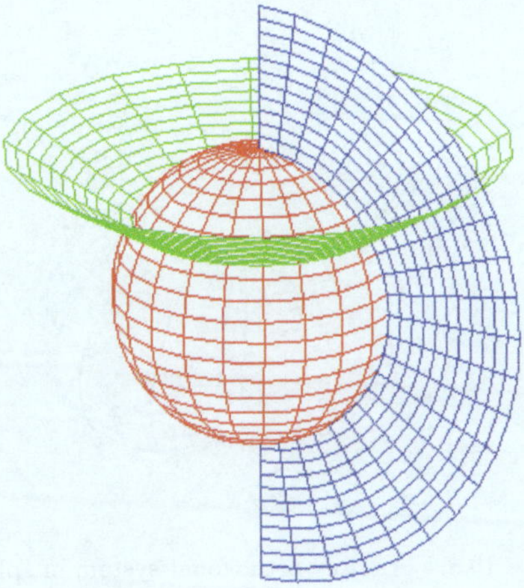

Figure 10.6. Spherical coordinates.

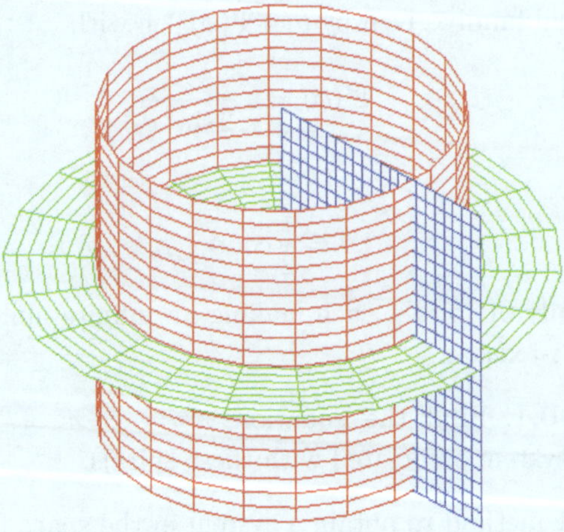

Figure 10.7. Cylindrical coordinates.

(c) Next, note that starting from orthogonal coordinates u_1, u_2, u_3 one can replace say u_2, u_3 by a new system v_2, v_3 simply choosing $v_2 = v_2(u_2, u_3)$ arbitrarily and then choosing $v_3 = v_3(u_2, u_3)$ so that the systems of curves $v_2 = c_2, v_3 = c_3$ in the u_2, u_3 surface are orthogonal. Since the tangent to $v_i(u_2, u_3) = c_i$ is $\partial_3 v_i \partial_2 - \partial_2 v_i \partial_3$, this means

$$H_2^2(\partial_3 v_2)(\partial_3 v_3) + H_3^2(\partial_2 v_2)(\partial_2 v_3) = 0.$$

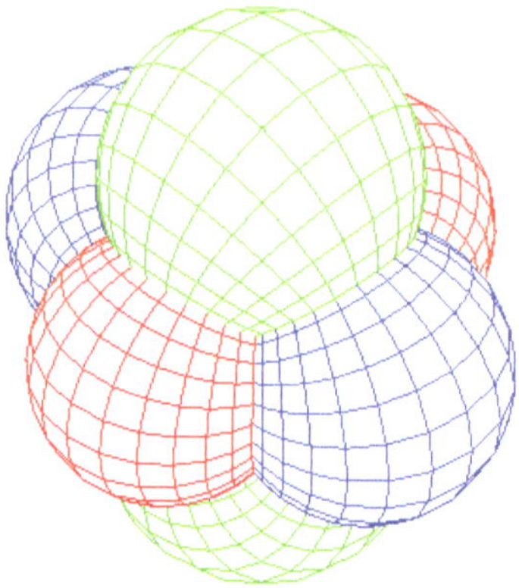

Figure 10.8. A triply orthogonal system of spheres.

For instance, in the sphere with spherical coordinates, the family of spirals $\theta = 2\phi + c$ (each making a full rotation around the z axis) has an orthogonal family given by $\theta = \Psi(\phi) + c$ with

$$\Psi'(\phi) = -\frac{1}{2\sin^2\phi},$$

that is

$$\theta = \frac{1}{2}\cot\phi + c,$$

another family of spirals, each making an infinite number of rotations around the z-axis.

Exercise 10.1. Find the equations $v = c$, $w = d$ of the triply orthogonal system associated with these spirals.

(d) Still another method to obtain a system in the space from a system in the plane is to rotate the latter around a symmetry axis of the plane system. Assuming this axis is the y-axis, and that

$$u(x^2, y) = c, \quad v(x^2, y) = d$$

is the plane system, this means replacing x^2 by $x^2 + z^2$,

$$u(x^2 + z^2, y) = c, \quad v(x^2 + z^2, y) = d.$$

These, together with the planes $z = \lambda x$, constitute a triple system in space. Applying this to the plane systems of circles described in the previous section one obtains systems in the space with spheres, planes and toruses.

Finally, we mention two versions of the space analogue of the plane confocal system of ellipses and hyperbolas seen in the previous section.

(e) Consider the equation

$$\frac{x^2}{a^2-\lambda}+\frac{y^2}{b^2-\lambda}+\frac{z^2}{c^2-\lambda}=1, \quad a^2<b^2<c^2. \tag{10.5}$$

If $\lambda < a^2$, this is an ellipsoid, if $a^2 < \lambda < b^2$, it is a one-sheet hyperboloid, and a two-sheet hyperboloid for $b^2 < \lambda < c^2$. Now, given (x, y, z), this is a function in λ with limit zero at $-\infty$, left-hand limit $+\infty$ at a^2, b^2, c^2 and right-hand limit $-\infty$ at a^2, b^2, c^2. On the other hand, the equation is cubic in λ. Therefore, it has a root $\lambda_1(x, y, z) < a^2$, another root $a^2 < \lambda_2 < b^2$ and a third root $b^2 < \lambda_3 < c^2$.

Thus, through each point there is exactly one surface of each family. We claim that they form a triply orthogonal system. Indeed, the three normals are proportional to

$$n_i=\left(\frac{x}{a^2-\lambda_i},\frac{y}{b^2-\lambda_i},\frac{z}{c^2-\lambda_i}\right), \quad i=1,2,3.$$

Then

$$\langle n_i,n_j\rangle=\frac{x^2}{(a^2-\lambda_i)(a^2-\lambda_j)}+\frac{y^2}{(b^2-\lambda_i)(b^2-\lambda_j)}+\frac{z^2}{(c^2-\lambda_i)(c^2-\lambda_j)},$$

which is seen to be zero subtracting (10.5) with $\lambda = \lambda_1, \lambda_2$. The functions λ_i are called the *ellipsoidal coordinates* in space.

(f) For the last example, with $a^2 < b^2$ as before, we consider the two elliptic cones

$$\frac{x^2}{\lambda_1}=\frac{y^2}{b^2-\lambda_1}+\frac{z^2}{a^2-\lambda_1}, \quad 0<\lambda_1<a^2, \qquad \frac{y^2}{b^2-\lambda_2}=\frac{x^2}{\lambda_2}+\frac{z^2}{\lambda_2-a^2}.$$

Again, through each point there is a unique cone in each family, and it is easily checked that together with the spheres they constitute a triply orthogonal system.

As a final remark, most of the triple systems mentioned here have singular points where some of the surfaces of the family are not defined.

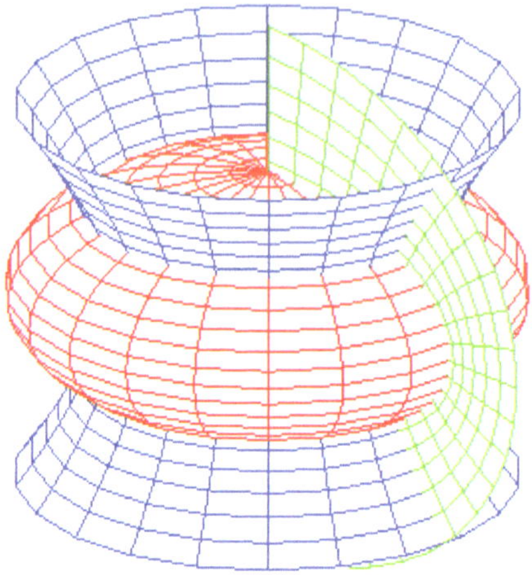

Figure 10.9. A triply orthogonal system with ellipsoids and cones.

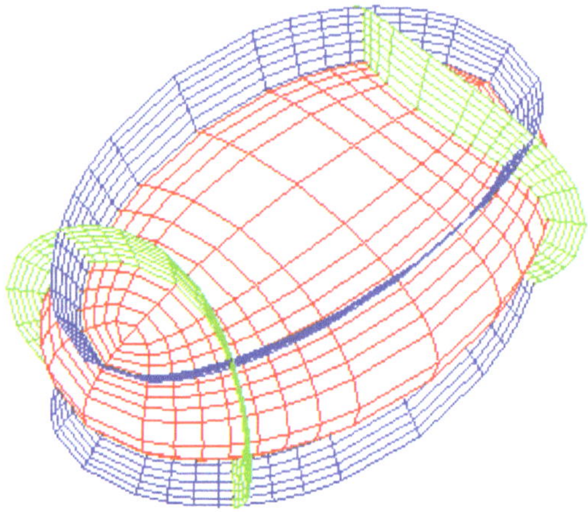

Figure 10.10. A triply orthogonal system with ellipsoids and elliptic cones.

Other examples are shown in Figures 10.9 and 10.10. These figures are taken from the interesting website [11] containing a handful of examples.

10.4 Rigidity of Conformal Maps in Space

In this section, we prove Liouville's theorem announced in the previous section. We have seen that the inversion is conformal; the inversion, together with dilations and rigid motions, generate a group of conformal maps, called *Möbius maps.* These are all conformal maps:

Theorem 10.1. *If $n \geq 3$, every C^1 conformal diffeomorphism $\Phi = (u_1, \ldots, u_n)$ defined in a domain U is the restriction of a Möbius transformation. Every conformal coordinate system is obtained transforming Cartesian coordinates by a Möbius map.*

This result holds even with weaker regularity assumptions on Φ. We will present for simplicity an adaptation of Darboux's proof [5] in case Φ is of class C^3.

Proof. With $\Phi = (u_1, \ldots, u_n)$ the hypothesis is that the column vectors $X_i = D_i\Phi$ satisfy

$$\langle X_j, X_k \rangle = h^2 \delta_{kj}, \tag{10.6}$$

where remember h is called the *dilation factor*. Now we differentiate this equation, that is, we consider the fields $D_iX_k = D_iD_k\Phi$, which are symmetric in i, k (in fact, $D_iX_k = (H\Phi)(D_i, D_k)$, where here we look at the hessian $H\Phi = (Hu_1, \ldots, Hu_n)$ as a symmetric vector-field valued bilinear map). We get for all i

$$\langle D_iX_j, X_k \rangle + \langle X_j, D_iX_k \rangle = 0, \quad j \neq k,$$

and

$$\langle D_iX_k, X_k \rangle = hD_ih.$$

If $i \neq j \neq k$ and we consider the first equation for the triplets $(i, j, k), (j, k, i)$ and (k, i, j), we conclude that $\langle D_iX_j, X_k \rangle = 0$ for $i \neq j \neq k$. This means that D_iX_j must be a linear combination of X_i, X_j with coefficients, using the second equation,

$$D_iX_j = \frac{D_jh}{h}X_i + \frac{D_ih}{h}X_j, \quad i \neq j.$$

We compute D_iX_i in the same way, obtaining

$$D_iX_i = \frac{D_ih}{h}X_i - \sum_{l \neq i} \frac{D_lh}{h}X_l.$$

At this point, we note that in case h is constant, then all fields X_i are constant and Φ is a linear orthogonal map followed by a translation.

As a consequence, if two conformal mappings Φ_1, Φ_2 have the same dilation factor $h_1 = h_2$, one is obtained from the other by a rigid motion, because $\Phi_1(\Phi_2)^{-1}$ has dilation factor one.

It is convenient to use the unitary fields $U_j = \frac{X_j}{h}$ instead, and introduce Darboux's notation $\beta_j = \frac{D_j h}{h}$. In terms of those, these relations become

$$D_i U_j = \beta_j U_i, \quad i \neq j, \quad D_i U_i = -\sum_{l \neq i} \beta_l U_l. \tag{10.7}$$

We look now at $D_k D_i U_j$ for $k \neq i \neq j$ (in the context of paragraph 8.1.3, this would correspond looking to the third covariant derivative $(d^\nabla)^3\Phi$, a symmetric trilinear map). Using (10.7), we obtain

$$D_k D_i U_j = D_k(\beta_j U_i) = (D_k\beta_j)U_i + \beta_j D_k U_i = (D_k\beta_j)U_i + \beta_j\beta_i U_k.$$

Since this is symmetric in i, k, we conclude that

$$D_i\beta_j = \beta_i\beta_j, \quad i \neq j. \tag{10.8}$$

Next we look at $D_k D_i U_i$ for $k \neq i$. On the one hand,

$$D_k(D_i U_i) = -D_k\left(\sum_{l\neq i}\beta_l U_l\right) = -\sum_{l\neq i}(D_k\beta_l)U_l - \sum_{l\neq i}\beta_l D_k U_l.$$

In both sums we distinguish whether $l = k$ or not and use (10.7), (10.8):

$$\begin{aligned} &= -(D_k\beta_k)U_k - \sum_{l\neq k, l\neq i}\beta_k\beta_l U_l + \sum_{p\neq k}\beta_k\beta_p U_p - \sum_{l\neq k, l\neq i}\beta_l^2 U_k \\ &= -D_k\beta_k U_k + \beta_k\beta_i U_i - \sum_{l\neq i, l\neq k}\beta_l^2 U_k. \end{aligned}$$

On the other hand,

$$D_i(D_k U_i) = D_i(\beta_i U_k) = (D_i\beta_i)U_k + \beta_i\beta_k U_i.$$

So we get

$$D_k\beta_k + D_i\beta_i + \sum_{l\neq i, l\neq k}\beta_l^2 = 0. \tag{10.9}$$

In terms of $\rho = \frac{1}{h}$, equation (10.8) becomes

$$D_{ij}\rho = 0, \quad i \neq j,$$

while equation (10.9) becomes

$$\rho(D_{ii}\rho + D_{jj}\rho) = \sum_l (D_l\rho)^2.$$

This implies that

$$\rho(x) = \sum_j \rho_j(x_j),$$

with all quantities $\rho_j''(x_j)+\rho_i''(x_i), i \neq j$ being equal. Since $n > 2$, all $\rho_j''(x_j)$ are equal, therefore a constant. Then ρ must have the form

$$\rho(x) = a\left(\sum_i x_i^2\right) + 2\sum b_i x_i + c.$$

When replaced in (10.9), we get that a, b_i, c must satisfy

$$\sum_i b_i^2 = ac.$$

Next, if $a = 0$, then ρ and h are constants and Φ is a composition of a dilation, an orthogonal linear transformation and a translation. If $a \neq 0$, then with $b_i = ac_i, c = a\sum_i c_i^2, p = (c_1, \ldots, c_n)$, we recognize in

$$h(x) = \frac{1}{a(x^2 + 2\sum_i c_i x_i + \sum_i c_i^2)} = \frac{1}{a}\frac{1}{|x+p|^2},$$

the dilation factor of a dilation followed by an inversion with pole at $-p$. □

10.5 The Lamé Surfaces*

10.5.1 In this section, we consider again triply orthogonal systems of surfaces in space, that is, three mutually orthogonal vector fields X_i satisfying the integrability conditions (10.2) or (10.3). Then $\lambda_i X_i = \nabla u_i$ for some integrating factors λ_i and functions u_i, so that the three systems of surfaces are given by $u_i = c_i$.

In dimension two, we saw that every family of curves can be completed to a bi-orthogonal system. The problem whether in space an arbitrary family of surfaces $u = c$ can be completed to a triple system was of central interest in the 19th century among geometers. Chasles stated incorrectly that this was always the case, later Bouquet gave a counterexample and

Lamé, Cayley, Darboux and others realized that there is a third-order partial differential equation on u governing this situation.

The systems of surfaces $u = c$ that can be completed to a triply orthogonal system are called *Lamé surfaces.* Studying them, we will encounter some notions pertaining to Differential Geometry of surfaces, such as curvature and umbilical points, approached from this perspective.

So we are given a family $u = c$ and consider $u_3 = u$ in the above situation. The question is to find out what condition must satisfy u, or rather $X = \nabla u$, so that there exist two vector fields X_1, X_2, so that

$$\langle X_1, X_2\rangle = 0, \quad \langle X_i, X\rangle = 0, \quad \langle [X_i, X], X_j\rangle = 0, \quad i \neq j. \tag{10.10}$$

All conditions are homogeneous in X_1, X_2; for the last one this is because

$$[\mu X_i, X] = \mu[X_i, X] - X(\mu)X_i.$$

The last two equations are equivalent to the complete integrability of both X_1, X_2

$$\langle \nabla \times X_i, X_i\rangle = 0, \quad i = 1, 2.$$

Thinking that both $X = \nabla u, X_1 = \nabla v$ are gradients, differentiating

$$\langle X, X_1\rangle = \sum_i (D_i u)(D_i v) = 0$$

yields

$$\sum_i (D_{ik}u)(D_i v) = -\sum_i (D_i u)(D_{ik}v), \quad k = 1, \ldots, n,$$

which means that

$$D_X X_1 = -D_{X_1}(X), [X, X_1] = -2D_{X_1}X,$$

and analogously for X_2. Therefore,

$$\langle X_1, X_2\rangle = 0, \langle X_i, X\rangle = 0, \langle D_{X_i}X, X_j\rangle = 0, \quad i \neq j,$$

which remain homogeneous on X_1, X_2. The last condition means that $D_{X_i}X$ is a linear combination of X, X_i.

10.5.2 Next, we will interpret these conditions geometrically, in terms of the surfaces $S : u = c$. First, we observe that

$$\langle D_{X_i} X, X_j \rangle = Hu(X_i, X_j) = 0,$$

with Hu the Hessian of u, is a symmetric condition, as it should be. We introduce the (intrinsic) unit normal

$$N = \lambda X, \quad \lambda = \frac{1}{|\nabla u|}.$$

For a vector field Z tangent to S, $\langle Z, N \rangle = 0$, differentiating $\langle N, N \rangle = 1$ yields $\langle D_Z N, N \rangle = 0$, that is $D_Z N$ is also tangent. The linear map in the tangent space to S

$$W : Z \to D_Z N,$$

is called the *Weingarten* endomorphism. One has

$$D_Z N = Z(\lambda) X + \lambda D_Z X, \tag{10.11}$$

showing that $\frac{1}{\lambda} D_Z N$ is the tangential component of $D_Z X$.

Thus, $D_{X_i} X$ is a linear combination of X, X_i if and only if X_i is a eigenvector of W. The eigenvectors of W are called *principal directions*; their integral curves are called *principal curvature lines*, the eigenvector being interpreted as a curvature. So, we can restate our analysis up to now in the following theorem of Dupin:

Theorem 10.2. *The surfaces of a triply orthogonal system meet along their principal curvature lines.*

If both Z_1, Z_2 are tangent to S, it follows from (10.11) that

$$\langle D_{Z_1} N, Z_2 \rangle = \lambda \langle D_{Z_1} X, Z_2 \rangle = \lambda Hu(Z_1, Z_2).$$

The left-hand side is called the *second fundamental form* of S, of main importance in the theory of surfaces. Our analysis shows how it arises too in relation to triply orthogonal systems.

10.5.3 We continue our analysis of equation (10.10) satisfied by a triply orthogonal system and its consequence

$$Hu(X_1, X_2) = 0. \tag{10.12}$$

The bilinear map Hu (or rather its restriction to the tangent space) is symmetric and so it diagonalizes in an orthonormal basis. Up to the factor λ the eigenvalues are those of W, the principal curvatures. If they are equal, then $Hu(Z_1, Z_2) = 0$ if Z_1, Z_2 are orthogonal. The points in S where the two principal curvatures agree are called *umbilical.* At those points, the condition

$$Hu(X_1, X_2) = 0$$

is then automatic. For instance, if S consists entirely of umbilical points, such as a sphere or a plane (in fact, such a surface must be either a sphere or a plane), then this condition is empty. At a non-umbilical point, Hu has well-defined principal directions corresponding to the two different eigenvalues. Both the eigenvalues and eigenvectors will depend smoothly on the coefficients of Hu (rather its restriction to the tangent space). Imposing the integrability condition on the principal directions will thus lead to a third-order partial differential equation on u (in fact, the above analysis shows that if one of the directions is completely integrable, then so is the other one). Finding explicitly this partial differential equation requires some algebraic work, first done by Cayley and Darboux, that we will describe in what follows.

Before doing so in general, we look now at a particular case,

$$u(x, y, z) = A_1(x) + A_2(y) + A_3(z),$$

first studied by Bouquet, for which the computations are easier.

We find first the principal directions $X_2 = (B_1, B_2, B_3)$, that is the tangent fields for which $D_{X_2}X - \lambda X_2$ is proportional to X, that is

$$(B_1(A_1'' - \lambda), B_2(A_2'' - \lambda), B_3(A_3'' - \lambda)) = \mu(A_1', A_2', A_3').$$

This means that

$$B_i = \mu \frac{A_i'}{A_i'' - \lambda}, \quad i = 1, 2, 3,$$

where λ must be chosen so that X_2 is tangent, that is, a root of the equation

$$\Delta = \sum_i \frac{(A_i')^2}{A_i'' - \lambda} = 0.$$

A smooth choice of λ is possible outside the umbilical points. The complete integrability condition for X_2 becomes

$$\langle \nabla \times X_2, X_2 \rangle = B_2 B_3 (D_1 \lambda) \left(\frac{1}{A_2'' - \lambda} - \frac{1}{A_2'' - \lambda} \right)$$

$$+ B_1 B_3 (D_2 \lambda) \left(\frac{1}{A_3'' - \lambda} - \frac{1}{A_1'' - \lambda} \right) + B_1 B_2 (D_3 \lambda) \left(\frac{1}{A_1'' - \lambda} - \frac{1}{A_2'' - \lambda} \right) = 0.$$

From the equation $\Delta = 0$, a computation shows that

$$|X_2|^2 D_i \lambda = B_i^2 A_i''' - 2 B_i A_i'',$$

which leads to

$$(A_3'' - A_2'')(A_1' A_1''' - 2A_1''(A_1'' - \lambda)) + (A_1'' - A_3'')(A_2' A_2''' - 2A_2''(A_2'' - \lambda)) + (A_2'' - A_1'')(A_3' A_3''' - 2A_3''(A_3'' - \lambda)) = 0,$$

and so to the third-order differential equation

$$(A_3'' - A_2'')(A_1' A_1''' - 2(A_1'')^2) + (A_1'' - A_3'')(A_2' A_2''' - 2(A_2'')^2) + (A_2'' - A_1'')(A_3' A_3''' - 2(A_3'')^2) = 0.$$

For example, for the family of ellipsoids

$$\frac{x^2}{a^2} + \frac{y^2}{b^2} + \frac{z^2}{c^2} = c,$$

the condition becomes

$$(a^2 - b^2)(b^2 - c^2)(c^2 - a^2) = 0,$$

so it is not a Lamé family unless they are of revolution.

10.5.4 We continue now discussing (10.10) and (10.12) in general. Loosely speaking, we have six unknown functions, the components of X_1, X_2, four equations, all linear in X_1, X_2 and not involving derivatives of X_i.

$$\langle X_1, X_2 \rangle = 0, \quad \langle X_i, X \rangle = 0, \quad Hu(X_1, X_2) = 0.$$

To these we should add the complete integrability condition on either X_1, X_2, but they involve derivatives of X_i. The equations above are simply those of the principal curvature directions, so they cannot contain all information. To obtain an additional linear equation on the components of X_i we will differentiate the last one with respect to X. This uses that not only a single surface $u = c$ is within a triple system, but the whole family. Assuming again that $X_1 = \nabla v, X_2 = \nabla w$, we thus consider

$$0 = X(Hu(X_1, X_2)) = X\left(\sum_{ij} D_{ij}uD_ivD_jw\right)$$

$$= \sum_{ij} X(D_{ij}u)D_ivD_jw + \sum_{ij} D_{ij}u(D_jwXD_iv + D_ivXD_jw).$$

The last two terms equal

$$\langle D_XX_1, D_{X_2}X\rangle + \langle D_XX_2, D_{X_1}X\rangle,$$

whence, using again that $D_XX_1 = -D_{X_1}(X), D_XX_2 = -D_{X_2}(X)$, we obtain another linear equation in the components of X_1, X_2.

$$X(Hu(X_1, X_2)) = \sum_{ij} X(D_{ij}u)D_ivD_jw - 2\langle D_{X_1}X, D_{X_2}X\rangle.$$

With the notation A, B for matrices

$$a_{ij} = D_{ij}u, b_{ij} = X(a_{ij}) - 2\sum_k a_{ik}a_{jk},$$

the last equation is

$$\langle BX_1, X_2\rangle = \sum_{ij} b_{ij}D_ivD_jw = 0.$$

Together with this we have the equations

$$\langle AX_1, X_2\rangle = \sum_{ij} a_{ij}D_ivD_jw = 0,$$

and the orthogonality conditions, which Darboux writes in the form

$$\langle X_1, X_2\rangle = 0, \quad \langle X_1, X\rangle X_2 + \langle X_2, X\rangle X_1 = 0.$$

These are six homogeneous equations in the six variables

$$\alpha_1 = (D_1v)(D_1w), \quad \alpha_2 = (D_2v)(D_2w), \quad \alpha_3 = (D_3v)(D_3w),$$
$$\alpha_{12} = (D_1v)(D_2w) + (D_2v)(D_1w), \quad \alpha_{31} = (D_1v)(D_3w) + (D_3v)(D_1w),$$
$$\alpha_{23} = (D_2v)(D_3w) + (D_3v)(D_2w).$$

Since $X_1, X_2 \neq 0$, not all α are zero, and we finally obtain the desired third-order equation that a Lamé family must satisfy: $\det M = 0$ where

$$M = \begin{pmatrix} b_{11} & b_{22} & b_{33} & b_{23} & b_{31} & b_{12} \\ a_{11} & a_{22} & a_{33} & a_{23} & a_{31} & a_{12} \\ 1 & 1 & 1 & 0 & 0 & 0 \\ 2D_1u & 0 & 0 & 0 & D_3u & D_2u \\ 0 & 2D_2u & 0 & D_3u & 0 & D_1u \\ 0 & 0 & 2D_3u & D_2u & D_1u & 0 \end{pmatrix}.$$

At a fixed point we can choose the coordinates so that the normal is the x-axis, $D_2u = D_3u = 0$, and the principal directions are those along the y, z-axis, that is $D_{23}u = 0$. Then the equation takes the form

$$2(D_1u)^3(D_{22}u - D_{33}u)(D_1uD_{123} - 2D_{12}uD_{13}u) = 0.$$

This shows that the equation is automatically fulfilled at the umbilical points, as expected.

10.5.5 It remains to show that conversely this equation guarantees that $u = c$ is a Lamé family. If the equation holds, this means that the homogeneous system has a solution with not all α equal to zero. Now we must solve the system

$$\alpha_1 = v_1w_1, \alpha_2 = v_2w_2, \alpha_3 = v_3w_3, \alpha_{12} = v_1w_2 + v_2w_1, \alpha_{31} = v_1w_3 + v_3w_1,$$

$$\alpha_{23} = v_2w_3 + v_3w_2,$$

to get the six variables $v_i, w_i, i = 1, 2, 3$, modulo a dilation. This requires

$$\alpha_{ij} \geq 4\alpha_i\alpha_j,$$

conditions that can be easily seen to follow from the last three equations in M. This gives us two non-zero vector fields $X_1 = (v_1, v_2, v_3), X_2 =$

(w_1, w_2, w_3), such that

$$\langle X, X_1 \rangle = 0, \quad \langle X, X_2 \rangle = 0, \quad \langle X_1, X_2 \rangle = 0,$$

and

$$A(X_1, X_2) = 0, \quad B(X_1, X_2) = 0.$$

We must prove now that these together imply the integrability conditions

$$\langle [X_i, X], X_j \rangle = 0.$$

The way we obtained the equation for B shows that

$$\begin{aligned} 0 = X(A(X_1, X_2)) - B(X_1, X_2) &= \langle D_X X_1, D_{X_2} X \rangle + \langle D_X X_2, D_{X_1} X \rangle \\ &\quad + 2\langle D_{X_1} X, D_{X_2} X \rangle. \end{aligned} \tag{10.13}$$

We will show that the orthogonality relations among $X = \nabla u, X_1, X_2$, (10.13) and (10.12) imply that X_1, X_2 are completely integrable. Assuming as we may that $|X_1| = |X_2| = 1$, we define coefficients

$$\begin{aligned} D_X X_i &= a_i X + b_i X_1 + c_i X_2, \quad & D_{X_i} X &= a_i' X + b_i' X_1 + c_i' X_2, \\ D_{X_1} X_2 &= dX + eX_1 + fX_2, \quad & D_{X_2} X_1 &= d' X + e' X_1 + f' X_2, \\ D_X X &= \alpha X + \beta X_1 + \gamma X_2. \end{aligned}$$

Differentiating the orthogonality relations among X, X_1, X_2 with respect to X, X_1, X_2 the following relations are seen to hold, with $\lambda = |X|$,

$$c_1 = -b_2, \quad \beta = -\lambda a_1, \quad \gamma = -\lambda a_2.$$

On the other hand, X being a gradient implies $\langle D_{X_i} X, X_j \rangle = \langle D_{X_j} X, X_i \rangle$ for all X_i, X_j, thus

$$\beta = \lambda a_1', \quad b_2' = c_1', \quad \gamma = \lambda a_2'.$$

So one has $c_1 = -b_2, a_1 = -a_1', a_2 = -a_2'$. Equation (10.12) reads $b_2' = c_1' = 0$, while equation (10.13) reads

$$a_1 a_2' + b_1 b_2' + c_1 c_2' + a_1' a_2 + b_1' b_2 + c_1' c_2 + 2(a_1' a_2' + b_1' b_2' + c_1' c_2') = 0.$$

Putting everything together we get

$$b_2(b_1' - c_2') = 0.$$

So, at a non-umbilical point it follows that $b_2 = c_1 = 0$, and so

$$\langle [X, X_2], X_1 \rangle = b_2 - b_2' = 0, \quad \langle [X, X_1], X_2 \rangle = c_1 - c_1' = 0,$$

showing that X_1, X_2 are completely integrable.

In conclusion, the differential equation is meaningful only at non-umbilical points and it is necessary and sufficient in order that $u = c$ can be completed to a triple orthogonal system, which is unique. Depending on the behavior of the principal directions near umbilical points, the two orthogonal families will be defined or not at those points.

Coming back to Bouquet's situation, an easy example of Lamé surface is obtained by solving

$$A'A''' = 2(A'')^2.$$

Its general solution is $A = \log x^\alpha$. Therefore,

$$x^\alpha y^\beta z^\gamma = c$$

is a Lamé family.

Chapter 11

Measuring Sets: The Riemann Integral

When confronted with an undergraduate course on integration theory, it is quite usual that the professor doubts between Riemann's and Lebesgue's theories. Both theories start answering a natural question: what is the measure (length, area, volume) of a set? Can all sets be measured? Riemann's theory is somehow more elementary, needs less machinery and is sufficient for all practical purposes of the integral. For instance, in numerical analysis integrals are viewed and approximated as Riemann integrals. However, the Riemann integral has some mathematical weak points, the most important being that it leads to non-complete spaces, and thus weak existence theorems. Lebesgue's theory is richer, there are more Lebesgue measurable sets than Riemann, leading to complete spaces, and having better properties regarding limits. In this and the next chapter, and for the teacher to decide, we present both theories and compare them. The present chapter deals with Riemann integration.

11.1 Measure of Sets

11.1.1 Based on three intuitive properties, in Section 1.3, we have studied measures of parallelepipeds. In this section, the goal is to *define* the n-dimensional measure $m(A)$ (or $m_n(A)$ when convenient) of a set $A \subset \mathbb{R}^n$.

The ideal goal would be to define $m(A) \geq 0$ for an arbitrary $A \subset \mathbb{R}^n$ and so that

(a) $m(R)$ is the product of the lengths of the sides when R is a rectangle.
(b) It is *countably additive*: if $A_1, A_2, \ldots, A_k, \ldots$ is a sequence of disjoint sets, then $m(\cup_i A_i) = \sum_i m(A_i)$.
(c) m is invariant by rigid motions: $m(T(A)) = m(A)$ for all rigid motions T.

The second property might be replaced by the *finitely additive property*, but this one would lead, as shown in what follows, to certain deficiencies of the theory.

Theorem 11.1. *It is not possible to define the measure $m(A)$ of an arbitrary set so that the three properties above hold. That is, if these properties are to hold, there must exist* non-measurable sets.

Proof. This was shown by Haussdorf in 1914. We reproduce instead the well-known example by Vitali of a non-measurable set.

Using the axiom of choice, we consider a set $E \subset [0, 1]$ that contains exactly one element of each coset $\mathbb{R}/\mathbb{Q}$. There are a non-countable number of cosets, each being countable and dense in $\mathbb{R}$. We claim that E is not measurable, we cannot speak about $m(E)$ and keeping the three properties above. If $y \in [0, 1]$, there exists a unique $x \in E$ such that $y - x \in \mathbb{Q}$, that is, $[0, 1] \subset \cup_q E + q$, where q runs on the rationals in $[-1, 1]$. So,

$$[0, 1] \subset \cup_q E + q \subset [-1, 2],$$

and all translates are disjoint. If E were measurable, we would have, since $m(E + q) = m(E)$,

$$1 \leq \sum_q m(E) \leq 3,$$

which is impossible, because the above sum is zero if $m(E) = 0$ and infinite if $m(E) > 0$. □

Another famous result in the same direction is the Banach–Tarski theorem (or paradox): if $n \geq 3$, a ball can be cut into a finite number of pieces, which using rigid motions can be reassembled into a ball of any desired volume. Of course, these pieces cannot be measurable. We point out that these constructions rely on the axiom of choice.

11.1.2 The goal is then to define $m(A)$, if not for all sets, for a sufficiently large class of sets. We will construct the so-called *Lebesgue measure* starting with elementary sets and moving gradually to more general sets, always guided by the three properties above. At a certain point it will appear the class of *Lebesgue measurable sets*, those sets for which we can speak properly about their measure.

We fix once for all a Cartesian coordinate system, say the canonical one. Of course, we want that the final result be independent of that choice, this is the meaning of the third property above. We first consider an interval $R = \prod[a_i, b_i]$ and define its measure as

$$m(R) = (b_1 - a_1) \cdots (b_n - a_n).$$

In particular, points have zero measure. Next we consider sets $A = \cup R_i$ which can be expressed as union of a finite number of intervals. In this case, the intervals can be chosen with non-overlapping interiors, because if $R_1, \ldots, R_N$ are intervals, by adding new edges we see that there are other intervals with non-overlapping interiors with the same union, see Figure 11.1.

Of course, for such a set we define

$$m(A) = \sum_i m(R_i).$$

Note that the above expression of A is not unique; it should be checked that if $A = \cup_j S_j$ is another one, then $\sum_i m(R_i) = \sum_j m(S_j)$, something that is left to the reader (not entirely trivial from an algebraic point of view).

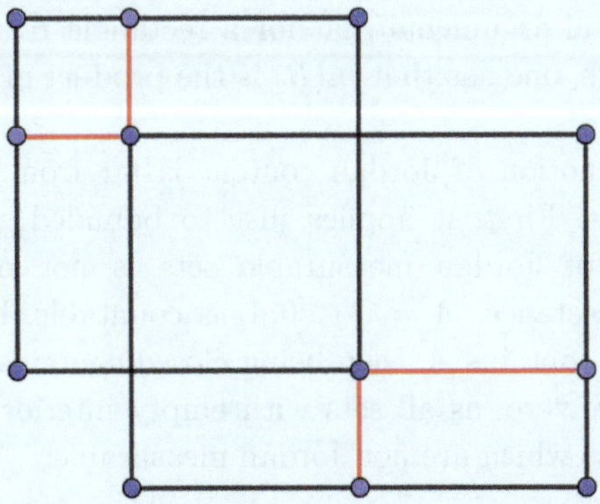

Figure 11.1. Adding edges.

The next natural step is to consider sets that are covered by a finite number of intervals, that is, bounded sets. If A is bounded, the number

$$\overline{c}(A) = \inf\left\{\sum_i m(R_i), A \subset \cup_i R_i\right\},$$

is called the *exterior Jordan content* of A. Again, in the definition we may assume that the R_i have non-overlapping interiors. Evidently, one has

$$\overline{c}(\cup_i A_i) \leq \sum_i \overline{c}(A_i).$$

One can consider in an analogous way approximations from within and the *inner Jordan content*

$$\underline{c}(A) = \sup\left\{\sum m(R_i), \cup R_i \subset A\right\},$$

where R_i have non-overlapping interiors. If these contents are equal, the set A is said to be *Jordan measurable* and its common value its *Jordan content* $m(A)$. Since $A \subset \cup_i R_i$ iff $\overline{A} \subset \cup_i R_i$, one has $\overline{c}(A) = \overline{c}(\overline{A})$; also, if $\cup R_i \subset A$, and we shrink each R_i a bit around its center, $R_i' = \lambda R_i$, then $\cup R_i' \subset \mathring{A}$ and so $\underline{c}(A) = \underline{c}(\mathring{A})$.

Exercise 11.1. A set A is Jordan measurable if and only if bA has zero exterior Jordan content. In fact,

$$\overline{c}(bA) = \overline{c}(A) - \underline{c}(A).$$

Since $b(A \cup B), b(A \cap B) \subset b(A) \cup b(B)$, finite unions and intersections of Jordan measurable sets are Jordan measurable. The Jordan content, defined on the class of Jordan measurable sets, is invariant by translations, and it is routinely checked that it is finitely additive. From this one can prove by easy geometrical arguments that for a rectangle R, which is of course Jordan measurable, one has that $m(R)$ is the product of the lengths of the sides.

However, the notion of Jordan content is far from our goal and has serious limitations. First, it applies just to bounded sets; secondly, the countable union of Jordan measurable sets is not necessarily Jordan measurable. For instance, $A = \mathbb{Q} \cap [0, 1]$ is countable, has outer measure 1 (because if $\cup R_i$ contains A, then being closed contains its closure $[0, 1]$), and inner measure zero, as all sets with empty interior. More than that, there are open sets which are not Jordan measurable:

Example 11.1. We construct the so-called *fat Cantor set* in $n = 1$. We start from $[0, 1]$ and delete an open central interval of length $\frac{1}{4}$; next we delete from the two remaining intervals their open central intervals of length $\frac{1}{16}$, leaving us with one interval of length $\frac{1}{4}$ deleted, two of length $\frac{1}{16}$ deleted and four closed intervals of length $\frac{5}{32}$ undeleted. Next, from each of the latter we delete a central interval of length $\frac{1}{64}$, so keeping 8 intervals of length $\frac{9}{128}$. After $k+1$ steps, we have removed $1+2+\cdots+2^k$ open intervals whose total length equals

$$\frac{1}{4}+2\frac{1}{16}+\cdots+2^k\frac{1}{4^{k+1}}=\frac{1}{2}-\frac{1}{2^{k+2}},$$

and keeping 2^{k+1} closed intervals each of length $\frac{1}{2^{k+2}}(1+\frac{1}{2^{k+1}})$ from which at next stage an open central interval of length $\frac{1}{4^{k+2}}$ will be deleted. The fat Cantor E set is the remaining set. Its complement is the union of all deleted open sets, so E is closed. Now, we claim that $\mathring{E}$ is empty, that is, E contains no open interval (these sets are called *nowhere dense*). Indeed, assume $(a,b)\subset E$, so that at the kth stage (a,b) lies in the undeleted part, that is, it must be contained in one of the 2^{k+1} intervals of length $\frac{1}{2^{k+2}}(1+\frac{1}{2^{k+1}})$, something not possible for k big enough. So $bE=E$. Now let us check that $\overline{c}(E)\geq\frac{1}{2}$. Assume that $I_1,\ldots,I_N$ are closed intervals with non-overlapping interiors such that $E\subset\cup_{i=1}^N I_i=I$. Then $[0,1]\setminus I$ is a finite union of intervals that must lie in the deleted part, whence it has total length at most $\frac{1}{2}$, and so $\sum_i |I_i|\geq\frac{1}{2}$. This shows that E is not Jordan measurable. Its complement, a countable union of disjoint open intervals, is not Jordan measurable either, in spite of the fact that the total length of these intervals is $\frac{1}{2}$, and that should be its length from an intuitive point of view.

The general Cantor set is obtained removing a central interval of length r_k from each remaining subinterval at the k-th step. The so-called *Cantor ternary set* occurs when $r_k=(\frac{1}{3})^k$, that is, replacing $\frac{1}{4}$ by $\frac{1}{3}$.

11.1.3 These limitations come from the fact that only finite coverings of intervals are used. If instead we use countable collections, then we are led to the Lebesgue measure.

Definition 11.1. The exterior Lebesgue measure $m^*(A)$ of an arbitrary set is defined

$$m^*(A)=\inf\left\{\sum_{i=1}^{\infty}m(R_i),A\subset\cup_{i=1}^{\infty}R_i\right\},$$

where the R_i are intervals.

Again, in the definition we may assume that the R_i have non-overlapping interiors. Note that $m^*(A)$ may be $+\infty$ and that m^* is obviously monotone: $m^*(A) \leq m^*(B)$ if $A \subset B$. Note too that now the set $A = \mathbb{Q} \cap [0,1]$, and in fact all countable sets, have zero exterior Lebesgue measure. The following facts are proposed as exercises:

Exercise 11.2.

(a) m^* is countably sub-additive: $m^*(\cup_{i=1}^{\infty} A_i) \leq \sum_{i=1}^{\infty} m^*(A_i)$.
(b) If A is compact, $m^*(A) = \overline{c}(A)$.
(c) $m^*(P) = m(P)$ if P is a parallelepiped.
(d) If A is a union of a countable family of intervals R_i with non-overlapping interiors, $m^*(A) = \sum_i m(R_i)$.

However, the exterior measure m^* is not finitely additive,

$$m^*(A \cup B) = m^*(A) + m^*(B), \quad A \cap B = \emptyset.$$

The point is that from a countable covering of $A \cup B$ by intervals R_i it is not possible in general to discriminate those that cover A from those that cover B. If the sets A, B are a positive distance apart, then this is possible:

Exercise 11.3. If $d(A, B) > 0$, then $m^*(A \cup B) = m^*(A) + m^*(B)$.

Regarding point (d) above, the following fact is relevant:

Proposition 11.1. *Every open set U is a countable union of closed cubes with non-overlapping interiors.*

Proof. Consider the *dyadic cubes* of size $2^{-i}, i \in \mathbb{N}$

$$Q = \left[\frac{k_1}{2^i}, \frac{k_1+1}{2^i}\right) \times \cdots \times \left[\frac{k_n}{2^i}, \frac{k_n+1}{2^i}\right), \quad k_1, \ldots, \ k_n = 1, 2, \ldots.$$

For fixed i, they have non-overlapping interiors, each cube of size 2^{-i} contains 2^n cubes of size 2^{-i-1}, each cube is included in just one of double size. Now we consider the collection $\mathcal{Q}$ of dyadic cubes included in U. For every point $x \in U$, there is a ball centered at x included in U, therefore, also a dyadic cube containing x and included in U. This means that the union of all cubes in $\mathcal{Q}$ is exactly U and selecting the maximal cubes in $\mathcal{Q}$ we are done. □

Exercise 11.4. $m^*(A) = \inf\{m^*(U), A \subset U, \text{U open}\}$. This property is known as *outer regularity*.

11.1.4 As m^* is not countably additive, we must restrict attention to some particular sets, the *Lebesgue measurable sets*. There are different but equivalent definitions, the most common being the so-called *Carathéodory method*. Here we have chosen a different one based on the idea that "a measurable set should be well approximated in measure by open sets". The presentation follows essentially that in [20].

Definition 11.2. A set A is said to be *Lebesgue measurable* if for every $\varepsilon > 0$ there is an open set U containing A such that $m^*(U \setminus A) < \varepsilon$. The restriction m of m^* to the class $\mathcal{L}$ of Lebesgue measurable sets is called *the Lebesgue measure.*

Obviously, every open set is measurable. It is also obvious that sets A with $m^*(A) = 0$ are also measurable. The following proposition ensures in particular that all sets defined in topological terms are measurable.

Theorem 11.2.

(a) *Closed sets are Lebesgue measurable.*
(b) *If A is Lebesgue measurable, so is A^c.*
(c) *If $A_k, k = 1, 2, \ldots$, is a sequence of Lebesgue measurable sets, $\cup_k A_k$ and $\cap_k A_k$ are measurable.*

Proof. We start proving the first part of the third point. Given $\varepsilon > 0$, there exists an open set U_k containing A_k such that $m^*(U_k \setminus A_k) \le \varepsilon 2^{-k}$. Then $U = \cup_k U_k$ is open, contains $\cup A_k$ and

$$m^*(U \setminus \cup A_k) = m^*(\cup_k (U_k \setminus A_k)) \le \sum_k m^*(U_k \setminus A_k) \le \sum_k \varepsilon 2^{-k} = \varepsilon.$$

Now, since every closed set is a countable union of compact sets, it is sufficient for point one to prove that a compact set K is measurable. First, note that $m^*(K) < +\infty$ because K is bounded. Given $\varepsilon > 0$ there exists an open set U, $K \subset U$, such that $m(U) \le m^*(K) + \varepsilon$. We will show that $m^*(U \setminus K) \le \varepsilon$. Being open, $U \setminus K = \cup_k Q_k$ with Q_k dyadic cubes with non-overlapping interiors, and $m^*(U \setminus K) = \sum_k m(Q_k)$, so it is enough to prove that $\sum_{k=1}^N m(Q_k) \le \varepsilon$ for every N. The sets $\cup_{k=1}^N Q_k$ and K are at positive distance, hence by Exercise 11.3

$$m^*(\cup_{k=1}^N Q_k) + m^*(K) = m^*(\cup_{k=1}^N Q_k \cup K) \le m^*(U) \le m^*(K) + \varepsilon,$$

and therefore $m^*(\cup_{k=1}^N Q_k) = \sum_k m^*(Q_k) \le \varepsilon$.

If A is measurable, and U_k is open, $A \subset U_k$, with $m^*(U_k \setminus A) \leq \frac{1}{k}$, the set $\cap_k U_k$ contains A and their difference is a set Z of zero outer measure. Therefore, $A^c = \cup_k U_k^c \cup Z$ is measurable because U_k^c, Z are measurable. The second part of the last point is an obvious consequence of the other properties. □

11.1.5 A class of sets in $\mathbb{R}^n$ (including the empty set and the whole space) enjoying properties two and three above is called a *σ-algebra*. Thus, the theorem states that the class of Lebesgue measurable sets is a σ-algebra containing all open and closed sets. Therefore, it contains the smallest σ-algebra containing all open and closed sets. The latter is called the *Borel class* and sets in this class are called *Borelians*. Intuitively, Borelian sets are those that can be obtained from open and closed sets using countable unions, countable intersections, etc. From the proof above we can extract other characterizations of measurability:

Theorem 11.3. *The following are equivalent:*

(a) *A is Lebesgue measurable.*
(b) *For every $\varepsilon > 0$, there is a closed set $F \subset A$ such that $m^*(A \setminus F) \leq \varepsilon$.*
(c) *For every $\varepsilon > 0$, there is a closed set $F \subset A$ and an open set $U, A \subset U$ such that $m^*(U \setminus F) \leq \varepsilon$.*
(d) *A is the union of a countable union of closed sets and a set of zero measure.*
(e) *A is obtained from a countable intersection of open sets deleting a set of zero measure.*

The easiest way to check that a set A is measurable is the following:

Corollary 11.1. *If for every $\varepsilon > 0$ there exists measurable sets $B_\varepsilon, C_\varepsilon$ such that $C_\varepsilon \subset A \subset B_\varepsilon$ and $m(B_\varepsilon \setminus C_\varepsilon) < \varepsilon$, then A is measurable and $m(A) = \lim m(B_\varepsilon) = \lim m(C_\varepsilon)$.*

Note that as a consequence of point two if A is measurable, then

$$m(A) = \sup\{m(F) : F \subset A, F \,\text{closed}\}. \tag{11.1}$$

Another consequence is that every Jordan measurable set is also Lebesgue measurable, because $\mathring{A} \subset A \subset \mathring{A} \cup b(A)$ and $b(A)$ has zero exterior measure.

11.1.6 We summarize now the main properties of the Lebesgue measure.

Theorem 11.4.

(a) *The Lebesgue measure is countably additive.*
(b) *If A_k is an increasing sequence of measurable sets and $A = \cup A_k$, then $m(A_k) \to m(A)$.*
(c) *If A_k is a decreasing sequence of measurable sets and some has finite measure, and $A = \cap A_k$, then $m(A_k) \to m(A)$.*
(d) *The Lebesgue measure is invariant by translations: if A is measurable, $A + x$ is measurable and $m(A + x) = m(A)$.*

Proof. Assume that $A_k, k \in \mathbb{N}$ are Lebesgue measurable and disjoint, and set $A = \cup_k A_k$. We want to show that $m(A) = \sum_k m(A_k)$. We already know that $m = m^*$ is sub-additive, so it is enough to prove that $\sum_{k=1}^{\infty} m(A_k) \le m(A)$, that is, $\sum_{k=1}^{N} m(A_k) \le m(A)$ for all N when $m(A)$ is finite. Suppose first that each A_k is bounded; then, by (11.1) given $\varepsilon > 0$ there is a compact $K_k \subset A_k$ such that $m(A_k) \le m(K_k) + \varepsilon 2^{-k}$. The compacts $K_k, k = 1, \ldots, N$ are disjoint, so at positive distance one from each other, and so using Exercise 11.3,

$$\sum_{k=1}^{N} m(A_k) \le \sum_{k=1}^{N} m(K_k) + \sum_{k=1}^{N} \varepsilon 2^{-k} = m(\cup_{k=1}^{N} K_k) + \sum_{k=1}^{N} \varepsilon 2^{-k} \le m(A) + \varepsilon.$$

Now we deal with the case of general A_k. Consider any sequence B_m of disjoint bounded sets with $\cup_m B_m = \mathbb{R}^n$. Then A is the countable disjoint union of $A_k \cap B_m, k, m \in \mathbb{N}$, and A_k is the disjoint union of the bounded sets $A_k \cap B_m, m \in \mathbb{N}$, whence by what has been proved,

$$m(A) = \sum_{k,m} m(A_k \cap B_m) = \sum_k \sum_m m(A_k \cap B_m) = \sum_k m(A_k).$$

In the situation of point (b), the sets $A'_k = A_k \setminus A_{k-1}$ are disjoint and $A = \cup A'_k$, so by point one,

$$m(A) = \lim_N \sum_{k=1}^{N} m(A'_k) = \lim_N m(A_N).$$

For point (c), we may assume that $m(A_1)$ is finite; since $A_1 \setminus A_k$ increases to $A_1 \setminus A$ point one implies that

$$m(A_1) - m(A_k) = m(A_1 \setminus A_k) \to m(A_1 \setminus A) = m(A_1) - m(A).$$

The last point is obvious. □

Since every closed set is the union of an increasing sequence of compact sets, (11.1) and point two above imply:

Corollary 11.2. *The Lebesgue measure is* inner regular, *meaning that for a measurable set A*

$$m(A) = \sup\{m(K) : K \, \text{compact}, K \subset A\}.$$

A set A is measurable if and only if for every $\varepsilon > 0$ there is an open set $U, A \subset U$ and a compact set $K \subset A$ such that $m(U \setminus K) < \varepsilon$.

The Lebesgue measure enjoys a very important unicity property as shown by the next theorem:

Theorem 11.5. *Assume that $A \to \mu(A)$ is another assignment on Lebesgue measurable sets, outer regular, countably additive and invariant by translations. Then there is a constant $c > 0$ such that $\mu(A) = cm(A)$.*

Proof. Set $c = \mu(Q)$, where Q is the unit cube $[0,1)^n$. Then using finite additivity, it follows that $\mu([0,p)^n) = cp^n$ and $\mu([0,\frac{p}{q})^n) = c(\frac{p}{q})^n$. It follows that $\mu(Q) = cm(Q)$ for all cubes. By Proposition 11.1, it follows that $\mu(U) = cm(U)$ for all open sets. As both are outer regular, it follows that $\mu = cm$. □

Theorem 11.6. *Assume $T \in GL(n)$; then if A is measurable, $T(A)$ is measurable and*

$$m(T(A)) = |\det T| \, m(A). \tag{11.2}$$

In particular, the Lebesgue measure is invariant by rigid motions.

Proof. By (1.10),

$$md(x,y) \leq d(Tx,Ty) \leq Md(x,y),$$

implying that the image by T of a ball centered at x contains a ball centered at Tx. Therefore, T maps open sets to open sets. This inequality implies as well that if R is an interval, there is an interval R' such that $T(R) \subset R'$ and $m(R') \leq Cm(R)$ for some constant C. From this it follows that $m^*(T(A)) \leq Cm^*(A)$. Now, if A is measurable and ε is given, there is an open set $U, A \subset U$, such that $m^*(U \setminus A) < \varepsilon$; then $T(U)$ is open, contains $T(A)$ and

$$m^*(T(U) \setminus T(A)) = m^*(T(U \setminus A)) \leq C\varepsilon,$$

showing that $T(A)$ is measurable.

In the previous section, (11.2) was shown for rectangles and parallelepipeds. Using Proposition 11.1 it holds too for open sets and hence for a general measurable set. □

Note that in fact one has $m^*(T(A)) = |\det T| m^*(A)$.

The fact that $m(T(A)) = m(A)$ if T is a rigid motion can be interpreted as follows. Consider rectangles $R' = T(R)$, with R an interval. Then

$$\inf\left\{\sum_m (R'_i) : A \subset \cup R'_i\right\} = \inf\left\{\sum m(R_i) : T^{-1}(A) \subset \cup R_i\right\}$$
$$= m^*(T^{-1}(A)) = m^*(A),$$

showing that the construction of the Lebesgue measure can be done with arbitrary rectangles.

11.2 The Riemann Integral

11.2.1 Riemann's integration theory applies to real functions $f : A \to \mathbb{R}$, with *both* f and $A \subset \mathbb{R}^n$ bounded. If R is an interval containing A and $\tilde{f}$ is defined on R as being equal to f on A and zero on $R \setminus A$, of course, $\int_A f(x)\, dm$ is to be equal to $\int_R \tilde{f}(x)\, dm$. It is therefore sufficient to consider bounded functions on intervals.

For positive f, the goal is to define the volume of the subgraph

$$G(R, f) = \{(x, z) : x \in R, 0 \leq z \leq f(x)\}.$$

The intuitive idea is to break R into many small intervals R_j with disjoint interiors, take $p_j \in R_j$, argue that R_j being very small f is approximately equal to $f(p_j)$ on R_j and approximate by the *Riemann sum*

$$\sum_j f(p_j)\, m(R_j).$$

Thus, $\int_R f(x)\, dm$, the volume of $G(R, f)$, should be in some sense the limit of these Riemann sums as the size of the R_j goes to zero. To formalize this idea it is convenient to use the language of *upper and lower Riemann sums.*

A *partition* of an interval $[a, b]$ is a finite number of ordered points

$$a = x_0 < x_1 < x_2 < \cdots < x_n = b.$$

A *partition* P of the interval $R = \prod[a_i, b_i]$ is the family of sub-intervals R_j with non-overlapping interiors determined by $P = P_1 \times \cdots \times P_n$, where P_i

is a partition of $[a_i, b_i]$, so that $R = \cup R_j$. Note that if $R = \cup S_j$ is a general decomposition of R in intervals with non-overlapping interiors, the S_j can in turn be decomposed so that the decomposition becomes associated to a partition (just define P_i to be the set of ith coordinates of the vertices of the S_j). Another partition Q is said to be *finer* than P if $P \subset Q$.

If f is bounded on R, for each R_j we consider

$$\alpha_j = \alpha_j(f) = \sup\{f(x), x \in R_j\}, \quad \beta_j = \beta_j(f) = \inf\{f(x), x \in R_j\},$$

and define the *upper and lower Riemann sum of f on P*

$$U(f, P) = \sum_j \alpha_j m(R_j), \quad L(f, P) = \sum_j \beta_j m(R_j),$$

to be thought as lower and upper approximations. See Figure 11.2.

Obviously, $L(f, P) \leq U(f, P)$. Now we claim that if Q is finer than P, then

$$L(f, P) \leq L(f, Q), \quad U(f, Q) \leq U(f, P).$$

It is enough to prove this if Q is obtained from P by adding a point c_i in each P_i. If so, one of the intervals R_j appears in Q split in 2^n sub-intervals. Since the supremum of f in each of those is not greater than α_j, the sum

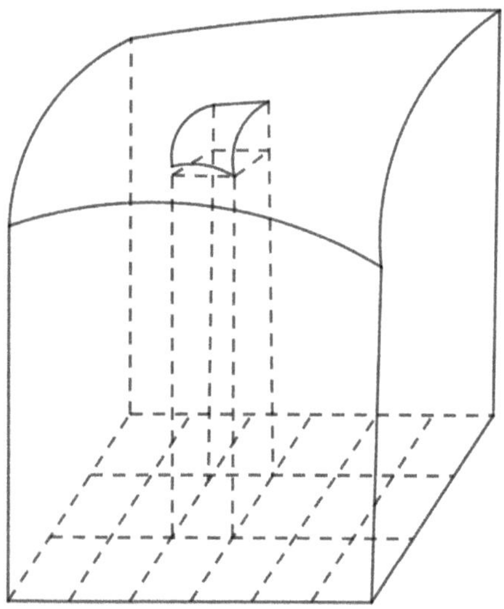

Figure 11.2. A partition.

of all these 2^n terms in $U(f,Q)$ is not greater than $\leq \alpha_j m(R_j)$. The other terms in $U(f,Q)$ are the same as in $U(f,P)$, therefore $U(f.Q) \leq U(f,P)$. In the same way, we see that $L(f,P) \leq L(f,Q)$.

If P, Q are arbitrary partitions, then $S = \prod(P_i \cup Q_i)$ is finer than both P, Q, whence

$$L(f,P) \leq L(f,S) \leq U(f,S) \leq U(f,Q),$$

as claimed. The *lower and upper Riemann integral of f on R* are then respectively defined by

$$L(f) = \sup_P L(f,P), \quad U(f) = \inf_P U(f,P).$$

They are well defined and obviously $L(f) \leq U(f)$.

Definition 11.3. If $L(f) = U(f)$, f is said to be *Riemann integrable on R*. This common value is called the *Riemann integral of f on R* and denoted

$$\int_R f(x)\, dm.$$

Since

$$U(f) - L(f) = \inf_P (U(f,P) - L(f,P)),$$

it follows that f is Riemann integrable if and only if for every $\varepsilon > 0$ there exists a partition P such that

$$U(f,P) - L(f,P) = \sum_j (\alpha_j - \beta_j) m(R_j) < \varepsilon. \tag{11.3}$$

The left-hand term is called *the oscillation of f on P* and denoted $O(f,P)$.

Instead of partitions, we could speak about the *step functions*

$$\sum_i \lambda_j 1_{R_j},$$

and the above criterion is about how well f is approximated by step functions.

11.2.2 Recall from paragraph 11.1.2 that if A is a bounded set, the number

$$\overline{c}(A) = \inf \left\{ \sum_{i=1}^{n} m(R_i), A \subset \cup_{i=1}^{n} R_i \right\},$$

is called the *exterior Jordan content* of A. In this definition we may evidently assume that the R_i have non-overlapping interiors, so that if $f = 1_A$, the characteristic function of A, and $A \subset R$, then $U(f) = \overline{c}(A)$. Thus, 1_A is Riemann integrable iff A is *Jordan measurable.*

If $f \geq 0$, the boundary of the subgraph

$$G(R, f) = \{(x, z) \in \mathbb{R}^{n+1} : x \in R, z = f(x)\},$$

has three parts:

$$A_1 = \{(x, z), x \in bR\}, \quad A_2 = \{(x, 0), x \in \mathring{R}\}, \quad A_3 = \{(x, f(x)) : x \in R\}.$$

Both A_1, A_2 are contained in n-dimensional linear sub-manifolds, so their $(n+1)$-Jordan content is zero. The condition (11.3) means that the graph A_3 has zero Jordan content too, that is, $G(R, f)$ is Jordan measurable. In fact,

$$\underline{c}(G(R, f)) = L(f), \quad \overline{c}(G(R, f)) = U(f),$$

so Riemann integrability of f amounts to Jordan measurability of $G(R, f)$, and in this case

$$m(G(R, f)) = \int_R f(x)\, dm,$$

as expected.

11.2.3 A *Riemann sum associated to a partition* P is defined by

$$\Sigma(f, P) = \sum_j f(\xi_j) m(R_j),$$

where $\xi_j \in R_j$. Note that the notation does not make explicit the choice of points ξ_j, $\Sigma(f, P)$ denotes all of them. As $\beta_j \leq f(\xi_j) \leq \alpha_j$, one has

$$L(f, P) \leq \Sigma(f, P) \leq U(f, P). \tag{11.4}$$

Theorem 11.7. *A bounded function f is Riemann integrable on R with integral L if and only if the Riemann sums have limit L in the following sense: for all $\varepsilon > 0$, there is a partition P such that if $P \subset Q$, then*

$$|L - \Sigma(f, Q)| < \varepsilon, \tag{11.5}$$

for every Riemann sum associated to Q.

Proof. If f is Riemann integrable, given ε there exists P such that

$$\int_R f(x)\,dm - \varepsilon < L(f,P) \leq \int_R f(x)\,dm \leq U(f,P) < \int_R f(x)\,dm + \varepsilon.$$

Since lower sums increase and upper sums decrease when refining a partition, the above holds too for Q, and (11.4) implies (11.5). In the converse direction, it is enough to note that $U(f,Q)$ (resp., $L(f,Q)$) is the supremum (resp., the infimum) of the Riemann sums $\Sigma(f,Q)$ through all possible choices $\xi_j \in R_j$. □

The theorem means that the Riemann integral is the limit of the Riemann sums as the partition gets finer. Intuitively, the number of terms in the sum increases to infinity, while each term $f(\xi_j)m(R_j)$ becomes *infinitesimal.* The existence of the limit represents thus a balance between these two opposite trends. One says that the integral is *an infinite sum of infinitesimal quantities.* In classical notation, the volume of an infinitesimal interval is denoted $dx = dx_1\,dx_2\,\cdots\,dx_n$, ξ_j is replaced by x and the infinite sum is denoted $\int$, leading to

$$\int_R f(x)\,dx.$$

Note however that we are using dm instead of dx, which we use just for one-dimensional integrals $\int_a^b f(x)\,dx$.

A slightly different way of formalizing the term *the partition gets finer* is as follows. For a partition P we set

$$\|P\| = \max_j \delta(R_j),$$

where $\delta(R_j) = \max\{|x-y|, x, y \in R_j\}$ is the diameter of R_j.

Exercise 11.5. Prove that f is Riemann integrable on R with integral L if and only if the Riemann sums have limit L in the following sense: for every $\varepsilon > 0$ there is $\delta > 0$ such that

$$|L - \Sigma(f,P)| < \varepsilon$$

whenever $\|P\| < \delta$.

11.2.4 If $f \geq 0$ is interpreted as a density mass distribution on a body A, $M = \int_A f(p)\,dm(p)$ represents the total mass. In case of a discrete mass distribution in space, with masses m_i at points p_i, the center of mass has coordinates $(\overline{x}, \overline{y}, \overline{z})$ with

$$\overline{x} = \frac{\sum_i x_i m_i}{M}, \quad M = \sum_i m_i,$$

and analogously with $\overline{y}, \overline{z}$. Thinking in $dM = f(p)\, dm(p)$ as an infinitesimal mass placed at p, we see that the center of mass of a mass distribution on A with density f in space has coordinates

$$\overline{x} = \frac{\int_A x\, dM}{M}, \quad \overline{y} = \frac{\int_A y\, dM}{M}, \quad \overline{z} = \frac{\int_A z\, dM}{M}, \quad M = \int_A dM.$$

11.2.5 Riemann sums, as they become finer, are infinite sums of infinitely small terms. A convergent numerical series $\sum_j a_j$, with $a_j \to 0$ is also an infinite sum of infinitely small terms, but their structure is different. Consider for example a Riemann sum of f defined on $[0, 1]$ corresponding to the partition $\frac{j}{n}, 0 \leq j \leq N$ of $[0, 1]$, say

$$\Sigma_N = \sum_{j=0}^{N-1} f\left(\frac{j}{N}\right) \frac{1}{N},$$

where for each interval $[\frac{j}{N}, \frac{j+1}{N}]$ we have chosen $\xi_j = \frac{j}{N}$. For the partial sums $S_N = \sum_{j=1}^{N} a_j$ of a series, when passing from S_N to S_{N+1} the number of terms increases by one but their jth term is the same, while for Σ_N also the number of terms increases by one but their jth term changes. The structure of Σ_N is

$$\Sigma_N = \sum_{j=0}^{N} a_{jN},$$

with $N a_{jN}$ depending on $\frac{j}{N}$. Similarly, a Riemann sum of $f(x, y)$ defined on $[0, 1] \times [0, 1]$ has the form

$$\Sigma_N = \sum_{i,j=0}^{N-1} a_{ijN},$$

with $N^2 a_{ijN}$ depending on $\frac{i}{N}, \frac{j}{N}$.

Example 11.2. The expression $\Sigma_N = \sum_{j=0}^{N} \frac{1}{j+N}$ is a Riemann sum of $f(x) = \frac{1}{1+x}$ because

$$\frac{1}{j+N} = f\left(\frac{j}{N}\right) \frac{1}{N}.$$

Similarly,

$$\sum_{i,j=0}^{N-1} \frac{1}{N^2 + iN + jN}$$

is a Riemann sum of $f(x,y) = \frac{1}{1+x+y}$ because

$$\frac{1}{N^2 + iN + jN} = \frac{1}{N^2} f\left(\frac{i}{N}, \frac{j}{N}\right).$$

11.2.6 An important interpretation of $\int_R f(x)\,dm$ is in terms of *mean value.* Assume that f is integrable on R and let P_N be the partition of R obtained considering N equally spaced points in each side. So P has $M = N^n$ intervals R_j with $m(R_j) = \frac{1}{M} m(R)$. If $\xi_j \in R_j$,

$$\frac{1}{m(R)} \int_R f(x)\,dm = \lim_{N\to+\infty} \frac{1}{M} \sum f(\xi_j),$$

so that $\int_R f(x)\,dm$ can be viewed as *the mean value of f over R.*

11.2.7 If f is uniformly continuous, the term $\alpha_j - \beta_j$ in (11.3) is arbitrarily small if $\|P\|$ is small; Proposition 2.3 in Chapter 2 implies:

Theorem 11.8. *Continuous functions are Riemann integrable.*

Exercise 11.6. Prove that in dimension $n = 1$ a monotone function in an interval $[a,b]$ is Riemann integrable.

The two last results are particular cases of next theorem characterizing Riemann integrability in terms of continuity. We want to understand when there are partitions with arbitrarily small oscillations. Now, the general term in (11.3) is the product of two terms, $(\alpha_j - \beta_j)m(R_j)$. Let us think in very fine partitions; the quantity $\alpha_j - \beta_j$ will be small for those R_j in which f is continuous, and large if f has discontinuities on R_j. However, if the total volume of the latter is arbitrarily small, the oscillation will be small, too. This is the essential idea in the proof of the next theorem: if the set of discontinuities of f can be covered by intervals with arbitrarily small volume, f will be integrable. In fact, we already know this for $f = 1_A$; in this case the set of discontinuities is bA, and it was already noted that 1_A is Riemann integrable if and only if A is Jordan measurable, that is, bA has zero Jordan content. In general, the set of discontinuities of f

$$D(f) = \{a \in R : f \text{ is not continuous at } a\},$$

is not compact, and the proof is more involved. To quantify the discontinuity of f at a we introduce

$$\omega_f(a) = \lim_{\delta \to 0} \sup\{|f(x) - f(y)|, x, y \in Q(a, \delta) \cap R\},$$

where $Q(a, \delta)$ is the cube centered at a of size δ. Thus, f is continuous at a if and only if $\omega_f(a) = 0$, $D(f) = \{a \in R : \omega_f(a) > 0\}$. Note that if $A \subset R$ and $a \in \mathring{A}$, then

$$\sup_{x \in A} f(x) - \inf_{x \in A} f(x) \geq \omega_f(a).$$

We consider

$$D_\tau = \{a \in R : \omega_f(a) \geq \tau\}.$$

This set is closed, hence compact: for if $a_k \in D_\tau, a_k \to a$ and it were $\omega_f(a) < \tau$, there would exist $\delta > 0$ such that

$$\sup\{|f(x) - f(y)|, x, y \in Q(a, \delta) \cap R\} < \tau.$$

Then for all points b in the interior of $Q(a, \delta) \cap R$ one has $\omega_f(b) < \tau$, excluding all a_k. These considerations done, we can address the characterization of Riemann integrable functions.

Theorem 11.9. *The following statements are equivalent:*

(a) *f is Riemann integrable.*
(b) *$D_\tau(f)$ has zero exterior Jordan for all τ.*
(c) *$D(f)$ has zero Lebesgue measure, that is, for every ε there are intervals R_j covering $D(f)$ with $\sum_j m(R_j) < \varepsilon$.*

Proof. Assume first that f is integrable; given $\varepsilon > 0$ let P be a partition of R such that

$$S(f, P) - s(f, P) = \sum_j (\alpha_j - \beta_j) m(R_j) \leq \varepsilon.$$

If $\mathring{R}_j$ meets $D_\tau(f)$, then $\alpha_j - \beta_j \geq \tau$, proving that

$$\tau \overline{c}(D_\tau(f)) \leq \varepsilon.$$

Therefore, $\overline{c}(D_\tau) = 0$.

Assume now that $\overline{c}(D_\tau(f)) = 0$ for all τ. With $\varepsilon > 0$ given we want to find P with $O(f, P) < \varepsilon$. First, we cover $D_\varepsilon(f)$ with a finite number of intervals R_j^1 with disjoint interiors such that $\sum_j m(R_j^1) \leq \varepsilon$. Let B be their union; replacing each R_j^1 by a dilated λR_j^1 (with respect to its center) we may assume that $D_\varepsilon(f) \subset \mathring{B}$. If $x \in R \setminus \mathring{B}$, then $\omega_f(x) < \varepsilon$, therefore there is a cube $Q(x, \delta_x)$ centered at x, which we choose disjoint from $D_\varepsilon(f)$, where the oscillation of f is smaller than ε. Since $R \setminus \mathring{B}$ is compact, a finite number of such cubes, denoted $W_1, \ldots, W_N$, cover what is not covered by R_j^1. The $W_1, \ldots, W_N$ determine intervals R_j^2 with disjoint interiors with the same union, and in each R_j^2 the oscillation of f is smaller than ε.

Altogether, R is covered by the R_j^1 and R_j^2. If $P_1, P_2, \ldots, P_n$ are respectively the projections on the coordinate axis of the vertices of all these intervals, then $P = P_1 \times P_2 \times \cdots P_n$ has intervals S of two types. Let $\mathcal{A}$ be the collection of all those included in some R_j^1; evidently

$$\sum_{S \in \mathcal{A}} m(S) = \sum_j m(R_j^1) < \varepsilon.$$

The remaining ones, call them of class $\mathcal{B}$, are included in some R_j^2, and so the oscillation of f there is $< \varepsilon$. We split the oscillation of f on P into two parts

$$O(f, P) = O(\mathcal{A}) + O(\mathcal{B}).$$

If $|f| \leq K$, in $O(\mathcal{A})$ we use $M(S) - m(S) \leq 2K$, so that

$$O(\mathcal{A}) \leq 2K \sum_{S \in \mathcal{A}} m(S) < 2K\varepsilon.$$

In $\mathcal{B}$, we use $M(S) - m(S) \leq \varepsilon$ so that

$$O(\mathcal{B}) \leq \varepsilon \sum_{S \in \mathcal{B}} |S| \leq \varepsilon |R|.$$

Thus, $O(f, P) \leq C\varepsilon$ and f is integrable.

Thus, we have proved that (a), (b) are equivalent. The last statement follows from the fact that $D(f)$ is the countable union of the compact sets $D_{\frac{1}{n}}(f), n \in \mathbb{N}$. □

We say that a certain property P relative to points *holds almost everywhere (a.e.) on* A if the set of points in A for which P is not true has zero Lebesgue measure. Thus, Riemann integrable functions are a.e continuous.

The theorem we just proved shows a deficiency of the Riemann integral. If A is the set of rationals in $[0, 1]$, trivially 1_A is not Riemann integrable (the upper sums are 1 and the lower sums are 0, the set of discontinuities is $[0, 1]$), but being countable is a set of zero Lebesgue measure.

11.2.8 As mentioned before, if f is a bounded function defined on a Jordan measurable set $A \subset R$, one says that *f is Riemann integrable on A* if the function $\tilde{f}$ on R which equals f on A and zero outside is integrable, and in this case we set

$$\int_A f(x)\,dm = \int_R \tilde{f}(x)\,dm.$$

Note that the upper integral is

$$\overline{\int_A} f(x)\,dm = \inf \sum_{R_j \cap A \neq \emptyset} \sup\{f(x), x \in R_j \cap A\} m(R_j), \tag{11.6}$$

and that the lower integral equals

$$\underline{\int_A} f(x)\,dm = \sup \sum_{R_j \subset A} \min\{f(x), x \in R_j\} m(R_j), \tag{11.7}$$

where in both cases the R_j are intervals with non-overlapping interiors covering A.

Evidently, if A has zero content, all bounded functions are integrable on A with integral zero. The points of discontinuity of $\tilde{f}$ are either points in $\mathring{A}$ where f is discontinuous or else points in bA. As bA has zero Jordan content, it follows from Theorem 11.9:

Proposition 11.2. *A bounded function f on a Jordan measurable set A is integrable if and only if*

$$\{x \in \mathring{A} : f \text{ is discontinuous at } x\},$$

has Lebesgue measure zero. Obviously, all bounded continuous functions on A are integrable.

Exercise 11.7. If f is positive and

$$G(f) = \{(x, z) \in \mathbb{R}^{n+1} : x \in A, z \leq f(x)\},$$

prove that f is integrable on A if and only if $G(f)$ is Jordan-measurable and in this case its integral on A equals the $(n+1)$-dimensional content of the subgraph.

Exercise 11.8. Check the following properties of the integral:

(a) If f, g are integrable on A and $\lambda \in \mathbb{R}$, $f+g, \lambda f$ are integrable on A and

$$\int_A (f+g)(x)\,dm = \int_A f(x)\,dm + \int_A g(x)\,dm, \int \lambda f = \lambda \int_A f(x)\,dm.$$

(b) If f, g are integrable on A and $f \leq g$, then $\int_A f(x)\,dm \leq \int_A g(x)\,dm$.
(c) If f is Riemann integrable on A and $|f| \leq K$, so is $|f|$ and

$$\left|\int_A f(x)\,dm\right| \leq \int_A |f(x)|\,dm \leq Km(A).$$

(d) If $A \subset B$ are Jordan measurable and f is integrable on B, then f is integrable on A.
(e) If A, B are disjoint Jordan measurable sets and f is integrable on $A \cup B$, then

$$\int_{A\cup B} f(x)\,dm = \int_A f(x)\,dm + \int_B f(x)\,dm.$$

(f) If f is integrable on A, $\int_A f(x)\,dm = \int_{\mathring{A}} f(x)\,dm = \int_{\overline{A}} f(x)\,dm$.
(g) If A, B are Jordan measurable sets with non-overlapping interiors, then

$$\int_{A\cup B} f(x)\,dm = \int_A f(x)\,dm + \int_B f(x)\,dm.$$

(h) If $f \geq 0$ is integrable on A and $\int_A f(x)\,dm = 0$, then $f = 0$ a.e. on A. If f is continuous on A, then $f(x) = 0, x \in \mathring{A}$.
(i) If f, g are Riemann integrable on A, so is the product fg.

In particular, in a Riemann integral $\int_A f(x)\,dm$ we may think that A is a bounded open set (replacing A by $\mathring{A}$), or a compact set (replacing A by $\overline{A}$), with bA of zero Jordan content, and that f is bounded and a.e. continuous in A. We can view

$$\rho = \frac{1}{m(A)} \int_A f(x)\,dm,$$

as the *mean value of* f *over* A.

Exercise 11.9. Prove that ρ is a value of f if A is connected and f is continuous in A.

11.2.9 Next, we consider *vector-valued integration.* A vector-valued function $f : A \to \mathbb{R}^m$ with components $f = (f_1, \ldots, f_m)$ is said to be *Riemann integrable on* A if every f_j is, in which case we define

$$\int_A f(x)\,dm = \left(\int_A f_1(x)\,dm, \ldots, \int_A f_m(x)\,dm\right).$$

All properties relative to scalar integration extend to this setting with obvious modifications. The only result that is not directly obtained considering the components is the next proposition.

Proposition 11.3. *If f is Riemann integrable on A, then so is $|f|$ and*

$$\left|\int_A f(x)\,dm\right| \leq \int_A |f(x)|\,dm.$$

Proof. It is enough to consider as A an interval R. The integrability of $|f|$ follows from

$$||f(x)| - |f(y)|| \leq |f(x) - f(y)| \leq \sum_{i=1}^{m} |f_i(x) - f_i(y)|,$$

implying that for the oscillations

$$O(|f|, P) \leq \sum_i O(f_i, P).$$

For $u = \int_A f(t)\,dt$, we can write

$$|u| = \max\{\langle u, v\rangle, |v| = 1\} = \max_v \int_A \langle f(t), v\rangle\,dt.$$

But $\langle f(t), v\rangle \leq |f(t)|$. □

Note that $|f|$ can be Riemann integrable without f being integrable; for instance, if A is not Jordan measurable, $f = 1_A - 1_{A^c}$.

11.2.10 The last proposition implies for real-valued f

$$\left|\int_A f(x)\,dm\right| \leq Cm(A), \tag{11.8}$$

if $|f(x)| \leq C, x \in A$. The best constant C is of course

$$\|f\|_\infty = \sup_{x\in A} |f(x)|.$$

The left-hand side is called the *supremum norm*. It is indeed a norm in the space of all bounded functions on A. Recall that the notion of convergence associated to this norm is the *uniform convergence*, stronger than pointwise convergence: $\|f_k - f\|_\infty \to 0$ means that for every $\varepsilon > 0$ there is k_0 such that for $k > k_0$ one has

$$|f_k(x) - f(x)| < \varepsilon, \quad x \in A.$$

Recall also that if $f_k \to f$ uniformly and all f_k are continuous at $p \in A$, then f is continuous at p, too.

Proposition 11.4. *Assume that f_k are Riemann-integrable on A and that $f_k \to f$ uniformly in $\mathring{A}$. Then f is Riemann integrable on A and*

$$\int_A f_k(x)\,dm \to \int_A f(x)\,dm.$$

Proof. We must show that

$$D = \{p \in \mathring{A} : f \text{ is not continuous at}\, p\},$$

has zero Lebesgue measure. But this set is contained in $\cup_k D_k$ where D_k is the set of discontinuities of f_k in $\mathring{A}$, which has zero measure because f_k is integrable.

We outline a proof independent of Proposition 11.9 when A is an interval R. Given $\varepsilon > 0$, we must find a partition P with oscillation $U(f,P) - L(f,P) < \varepsilon$. But this follows from, with the notations in paragraph 11.2.1,

$$\alpha_j(f) \le \alpha_j(f_k) + |f - f_k|, \quad \beta_j(f) \ge \beta_j(f_k) - |f - f_k|,$$

implying

$$U(f,P) - L(f,P) \le U(f_k,P) - L(f_k,P) + 2m(R)|f - f_k|.$$

The last assertion follows from (11.8). □

The above is the only limit theorem naturally stated in the context of the Riemann integral. Somehow it is unsatisfactory as it does not apply to situations as in the following example:

Example 11.3. Let $f_k(x) = x^k$ in $[0,1]$. Then $f_k(x)$ has point-wise limit f equal to zero for $x < 1$, $f(1) = 1$. The convergence is not uniform, but still

$$\int_0^1 f_k(x)\,dx = \frac{1}{k+1} \to 0 = \int_0^1 f(x)\,dx.$$

The example suggests that the following result should hold:

Proposition 11.5. *Assume that f_k are uniformly bounded, Riemann-integrable on A, and $f_k(x) \to f(x), x \in A$, with f Riemann-integrable. Then*

$$\int_A f_k(x)\,dm \to \int_A f(x)\,dm.$$

This result, which is true, is best proved in the context of Lebesgue integration. A formal proof in the context of Riemann integration is

$$\int_A f\,dm = \lim_P S(f,P) = \lim_P \lim_k S(f_k,P) = \lim_k \lim_P S(f_k,P) = \lim_k \int_A f_k\,dm,$$

but interchanging the limits needs justification. This is an example indicating that Riemann integral has bad properties regarding limiting processes.

Chapter 12

The Lebesgue Integral

Even though we are mainly interested in the Euclidean setting, the Lebesgue integral is presented in the more general context of *measure spaces*, including *probability spaces*. Then, the specific aspects of Lebesgue integral in $\mathbb{R}^n$ are described, together with its relation with multi-dimensional improper Riemann integrals and continuous functions. Last section provides multidimensional versions of the fundamental theorem of calculus in terms of *densities and set functions* defined on intervals, both for Riemann integrals and Lebesgue integrals. These results are of central importance in this book, as they are the main ingredient in the proof on the change of variable formula and the main theorems in vector analysis. As far as the author knows, these are original unpublished results.

12.1 Measure Spaces

12.1.1 In retrospective, the Riemann integral is a theory closely related to continuity, and Riemann integrable functions are not far from being continuous functions. Although more than sufficient for practical applications, from a mathematical point of view the Riemann integral is too rigid and has some inconveniences, especially regarding limiting processes, completeness, iterated integrals and others. In this section, we introduce the Lebesgue integral, a more satisfactory theory from a mathematical point of view.

To obtain a more flexible theory, Lebesgue's main idea is to replace the partitions in the domain of definition by partitions in the *range* of the function. A well-known analogous setting is the following: assume you have a number of coins on a table, with different values, and you want to count the total value. According to Riemann, we would start say from left to right, from bottom to top, and keep adding the values found along the way. According to Lebesgue, we would first gather the coins according to its value.

To concretize this idea in mathematical terms, let f be a real function on some set $A \subset \mathbb{R}^n$, and assume for simplicity that $0 \le f \le M$; we consider a partition P of $[0, M]$

$$P : 0 = \lambda_0 < \lambda_1 < \cdots < \lambda_N = M.$$

In the set

$$E_i = f^{-1}([\lambda_i, \lambda_{i+1})) = \{x \in A : \lambda_i \le f(x) < \lambda_{i+1}\},$$

we replace $f(x)$ by the smaller value λ_i and define the *lower Lebesgue sum*

$$L(f, P) = \sum_i \lambda_i m(E_i). \tag{12.1}$$

The Lebesgue integral would then be defined as the limit of these sums as P gets finer.

Now, the set E_i is no longer an interval, it might be quite general. So, for this to make sense, we need to know the meaning of $m(E_i)$, the set E_i should be *measurable.* Put differently, considering partitions by intervals in the range of the function leads to partitions in the domain of definition by general measurable sets. That's why the construction of the Lebesgue measure, developed in Section 11.1, is a required previous step.

12.1.2 In defining the Lebesgue integral from the Lebesgue measure, we will consider a more general abstract setting, a *measure space.* This setting includes the *probability spaces* for which the integral of a function becomes the expected value of a random variable.

We replace $\mathbb{R}^n$ by a general set X, and the class of Lebesgue measurable sets by a collection $\mathcal{M}$ of subsets of X satisfying the same properties:

(a) $X \in \mathcal{M}$.
(b) If $A \in \mathcal{M}$, then $A^c \in \mathcal{M}$.
(c) If $A_k \in \mathcal{M}$ is a sequence of sets in $\mathcal{M}$, then $\cup_k A_k \in \mathcal{M}$.

We say that $\mathcal{M}$ is a *σ-algebra on X*.

Next we consider a *positive measure* μ on X replacing the Lebesgue measure m. This means a set function

$$\mu : \mathcal{M} \longrightarrow [0, +\infty],$$

($+\infty$ is an allowed value, there are Lebesgue measurable sets of infinite measure), that is *countably additive*:

$$\mu(\cup_k A_k) = \sum_k \mu(A_k), \quad A_k \in \mathcal{M}, \quad A_k \cap A_j = \emptyset, \quad k \neq j.$$

Note that this property does not depend on the ordering of the sets A_k.

The triple $(X, \mathcal{M}, \mu)$ is called a *measure space*. If $A \in \mathcal{M}$, restricting μ to the sets $B \in \mathcal{M}, B \subset A$ we obtain another measure space.

The following properties of $\mathcal{M}$ and μ are straightforward consequences of the definitions and are proposed as exercises to the reader.

Exercise 12.1. If $\mathcal{M}$ is a σ-algebra on a set X, then,

(a) $\emptyset \in \mathcal{M}$.
(b) If $A_k \in \mathcal{M}, \cap_k A_K \in \mathcal{M}$.
(c) Finite unions and intersections of sets in $\mathcal{M}$ are in $\mathcal{M}$.
(d) If $A, B \in \mathcal{M}$, then $A \setminus B \in \mathcal{M}$.

Exercise 12.2. For a measure space, prove:

(a) $\mu(\emptyset) = 0$.
(b) If $A, B \in \mathcal{M}, A \subset B$, then $\mu(A) \leq \mu(B)$. If $\mu(B)$ is finite, $\mu(B \setminus A) = \mu(B) - \mu(A)$.
(c) If (A_k) is an increasing sequence of sets in $\mathcal{M}$ and $A = \cup_k A_k$, then $\mu(A_k) \to \mu(A)$.
(d) If (A_k) is a decreasing sequence of sets in $\mathcal{M}$ and some $\mu(A_k)$ is finite, $\mu(A_k) \to \mu(A)$.

A measure space is called *complete* if subsets of measurable sets of zero measure are measurable (whence of zero measure, too): $A \in \mathcal{M}, \mu(A) = 0, E \subset A$ implies $E \in \mathcal{M}, \mu(E) = 0$.

If a measure space is not complete, it can be completed as outlined in the following exercise:

Exercise 12.3. Let $\mathcal{M}^*$ be the class obtained by adding to $\mathcal{M}$ all subsets of sets of zero measure. Equivalently, $E \in \mathcal{M}^*$ if there exist two measurable sets A, B with $A \subset E \subset B, \mu(B \setminus A) = 0$. Defining in this situation $\mu(E) = \mu(A)$, prove that $(X, \mathcal{M}^*, \mu)$ is a complete measure space.

We will accordingly assume from now on that the measure space is complete. The Lebesgue measure constructed in Section 11.1 is already complete.

Besides Lebesgue measure and probability spaces, another example to have in mind is the *counting measure* defined on all subsets $A \subset \mathbb{N}$ defining $\mu(A)$ as the number of elements of A if A is finite and $+\infty$ otherwise.

12.1.3 To define the integral, according to Lebesgue idea we must consider real functions f defined on X such that the pre-images $f^{-1}(I)$ of intervals are measurable. Another important feature of the Lebesgue integral is that functions are allowed to take infinite values. This will make it unnecessary to talk about improper Lebesgue integrals. On the other hand, since subsets of zero measure are measurable, the values of f on those sets are irrelevant. More than that, we do not need f to be defined there.

Altogether, and with some abuse of notation, one considers functions

$$f : X \longrightarrow [-\infty, +\infty],$$

defined a.e. on X. We define f as *measurable* if $\{x : f(x) > \lambda\} \in \mathcal{M}$ for all $\lambda \in \mathbb{R}$. Note that all these sets are defined modulo a set of zero measure. Then $\{f \geq \lambda\}, \{f < \lambda\}, \{f \leq \lambda\}, \{f = +\infty\}, \{f = -\infty\}$ are measurable, too. In fact, this is equivalent to requiring that $f^{-1}(A) \in \mathcal{M}$ for every open set or closed set A.

In case of $(\mathbb{R}, \mathcal{L}, m)$, continuous functions are measurable. It is important to note that if one knows a continuous function f a.e., then $f(x)$ is completely known for all x, because $f(x) = \lim_{y \to x} f(y)$ and obviously the complement of a set of zero measure cannot contain balls and so is a dense set.

For the counting measure measurability is a void condition, any sequence (x_k) is measurable.

The following are stability properties of measurable functions:

(a) The measurable functions taking finite values form a linear space. It is clear that λf is measurable if f is. If both f, g are measurable and I is an open interval, $f(x) + g(x) \in I$ means that the point $(f(x), g(x))$ belongs to the open set U in the plane

$$U = \{(x, y) : x + y \in I\}.$$

U is a union of countably many open intervals R_i, so it is enough to prove that $(f(x), g(x)) \in R$ defines a measurable set for every interval R. But if $R = J \times K, J, K$ open intervals, this set is $f^{-1}(J) \cap g^{-1}(K)$ and therefore it is measurable.

(b) The same argument shows that if $f_1, \ldots, f_N$ are measurable functions taking finite values and g is a continuous function on $\mathbb{R}^N$, then $g(f_1, \ldots, f_N)$ is measurable.

(c) Later on we will be considering complex-valued or vector-valued functions $f = (f_1, \ldots, f_N)$; f is said to be measurable if all components are. This is equivalent to requiring that $f^{-1}(U)$ is open for every open set $U \subset \mathbb{R}^N$.

(d) For measurable f, g the functions

$$\max(f,g)(x) = \max(f(x), g(x)), \quad \min(f,g)(x) = \min(f(x), g(x)),$$

are easily seen to be measurable. In particular, the *positive and negative parts* of a real function f

$$f^+ = \max(f, 0), \quad f^- = f - f^+ = \max(-f, 0),$$

are measurable. Note that $f^+ = f$ in the set A where $f \geq 0$ and zero outside and

$$f = f^+ - f^-, \quad |f| = f^+ + f^-.$$

For next properties involving a sequence f_k of measurable functions, note that if f_k is defined on D_k with $\mu(X \setminus D_k) = 0$, then all f_k are defined on the common set $D = \cap D_k$ for which $\mu(X \setminus D) = 0$.

(a) The class of measurable functions is closed under point-wise limits, that is, if f_k are measurable and the point-wise limit

$$f(x) = \lim_k f_k(x),$$

exists for a.e $x \in X$, then f is measurable. This is because

$$\{f > \lambda\} = \cup_k \{f_k > \lambda\}.$$

(b) More generally, for measurable functions f_k, the functions

$$\limsup f_k(x) = \inf_m \sup_{k \geq m} f_k(x), \quad \liminf f_k(x) = \sup_m \inf_{k \geq m} f_k(x),$$

are also measurable.

The easiest measurable functions are the simple functions. A *simple function* is a measurable one defined everywhere on X and taking only a finite number of values $-\infty < \lambda_1 < \cdots < \lambda_N < +\infty$, that is,

$$s = \sum_i \lambda_i 1_{A_i}, \quad A_i = \{s = \lambda_i\}.$$

Here $A_i \in \mathcal{M}$ and 1_A denotes the characteristic function of A. The sets A_i constitute an a.e. partition of X. These simple functions replace the step functions in the Riemann theory. It is straightforward that linear combinations of simple functions are simple; also, $\max(s_1, \ldots, s_m)$ and $\min(s_1, \ldots, s_m)$ are simple if the s_i are.

Proposition 12.1. *A function $f : X \to [-\infty, +\infty]$ is measurable if and only if it is a.e. the point-wise limit of a sequence s_N of simple measurable functions. If f takes values in $[0, +\infty]$, the sequence s_N can be taken non-decreasing.*

Proof. We may assume f defined everywhere. Decomposing f into its positive and negative parts, it is enough to prove the last statement.

For a given measurable function $f : X \longrightarrow [0, +\infty]$, there is a one-to-one correspondence between partitions P of $[0, +\infty)$

$$P : 0 = \lambda_1 < \lambda_2 < \cdots < \lambda_N < +\infty,$$

and positive simple functions s with $s \leq f$. A partition defines the simple function

$$s = \sum_i \lambda_i 1_{E_i},$$

with

$$E_i = f^{-1}([\lambda_i, \lambda_{i+1})), \quad i = 1, \ldots, N-1, \quad E_N = \{f \geq \lambda_N\}.$$

Conversely, if

$$s = \sum_i \lambda_i 1_{E_i}, \quad \lambda_i < \lambda_{i+1}$$

and $s \leq f$, then $E_i = \{\lambda_i \leq f < \lambda_{i+1}\}$.

For each $N \in \mathbb{N}$, consider the partition P_N with points $k2^{-N}, 0 \leq k \leq M = N2^N$. If s_N is the simple function corresponding to P_N,

$$s_N = \sum_k k2^{-N} 1_{E_k}, E_k = \{k2^{-N} \leq f < (k+1)2^{-N}\}, k < M,$$

$$E_M = \{N \leq f\},$$

then s_N is increasing, because $s_{N+1}(x) = s_N(x)$ if $f(x)$ is finite and in the first half of $[k2^{-N}, (k+1)2^{-N})$ and $s_{N+1}(x) > s_N(x)$ if $f(x)$ is on the second half. If $f(x) = +\infty$, then $s_N(x) = N$. Since $f(x) - s_N(x) \leq 2^{-N}$ if $f(x)$ is finite, it follows that $s_N(x) \to f(x)$ for all $x \in X$. □

12.2 Lebesgue Integrable Functions

12.2.1 We proceed now to define Lebesgue integrability. A main characteristic of Lebesgue integration is that the integrability of a function depends just on *size*, more precisely on the measure of the set where the function takes big values, nothing else. It is also desirable that whenever f is integrable on X, it should be integrable on every measurable set A. Considering $A = \{f \geq 0\}$, the conclusion is that in Lebesgue's theory, for a measurable function f, integrability of $f, |f|$ should be equivalent. That's why it is enough to restrict attention to positive functions.

For a simple non-negative function $s = \sum_i \lambda_i 1_{A_i}$ and $A \in \mathcal{M}$ we define its Lebesgue integral

$$\int_A s\,d\mu = \sum_i \lambda_i \mu(A \cap A_i).$$

Here, if $\mu(A \cap A_i) = +\infty$ and $\lambda_i = 0$, we assign value zero to the product (if there are no coins in an infinite table, the total value is zero). That is, $0 \cdot +\infty = 0$.

The value of the integral can be $+\infty$, exactly when the *support* has infinite measure

$$\mu(\{x : s(x) > 0\} = +\infty.$$

The following properties are then easy consequences of the definitions:

(a) $\displaystyle\int_A s\,d\mu \leq \int_B s\,d\mu, \quad A \subset B, \quad s \geq 0.$

(b) $\displaystyle\int_A s_1\,d\mu \leq \int_A s_2 d\mu, \quad s_1 \leq s_2, \quad$ a.e.

(c) For non-negative simple $s_1, \ldots, s_m$ and constants $c_1, \ldots, c_m \geq 0$,

$$\int_A \left(\sum_j c_j s_j\right) d\mu = \sum_j c_j \int_A s_j\,d\mu.$$

(d) For mutually disjoint $A_k \in \mathcal{M}, k \in \mathbb{N}$,

$$\int_{\cup_k A_k} s\,d\mu = \sum_k \int_{A_k} s\,d\mu.$$

The last property says that

$$\overline{s}(A) = \int_A s\,d\mu$$

is a measure, too.

12.2.2 For a non-negative measurable function f, we define its Lebesgue integral on $A \in \mathcal{M}$ as

$$\int_A f\,d\mu = \sup \int_A s\,d\mu,$$

where s is simple and $0 \leq s \leq f$ a.e. on A. Of course, if f itself is simple, both definitions agree.

In terms of partitions P of $[0, +\infty)$

$$P : 0 = \lambda_0 < \lambda_1 < \cdots < \lambda_N,$$

the integral of the associated simple function, as in the proof of Proposition 12.1, is the lower Lebesgue sum

$$L(f, P; A) = \sum_i \lambda_i \mu(A \cap A_i),$$

where $A_i = \{x \in A : \lambda_i \leq f(x) < \lambda_{i+1}\}, A_N = \{\lambda_N \leq f\}$. Thus, an equivalent definition is

$$\int_A f\,d\mu = \sup L(f, P, A).$$

For the counting measure, the integral of a positive sequence (x_k) is of course $\sum_k x_k$.

The integral can be infinite. For instance, for $f(x) = x^{-\alpha}$ in $A = [0, 1]$ and the partition $0 < 1 < 2^\alpha < \cdots < N^\alpha$, the lower Lebesgue sum is

$$\sum_{k=1}^{N-1} (k+1)^\alpha \left(\frac{1}{k} - \frac{1}{k+1} \right) + N^{\alpha - 1},$$

which is arbitrarily large if $\alpha > 1$. Analogously, if f is decreasing in $A = [1, +\infty)$ and P is the partition $f(N), f(N-1), \ldots, f(1)$, the lower Lebesgue sum is $\sum_{k=1}^{N} f(k) \geq N f(N)$ and so arbitrarily large if $x f(x) \to +\infty$ as $x \to +\infty$.

A monotone function φ on an interval I is measurable. For positive and continuous φ, a Lebesgue sum is a Riemann sum, and conversely.

12.2.3 The following properties are again easy consequences of the definitions, based on the same properties for non-negative simple functions:

(a) $\displaystyle\int_A f\,d\mu \le \int_B f\,d\mu, A \subset B, \quad f \ge 0.$

(b) $\displaystyle\int_A f_1\,d\mu \le \int_A f_2 d\mu, \quad f_1 \le f_2, \quad$ a.e.

(c) For non-negative measurable $f_1, \ldots, f_m$ and constants $c_1, \ldots, c_m \ge 0$,

$$\int_A \left(\sum_j c_j f_j \right) d\mu = \sum_j c_j \int_A f_j\,d\mu.$$

(d) For mutually disjoint $A_1, \ldots, A_m \in \mathcal{M}$,

$$\int_{\cup_j A_j} f\,d\mu = \sum_j \int_{A_j} f\,d\mu.$$

Next, we discuss sets of zero measure in integrals. Note that $\int_A f\,d\mu = 0$ in two cases: if $\mu(A) = 0$, for all f, and if $f(x) = 0, x \in A$. As a consequence, if f, g are non-negative measurable functions and $f = g$ a.e. on A, that is,

$$\mu(E) = 0, E = \{x \in A : f(x) \ne g(x)\}) = 0,$$

(note that E is measurable), then

$$\int_A f\,d\mu = \int_E f\,d\mu + \int_{A\setminus E} f\,d\mu = \int_{A\setminus E} f\,d\mu = \int_{A\setminus E} g\,d\mu = \int_A g\,d\mu.$$

Thus, as well as in the definition of measurability, the values of f on sets of zero measure do not matter regarding integration.

12.2.4 According to what was explained in paragraph 12.2.1, a measurable function $f : X \to [-\infty, +\infty]$ is said to be *Lebesgue integrable* if $\int_X |f|\,d\mu < +\infty$. Then both $\int_X f^+\,d\mu, \int_X f^-\,d\mu$ are finite and we can define

$$\int_A f\,d\mu = \int_A f^+\,d\mu - \int_A f^-\,d\mu.$$

More generally, if f^+ is integrable and f^- is not, we set $\int_X f\,d\mu = -\infty$, and $\int_X f\,d\mu = +\infty$ if f^- is integrable and f^+ is not.

An a.e. defined complex or vector-valued function is said to be *Lebesgue integrable* if every component is, which amounts to

$$\int_X |f|\, d\mu < +\infty.$$

The integral is defined component-wise. For instance, for a complex-valued function $f = u + iv$,

$$\int_A f\, d\mu = \int_A u^+\, d\mu - \int_A u^-\, d\mu + i\left(\int_A v^+\, d\mu - \int_A v^-\, d\mu\right).$$

It is straightforward to see that linear combinations $af + bg$ of integrable functions are integrable and

$$\int_A (af + bg)\, d\mu = a\int_A f\, d\mu + b\int_A g\, d\mu.$$

The same proof as in Proposition 11.3 shows that for integrable functions

$$\left|\int_A f\, d\mu\right| \le \int_A |f|\, d\mu.$$

Again, if f, g are complex-valued integrable and $f(x) = g(x)$ a.e. on X, then $\int_A f\, d\mu = \int_A g\, d\mu$ for all measurable sets A. From the point of view of integration, f, g can be identified. Both measurable and integrable functions are considered as defined modulo sets of zero measure.

Proposition 12.2. *Let $E_\lambda = \{|f| \ge \lambda\}$. If f is integrable, then*

$$\mu(E_\lambda) \le \frac{1}{\lambda}\int_X |f|\, d\mu.$$

In particular, an integrable $f : X \to [-\infty, +\infty]$ is finite a.e.

$$\mu(\{x : |f(x)| = +\infty\}) = 0.$$

Proof. For the first statement,

$$\int_X |f|\, d\mu \ge \int_{E_\lambda} |f|\, d\mu \ge \lambda\mu(E_\lambda).$$

This implies $\mu(E_k) \to 0$. Since E_k increases to $\{|f| = +\infty\}$, the result follows. □

This means that we could have assumed from the very beginning that measurable and integrable functions are defined a.e. and finite. The inequality in the proposition is called the *Markov inequality*.

12.3 Integrals and Limits

12.3.1 We deal now with limiting processes with non-negative functions. In this aspect, the Lebesgue integration turns out to be a very flexible tool. The following is *Lebesgue's monotone convergence theorem.*

Theorem 12.1. *Let f_k be a sequence of non-negative measurable functions such that $f_k(x) \le f_{k+1}(x)$ for a.e. $x \in X$. Then $f(x) = \lim_k f_k(x)$ (which is defined a.e. on X) is measurable and*

$$\lim_k \int_X f_k \, d\mu = \int_X f \, d\mu.$$

Proof. We already know that f is measurable. Clearly, the left-hand limit L exists and $L \le \int_X f\,d\mu$, because $f_k \le f$. To prove the reverse inequality, we must show that for every simple function $s, 0 \le s \le f$, one has $\int_X s\,d\mu \le L$. Given $0 < \lambda < 1$, we consider the measurable sets $A_k = \{f_k \ge \lambda s\}$. Since $f = \lim_k f_k$, $X = \cup_k A_k$. Then, in the notation of paragraph 12.2.1

$$\int_X f_k \, d\mu \ge \int_{A_k} f_k \, d\mu \ge \lambda \int_{A_k} s \, d\mu = \lambda \overline{s}(A_k).$$

Since $\overline{s}$ is a measure, the right-hand side has limit $\lambda \overline{s}(X) = \lambda \int_X s\,d\mu$. Taking now the limit as $\lambda \to 1$, the proof is finished. □

Applying the theorem to the sequence of partial sums, we obtain:

Corollary 12.1. *If $f_k, k \in \mathbb{N}$, are non-negative measurable functions, then*

$$\int_X \left(\sum_k f_k(x) \right) d\mu = \sum_k \int_X f_k(x) \, d\mu.$$

In particular, for mutually disjoint $A_k, k \in \mathbb{N}$,

$$\int_{\cup A_k} f \, d\mu = \sum_k \int_{A_k} f \, d\mu.$$

For a double sequence $a_{jk} \ge 0, j, k \in \mathbb{N}$,

$$\sum_j \sum_k a_{jk} = \sum_k \sum_j a_{jk}.$$

Corollary 12.2 (Fatou's lemma). *If $f_k, k \in \mathbb{N}$ are non-negative measurable functions, then*

$$\int_X (\liminf f_k) \, d\mu \le \liminf \int_X f_k \, d\mu.$$

Proof. It is enough to apply the theorem to $g_k = \inf_{i\geq k} f_i$. □

Exercise 12.4. For a positive measurable function f its *distribution function* is defined

$$\varphi_f(\lambda) = \mu(\{x : f(x) > \lambda\}), \quad \lambda \geq 0.$$

(a) Prove that φ_f is non-increasing, with lateral limits

$$\varphi(\lambda-) = \mu(\{x : f(x) \geq \lambda\}), \quad \varphi(\lambda+) = \varphi(\lambda).$$

(b) Prove that a monotone function has a countable set of discontinuities and that it is a measurable function.

(c) Prove that if $f = s$ is simple, then

$$\int_X s\, d\mu = \int_{[0,+\infty)} \varphi_s(\lambda)\, d\lambda.$$

Using the monotone convergence theorem and Proposition 12.1, prove that this holds for f measurable and positive.

(d) Prove that the same holds replacing φ_f by

$$\mu(\{x : f(x) \geq \lambda\}).$$

12.3.2 Now we look at limiting processes for general functions.

Theorem 12.2 (Lebesgue's dominated convergence theorem). *Let f_k be a sequence of complex measurable functions such that $f(x) = \lim_k f_k(x)$ exists a.e. and $|f_k(x)| \leq g(x)$ a.e. for some integrable g. Then f is integrable and*

$$\lim_k \int_X |f_k - f|\, d\mu \to 0, \quad \lim_k \int_X f_k\, d\mu = \int_X f\, d\mu.$$

Proof. As $|f| \leq g$, f is integrable. Applying Fatou's lemma to $2g - |f_k - f|$, we get

$$\begin{aligned}\int_X 2g\, d\mu &\leq \liminf_k \int_X (2g - |f_k - f|)\, d\mu \\ &= \int_X 2g\, d\mu - \limsup_k \int_X |f_k - f|\, d\mu.\end{aligned}$$

Therefore, $\limsup_k \int_X |f_k - f|\, d\mu = 0$. Finally,

$$\left| \int_X f_k\, d\mu - \int_X f\, d\mu \right| \leq \int_X |f_k - f|\, d\mu \to 0.$$ □

The rule

$$\int_X \lim_k f_k(x)\, d\mu(x) = \lim_k \int_X f_k(x)\, d\mu(x),$$

does not hold without further hypothesis. For instance, if f is an integrable function on $\mathbb{R}$ vanishing at infinity, then $f_k(x) = f(x+k) \to 0$ as $k \to +\infty$ while $\int_{\mathbb{R}} f_k\, dx = \int_{\mathbb{R}} f\, dx$ is constant.

In the next section, we will see that Riemann integrable functions are Lebesgue integrable with the same integral. Thus, the previous theorem implies Proposition 11.5.

Corollary 12.3. *Assume f_k is a sequence of complex measurable functions such that*

$$\sum_k \int_X |f_k|\, d\mu < +\infty.$$

Then $f(x) = \sum_k f_k(x)$ converges a.e., f is integrable, and

$$\int_X f\, d\mu = \sum_k \int_X f_k\, d\mu.$$

Proof. By Corollary 12.1,

$$\int_X \sum_k |f_k|\, d\mu < +\infty,$$

and by Proposition 12.2, $\sum_k |f_k(x)| < +\infty$ a.e. Therefore, the series $\sum_k f_k(x)$ converges absolutely whence is convergent a.e., too. Applying the theorem to its partial sums finishes the proof. □

For the counting measure, the corollary states that if a double sequence (a_{jk}) is absolutely convergent

$$\sum_j \sum_k |a_{jk}| < +\infty,$$

(note that the order of summation does not matter, by Corollary 12.1), then the sum of the double series does not depend on the order of summation. For a general double series the formula

$$\sum_j \sum_k a_{jk} = \sum_k \sum_j a_{jk},$$

does not hold, even if the four sums involved are defined.

If f is integrable and

$$f_\lambda(x) = \begin{cases} f(x), & |f(x)| < \lambda, \\ 0, & |f(x)| \geq \lambda, \end{cases}$$

then $f_\lambda(x) \to f(x)$ a.e.; application of Lebesgue's dominated convergence theorem gives that $\int_X |f - f_\lambda|\, d\mu \to 0$ as $\lambda \to +\infty$, that is,

$$\int_{|f|\geq\lambda} |f|\, d\mu \to 0,$$

that implies that

$$\lambda\, \mu\{|f| \geq \lambda\} \to 0,$$

improving Proposition 12.2.

Another application of the theorem, in the Euclidean setting, is obtained by truncation, with

$$f_\lambda(x) = \begin{cases} f(x), & |x| < \lambda, \\ 0, & |x| \geq \lambda. \end{cases}$$

We obtain that for integrable f,

$$\int_{|x|>\lambda} |f(x)|\, dm \to 0.$$

For the counting measure, this is the statement that for an absolutely convergent series $\sum_j a_j$, one has

$$\sum_{j>k} |a_j| \to 0, \quad k \to +\infty.$$

12.4 Functions Defined by Integrals

In the framework of Lebesgue integration, continuity and differentiability of functions defined by integrals are stated as follows. Assume that $F(x, y)$ is defined on $X \times U$, U is a domain in $\mathbb{R}^n$, and that it is integrable in x for all y. We can then define

$$f(y) = \int_X F(x, y)\, d\mu(x), \quad y \in U.$$

Proposition 12.3. *Assume that F is continuous in y for a.e. x and that there is an integrable function g such that $|F(x, y)| \leq g(x)$ for a.e. x. Then*

f is continuous in U. If, moreover, F is of class C^1 in y for a.e. x and

$$|\nabla_y F(x, y)| \le h(x),$$

for a.e. x, and an integrable function h, then F is of class C^1 in U and

$$\nabla f(y_0) = \int_X \nabla_y F(x, y_0)\, d\mu(x).$$

Proof. For an arbitrary sequence $y_k \to y$, the functions $f_k(x) = F(x, y_k)$ satisfy for k big enough the hypothesis of Lebesgue's dominated convergence theorem, so

$$f(y_k) = \int_X f_k\, d\mu \to \int_X F(x, y)\, d\mu = f(y).$$

Similarly, for a direction v and a sequence $t_k \to 0$

$$\frac{f(y + t_k v) - f(y)}{t_k} = \int_X \frac{F(x, y + t_k v) - F(x, y)}{t_k}\, d\mu(x).$$

By the mean-value theorem and the hypothesis, for a.e. x

$$\left| \frac{F(x, y + t_k v) - F(x, y)}{t_k} \right| \le h(x),$$

so we can apply the theorem again and conclude that

$$\lim_k \frac{f(y + t_k v) - f(y)}{t_k} = \int_X D_v F(x, y)\, d\mu(x).$$

Since this is true for every sequence t_k, we conclude that $D_v f(y)$ exists and equals the right-hand side, which is continuous in y by the first part of the theorem. □

Note from the proofs that it is enough to assume that the uniform bound hypothesis holds locally around each point of U.

Example 12.1. Our aim is to compute

$$I = \int_0^{+\infty} e^{-x^2}\, dx.$$

No explicit anti-derivative of the Gaussian function is available, so we need an indirect method. Introduce

$$f(t) = \int_0^{+\infty} \frac{e^{-t^2(1+x^2)}}{1 + x^2}\, dx, \quad t > 0.$$

The function $F(x,t)$ inside the integral satisfies

$$|F(x,t)| \le g(x) = \frac{1}{1+x^2}, \quad |D_t F(x,t)| = 2te^{-t^2(1+x^2)}.$$

For $t_0 > 0$ fixed and $2t_0 > t > t_0/2$,

$$|D_t F(x,t)| \le h(x) = 4t_0 e^{-t_0 x^2/4},$$

so we can apply the theorem and get

$$f'(t) = -2te^{-t^2} \int_0^{+\infty} e^{-t^2 x^2}\, dx.$$

Making the substitution $y = tx, dy = tdx$,

$$f'(t) = -2e^{-t^2} \int_0^{+\infty} e^{-y^2}\, dy = -2Ie^{-t^2}.$$

This implies, for $a < b$,

$$f(b) - f(a) = -2I \int_a^b e^{-t^2}\, dt.$$

By the dominated convergence theorem,

$$\lim_{a\to 0} f(a) = \int_0^{+\infty} \frac{1}{1+x^2}\, dx = \frac{\pi}{2}, \quad \lim_{b\to +\infty} f(b) = 0,$$

whence $I = \frac{\sqrt{\pi}}{2}$. A different method is explained in Example 13.16.

The differentiation rule in the proposition does not hold in general, even if both terms make sense. On the other hand, f may be continuous at one point even if the hypothesis does not hold locally.

Example 12.2. Consider

$$f(t) = \int_0^{+\infty} F(x,t)\, dx, \quad F(x,t) = \frac{t^\alpha e^{-x}}{(t^2 + e^{-x})^2}, \quad t \in \mathbb{R}.$$

A computation shows that

$$f(t) = \frac{t^{\alpha-2}}{1+t^2}.$$

If $\alpha \ge 2$, this is a continuous function of t; if $\alpha > 2$, $F(x,t) \le Ce^{-\beta x}$ for t close to zero, for some $\beta > 0$. But for $\alpha = 2$, $F(x,t) \le g(x)$ cannot hold for t close to zero for an integrable function g because $g(x) \ge F(x, e^{-x/2}) = \frac{1}{4}$. For $\alpha = 3$, f is differentiable with $f'(0) = 1$, but $D_t F(x, 0) = 0$. Similarly as F with $\alpha = 2$, $D_t F$ does not satisfy an estimate $D_t F(x,t) \le h(x)$ with h integrable for t close to zero.

12.5 Probability Spaces

We briefly comment on Lebesgue's theory in the context of probability spaces. As said before, a measure space with $\mu(X) = 1$ is called a *probability space*, in which case the notation $(\Omega, \mathcal{E}, p)$ is used instead of $(X, \mathcal{M}, \mu)$. Elements of Ω are all possible outcomes of a random experiment, and sets $A \in \mathcal{E}$ are called *observable events* of the random experiment at hand. In this context, events A, B with $A \cap B = \emptyset$ are called *incompatible.* Intuitively, $p(A)$ is the limit of the observed *relative frequency* of A,

$$p(A) = \lim_{N\to\infty} \frac{M}{N}, \tag{12.2}$$

where M is the number of times that A has been observed in N repetitions of the experiment in identical circumstances. Instead of trying to justify this as a definition, probability theory proceeds axiomatically. One accepts that events have probabilities obeying the axiom $p(\cup_k A_k) = \sum_k p(A_k)$ whenever the events A_k are pair-wise incompatible. Within the theory it is then proved that in an appropriate sense (12.2) holds. This is the content of the so-called *laws of large numbers.*

Measurable functions in a probability space are called *random variables.* Measurability of $X : \Omega \to \mathbb{R}$ means that $X(\omega)$ is an *observable numerical characteristic* of the output ω. Simple functions are here random variables X taking a finite number of values $x_1, \ldots, x_k$. If $A_i = \{X = x_i\}$, the definition of the integral

$$I = \sum_i x_i p(X = x_i)$$

is intuitively a mean-value or *expected value* of X because of the following. Assume that in N repetitions of the random experiment, the observed values of X are $y_1, \ldots, y_N$ and consider the observed mean value

$$\overline{y} = \frac{1}{N}(y_1 + \cdots + y_N).$$

If x_i appears m_i times in this set of values, $\sum_{i=1}^k m_i = N$, then

$$\overline{y} = \sum_i x_i \frac{m_i}{N},$$

which by (12.2) has limit I. Of course, this interpretation goes over to a general random variable by approximation.

12.6 Lebesgue Integral in $\mathbb{R}^n$

Now we specialize to the Euclidean measure space $(\mathbb{R}^n, m_n, \mathcal{L})$. The invariance by rigid motions T of measurable sets, $m_n(T(A)) = m_n(A)$, implies the invariance of the integral in the following sense. Given f defined on $\mathbb{R}^n$, define

$$f_T(x) = f(T^{-1}x).$$

Then f is integrable if and only if f_T is integrable with the same integral. Applying the definitions it is enough to prove this for a simple function $s = \sum \lambda_i 1_{A_i}$; then $s_T = \sum \lambda_i 1_{T(A_i)}$ and so

$$\int_{\mathbb{R}^n} s_T \, dm = \sum_i \lambda_i m(T(A_i)) = \sum_i \lambda_i m(A_i) = \int_{\mathbb{R}^n} s \, dm.$$

Since $m(T(A)) = |\det T| m(A)$ for a general invertible linear map T, the same argument shows that

$$\int_{\mathbb{R}^n} f_T \, dm = |\det T| \int_{\mathbb{R}^n} f \, dm.$$

In Section 13.3, we will study the general change of variable formula.

As shown in Theorem 11.3, Lebesgue measurable sets are strongly related to Borel sets, they are arbitrarily close in measure to open and compact sets. In an analogous way, Lebesgue measurable and integrable functions are related to continuous functions, as shown by the following results.

Theorem 12.3. *If f is Lebesgue integrable on U and $\varepsilon > 0$, there is a continuous function g with compact support in U such that $\int_U |f-g| \, dm < \varepsilon$.*

Proof. To prove this we may assume that f is non-negative. By definition of the integral, there is a simple function s with $\int_U f \, dm - \int_U s \, dm = \int_U (f - s) \, dm < \varepsilon/2$, so it is enough to show that a simple integrable non-negative function can be approximated by functions in $C_c(U)$. For this, in turn, it suffices to show that 1_A, with $A \subset U$ measurable and $m(A)$ finite, can be arbitrarily approximated by a function in $C_c(U)$. Given $\varepsilon > 0$, there is an open set $V, A \subset V$ and a compact $K \subset A$ such that $m(V \setminus K) < \varepsilon$. Replacing V by $V \cap U$ we may assume $V \subset U$. By Theorem 5.4 there is $\phi \in C_c(V), 0 \leq \phi \leq 1, \phi(x) = 1, x \in K$. Then $1_A - \phi$ is supported in $V \setminus A \subset V \setminus K$ whence

$$\int_U |1_A - \phi| \, dm \leq 2m(V \setminus K) \leq 2\varepsilon. \qquad \square$$

We will need the next result in paragraph 12.8.3. In its statement, a *lower semi-continuous function* v in U is a real-valued function possibly taking infinite values, $-\infty \leq v \leq +\infty$, such that for every $p \in U$,

$$\liminf_{x \to p} v(x) \geq v(q).$$

Recall that $\liminf_{x \to p} v(x) = \sup_{\varepsilon > 0} \inf\{v(x), x \in B(p, \varepsilon)\}$. This amounts to saying that $\{x : v(x) > a\}$ is an open set for every $a \in \mathbb{R}$. The function u is called upper semi-continuous if $-u$ is lower semi-continuous.

Exercise 12.5. Prove that finite linear combinations $\sum_i c_i v_i, c_i > 0$ of lower (resp., upper) semi-continuous functions are lower (resp., upper) semi-continuous. Prove that the point-wise increasing (resp., decreasing) limit of lower (resp., upper) semi-continuous functions is also lower (resp., upper) semi-continuous.

The characteristic functions $1_V, 1_K$ of an open set V and compact set K are respectively lower and upper semi-continuous. With this in mind, the following result, the *Vitali–Carathéodory theorem*, can be seen as the analogue of the structure result Corollary 11.2 of Lebesgue measurable sets:

Theorem 12.4. *If f is real-valued, Lebesgue integrable on U and $\varepsilon > 0$, there exists u, v such that $u \leq f \leq v$, u is upper semi-continuous, v is lower semi-continuous, and $\int_U (v - u)\, dm < \varepsilon$.*

Proof. We take it from Rudin's book [14]. We may assume $f \geq 0$. Note that if f is the characteristic function of a measurable set, the result follows from Corollary 11.2. We will thus write f as an (infinite) linear combination of such characteristic functions. By Corollary 12.1, there are simple functions s_n increasing to f; then $f = \sum_n (s_n - s_{n-1}) = \sum_{j=1}^{\infty} c_j 1_{A_j}$. By Corollary 12.1, there exist open sets V_j and compact sets K_j such that $K_j \subset E_j \subset K_j$ and $c_j m(V_j \setminus K_j) < 2^{-j-1}\varepsilon$. Then $v = \sum_j c_j 1_{V_j}$ is lower semi-continuous and $f \leq v$. The function $\sum_j c_j 1_{K_j}$ is not necessarily upper semi-continuous, but a finite sum $u = \sum_{j=1}^{N} c_j 1_{K_j}$ is; since

$$\int_U f\, dm = \sum_j c_j m(V_j),$$

we can take N big enough so that $\sum_{j=N+1}^{+\infty} c_j 1_{K_j} < \frac{\varepsilon}{2}$. Then

$$v - u = \sum_{j=1}^{N} c_j (1_{V_j} - 1_{K_j}) + \sum_{j=N+1}^{+\infty} 1_{V_j} \leq \sum_{j=1} c_j (1_{V_j} - 1_{K_j}) + \sum_{j=N+1}^{+\infty} 1_{E_j},$$

whence

$$\int_U (v-u)\,dm \le \sum_{j=1}^{+\infty} 2^{-j-1}\varepsilon + \frac{\varepsilon}{2} = \varepsilon. \qquad \square$$

12.7 Relations Between Riemann and Lebesgue Integration

12.7.1 We have seen two integration theories, Riemann's and Lebesgue's. In this section, we analyze the relations between them and the main differences.

Theorem 12.5. *A Riemann integrable function is Lebesgue integrable with the same integral.*

Proof. Riemann integrability of a bounded function f defined on an interval R implies the existence of a sequence of partitions $P_k \subset P_{k+1}, k \in \mathbb{N}$, of R such that $U(f,P_k) - L(f,P_k) \to 0$, and

$$I = \lim_k U(f,P_k) = \lim_k L(f,P_k),$$

being the Riemann integral. If R_{jk} are the intervals defined by P_k, we consider the step functions

$$s_k = \sum_j \beta_{jk} 1_{\mathring{R}_{jk}}, \quad t_k = \sum_j \alpha_{jk} 1_{\mathring{R}_{jk}},$$

with $\alpha_{jk} = \sup_{R_{jk}} f, \beta_{jk} = \inf_{R_{jk}} f$. We can consider s_k, t_k as a.e. defined simple functions satisfying for a.e. x

$$\begin{aligned} &s_k(x) \le s_{k+1}(x) \le f(x) \le t_{k+1}(x) \le t_k(x), \\ &\textstyle\int_R s_k\,dm = L(f,P_k), \quad \int_R t_k\,dm = U(f,P_k)\,. \end{aligned}$$

Then $s(x) = \lim_k s_k(x), t(x) = \lim_k t_k(x)$ exist a.e., are Lebesgue integrable, $s \le f \le t$ and

$$\int_R (t-s)\,dm = \lim_k \int_R (t_k - s_k)\,dm = 0,$$

implying $s = f = t$ a.e. It follows that f is Lebesgue integrable and its Lebesgue integral is

$$\int_R f(x)\,dm = \lim_k \int_R s_k\,dm = \lim_k L(f,P_k) = I. \qquad \square$$

12.7.2 Improper Riemann integrals. The Riemann integral deals with bounded functions defined on bounded sets. The concept of *improper Riemann integral* $\int_A f\,dm$ arises when either A is not bounded or f is not bounded in A, or both.

The setting is as follows. We assume that U is an open set in $\mathbb{R}^n$, f is a real function defined in U such that its restriction to every compact Jordan measurable $K \subset U$ is Riemann integrable. By Theorem 11.9, and since U is a countable union of cubes (Proposition 11.1) , this means f is bounded on every compact $K \subset U$ and that the set of discontinuities of f has zero Lebesgue measure. Of course we are interested in the case when either U is unbounded or f is not bounded near some points of bU, or both.

The obvious idea is to consider *exhausting sequences* K_k of compact Jordan measurable subsets, that is, $K_k \subset K_{k+1}, U = \cup_k K_k$, for instance using again Proposition 11.1. The natural definition is then:

Definition 12.1. We say that $\int_U f(x)\,dm$ is convergent (or that f is Riemann-integrable on U) with integral I if whenever K_k is an exhausting sequence for U,

$$\lim_k \int_{K_k} f(x)\,dm = I.$$

In fact, it is easy to see that if one requires that the limit exists and is finite for every exhausting sequence, then all limits are the same. We point out that when $n = 1$, this is not the classical definition. For instance, if $U = (0, +\infty)$, the classical definition is

$$\int_0^{+\infty} f(x)\,dx = \lim_{a\to 0, b\to+\infty} \int_a^b f(x)\,dx,$$

that amounts to considering exhausting compacts only of type $K_m = (a_m, b_m)$. The possibility to restrict to particular sequences of exhausting sets, using the order relation, is what makes a difference between $n = 1$ and $n > 1$, as will be explained shortly.

Three remarks are in order. First, if U is bounded and f is Riemann integrable on U, then of course $I = \int_U f\,dm$, because

$$\left|\int_U f\,dm - \int_{K_k} f\,dm\right| = \left|\int_{U\setminus K_k} f\,dm\right| \le \int_{U\setminus K_k} |f|\,dm \le Cm(U \setminus K_k) \to 0.$$

Secondly, if $f \geq 0$, then for an arbitrary exhausting sequence K_k

$$\int_{K_k} f(x)\, dm \to I = \sup_K \int_K f(x)\, dm,$$

so convergence simply means I finite. In fact, when $f \geq 0$, we consider that the integral is always defined

$$\int_U f(x)\, dm = \sup_K \int_K f(x)\, dm \leq +\infty.$$

The third remark is that if

$$\int_U |f|\, dm < +\infty,$$

then $\int_U f(x)\, dm$ is convergent. To see this, write $f = f^+ - f^-$, $|f| = f^+ + f^-$, then both $\int_U f^+\, dm$, $\int_U f^-\, dm$ are convergent and so is $\int_U f\, dm$. When $\int_U |f|\, dm < +\infty$, we say that $\int_U f\, dm$ is *absolutely convergent.*

Exercise 12.6. Show that the absolute convergence or divergence of an improper integral depends on the behavior of f near the boundary bU:

(a) If $|f(x)| \leq C|g(x)|$ for x close to bU, that is

$$\limsup_{x \to bU} \frac{|f(x)|}{|g(x)|} < +\infty,$$

and $\int_U g\, dm$ is convergent, so is $\int_U f\, dm$.

(b) If

$$0 < \lim_{x \to bU} \frac{|f(x)|}{|g(x)|} < +\infty,$$

then $\int_U f\, dm$, $\int_U g\, dm$ have the same character.

We noticed that absolute convergence implies convergence. In fact, the converse holds, too.

Theorem 12.6. *If $\int_U f\, dm$ is convergent, then it is absolutely convergent.*

Proof. Assume $\int_U |f|\, dm = +\infty$; then either $\int_U f^+\, dm = +\infty$ or $\int_U f^-\, dm = +\infty$, or both. We assume without loss of generality that $\int_U f^+\, dm = +\infty$ and will show that there is an exhausting sequence K_k of finite unions of intervals such that $\int_{K_k} f\, dm \to +\infty$. First, we define an exhausting family K'_k inductively as follows. Given K'_k, since $\int_{U\setminus K'_k} f^+\,dm = +\infty$, we can choose $K'_{k+1} \subset U$ with

$$\int_{K'_{k+1}\setminus K'_k} f^+\,dm > k + \int_{K'_k} |f|\,dm.$$

Now we will modify K'_k to a suitable K_k. With $A = \{f > 0\}$ we have

$$\int_{(K'_{k+1}\setminus K'_k)\cap A} f\,dm > k + \int_{K'_k} |f|\,dm.$$

Since f is a.e. continuous, we can replace this last set by its interior, and can choose in it a finite union of intervals L_k such that still

$$\int_{L_k} f\,dm > k + \int_{K'_k} |f|\,dm.$$

Now we define $K_k = K'_k \cup L_k$. This is an exhausting sequence, and by construction,

$$\int_{K_k} f\,dm = \int_{K'k} f\,dm + \int_{L_k} f\,dm > k.$$

□

Thus, no conditional convergent improper integrals (convergent, not absolutely convergent) exist using the definition above.

Exercise 12.7. In dimension one, for the well-known example $f(x) = \sin x/x$ one has

$$\lim_{b\to+\infty} \int_0^b f(x)\,dx = \frac{\pi}{2}, \int_0^{+\infty} |f|\,dx = +\infty.$$

Make explicit K_m, each a union of closed intervals and exhausting $(0, +\infty)$ such that $\int_{K_m} f\,dx \to +\infty$.

Thus, we can state:

Theorem 12.7. *A function f defined in an open set U is Riemann integrable in the improper sense if and only if it is Lebesgue integrable, bounded on every compact set $K \subset U$ and has a set of discontinuities $D(f)$ of zero measure. In this case, the Riemann and Lebesgue integrals are the same.*

Proof. If f is Riemann integrable on every Jordan measurable compact $K \subset U$, it follows from Theorem 12.5 that f is measurable. Exhausting U by a sequence K_k, since $D(f) = \cup_k (D(f) \cap \mathring{K}_k)$, Theorem 11.9 implies that $D(f)$ has zero Lebesgue measure. Finally, by Theorem 12.6 and the monotone convergence theorem, convergence of the improper Riemann integral is equivalent to Lebesgue integrability. □

To exhibit a Lebesgue integrable function which is not Riemann integrable we need functions with a set of discontinuities of positive measure. This cannot be accomplished by closed, explicit, formulas in terms of elementary functions. The easiest example is the characteristic function 1_A of a bounded measurable set with $m(bA) > 0$, as the irrationals in $[0, 1]$ or the fat Cantor set in Example 11.1. Thus, for practical applications, Riemann's theory is sufficient and, in fact, integrals are in practice evaluated or estimated numerically approaching them by Riemann sums, not Lebesgue sums.

12.7.3 What is then the point of the Lebesgue integral and what makes it an important tool in Analysis? The key word to answer this question is *completeness*. If $R(U), L^1(U)$ denote the space of Riemann and Lebesgue integrable functions in U, respectively, the relation between both is much like the one between the system of rational numbers $\mathbb{Q}$ and the system of real numbers.

We will prove two results: first, that $L^1(U)$ is complete and, secondly, that $R(U)$ is not by proving that the space $C_c(U)$ of continuous compactly supported functions in U is dense in $L^1(U)$. The distance considered in both $R(U)$ and $L^1(U)$ is the one defined by the L^1-norm

$$d(f, g) = \|f - g\|_1 = \int_U |f(x) - g(x)| \, dm.$$

For d to be a distance, we need, again, identifying functions which are equal a.e.

Theorem 12.8. *$L^1(U)$ is complete.*

Proof. Let $f_k, k = 0, 1, \ldots$ be a Cauchy sequence in $L^1(U)$,

$$\int_U |f_i(x) - f_j(x)| \, dm \to 0, \quad i, j \to +\infty.$$

Let k_l be an increasing sequence such that

$$\int_U |f_i(x) - f_j(x)|\, dm \leq 2^{-l}, \quad i,j \geq k_l.$$

By Corollary 12.3, the series

$$\sum_l (f_{k_{l+1}}(x) - f_{k_l}(x)),$$

is absolutely convergent a.e. and defines $\phi \in L^1(U)$. Since

$$\phi = f_{k_l} - f_{k_0} + \sum_{j \geq l} (f_{k_{j+1}}(x) - f_{k_j}(x)),$$

and

$$\sum_{j \geq l} |f_{k_{j+1}} - f_{k_j}| \leq \sum_{j \geq l} 2^{-j} = 2^{1-j},$$

we see that f_{k_l} is convergent with limit $f = \phi + f_{k_0}$. But from

$$|f - f_k| \leq |f - f_{k_l}| + |f_{k_l} - f_k|,$$

it follows that the whole sequence is also convergent to f. □

On the other hand, Theorem 12.3 can be restated as

Theorem 12.9. *The space $C_c(U)$ of compactly supported continuous functions in U is dense in $L^1(U)$.*

The importance of function spaces being complete has already been pointed out in Theorem 8.5. The following example considers a different kind of integral equation.

Example 12.3. We want to investigate continuous solutions of an integral equation

$$f(x) - Tf(x) = g(x), \quad Tf(x) = \int_{-\infty}^{+\infty} K(x,y)\, f(y)\, dy, \quad x \in \mathbb{R},$$

for a given continuous g. We can consider T acting in $L^1(\mathbb{R})$; this is so if

$$\sup_y \int |K(x,y)|\, dx = C < +\infty,$$

for then $\|Tf\|_1 \leq C\|f\|_1$. If $C < 1$, then T is contractive, $T'f = g + Tf$, too, and a fixed point of T' is a solution f. Thus, for every integrable g there is a unique solution f. But under mild conditions on K, Tf is continuous so $f = g + T(f)$ is also continuous if g is. This is the case for instance if $K(x,y) = e^{-2x(1+y^2)}$. For every continuous and integrable g, there is a unique continuous f such that $f - Tf = g$. The complete space $L^1(\mathbb{R})$ has been auxiliary in the proof. We might try using instead the complete space X of bounded continuous functions in $\mathbb{R}$. But as it happens, T is no longer a contraction in X.

Another important aspect is that Lebesgue theory is a more flexible tool that applies to other contexts. As shown in Section 12.5, Lebesgue theory is the appropriate tool in Probability theory to define the expectation or mean value of a random variable.

12.8 The Fundamental Theorem of Calculus for Multiple Integrals

12.8.1 The fundamental theorem of calculus in dimension one, viewed as describing relationships between differentiation and integration, can be generalized to several variables in a number of ways. Some will be analyzed later in this text in the context of vector analysis. In this section, we establish other versions in terms of *set functions*. The classical one-variable fundamental theorem of calculus can be stated in two parts:

Theorem 12.10.

(a) *If F is differentiable at every point $x \in [a,b]$ and $F' = f$ is Riemann integrable in $[a,b]$, then*

$$F(x) - F(a) = \int_a^x f(t)\,dt, \quad a \leq x \leq b, \tag{12.3}$$

whence the indefinite integral

$$G(x) = \int_a^x f(t)\,dt,$$

is also an anti-derivative of f.

(b) *For a Riemann integrable function f,*

$$G'(x) = \lim_{h \to 0} \frac{G(x+h) - G(x)}{h} = f(x),$$

at all continuity points of f, that is, for almost all points.

Proof. The second part is obvious. For the first part, if $P : a = x_0 < x_1 < \cdots < x_N = x$ is a partition of $[a, x]$, then, by the mean-value theorem,

$$F(x) - F(a) = \sum_i F(x_{i+1}) - F(x_i) = \sum_i f(\xi_i)(x_{i+1} - x_i),$$

is a Riemann sum associated to P, so the result follows. □

In terms of f, the first part states that if f is known to have an antiderivative (which is not always the case), then $G' = f$ everywhere. For continuous f, antiderivative and indefinite integral are equivalent concepts.

The fundamental theorem of calculus also holds for Lebesgue integrals:

Theorem 12.11.

(a) *If F is differentiable at all points of $[a, b]$ and $F' = f$ is Lebesgue integrable on $[a, b]$, then* (12.3) *holds.*
(b) *If f is Lebesgue integrable, $G'(x)$ exists a.e. and equals $f(x)$.*

A proof of (a) can be found in [14], and will be proved in what follows in general dimension. The second part implies that

$$\lim_{h \to 0} \frac{1}{2h} \int_{x-h}^{x+h} f(t)\, dt = f(x),$$

for a.e. x, see paragraph 12.8.4.

12.8.2 The generalization of statement (a) in Theorems 12.10 and 12.11 to general dimension is stated in terms of set functions. First, we consider the context of Riemann integration. A *set function* Φ on a domain U is a map defined on cubes $Q \subset U$ assigning to each Q a real or complex number $\Phi(Q)$ with the property that

$$\Phi(Q) = \sum_i \Phi(Q_i),$$

if (Q_i) is a finite partition of Q.

In case $U = [0, b_1] \times \cdots \times [0, b_n]$, there is a one-to-one correspondence between set functions and functions F in U through

$$F(x_1, \ldots, x_n) = \Phi([0, x_1] \times \cdots \times [0, x_n]).$$

The inverse transformation is

$$\Phi([c,d]) = F(d) - F(c),$$

if $n = 1$, and

$$\Phi([a,b] \times [c,d]) = F(b,d) - F(a,d) - F(b,c) + F(a,b),$$

if $n = 2$ with analogous expressions with 2^n terms for general n. The function F is called the *distribution function* of Φ.

A Riemann integrable function f defines a set function

$$\Phi_f(Q) = \int_Q f(x)\, dm.$$

Moreover, if $|f| \leq C$,

$$|\Phi_f(Q)| \leq Cm(Q).$$

Note that if f, g differ in a set of zero Jordan content, then $\Phi_f = \Phi_g$, so in general Φ_f cannot determine f point-wise. At all continuity points x of f

$$|f(x) - \frac{1}{m(Q)} \int_Q f(y)\, dm| \leq \frac{1}{m(Q)} \int_Q |f(y) - f(x)|\, dm,$$

has limit zero as Q shrinks to x. Therefore, by Theorem 11.9, Φ_f determines f almost everywhere, everywhere if f is continuous.

For a set function Φ, we define its *upper density*

$$\overline{D}_\Phi(x) = \limsup_{x \in Q} \frac{\Phi(Q)}{m(Q)} = \inf_{\varepsilon} \sup_{\delta(Q) \leq \varepsilon} \frac{\Phi(Q)}{m(Q)}, \tag{12.4}$$

where $Q \subset U$ are cubes. Analogously, the lower density is defined

$$\underline{D}_\Phi(x) = \liminf_{x \in Q} \frac{\Phi(Q)}{m(Q)} = \sup_{\varepsilon} \inf_{\delta(Q) \leq \varepsilon} \frac{\Phi(Q)}{m(Q)}.$$

In case both are finite and equal, we say that Φ has a *finite density* at x. Obviously, for a continuous function f, Φ_f has density f. In terms of the distribution function F of Φ, the existence of a density means that F has derivative f at all points if $n = 1$. Then, the following is a direct generalization of statement (a) in Theorem 12.10.

Theorem 12.12.

(a) *If a set function* Φ *has an upper Riemann integrable density* $\overline{D}_\Phi$ *at every point, then for every cube* $Q \subset U$,

$$\Phi(Q) \leq \int_Q \overline{D}_\Phi(x)\, dm.$$

(b) *If a set function* Φ *has a lower Riemann integrable density* $\underline{D}_\Phi$ *at every point, then for every cube* $Q \subset U$,

$$\Phi(Q) \geq \int_Q \underline{D}_\Phi(x)\, dm.$$

(c) *Consequently, if a set function* Φ *has a Riemann integrable density* f *at every point, then for every cube* $Q \subset U$,

$$\Phi(Q) = \int_Q f(x)\, dm.$$

Proof. Let $Q \subset U$ be a cube and let us break it into 2^n cubes S_i of equal measure. Since $\Phi(Q) = \sum_\Phi (S_i)$, one has

$$\Phi(S_i) \geq \frac{\Phi(Q)}{2^n},$$

whence

$$\frac{\Phi(S_i)}{m(S_i)} \geq \frac{\Phi(Q)}{m(Q)},$$

for at least one i. Repeating the argument we find a sequence Q_k of cubes, $Q_k \subset Q$, shrinking to some point $p \in Q$ such that

$$\frac{\Phi(Q_k)}{m(Q_k)} \geq \frac{\Phi(Q)}{m(Q)}.$$

Therefore, $\overline{D}_\Phi(p) \geq \frac{\Phi(Q)}{m(Q)}$. So $\Phi(Q) \leq \overline{D}_\Phi(p) m(Q)$ for some point $p \in Q$. This holds for all cubes. Now let (Q_i) be a partition of Q; then

$$\Phi(Q) = \sum_i \Phi(Q_i) \leq \sum_i \overline{D}_\Phi(p_i) m(Q_i),$$

which proves (a). The proof of (b) is similar and (c) follows from (a),(b).

□

As a consequence, we may state that for a set function Φ and a continuous function f, Φ has density f (a local condition) if and only if $\Phi(Q) = \int_Q f\,dm$ for all Q (a global condition).

If $n = 2$, in terms of the distribution function F, the result states that if

$$\lim_{a,b\to\alpha,c,d\to\beta} \frac{F(b,d) - F(a,d) - F(b,c) + F(a,b)}{(c-a)(d-b)} = f(\alpha,\beta)$$

exists at every point and f is Riemann integrable, then

$$F(x,y) = F(0,0) + \int_0^x \int_0^y f(s,t)\,ds\,dt.$$

In case F is of class C^2, the limit equals $f = D_{12}F$ and the latter formula follows by iteration of the fundamental theorem of calculus.

12.8.3 In this paragraph, we prove the analogue of Theorem 12.12 for Lebesgue integrable densities, the generalization to several variables of statement (a) in Theorem 12.11.

Theorem 12.13. *Let U be a domain in $\mathbb{R}^n$ and Φ a set function defined on cubes $Q \subset U$. Assume that Φ has a density $f(x)$ at every point $x \in U$ and that f is Lebesgue integrable on U. Then,*

$$\Phi(Q) = \int_Q f\,dm, Q \subset U.$$

Proof. We may assume Φ, f real-valued. Let $\varepsilon > 0$. By Theorem 12.4, there is a lower semi-continuous function v such that $f \le v$ and $\int_U (v - f)\,dm < \varepsilon$. Define

$$\Psi(Q) = \int_Q v\,dm - \Phi(Q).$$

Then, v being lower semi-continuous,

$$\liminf_{x\in R} \frac{\Psi(Q)}{m(Q)} = \liminf_{x\in Q} \left(\frac{1}{m(Q)} \int_Q v\,dm - \frac{\Phi(Q)}{m(Q)}\right) \ge v(x) - f(x) \ge 0.$$

Therefore, by Theorem 12.12, $\Psi(Q) \ge 0$, whence

$$\Phi(Q) \le \int_Q v\,dm = \int_Q f\,dm + \int_Q (v - f)\,dm < \int_Q f\,dm + \varepsilon.$$

This shows that $\Phi(Q) \le \int_Q f\,dm$ and applying the same argument to $-\Phi$ we are done. □

12.8.4 We point out some remarks. First, it is essential in the above proofs that the density is assumed to exist at *every* point. If it exists a.e., then the theorem does not hold.

Secondly, in other type of results the a.e. existence of the density is *proved.* This is the case for *Lebesgues' differentiation theorem*, the generalization of statement (b) in Theorem 12.11, establishing that

$$\lim_{r \to 0} \frac{1}{m(B(x,r))} \int_{B(x,r)} f(y)\, dm(y) = f(x), \tag{12.5}$$

holds for a.e. $x \in U$ if f is Lebesgue integrable in U. This theorem holds for cubes instead of balls, but for general intervals some assumption must be made on their excentricity.

Third, it is possible to characterize the set functions of the type $\Phi(Q) = \int_Q f\, dm$ with f Lebesgue integrable:

Theorem 12.14. *If* $\Phi = \Phi_f$ *with* f *Lebesgue integrable,* Φ *is* absolutely continuous, *meaning that for every* $\varepsilon > 0$ *there exists* $\delta > 0$ *such that* $\sum_i |\Phi(Q_i)| < \varepsilon$ *whenever* Q_i *are non-overlapping cubes and* $\sum_i m(Q_i) < \delta$*. Conversely, an absolutely continuous set function has at a.e. point a Lebesgue integrable density* f *and* $\Phi = \Phi_f$*.*

A reference for all these results is [10].

Chapter 13

Fubini's Theorem and Change of Variables

Up to now our presentation of the Riemann and Lebesgue integral has been very conceptual and has not said much about computation of integrals, and volumes in particular. In practice, computers implement numerical analysis methods viewing integrals as Riemann ones, that is, partitions are done in the domain of definition. This is an aspect not covered in this text. Still, it is convenient that the student learns some techniques to compute multiple integrals and volumes with no computer help. The definition of both the Riemann and the Lebesgue integral is very much based on Cartesian coordinates. The main result to compute multiple integrals in Cartesian coordinates is Fubini's theorem, or Cavalieri's principle in geometrical terms. The *change of variable formula* shows how to compute integrals when the domain of integration or the function to integrate, or both, are best described in a non-Cartesian coordinate system. In its proof we use the n-dimensional version of the fundamental theorem of calculus of the previous chapter, just identifying the Jacobian as a density.

13.1 Computing Multiple Integrals with Cartesian Coordinates

13.1.1 Using the definition, computing an integral is possible only in very special cases. Let us look at $f(x) = x^2$ in $[0, 1]$. Using the partition $P_N : 0 < \frac{1}{N} < \frac{2}{N} < \cdots < \frac{k}{N} < \cdots < 1$ the Riemann integral is the limit of the Riemann sums

$$\frac{1}{N}\sum_{k=0}^{N-1}\frac{k^2}{N^2}=\frac{1}{N^3}(1^2+2^2+\cdots+(N-1)^2).$$

Since the last sum equals $N(N+1)(2N+1)/6$, the integral is $\frac{1}{3}$. The Lebesgue sum for P_N is

$$\sum_{k=0}^{N-1}\frac{k}{N}\left(\sqrt{\frac{k+1}{N}}-\sqrt{\frac{k}{N}}\right)=N^{-\frac{3}{2}}\sum_{k=0}^{N-1}k(\sqrt{k+1}-\sqrt{k}),$$

which is not explicit. Still, we may state that

$$\lim_N N^{-\frac{3}{2}}\sum_{k=1}^{N-1}k(\sqrt{k+1}-\sqrt{k})=\frac{1}{3}.$$

In general, Riemann and Lebesgue sums are not explicit. For $n=1$, in first-year calculus we learned Barrow's rule

$$\int_{[a,b]}f\,dx=F(b)-F(a),\quad F'=f,$$

which reduces computation of integrals to computation of anti-derivatives. Exact values of the integral can be obtained in some cases combining two results, Fubini's theorem and the change of variables formula. In this section, we address the first one, starting in the context of the Riemann integral.

For $n>1$, at first glance one can realize that it is possible to compute an n-dimensional integral with n one-dimensional integrals in some particular cases. For instance, when f is *separable* on an interval R, meaning that

$$f(x_1,x_2,\ldots,x_n)=f_1(x_1)f_2(x_2)\cdots f_n(x_n),$$

with f_i bounded in $[a_i,b_i]$. If $P=P_1\times\cdots\times P_n$ is a partition of R, P_i a partition of $[a_i,b_i]$, it is clear for the Riemann sums that

$$\Sigma(f,P)=\Sigma(f_1,P_1)\cdots\Sigma(f_n,P_n),$$

proving that f is integrable if each f_i is, in which case

$$\int_R f(x)\,dm=\left(\int_{[a_1,b_1]}f_1\,dx_1\right)\cdots\left(\int_{[a_n,b_n]}f_n\,dx_n\right).$$

For a general function the key idea is given by geometric intuition. Let us assume $n=2$ with coordinates x,y and assume that $f\geq 0$ is continuous on

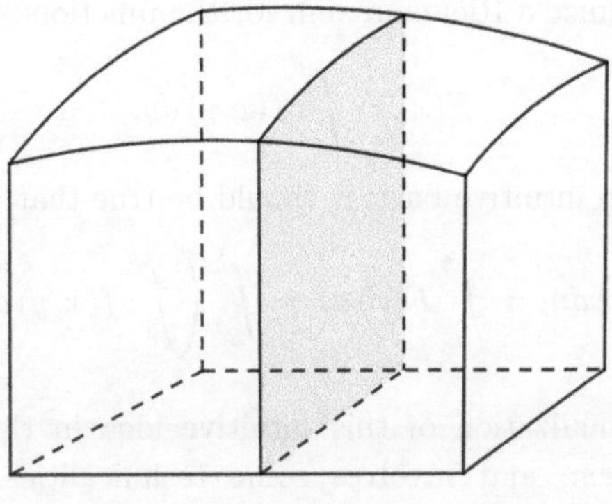

Figure 13.1. Cavalieri's principle.

$R = [a, b] \times [c, d]$, so that $\int_R f(x)\, dm$ is the volume of the three-dimensional subgraph $G(R, f)$. The idea, known as *Cavalieri's principle* is that we can compute this volume *using slices*, as illustrated in Figure 13.1.

In fact, Cavalieri's principle states that if two bodies A, B in space have the property that their slices

$$A_z = \{(x, y) \in \mathbb{R}^2 : (x, y, z) \in A\}, \quad B_z = \{(x, y) \in \mathbb{R}^2 : (x, y, z) \in B\},$$

have the same area for all z, then A, B have the same volume. Of course, one could use any other axes instead of the z-axis. The meaning is that we can compute the volume of $G(R, f)$ *by slices.* We consider a partition $a = x_0 < x_1 < \cdots < x_n = b$ and the slices

$$S_i = \{(x, y, f(x, y)) : x_i \leq x \leq x_{i+1}\}.$$

This slice has thickness $x_{i+1} - x_i$ and two faces, the subgraphs of $f_x(y) = f(x, y)$ for $x = x_i$ and $x = x_{i+1}$; if $\xi_i \in [x_i, x_{i+1}]$, the volume is approximately the thickness times the area of the subgraph of f_{ξ_i}, that is,

$$(x_{i+1} - x_i) \int_{[c,d]} f_{\xi_i}\, dy.$$

Incidentally, it is at these situations that making explicit the variable of integration in dy is useful. The volume of $G(R, f)$ is then approximately the sum of volumes of the slices,

$$\sum_i \left(\int_c^d f(\xi_i, y)\, dy \right) (x_{i+1} - x_i),$$

in which we recognize a Riemann sum for the function

$$F(x) = \int_c^d f(x,y)\,dy,$$

on $[a,b]$. So, on an intuitive basis it should be true that

$$\int_R f(x)\,dm = \int_a^b F(x)\,dx = \int_a^b \left(\int_c^d f(x,y)\,dy\right) dx.$$

13.1.2 The formalization of this intuitive idea in the general context is *Fubini's theorem*, and involves some technicalities, mainly because integrability of f in R does not imply integrability of all f_x in $[c,d]$. For instance,

$$f(x,y) = 0,\ \ x \neq \frac{1}{2}, \quad f\left(\frac{1}{2}, y\right) = 0,\ \ y \in \mathbb{Q}, \quad f\left(\frac{1}{2}, y\right) = 1,\ \ y \notin \mathbb{Q},$$

is trivially integrable with integral zero, but $f_{\frac{1}{2}}$ is not integrable. That's why one must deal with the upper and lower integral of f_x; in Fubini's theorem we use the notation

$$F_-(x) = L(f_x), \quad F^+(x) = U(f_x), \quad x \in [a,b].$$

Theorem 13.1. *For a bounded function f on $R = [a,b] \times [c,d]$, one has*

$$L(f) \le L(F_-) \le L(F^+) \le U(F^+) \le U(f),$$
$$L(f) \le L(F_-) \le U(F_-) \le U(F^+) \le U(f).$$

In particular, if f is Riemann integrable, then both F_-, F^+ are Riemann integrable with the same integral as f and f_x is Riemann integrable for a.e. x.

Proof. We prove $U(F^+) \le U(f)$, the other inequalities being analogous. Let P_1, P_2 be partitions of $[a,b], [c,d]$, respectively, and $P = P_1 \times P_2$. We denote by I_j the intervals of P_1 and by J_i those of P_2, so $I_j \times J_i$ are the intervals of P. Let M_{ij} be the supremum of f in $I_j \times J_i$, and consider

$$U(F^+, P_1) = \sum_j (\sup_{I_j} F^+)|I_j|.$$

By definition, $F^+(x) \le U(f_x, P_2)$. Since $\sup_{J_i} f_x \le M_{ij}$ for $x \in I_j$ one has

$$F^+(x) \le \sum_i M_{ij}|J_i|, \quad x \in I_j,$$

therefore,

$$U(F^+) \leq U(F^+, P_1) \leq \sum_{ij} M_{ij}|I_j||J_i| = U(f, P),$$

which gives $U(F^+) \leq U(f)$ minimizing in P. If f is Riemann integrable, the inequalities imply $L(F^+) = U(F^+), L(F_-) = U(F_-)$, proving that F_-, F^+ are Riemann integrable with the same integral as f. Then

$$\int_a^b (F^+ - F_-)\, dx = 0,$$

and therefore $F^+(x) = F_-(x)$ for a.e. x (Exercise 11.8, part (h)), that is, f_x is integrable for a.e. x. □

Evidently, one can exchange the roles of x, y and consider the functions $f^y(x) = f(x, y)$, etc.

If f is continuous, then f_x is continuous in y, whence integrable, so

$$F_-(x) = F_+(x) = \int_c^d f(x, y)\, dy.$$

By the results in paragraph 4.7.2, this function of x is continuous, and so the theorem states that the computation of a double integral can be reduced to the computation of the *iterated integrals* involving two one-dimensional integrals

$$\int_{[a,b]\times[c,d]} f(x, y)\, dm = \int_a^b \left(\int_c^d f(x, y)\, dy \right) dx = \int_c^d \left(\int_a^b f(x, y)\, dx \right) dy.$$

For a triple integral, there are six different iterated integrals, all with the same value.

13.1.3 Fubini's theorem is best stated in the context of Lebesgue theory. For simplicity we deal just with the Euclidean context, and assume that $f(x, y)$ is defined in $\mathbb{R}^2$. With the same notations, $f_x(y) = f(x, y) = f_y(x)$, the version of Fubini's theorem is as follows. We omit the proof (see [14]).

Theorem 13.2.

(a) *If f is measurable, then f_x is measurable for a.e. x and similarly for f_y.*

(b) *If* $f \geq 0$, *both*

$$F(x) = \int_{\mathbb{R}} f_x \, dy, \quad G(y) = \int_{\mathbb{R}} f_y \, dx,$$

(which are defined a.e. by the previous point) are measurable, and

$$\int_{\mathbb{R}} F \, dx = \int_{\mathbb{R}^2} f \, dm = \int_{\mathbb{R}} G \, dy.$$

(c) f *is Lebesgue integrable if and only if an iterated integral of* $|f|$ *is finite.*

(d) *If* f *is Lebesgue integrable both iterated integrals make sense and equal the integral of* f.

So, if one of the iterated integrals of $|f|$ is finite, then f is integrable and its integral equals the two iterated integrals. Note too that in particular the theorem states that if $A \subset \mathbb{R}^2$ has zero area, then for almost all x the section A_x has zero length, and similarly for A_y.

13.1.4 With Fubini's theorem one can explicitly compute multiple integrals in some cases.

Example 13.1. We saw in Example 11.2 that

$$\sum_{j=0}^{n-1} \frac{1}{j+n}$$

is a Riemann sum of $f(x) = \frac{1}{1+x}$ on $[0, 1]$. Since f is a continuous function on $[0, 1]$, it is integrable and hence

$$\lim_n \sum_{j=0}^{n-1} \frac{1}{j+n} = \int_0^1 \frac{1}{1+x} \, dx.$$

An anti-derivative of f is $\log(1 + x)$ whence the limit equals $\log 2$. In the same way,

$$\sum_{i,j=0}^{n-1} \frac{1}{n^2 + in + jn},$$

as a Riemann sum of the continuous function $f(x, y) = \frac{1}{1+x+y}$ on $Q = [0, 1] \times [0, 1]$ has limit

$$\int_Q \frac{1}{1+x+y} \, dm.$$

Since

$$\int_0^1 \frac{1}{1+x+y}\,dy = [\log(1+x+y)]_{y=0}^{y=1} = \log(2+x) - \log(1+x),$$

and

$$\int_0^1 \log(a+x)\,dx = [(a+x)\log(a+x)-x]_{x=0}^{x=1} = (1+a)\log(1+a)-1-a\log a,$$

the above limit equals

$$3\log 3 - 4\log 2.$$

Example 13.2. If $Q = [0,1] \times [0,1]$ is the unit square in the plane,

$$\begin{aligned}\int_Q xe^{xy}\,dm &= \int_0^1 \left(\int_0^1 xe^{xy}\,dy\right)dx = \int_0^1 [e^{xy}]_{y=0}^{y=1}\,dx \\ &= \int_0^1 (e^x - 1)\,dx = e - 2.\end{aligned}$$

Example 13.3. If $Q = [0,1] \times [0,1] \times [0,1]$ is the unit cube in space

$$\begin{aligned}\int_Q x^2 y \cos \pi xyz\,dm &= \frac{1}{\pi}\int_0^1 \left(\int_0^1 x \sin \pi xy\,dy\right)dx \\ &= \frac{1}{\pi^2}\int_0^1 (1 - \cos \pi x)\,dx = \frac{1}{\pi^2}.\end{aligned}$$

Example 13.4. Let us compute the volume of the cupola A defined by $-1 \leq x, y \leq 1, 0 \leq z \leq 2 - x^2 - y^2$. It equals

$$\begin{aligned}m(A) &= \int_{[-1,1]} \left(\int_{[-1,1]} (2 - x^2 - y^2)dy\right)dx \\ &= \int_{[-1,1]} \left(4 - 2x^2 - \frac{2}{3}\right)dx = \frac{16}{3}.\end{aligned}$$

Example 13.5. We analyze convergence of

$$\int_{\mathbb{R}^2} \frac{1}{1+|x|^\alpha + |y|^\beta}\,dm, \quad \alpha, \beta > 0.$$

In the dy integral, we use the change of variable $y = (1+|x|^\alpha)^{1/\beta} z$ to get

$$(1+|x|^\alpha)^{1/\beta - 1} \int_{\mathbb{R}} \frac{1}{1+|z|^\beta}\, dz.$$

The integral in z is finite iff $\beta > 1$. Then

$$\int_{\mathbb{R}} (1+|x|^\alpha)^{1/\beta-1}\, dx$$

is finite iff $\int_{|x|>1}$ is, that is

$$\int_1^{+\infty} x^{\alpha(1/\beta-1)}\, dx < +\infty.$$

Therefore, the integral is convergent iff

$$\frac{1}{\alpha} + \frac{1}{\beta} < 1.$$

Example 13.6. For

$$I = \int\int_{\mathbb{R}^2} \frac{|x|^\alpha |y|^\beta}{(1+x^2+y^4)^\gamma}\, dm,$$

in the y integral we use $y = (1+x^2)^{\frac{1}{4}} t$ to find

$$I = 2|x|^\alpha (1+x^2)^{\frac{1+\beta}{4}-\gamma} \int_{\mathbb{R}} \frac{t^\beta}{(1+t^4)^\gamma}\, dt.$$

Convergence occurs iff $4\gamma - \beta > 1, \beta > -1, \alpha > -1, 4\gamma - \beta - 2\alpha > 3$.

13.1.5 For a double integral $\int_A f\, dm$ on a measurable set, we identify the projection $\Pi_1(A)$ of A on the x-axis and the sections $A_x = \{y : (x,y) \in A\}$. Writing

$$A = \{(x,y) : x \in \Pi_1(A), y \in A_x\},$$

means describing A *by slices*. Then $\int_A f\, dm$ equals

$$\int_{\Pi_1(A)} \left(\int_{A_x} f(x,y)\, dy \right) dx,$$

with analogous expressions for higher-order integrals. In particular,

$$m_2(A) = \int_{\Pi_1(A)} l(A_x)\, dx,$$

and in general $m_n(A) = \int_{\Pi_1(A)} m_{n-1}(A_{x_1})\, dx_1$.

Let us consider some examples. First, we remind the student that, by definition, 2π is the length of a circle of radius 1. Archimedes proved by the method of exhaustion that the area of the disk equals the area of a rectangular triangle whose basis is the length of the circle and the height its radius:

$$A = \frac{1}{2} 2\pi = \pi.$$

In modern language, with Riemann sums, if P_N is the Nth circumscribed regular polygon and ε_N its side-length, then

$$N\varepsilon_N = \sum_{n=1}^{N} \varepsilon_N \to L,$$

while

$$\sum_{n=1}^{N} \frac{1}{2}\varepsilon_N \to A,$$

because $\frac{1}{2}\varepsilon_N$ is the area of the triangle with vertex at the origin and base a side of P_N. Therefore, the area of a disc of radius r is πr^2.

In spite of the fact that the length of an ellipse with semi-axis a, b is not expressible by elementary functions, its area can be computed explicitly:

$$\int_{-a}^{a} 2b\sqrt{1 - \frac{x^2}{a^2}}\, dx = ab \int_{-1}^{1} \sqrt{1 - x^2}\, dx = \pi ab,$$

because the last integral is the area of the unit disk.

If say A is the region in $\mathbb{R}^2$ between the graphs $y = \varphi_1(x), y = \varphi_2(x), x \in [a, b]$ with $\varphi_1(x) \leq \varphi_2(x)$, then

$$\int_A f(x, y)\, dm = \int_a^b \left(\int_{\varphi_1(x)}^{\varphi_2(x)} f(x, y)\, dy \right) dx.$$

Example 13.7. For $A = \{(x, y) : 0 \leq x \leq 1, 0 \leq y \leq x\}, f = e^{x^2}$,

$$\int_A f\, dm = \int_0^1 \left(\int_0^x e^{x^2}\, dy \right) dx = \int_0^1 x e^{x^2}\, dx = \frac{1}{2}(e - 1).$$

If instead we integrate first in x,

$$= \int_0^1 \left(\int_y^1 e^{x^2}\, dx \right) dy,$$

we find an integral in x that cannot be evaluated in terms of elementary integrals. This example shows that the order of integration matters when computing multiple integrals, regarding difficulty.

Example 13.8. We analyze convergence of

$$\int_{|x|\leq|y|} \frac{\log(1+|x|^\beta)}{1+|xy|^\alpha}\, dm.$$

In the dy integral, we change to $z = |x|y$ to obtain

$$\frac{\log(1+|x|^\beta)}{|x|} \int_{|z|\geq|x|^2} \frac{1}{1+|z|^\alpha}\, dz.$$

The dz integral is infinite if $\alpha \leq 1$, while if $\alpha > 1$ is bounded for $|x| < 1$ and if $|x| > 1$ of the order of

$$\int_{|z|\geq|x|^2} |z|^{-\alpha}\, dz = c|x|^{2(1-\alpha)}.$$

The integral in x has then two parts,

$$\int_0^1 \frac{\log(1+x^\beta)}{x}\, dx,$$

and

$$\int_1^{+\infty} \frac{\log(1+x^\beta)}{x^{2\alpha-1}}\, dx.$$

In the first one, $\log(1+x^\beta)$ is comparable to x^β, so it is convergent iff $\beta > 0$. In the second one, the log term does not play a role, and it is convergent iff $2\alpha - 1 > 1$. Altogether, the integral is convergent iff $\beta > 0, \alpha > 1$.

13.1.6 For a triple integral $\int_A f\, dm$ we proceed in the same way to describe it in cartesian coordinates. We first describe say $B = \Pi_{xy}(A)$, the projection on the xy-plane, and for $(x, y) \in B$, the slice $A_{x,y} = \{z : (x, y, z) \in A\}$. Then

$$\int_A f\, dm = \int_B \left(\int_{A_{xy}} f(x,y,z)\, dz \right) dm(x,y),$$

and compute the resulting double integral as before.

Example 13.9. Let A be the intersection of the ball with center $(0, 0, 0)$ and radius 1 and the ball with center $(0, 0, 1)$ and radius 1,

$$x^2 + y^2 + z^2 \leq 1, \quad x^2 + y^2 + z^2 \leq 2z.$$

The balls meet along $z = \frac{1}{2}, x^2 + y^2 \leq \frac{3}{4}$, so the projection on the xy plane is this disk. The description of A in cartesian coordinates is

$$-\frac{\sqrt{3}}{2} \leq x \leq \frac{\sqrt{3}}{2}, \quad -\sqrt{\frac{3}{4} - x^2} \leq y \leq \sqrt{\frac{3}{4} - x^2},$$
$$1 - \sqrt{1 - x^2 - y^2} \leq z \leq \sqrt{1 - x^2 - y^2}.$$

This is not an easy description. In the next section, we will learn how to describe sets using other systems of coordinates.

Example 13.10. Let us compute the volume of the ellipsoid

$$A = \left\{(x, y, z) : \frac{x^2}{a^2} + \frac{y^2}{b^2} + \frac{z^2}{c^2} \leq 1\right\},$$

using Cavalieri's principle: at level z the section is an ellipse of semi-axes

$$a\sqrt{1 - \frac{z^2}{c^2}}, \quad b\sqrt{1 - \frac{z^2}{c^2}},$$

therefore,

$$V = \pi ab \int_{-c}^{c} \left(1 - \frac{z^2}{c^2}\right) dz = \frac{4}{3}\pi abc.$$

In particular, the ball of radius r has volume $\frac{4}{3}\pi r^3$.

13.1.7 Next, we compute the volume of a ball of radius r in general dimension n, $\omega_n r^n$. By Cavalieri's principle, as before

$$\omega_n = \omega_{n-1} \int_{-1}^{1} (1 - t^2)^{\frac{n-1}{2}}\, dt = \omega_{n-1} \int_{0}^{1} (1 - x)^{\frac{n-1}{2}} x^{-\frac{1}{2}}\, dx.$$

The right-hand side is the beta function $B(\frac{n+1}{2}, \frac{1}{2})$, which in terms of the Euler gamma function

$$\Gamma(x) = \int_{0}^{+\infty} t^{x-1} e^{-t}\, dt,$$

equals

$$\frac{\Gamma(\frac{n+1}{2})\Gamma(\frac{1}{2})}{\Gamma(\frac{n}{2} + 1)}.$$

In particular, $\Gamma(\frac{1}{2}) = \sqrt{\pi}$. Using twice the recursion formula and that $\Gamma(x+1) = x\Gamma(x)$, it follows that

$$\omega_n = \omega_{n-2} \frac{\Gamma(\frac{n+1}{2})\Gamma^2(\frac{1}{2})\Gamma(\frac{n}{2})}{\Gamma(\frac{n+1}{2})\Gamma(\frac{n}{2}+1)} = \omega_{n-2}\frac{2\pi}{n}.$$

So, if $n = 2k$ is even,

$$\omega_n = \frac{(2\pi)^k}{n(n-2)\cdots 2} = \frac{\pi^k}{k!},$$

and if $n = 2k+1$ is odd,

$$\omega_n = 2\frac{(2\pi)^k}{n(n-2)\cdots 3}.$$

Both can be written

$$\omega_n = \frac{\pi^{\frac{n}{2}}}{\Gamma(\frac{n}{2}+1)}.$$

Note that ω_n increases while $n < 2\pi$ and decreases for $n > 2\pi$, and tends to zero very fast.

13.1.8 If f is C^1 with compact support in a domain U, then combining Fubini's theorem with the fundamental theorem of calculus one has for all directions v

$$\int_U D_v f \, dm = 0.$$

As a consequence, we have an *integration by parts formula*,

$$\int_U f D_v g \, dm = -\int_U g D_v f \, dm, \tag{13.1}$$

valid for C^1 functions, one of them compactly supported. In equation (18.5), we will see its complete version. Iterating, for a multi-index α,

$$\int_U f D^\alpha g \, dm = (-1)^\alpha \int_U g D^\alpha f \, dm. \tag{13.2}$$

This equation is the basis of an important concept, the *weak derivative*. It allows to consider derivatives of general functions which are not differentiable.

Definition 13.1. If f, h are Lebesgue integrable functions in a domain U, we say that $D^\alpha f = h$ *in the weak sense* if for all $g \in C_c^\infty(U)$ one has

$$\int_U f D^\alpha g \, dm = (-1)^\alpha \int_U gh \, dm.$$

If f is of class C^k, $|\alpha| = k$, (13.2) means that the weak derivative equals the classical one. This concept is of fundamental importance in the theory of partial differential equations. We will use it in the last chapters, too, in the context of vector analysis.

13.2 Convolution

13.2.1 In this section, we describe *convolution*, one of the basic operations in harmonic analysis. We denote by $L^1(\mathbb{R}^n)$ the space of complex-valued Lebesgue integrable functions provided with the norm

$$\|f\|_1 = \int_{\mathbb{R}^n} |f(x)|\, dm.$$

As mentioned before, we consider that functions are defined a.e., and identify those which are equal a.e.

This is an example of a *Banach space*, a linear space equipped with a norm, and so with a distance, and complete. In exactly the same way as for Euclidean space, a linear map T from $L^1(\mathbb{R}^n)$ to itself is continuous if and only if

$$\|Tf\|_1 \leq C\|f\|_1, \quad f \in L^1(\mathbb{R}^n),$$

for some constant C.

The *translation operator* $\tau_x, x \in \mathbb{R}^n$ acts on functions as

$$(\tau_x f)(y) = f(y - x), \quad y \in \mathbb{R}^n.$$

Clearly, $\tau_x f \in L^1(\mathbb{R}^n)$ if $f \in L^1(\mathbb{R}^n)$ and $\|\tau_x f\|_1 = \|f\|_1$. An important point is that $x \to \tau_x f$ is continuous, that is,

$$\lim_{y \to x} \|\tau_y f - \tau_x f\|_1 = 0.$$

This is clear if $f \in C_c(\mathbb{R}^n)$, by uniform continuity. Since

$$\begin{aligned}\|\tau_y f - \tau_x f\|_1 &\leq \|\tau_y f - \tau_y g\|_1 + \|\tau_y g - \tau_x g\|_1 + \|\tau_x g - \tau_x f\|_1 \\ &\leq 2\|f - g\|_1 + \|\tau_y g - \tau_x g\|_1,\end{aligned}$$

and $C_c(\mathbb{R}^n)$ is dense in $L^1(\mathbb{R}^n)$ (Theorem 12.9), this holds for all f.

The *convolution* of two functions f, g is defined as

$$(f * g)(x) = \int_{\mathbb{R}^n} f(y)g(x - y)\, dm(y).$$

This operation among functions appears in different contexts. One is related to probability theory. Assume X, Y are independent random variables with densities f_X, f_Y, respectively; their joint density function is then $f_X(x)f_Y(y)$, that is

$$p((X,Y) \in A) = \int_A f_X(x)f_Y(y)\,dm.$$

Note that indeed

$$p(X \in A_1, Y \in A_2) = p(X \in A_1)p(Y \in A_2).$$

We are interested in the law of the sum $Z = X+Y$. Its distribution function is

$$F_Z(z) = p(X+Y \leq z) = \int_A f_X(x)f_Y(y)\,dm,$$

with $A = \{(x,y) : x+y \leq z\}$. By Fubini's theorem,

$$\begin{aligned} F_Z(z) &= \int_{\mathbb{R}} f_X(x)\left(\int_{-\infty}^{z-x} f_Y(y)\,dy\right)dx = \int_{\mathbb{R}} f_X(x)\left(\int_{-\infty}^{z-x} g(y)\,dy\right)dx \\ &= \int_{\mathbb{R}} f_X(x)F_Y(z-x)\,dx = (f_X * F_Y)(x), \end{aligned}$$

where F_Y is the distribution function of Y. Differentiation yields

$$f_Z = f_X * f_Y.$$

If in an informal way we think of the density $f_X(x)$ as being $P(X = x)$, then the rule $f_Z = f_X * f_Y$ just proved reads

$$p(X+Y=z) = \int p(X=x, Y=z-x)\,dx = \int p(X=x)p(Y=z-x)\,dx,$$

and can be seen as a continuous version of the total probability formula.

To precise the definition of the convolution, we use Fubini's theorem:

Proposition 13.1. *If $f, g \in L^1(\mathbb{R}^n)$, $(f*g)(x)$ is defined for a.e. x, $f*g \in L^1(\mathbb{R}^n)$ and*

$$\int_{\mathbb{R}^n} (f*g)\,dm = \left(\int_{\mathbb{R}^n} f\,dm\right)\left(\int_{\mathbb{R}^n} g\,dm\right).$$

One has $f*g = g*f$ and $\|f*g\|_1 \leq \|f\|_1\|g\|_1$.

Proof. It is enough to apply Fubini's Theorem 13.2 to the function $F(x, y) = f(y)g(x-y)$, because the iterated integral

$$\int_{\mathbb{R}^n}\left(\int_{\mathbb{R}^n} |f(y)g(x-y)|\, dm(y)\right) dm(x)$$

equals

$$\begin{aligned}\int_{\mathbb{R}^n} |f(y)|\left(\int_{\mathbb{R}^n} |g(x-y)|\, dm(x)\right) dm(y) &= \int_{\mathbb{R}^n} |f(y)|\|\tau_y g\|_1 dm(y)\\ &= \|f\|_1\|g\|_1.\end{aligned}$$

□

In an analogous way, it can be shown that if f is integrable with compact support K and g is locally integrable, that is, g is integrable on every compact set, then $f*g$ is also locally integrable:

$$\begin{aligned}&\int_L\left(\int_K |f(y)||g(x-y)|\, dm(y)\right) dm(x)\\ &= \int_K |f(y)|\left(\int_L |g(x-y)|\, dm(x)\right) dm(y)\\ &\le \int_K |f(y)|\left(\int_{L-K} |g(z)|\, dm(z)\right) dm(y) < +\infty.\end{aligned}$$

The convolution $f*g$ is defined for other couplings as well. For instance, if $f \in L^\infty(\mathbb{R}^n), g \in L^1(\mathbb{R}^n)$, then $f*g$ is defined at all points and continuous, for if $|f| \le C$,

$$\begin{gathered}(f*g)(x) - (f*g)(z) = \textstyle\int_{\mathbb{R}^n} f(y)(g(x-y) - g(z-y))\, dm(y),\\ |(f*g)(x) - (f*g)(z)| \le C\|\tau_x g - \tau_z g\|_1.\end{gathered}$$

Example 13.11. The exponential random variable $E(\lambda)$ with parameter λ has density $\lambda e^{-\lambda x}$ in $x \ge 0$. The density of $E(\lambda)+E(\mu)$, if independent, is

$$\lambda\mu\int_0^z e^{-\lambda x}e^{-\mu(z-x)}\, dx = \lambda\mu e^{-\mu z}\int_0^z e^{(\mu-\lambda)x}\, dx.$$

If $\lambda = \mu$, the density is $\lambda^2 z e^{-\lambda z}$. If $\lambda \ne \mu$, the density is

$$\frac{\lambda\mu}{\mu-\lambda}(e^{-\lambda z} - e^{-\mu z}), \quad z > 0.$$

Writing

$$(f * g)(x) = \int_{\mathbb{R}^n} f(y)\tau_y g(x)\,dm(y), \quad f * g = \int_{\mathbb{R}^n} f(y)\tau_y g\,dm(y),$$

exhibits $f * g$ as a linear combination of translates of g. If $f \geq 0$ has integral one, it can be thought as a mean of g. For instance, for $f = \frac{1}{2h}1_{[-h,h]}$,

$$(f * g)(x) = \frac{1}{2h}\int_{x-h}^{x+h} g(y)\,dy.$$

An important aspect of convolution is in connection to *translation-invariant operators.* A linear operator $T : L^1(\mathbb{R}^n) \to L^1(\mathbb{R}^n)$ is called translation-invariant if it commutes with translations:

$$T(\tau_x f) = \tau_x(Tf), \quad x \in \mathbb{R}^n, \ f \in L^1(\mathbb{R}^n).$$

If we further assume that T is continuous, then it commutes with infinite sums (integrals) whence

$$\begin{aligned} T(f * g) &= T\left(\int_{\mathbb{R}^n} f(y)\tau_y g\,dm(y)\right) = \int_{\mathbb{R}^n} f(y)T(\tau_y g)\,dm(y) \\ &= \int_{\mathbb{R}^n} f(y)\tau_y Tg\,dm(y) = f * Tg. \end{aligned}$$

That is, T commutes with convolution.

Convolution with a fixed f is obviously translation-invariant: $f*(\tau_x g) = \tau_x(f*g)$. It can be proved that conversely, a continuous translation-invariant operator T is of the form $Tg = \mu * g$ where $f \in L^1(\mathbb{R}^n)$ is replaced by a finite measure μ:

$$\mu * g(x) = \int g(x-y)\,d\mu(y).$$

Exercise 13.1. Prove the discrete analogue of the last statement: the general form of a continuous translation-invariant operator T in the space of summable sequences

$$L^1(\mathbb{Z}) = \left\{a = (a_k)_{k\in\mathbb{Z}}, \sum_k |a_k| < +\infty\right\},$$

is $Ta = a * b$ with $b \in L^1(\mathbb{Z})$ where,

$$(a * b)_k = \sum_m a_m b_{k-m}.$$

13.2.2 An important application of convolution is to regularize and approximate functions. Assume that f is locally integrable and let ϕ be a smooth function with compact support (Section 5.4). Then the convolution $f * \phi$ is a smooth function with

$$D^\alpha(f * \phi) = f * D^\alpha \phi.$$

It is enough to prove this for D_i:

$$\left| \frac{(f*\phi)(x+he_i) - (f*\phi)(x)}{h} - f * D_i\phi(x) \right|$$
$$\leq \int |f(y)| \left| \frac{\phi(x+he_i-y) - \phi(x-y)}{h} - D_i\phi(x-y) \right| dm(y).$$

By the mean-value theorem, the last term equals $|D_i\phi(x-y-h') - D_i\phi(x-y)|$ for some $h', |h'| \leq |h|$. By the uniform continuity of $D_i\phi$ it is small, uniformly in y, for $|h|$ small, and the result follows.

Now we take ϕ supported in $B(0,1)$ and $\int \phi\, dm = 1$. For each $\varepsilon > 0$, we consider a scaled version $\phi_\varepsilon(x) = \varepsilon^{-n}\phi(x/\varepsilon)$, supported in $B(0,\varepsilon)$, $\int \phi_\varepsilon\, dm = 1$. The convolution

$$f_\varepsilon(x) = \int f(y)\phi_\varepsilon(x-y)\, dm(y),$$

being at each point x an average of f in $B(x,\varepsilon)$, is a regularized version of f. Now we will prove that if f is continuous, then f_ε converges to f uniformly on compact sets K, that is

$$\sup_{x \in K} |f_\varepsilon(x) - f(x)| \to 0, \quad \varepsilon \to 0.$$

Indeed,

$$f_\varepsilon(x) - f(x) = \int (f(x-y) - f(x))\phi_\varepsilon(y)\, dm(y)$$
$$= \int (f(x-\varepsilon y) - f(x))\phi(y)\, dm(y),$$

and the result follows using the uniform continuity of f on compacts. If f is continuous in a domain U, the same proof shows that f_ε is smooth in $U_\varepsilon = \{x \in U : d(x, U^c) \geq \varepsilon\}$ and converges to f uniformly on compact sets in U.

Analogous arguments work if f is a continuous function with compact support K and ϕ is a smooth function with integral one. Choosing $\phi(x) =$

$c_n e^{-|x|^2}$ leads to a proof of *Weierstrass' approximation theorem*, stating that f can be uniformly approximated by polynomials. Indeed, it is enough to approximate uniformly on compacts a fixed $f * \phi_\varepsilon$; assuming as we may that $\varepsilon = 1$, we consider the partial sum of the exponential

$$P_N(x) = \sum_{n=0}^{N} \frac{(-1)^n |x|^{2n}}{n!}.$$

Then $f * P_N$ is a polynomial, and

$$f * \phi(x) - f * P_N(x) = \int_K f(y)(\phi(x-y) - P_N(x-y))\, dm(y).$$

For x in a compact K', $x - y$ ranges in the compact $K' \setminus K$. Since $P_N \to \phi$ uniformly on compacts, it follows that $f * P_N \to f * \phi$ uniformly on K'.

Formally, ϕ_ε converges to the Dirac delta at the origin. The family ϕ_ε is thus called an *approximation of the identity.*

Exercise 13.2. Using an approximation of the identity and the formal identity $\tau_x f = f * \delta_x$, prove that if $T : L^1(\mathbb{R}^n) \to L^1(\mathbb{R}^n)$ commutes with convolutions, then it also commutes with translations.

13.3 Change of Variables in Multiple Integrals

13.3.1 By its definition, the Riemann integral distinguishes a particular coordinate system in $\mathbb{R}^n$, the canonical one. We might have used a general Cartesian coordinate system, sets which are intervals in this coordinate system, partitions, etc., to define the integral. On an intuitive basis, this should lead to the same class of integrable functions, with the same integral. Lebesgue integral depends on Lebesgue measure, which in its definition relies also on a choice of Cartesian coordinates.

In this section, we consider a general coordinate system, not necessarily linear, and study how measure and integral are expressed in those. A main motivation is that in an integral $\int_A f\, dm$ it might be that either A, f or both are best described in a different coordinate system.

13.3.2 We consider a smooth coordinate system $u_1, u_2, \ldots, u_n$ in an open set $V \subset \mathbb{R}^n$, that is, $h = (u_1, u_2, \ldots, u_n)$ is a diffeomorphism of class C^1 from V to an open set U, with a C^1 inverse map denoted by g. For $A \subset V$, we think of $h(A) = B \subset U$ as its description in the u-coordinates.

First, we claim that g maps a compact set $C \subset U$ of zero Jordan content to a compact set $g(C)$ of zero Jordan content, and the same applies to h. Since g is a homeomorphism, $g(C)$ is compact. Let

$$K = \left\{ u \in U : d(u, C) \leq \frac{1}{2} d(C, U^c) \right\}.$$

The set K is also compact and so

$$M = \sup_{u \in K} |\nabla g(u)| < +\infty.$$

By the mean-value theorem, if B is a ball inside K of radius r, $g(B)$ is included in a ball B' of radius Mr, whence $m(B') \leq M^n m(B)$. Given ε, there are balls B_i covering C with $\sum_i m(B_i) < \varepsilon$, and we may assume that $B_i \subset K$. Then $g(B_i) \subset B'_i$, with $\sum_i m(B'_i) \leq M^n \varepsilon$, and the B'_i cover $g(C)$, so $g(C)$ has zero Jordan content. The map g being a homeomorphism, one has $b(g(B)) = b(g(B))$ for $B \subset U$; applying the above to $C = bA$ we see that g maps Jordan measurable sets to Jordan measurable sets, and so does h. The above also shows that

$$m(g(E)) \leq Cm(E),$$

for $E \subset K$ compact Jordan measurable. Since

$$m(E) = \sup_{R_i \subset E} \sum_i m(R_i),$$

it follows that

$$m(g(E)) = \sup_{R_i \subset E} \sum_i m(g(R_i)). \tag{13.3}$$

If $E \subset U$ has zero Lebesgue measure and K_k is the compact set $K_k = \{u \in U : d(u, U^c) \geq \frac{1}{k}, |u| \leq k\}$, then $E = \cup_k (E \cap K_k), g(E) = \cup_k g(E \cap K_k)$, the same argument applies to $E \cap K_k$ to show that $g(E \cap K_k)$ has zero Lebesgue measure, implying $m(g(E)) = 0$. Thus, g maps sets of zero measure to sets of zero measure. Note that this holds for an arbitrary C^1 map.

Now, if f is defined on $A \subset V, B = h(A)$, the set of discontinuities of $f(g(u))$ in $\mathring{B} = h(\mathring{A})$ is mapped by g onto the set of discontinuities of f in $\mathring{a}$. By Proposition 11.9, it follows that $f(x)$ is Riemann integrable on A if and only if $F(u) = f(g(u))$ is Riemann integrable on B.

13.3.3 With these preliminaries covered we can now state the *change of variable* theorem for Riemann integrals. In the following statement, $f(g(u))|Jg(u)|$ is Riemann integrable, by statement (i) in Exercise 11.8.

Theorem 13.3. *The following* change of variable formula *holds if $B \subset U$ is compact Jordan measurable and f is Riemann integrable on $g(B)$:*

$$\int_{g(B)} f(x)\, dm(x) = \int_B f(g(u))|Jg(u)|\, dm(u), \tag{13.4}$$

where $Jg(u)$ is the Jacobian of g, *the determinant of $dg(u)$. In particular, the Riemann integral is independent of the Cartesian coordinates used to define it.*

Proof. Before filling the details we explain the intuitive idea of the proof. If Q is an infinitesimal cube centered at u of measure $dm(u)$, then $g(Q)$ is essentially the same as $dg(u)(Q)$ and so has measure $|Jg(u)|\, dm(u)$ by Theorem 1.7. By adding on all infinitesimal cubes in B, we get (13.4) for $f = 1$. This is the essential step, the proof for general f using just the definition of the integral. To make rigorous this argument the proof follows three steps:

(a) Proving (13.4) for $f = 1$ and B a (big) cube using Theorem 12.12.
(b) Proving (13.4) for $f = 1$ and B compact Jordan measurable set.
(c) Proving (13.4) in general.

We will also show that it is enough to prove inequalities.

We will prove that for a cube $Q \subset U$,

$$m(g(Q)) \leq \int_Q |Jg(u)|\, dm(u). \tag{13.5}$$

If this holds, then by (13.3) it holds too for $B \subset U$ compact Jordan measurable,

$$m(g(B)) \leq \int_B |Jg(u)|\, dm(u).$$

Thus,

$$m(S) \leq \int_{h(S)} |Jg(u)|\, dm(u),$$

for Jordan measurable compact $S \subset V$.

Then, to prove (13.4), we use (11.7); for intervals $S_i \subset V$,

$$\begin{aligned}
\int_{g(B)} f(x)\,dm(x) &= \sup \sum_{S_i \subset g(B)} \min\{f(x), x \in S_i\} m(S_i) \\
&\le \sup \sum_{S_i \subset g(B)} \min\{f(x), x \in S_i\} \int_{h(S_i)} |Jg(u)|\,dm(u) \\
&= \sup \sum_{S_i \subset g(B)} \min\{f(g(u)), u \in h(S_i)\} \int_{h(S_i)} |Jg(u)|\,dm(u) \\
&\le \sup \sum_{S_i \subset g(B)} \int_{h(S_i)} f(g(u))|J_g(u)|\,dm(u) \\
&\le \int_B f(g(u))|J_g(u)|\,dm(u).
\end{aligned}$$

Since the same inequality holds reversing the roles of g, h and

$$|J_g(h(x))||J_h(x)| = 1,$$

equation (13.4) follows.

So altogether it is enough to prove (13.5). When g is linear, we already know that

$$m(g(Q)) = |\det g|\, m(Q),$$

by Theorem 1.7. To prove it in general, we apply Theorem 12.12 to

$$\Phi(Q) = m(g(Q)).$$

The previous remarks ensure that Φ is indeed a set function. By Theorem 12.12, it is enough to prove that its upper density is dominated by $|Jg(u)|$:

$$\limsup_{u \in Q} \frac{m(g(Q))}{m(Q)} \le |Jg(u)|, \tag{13.6}$$

as the cube Q shrinks to u. To see this we will approximate g around u using the linear map $dg(u)$, for which we already know the result.

Before that it is worth considering two specific examples, polar coordinates in the plane and spherical coordinates in space. Using r, θ instead of u_1, u_2,

$$x = r\cos\theta, \quad y = r\sin\theta,$$

the Jacobian is $Jg = r$. Let $Q = [r, r+\delta] \times [\theta, \theta+\delta]$ be a square in polar coordinates, $A = g(Q)$ is a sector in the corona of radius $r, r+\delta$ and angular

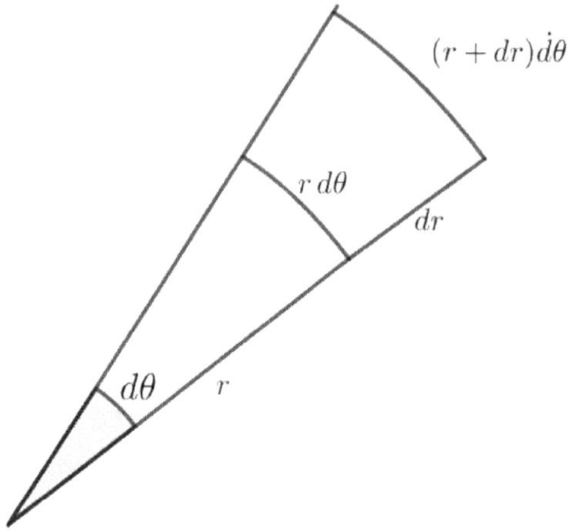

Figure 13.2. A sector, a square in polar coordinates.

amplitude δ, so its area is exactly

$$m(g(Q)) = \pi((r+\delta)^2 - r^2)\frac{\delta}{2\pi} = r\delta^2 + \frac{1}{2}\delta^3,$$

so (13.6) follows. Note that $rd\theta$ is the length of the arc joining (r,θ) with $(r,\theta+d\theta)$ and dr is the radial size of the sector, see Figure 13.2.

Similarly, in spherical coordinates (ρ,θ,ϕ) an infinitesimal cube $[\rho,\rho+\delta]\times[\theta,\theta+\delta]\times[\phi,\phi+\delta]$ is an infinitesimal shell with radial size δ, size $\rho\,\delta$ along meridians, and size $\rho\,\sin\phi\,\delta$ along parallels, whence with volume

$$\rho^2(\sin\phi)\,\delta^3 + o(\delta^3).$$

These examples being understood, to prove (13.6) in general,

$$m(g(Q)) \le |Jg(u)|m(Q) + o(m(Q)),$$

we use the approximation of g around u using $dg(u)$:

$$g(u+h) = L + E, L(u+h) = g(u) + dg(u)(h), |E| \le \tau(|h|)|h|, \tau(h) \to 0. \tag{13.7}$$

Let $v_j = \frac{\partial g}{\partial u_j}(u), j = 1,\ldots,n$ be the columns of $dg(u)$. If Q has side δ, L maps Q onto the parallelepiped $P = P(g(u),\delta v_1,\ldots,\delta v_n)$, whose measure is by Theorem 1.5.

$$m(P) = |J_g(u)|\delta^n = |J_g(u)|m(Q),$$

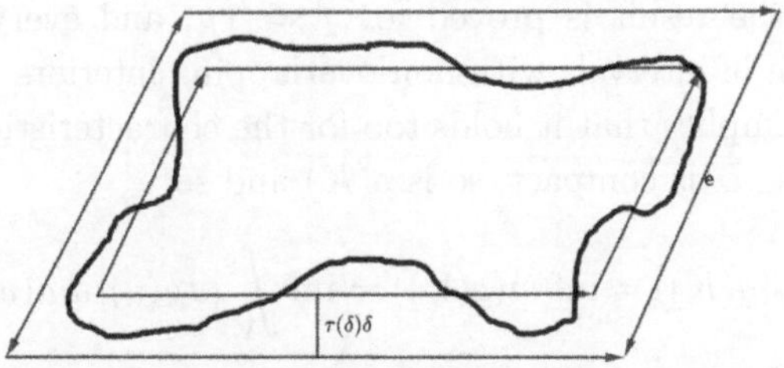

Figure 13.3. Comparing areas.

so let us compare $g(Q)$ with $P = L(Q)$. Since $|g - L| = |E| \leq \tau(|h|)|h|$, $g(Q)$ is included in the set

$$P_1 = \bigcup_{x \in P} B(x, \tau(\delta)\delta),$$

and so $m(g(Q)) \leq m(P_1)$. Now, $P_1 \setminus P$ is a union of $2n$ parallelepipeds, each having size δ in $n-1$ directions and size $\tau(\delta)\delta$ in the remaining direction (see Figure 13.3 for $n = 2$) whence it has measure bounded by $c\tau(\delta)\delta^n$, and (13.6) follows. □

In fact, the theorem is equivalent to (13.6). If we interpret g dynamically, it means that the Jacobian $J(g)$ is the distortion factor for infinitesimal volumes. If we interpret g as a change of coordinates, the meaning is that the measure of an infinitesimal rectangle in the u coordinates with size $du_1, \ldots, du_n$ is not the product $du_1 \cdots du_n$ but $Jg(u)du_1 \cdots du_n$, as shown explicitly for polar and spherical coordinates.

13.3.4 Now, we deal with the change of variables in Lebesgue integrals. We have seen that g transforms sets of zero measure into sets of zero measure. Since g is a homeomorphism, it preserves Borel sets, too, so by Theorem 11.3 g preserves Lebesgue measurable sets. Also, since $F^{-1}(I) = h(f^{-1}(I))$, f is measurable if and only if $F(u) = f(g(u))$ is. The statement is then:

Theorem 13.4. *If f is a non-negative measurable function, and B is measurable,*

$$\int_{g(B)} f(x)\, dm(x) = \int_B f(g(u))|Jg(u)|dm(u).$$

Therefore, a measurable function f is Lebesgue integrable on $g(B)$ if and only if $f(g(u))|Jg(u)|$ is integrable in B, with the same integral.

Proof. Since the result is proved for $f = 1_R$, and every open set is a countable union of intervals with non-overlapping interiors (Theorem 11.1) Corollary 12.1 implies that it holds too for the characteristic function of an open set. If $K \subset U$ is compact, so is $g(K)$ and so

$$m(g(K)) = \inf_V m(g(V)) = \inf_V \int_V |Jg(u)|\, dm(u),$$

where V runs over all open sets $K \subset V \subset U$. Since V can be assumed to be included in a fixed compact, the dominated convergence theorem implies that the result holds for compact sets. The monotone convergence theorem implies then that it holds for countable unions of closed sets, whence by Theorem 11.3 it holds for a measurable set. This means that the theorem holds for $f = 1_A$, A measurable, and so for simple non-negative functions as well. Another application of the monotone convergence theorem and Proposition 12.1 finishes the proof. □

13.3.5 It is worth pointing out some comments in relation to the familiar substitution rule in dimension $n = 1$: if $g : [a, b] \to \mathbb{R}$ is C^1, then

$$\int_{g(a)}^{g(b)} f(x)\, dx = \int_a^b f(g(u))g'(u)\, du.$$

This is a rule regarding *oriented* integrals $\int_a^b f\, dx$ defined as

$$\int_a^b f\, dx = \int_{[a,b]} f\, dx, \; a \leq b, \quad \int_a^b f\, dx = -\int_{[a,b]} f\, dx, \; a \geq b.$$

It holds because both terms, as functions of b, equal zero at $b = a$ and have derivative $f(g(b))g'(b)$. This substitution rule is more general than (13.4) because g is not supposed to be one-to-one. If g is one-to-one, then g is either non-decreasing or non-increasing, and both formulas are equivalent.

13.3.6 Recall from Corollary 6.1 that a diffeomorphism is simply a one-to-one map with non-vanishing Jacobian at every point. As it is the case for polar and spherical coordinates, the assumption $Jg(u) \neq 0, u \in U$ can be deleted in (13.4), both for the Riemann and Lebesgue versions. This is a consequence of Sard's theorem that follows.

In paragraph 13.3.2, it was pointed out that C^1 maps $g : U \subset \mathbb{R}^n \to \mathbb{R}^n$ transform sets of zero measure into sets of zero measure. This implies that for a C^1 map $g : U \subset \mathbb{R}^n \to \mathbb{R}^m$ with $n < m$, the whole $g(U)$ has zero measure, because we can look at g as the restriction to $\mathbb{R}^n$, a set of zero measure in $\mathbb{R}^m$, of a function of m variables. The situation for $n \geq m$ is the one considered in Sard's theorem.

Theorem 13.5. *Let $g : U \to \mathbb{R}^m$ of class C^1 in a domain $U \subset \mathbb{R}^n, n \geq m$ and consider the set*

$$S = \{u \in U : \text{rank} \quad dg(u) < m\}.$$

Then $g(S)$ has m-dimensional Lebesgue measure zero if $n = m$. If $n > m$, the conclusion holds if g is of class $C^k, k \geq \frac{n}{m}$.

Note that for linear maps the statement is obvious. The intuition is that since $g(u)$ is well approximated near a point a by $g(a) + dg(a)(u - a)$, g inherits this property of dg.

Proof. We prove just the statement for $n = m$. A proof for $n > m$ can be found in [12]. We prove that for every compact $K \subset S$, the set $g(K)$ has zero Jordan content. Given $\varepsilon > 0$, by the uniform continuity of dg on compacts we may choose $\delta > 0$ such that for any ball B of radius δ centered at $a \in K$,

$$|g(u) - g(a) - dg(a)(u - a)| \leq \varepsilon|u - a| \leq \varepsilon\delta, \quad u \in B.$$

Since $Jg(a) = 0$, as u ranges over B the point $g(a) + dg(a)(u - a)$ lies in a ball of radius $M\delta$ within a linear sub-manifold of dimension $n - 1$, and $g(u)$ is at distance less than $\varepsilon\delta$ from it. Therefore,

$$m(g(B)) \leq C\delta^{n-1}\varepsilon\delta = C\varepsilon m(B).$$

A family B_i of such balls covers K with $\sum_i m(B_i)$ comparable to $m(K)$ whence

$$m(g(K)) \leq \sum_i m(g(B_i)) \leq C\varepsilon m(K).$$

Since ε is arbitrary, this implies $m(g(K)) = 0$. □

For a general C^1 map $g : U \to \mathbb{R}^n$, we denote by $N(x)$ the *multiplicity function*, defined as the number of points in $g^{-1}(x)$ if this set is finite, and $+\infty$ otherwise.

Theorem 13.6. *For h integrable on U, one has*

$$\int_{g(U)} \left(\sum_{u \in g^{-1}(x)} h(u) \right) dm(x) = \int_U h(u)|Jg(u)|\, dm(u), \tag{13.8}$$

and for f integrable on $g(U)$,

$$\int_{g(U)} N(x)f(x)\,dm(x) = \int_U f(g(u))|Jg(u)|\,dm(u). \tag{13.9}$$

Proof. We may assume $h \geq 0$. It is worth proving it first for $n = 1$. If S is the (closed) set of critical points, $U = (a,b) \setminus S$ is open, whence a union of a finite or countable number of open intervals (a_i, b_i). Since $g' = 0$ on S,

$$\int_{[a,b]} h(u)|g'(u)|\,du = \sum_i \int_{(a_i,b_i)} h(u)|g'(u)|\,du.$$

In each of those $g = g_i$ is strictly increasing, so the above equals

$$\sum_i \int_{g((a_i,b_i))} h(g_i^{-1})(x)\,dx.$$

Let I_i denote the characteristic function of $g((a_i,b_i))$. By Corollary 12.1, this equals

$$\int_{g([a,b])} \left(\sum_i I_i(x)h(g_i^{-1})(x)\right) dx.$$

Now, for $x \notin g(S)$, and thus a.e. by Theorem 13.5, one has

$$\sum_i I_i(x)h(g_i^{-1})(x) = \sum_{g(u)=x} h(u),$$

and so (13.8) is proved for positive h.

To prove it in general, as before using Sard's theorem we may assume that $Jg \neq 0$ in U. By the inverse function theorem, each fiber $g^{-1}(x)$ is a discrete set, whence finite on every compact. Using an exhausting sequence of compact sets and monotone convergence it is enough to prove (13.8) for $h \geq 0$ compactly supported in $K \subset U$. By the inverse function theorem, for each $p \in K$ there is a ball $B(p)$ where g is a diffeomorphism $g = g_i$. Let $B_1, \ldots, B_N$ a finite covering of K and $\varphi_i, i = 1, \ldots, N$ a partition of unity for that covering as in Theorem 5.5. Then, with $h_i = \varphi_i h$,

$$\begin{aligned}\int_U h_i(u))|Jg(u)|dm(u) &= \sum_i \int_{B_i} h_i(u))|Jg(u)|dm(u)\\ &= \sum_i \int_{g(B_i)} h_i(g_i^{-1}(x))dm(x).\end{aligned}$$

If I_i denotes the characteristic function of $g(B_i)$, we finish as before. □

For $f = 1$, it follows that

$$\int_{g(U)} N(x)\, dm(x) = \int_U |Jg(u)|\, dm(u),$$

and so N is finite a.e. if Jg is integrable on U. In dimension $n = 1$, a theorem by S. Banach shows that this result holds for more general functions, the functions of *bounded variation*, for which the right-hand side is replaced by

$$\sup_P \sum_i |g(x_{i+1} - g(x_i)|,$$

P denoting a partition $a = x_0 < \cdots < x_N = b$ of $[a, b]$.

Formulas (13.8) and (13.9) can be seen as special cases of the co-area formulas to be found in Section 15.1. Theorem 13.6 holds for more general maps, Lipschitz maps, and can be found in full generality in [7].

13.3.7 In practice, we use the change of variable formula as follows. We wish to compute an integral $\int_A f(x)\, dm$ and realize that A, f have a simpler expression in some system of coordinates $u_1; \ldots, u_n$ defined by $x = g(u)$. We work in this coordinate system, describing A in the u-coordinates and verbalize that the expression of the volume element $dm(x)$ is $|J_g(u)|\, dm(u)$ in the new variables.

For instance, an ellipsoid A with semi-axes a, b, c, in the variables u, v, w defined by $x = au, y = bv, z = cw$ becomes the unit ball B, whence

$$m(A) = \int_B |Jg| du\, dv\, dw = \int_B abc\; du\, dv\, dw = \frac{4}{3}\pi abc.$$

Let us look in some detail to polar, cylindrical and spherical coordinates. The map

$$g(r, \theta) = (r\cos\theta, r\sin\theta),$$

has Jacobian r and is one-to-one only if θ ranges on an open interval of length at most 2π. So, strictly speaking it is a change of variables only in a domain V which omits a half-line. However, since both a half-line and its description in polar coordinates (θ constant) have zero area, both terms of the formula become unaltered if we delete these exceptional points. This amounts to saying that we can use polar coordinates as if they were coordinates in the whole plane, the area element dm being $r\, dr\, d\theta$. The same remark applies to the spherical and cylindrical coordinates in space.

Example 13.12. Let us compute again the volume of the unit ball A. In spherical coordinates (ρ, θ, ϕ), the ball is the interval $[0,1] \times [0, 2\pi] \times [0, \pi]$, up to sets of zero volume. Since the Jacobian equals $\rho^2 \sin\phi$,

$$m(A) = \int_{[0,1]} \int_{[0,2\pi]} \int_{[0,\pi]} \rho^2 \sin\phi \, dr \, d\theta \, d\phi = \frac{4}{3}\pi.$$

Example 13.13. We compute the volume of

$$A = \{(x, y, z) : z \geq 0, z^2 \geq x^2 + y^2 : x^2 + y^2 + z^2 \leq 1\},$$

the part of the unit ball inside the cone $z \geq 0, z^2 = x^2 + y^2$. The cone and the sphere meet at $z = \frac{1}{\sqrt{2}}, x^2 + y^2 = \frac{1}{2}$. In cylindrical coordinates (r, θ, z), A is described by $0 \leq r \leq \frac{1}{\sqrt{2}}, 0 \leq \theta \leq 2\pi, r \leq z \leq \sqrt{1 - r^2}$, so

$$m(A) = 2\pi \int_0^{\frac{1}{\sqrt{2}}} (\sqrt{1 - r^2} - r) \, dr = \frac{2\pi}{3}\left(1 - \frac{1}{\sqrt{2}}\right).$$

The computation is simpler in spherical coordinates; in those, A is the interval $0 \leq \rho \leq 1, 0 \leq \theta \leq 2\pi, 0 \leq \phi \leq \frac{\pi}{4}$, so

$$\begin{aligned} m(A) &= \int_{[0,1]} \int_{[0,2\pi]} \int_{[0,\frac{\pi}{4}]} \rho^2 \sin\phi \, d\rho \, d\theta \, d\phi = \frac{2}{3}\pi \int_{[0,\frac{\pi}{4}]} \sin\phi \, d\phi \\ &= \frac{2\pi}{3}\left(1 - \frac{1}{\sqrt{2}}\right). \end{aligned}$$

Example 13.14. Given two points F_1, F_2 at distance $2c$ (focci), Bernouilli's lemniscate is the locus of points P in the plane such that $d(p, F_1)d(p, F_2) = c^2$. Assuming $F_1 = (c, 0), F_2 = (-c, 0)$, its equation is

$$(x^2 + y^2)^2 = 2c^2(x^2 - y^2).$$

In polar coordinates, the region bounded by the curve is $r^2 \leq 2c^2(\cos^2\theta - \sin^2\theta) = 2c^2 \cos 2\theta$. The right-hand side is described by $-\frac{\pi}{4} \leq \theta \leq \frac{\pi}{4}$, $0 \leq r^2 \leq 2c^2 \cos 2\theta$, so its area is

$$A = \int_{-\frac{\pi}{4}}^{\frac{\pi}{4}} \left(\int_0^{c\sqrt{2\cos 2\theta}} r \, dr \right) d\theta = c^2 \int_{-\frac{\pi}{4}}^{\frac{\pi}{4}} \cos 2\theta \, d\theta = c^2.$$

Example 13.15. We analyze for what values of $\alpha > 0$ the following double integral is convergent

$$I = \int_{\mathbb{R}^2} \frac{1}{(1+x^2+y^2)^\alpha}\, dm.$$

When integrating in y we change variables, $y = \sqrt{1+x^2}\, t$, to obtain

$$(1+x^2)^{\frac{1}{2}-\alpha} \int_{\mathbb{R}} \frac{1}{(1+t^2)^\alpha}\, dt.$$

If $2\alpha > 1$, this converges with value $c_\alpha(1+x^2)^{\frac{1}{2}-\alpha}$. The integral in x, and so I, converges iff $2\alpha - 1 > 1, \alpha > 1$.

We can use alternatively polar coordinates to compute exactly its value:

$$I = \int_0^\infty \int_0^{2\pi} \frac{r}{(1+r^2)^\alpha}\, dr\, d\theta = 2\pi \int_0^\infty \frac{r}{(1+r^2)^\alpha}\, dr$$

$$= \pi \int_0^\infty \frac{1}{(1+t)^\alpha}\, dt = \frac{\pi}{\alpha - 1}.$$

Next is an alternative for Example 12.1:

Example 13.16. The integral

$$I = \int_{\mathbb{R}} e^{-\frac{1}{2}x^2}\, dx,$$

cannot be computed using Barrow's rule. But

$$I^2 = \left(\int_{\mathbb{R}} e^{-\frac{1}{2}x^2}\, dx\right)\left(\int_{\mathbb{R}} e^{-\frac{1}{2}y^2}\, dy\right) = \int_{\mathbb{R}^2} e^{-\frac{1}{2}(x^2+y^2)}\, dm,$$

that, using polar coordinates, equals

$$\int_0^{+\infty} e^{-\frac{1}{2}r^2}\, r\, dr\, d\theta = 2\pi \int_0^\infty e^{-t}\, dt = 2\pi.$$

The function

$$\frac{1}{\sqrt{2\pi}} e^{-\frac{1}{2}x^2},$$

is the *Gaussian function*, the density of the standard normal law, denoted $N(0,1)$.

The general *normal law* $N(\mu, \sigma)$ with center μ and variance σ^2 is obtained by translation and rescaling:

$$\frac{1}{\sigma\sqrt{2\pi}}e^{-\frac{1}{2}(\frac{x-\mu}{\sigma})^2}.$$

Example 13.17. To compute

$$I = \int_S e^{-|x+y|^2-|x-y|}\,dm,$$

on the strip S defined by $|x-y| \le 2$ we change variables $u = x+y, v = x-y$. The Jacobian of the transformation is $\frac{1}{2}$, therefore,

$$\begin{aligned} I &= \frac{1}{2}\int_{|v|\le 2} e^{-u^2-|v|}\,du\,dv = \frac{1}{2}\left(\int_{\mathbb{R}} e^{-u^2}\,du\right)\left(\int_{|v|\le 2} e^{-|v|}\,dv\right) \\ &= \sqrt{\pi}(1-e^{-2}). \end{aligned}$$

Many physical quantities (mass, force, pressure, etc.) are described in terms of densities. Also, the probability distribution of a random variable X can be described in terms of a *probability density* f_X.

Example 13.18. Let us find the total mass and center of mass of the body

$$A = \{(x, y, z) : z \ge 0, z^2 \ge x^2 + y^2 : x^2 + y^2 + z^2 \le 1\},$$

assuming that it has density proportional to the distance to the z-axis, that is, $f = \lambda\sqrt{x^2+y^2}$. We compute the mass using spherical coordinates, in which $f = \lambda\,\rho\,\cos\phi$

$$\begin{aligned} M &= \lambda\int_A \sqrt{x^2+y^2}\,dm = \lambda\int_{[0,1]}\int_{[0,2\pi]}\int_{[0,\frac{\pi}{4}]} \rho^3\cos\phi\sin\phi\,d\rho\,d\theta\,d\phi \\ &= \lambda\frac{1}{4}2\pi\int_0^{\frac{\pi}{4}}\cos\phi\sin\phi\,d\phi = \lambda\frac{1}{4}2\pi\frac{1}{4} = \lambda\frac{\pi}{8}. \end{aligned}$$

It is clear from the symmetry that $\overline{x} = \overline{y} = 0$. Finally,

$$\begin{aligned} \int_A z\,f\,dm &= \lambda\int_{[0,1]}\int_{[0,2\pi]}\int_{[0,\frac{\pi}{4}]} \rho^4\cos\phi\sin^2\phi\,d\rho\,d\theta\,d\phi \\ &= \lambda\frac{1}{5}2\pi\frac{1}{3}\frac{1}{2\sqrt{2}} = \lambda\pi\frac{1}{15\sqrt{2}}, \end{aligned}$$

and so $\overline{z} = \frac{8}{15\sqrt{2}}$.

13.3.8 The *bi-variant Normal* density for a random pair X, Y has five parameters

$$f_{X,Y}(x,y)$$
$$= c\exp\left[-\frac{1}{2(1-\rho^2)}\left(\frac{(x-\mu_x)^2}{\sigma_x^2}+\frac{(y-\mu_y)^2}{\sigma_y^2}-2\rho\frac{(x-\mu_x)(y-\mu_y)}{\sigma_x\sigma_y}\right)\right],$$

$$c = \frac{1}{2\pi\sigma_x\sigma_y(1-\rho^2)^{\frac{1}{2}}},$$

with $|\rho| < 1$. The standard one corresponds to $\mu_x = \mu_y = 0, \sigma_x = \sigma_y = 1$, $\rho = 0$,

$$f_{X,Y}(x,y) = \frac{1}{2\pi}\exp\left[-\frac{1}{2}(x^2+y^2)\right].$$

Recall that the meaning is that

$$\int_B f_{X,Y}(x,y)\,dm$$

is the probability that (X, Y) takes value in B, $p((X,Y) \in B)$.

Let us compute the so-called *marginal* densities

$$f_X(x) = \int_{\mathbb{R}} f(x,y)\,dy,\ f_Y(y) = \int_{\mathbb{R}} f(x,y)\,dx.$$

With $x' = x - \mu_x, y' = y - \mu_y$, and completing squares in the exponent

$$f_Y(y) = \frac{1}{2\pi\sigma_x\sigma_y(1-\rho^2)^{\frac{1}{2}}}\exp\left[-\frac{1}{2}\frac{y'^2}{\sigma_y^2}\right]$$
$$\times\int_{\mathbb{R}}\exp\left[-\frac{1}{2}\frac{1}{(1-\rho^2)}\left(\frac{x'}{\sigma_x}-\frac{\rho y'}{\sigma_y}\right)^2\right]dx'.$$

With the change of variable

$$\frac{1}{(1-\rho^2)^{\frac{1}{2}}}\left(\frac{x'}{\sigma_x}-\frac{\rho y'}{\sigma_y}\right) = t, \quad dx' = \sigma_x(1-\rho^2)^{\frac{1}{2}}\,dt,$$

we get

$$f_Y(y) = \frac{1}{2\pi\sigma_y}\exp\left[-\frac{1}{2}\frac{(y-\mu_y)^2}{\sigma_y^2}\right]\int_{\mathbb{R}} e^{-\frac{1}{2}t^2}\,dt = \frac{1}{\sqrt{2\pi}\sigma_y}\exp\left[-\frac{1}{2}\frac{(y-\mu_y)^2}{\sigma_y^2}\right].$$

This shows that f_Y is $N(\mu_y, \sigma_y)$; analogously, f_X is $N(\mu_x, \sigma_x)$. By Fubini's theorem, $f_{X,Y}$ has indeed integral one.

Let Σ be the co-variance matrix with entries

$$\Sigma = \begin{pmatrix} \text{Var}X & \text{Cov}(X,Y) \\ \text{Cov}(X,Y) & \text{Var}Y \end{pmatrix}.$$

A computation shows that

$$\Sigma = \begin{pmatrix} \sigma_x^2 & \rho\sigma_x\sigma_y \\ \rho\sigma_x\sigma_y & \sigma_y^2 \end{pmatrix}.$$

One has $\det\Sigma = (1-\rho^2)\sigma_x^2\sigma_y^2$ and

$$\Sigma^{-1} = \frac{1}{(1-\rho^2)}\begin{pmatrix} \frac{1}{\sigma_x^2} & -\frac{\rho}{\sigma_x\sigma_y} \\ -\frac{\rho}{\sigma_x\sigma_y} & \frac{1}{\sigma_y^2} \end{pmatrix}.$$

If R is the column vector $(X,Y)^t$, and $\Lambda = (\mu_x, \mu_y)^t$, the density is written in matrix form

$$\frac{1}{2\pi|\det\Sigma|^{\frac{1}{2}}}\exp\left[-\frac{1}{2}(R-\Lambda)^t\Sigma^{-1}(R-\Lambda)\right].$$

This means that Σ plays the role of σ^2 in the univariate normal distribution.

The general multivariate version is as follows: a random column vector $X = (X_1, X_2, \ldots, X_n)^t$ is said to have a *multivariant Normal distribution* $N(\Lambda, \Sigma)$, where Λ is a column vector and Σ a positive definite symmetric matrix, if the joint density function is

$$f_X(x) = \frac{1}{\sqrt{(2\pi)^n|\det\Sigma|}}\exp\left[-\frac{1}{2}(X-\Lambda)^t\Sigma^{-1}(X-\Lambda)\right].$$

As before, each X_i is then a normal variable centered at μ_i and Σ is the covariance matrix with entries

$$\text{Cov}(X_i, X_j) = \int\int (x_i - \mu_i)(x_j - \mu_j) f_X(x)\, dm.$$

If Σ is diagonal, that is the variables X_j are uncorrelated, then they are independent. The law $N(0, I)$ with I the identity matrix, that is $X = (X_1, \ldots, X_n)$ with the X_j independent standard normal law, is called the *multivariate standard normal law.*

The expression

$$P(x_1, \ldots, x_n) = \frac{1}{2}(X-\Lambda)^t\Sigma^{-1}(X-\Lambda),$$

is a homogeneous polynomial in $x_1, \ldots, x_n$. P is a strictly convex function, Λ is the only critical point of P and Σ^{-1} is the hessian HP.

13.3.9 The following remark is related to the co-area formula in Section 15.1. In Fubini's theorem for rectangular coordinates, say in $n = 2$ in the whole plane

$$I = \int \left(\int f(x,y)\,dy \right) dx,$$

dy is arc-length on the line $x = c$ and dx can be thought as arc-length on any line $y = c$, say $y = 0$, so it can be interpreted that

$$I = \int_{y=0} \phi\, ds, \quad \phi(c) = \int_{x=c} f\, ds.$$

If u, v is a coordinate system with $|J| = 1$, also

$$I = \int \left(\int f(u,v)\,dv \right) du,$$

but these integrals are not in general with respect to arc-length of the level curves of u, v. These are integrals in the u, v plane and are not integrals along the u, v axis.

For a general coordinate system u, v, consider

$$J = \int_{v=0} \phi\, ds, \quad \phi(c) = \int_{u=c} f ds.$$

Since on $u = c$,

$$ds = \sqrt{\left(\frac{\partial x}{\partial v}(c,v)\right)^2 + \left(\frac{\partial y}{\partial v}(c,v)\right)^2}\, dv,$$

and on $v = 0$,

$$ds = \sqrt{\left(\frac{\partial x}{\partial u}(u,0)\right)^2 + \left(\frac{\partial y}{\partial u}(u,0)\right)^2}\, du,$$

one has $I = J$ only if

$$|Jg(u,v)|^2 = \left(\left(\frac{\partial x}{\partial v}(u,v)\right)^2 + \left(\frac{\partial y}{\partial v}(u,v)\right)^2\right) \times \left(\left(\frac{\partial x}{\partial u}(u,0)\right)^2 + \left(\frac{\partial y}{\partial u}(u,0)\right)^2\right).$$

In an orthogonal system of coordinates, one has

$$|Jg(u,v)| = |\partial_u||\partial_v|,$$

(that is, the area of an interval is the product of lengths!). If, moreover, $|\partial_u|$ does not depend on v, then the above holds.

This is the case for polar coordinates, for which $|\partial_r| = 1$

$$\int_{\mathbb{R}^2} = \int_0^\infty \left(\int_0^{2\pi} r\, d\theta \right) dr.$$

The inner integral is with respect to arc-length in $|x| = r$, and the dr integral is with respect to arc-length in an arbitrary ray $\theta = c$.

Requiring

$$I = \int \left(\int f ds(v) \right) ds(u) = \int \left(\int f ds(u) \right) ds(v),$$

means that $\partial u, \partial v$ are orthogonal and depend only on u, v, respectively. The latter condition means (assuming u, v of class C^2) that

$$x_u x_{uv} + y_u y_{uv} = 0, \quad x_v x_{vu} + y_v y_{vu} = 0.$$

This implies $x_{uv} = y_{uv} = 0$, so $x = A(u)+B(v), y = C(u)+D(v)$. Requiring that ∂_u, ∂_v are orthogonal leads to $A' = kC', D' = -kB'$ for some constant k, so

$$\partial_u = C'(k, 1), \quad \partial_v = B'(1, -k).$$

This means that the change of variables is a linear orthonormal one followed by a rescaling in each direction, so the new axes are necessarily Cartesian.

13.3.10 In case $|Jg| = 1, |Jh| = 1$, if we think of g, h dynamically, that is as transformation moving points, they are *measure-preserving*, $m(h(A)) = m(A)$ for all compact Jordan measurable sets A. All affine transformations $g(u) = c + Mu$, with M an invertible matrix, $|\det M| = 1$, are measure preserving. For a C^1 measure-preserving transformation on a domain U, since J_h is continuous, either $J_h = 1$ or $J_h = -1$. In dimension $n = 2$, $h = (u, v)$, they are described by the PDE equation

$$u_x v_y - u_y v_x = 1.$$

To find non-affine examples, assume $u = x$, so that the equation becomes $v_y = 1$; then $v = y + f(x)$ with an arbitrary f. The maps

$$(x, y) \to (x, y + f(x))$$

are called *shears*. The fact that these maps preserve area can be seen geometrically, because vertical segments lying on $x = a$ are translated vertically by an amount $f(a)$ and so by Cavalieri's principle the area is preserved.

13.3.11 As an application of the change of variables formula, we will consider how a multivariant normal distribution transforms under a linear change. Let $A = (a_{ij})$ be an $n \times n$ non-singular matrix, and set $Y = AX$, so that each Y_i is a linear combination of X_j. Then

$$p(Y \in B) = p(X \in A^{-1}B) = \int_{A^{-1}B} f_X(x)\, dm(x)$$
$$= \int_B f_X(A^{-1}y)|\det A^{-1}|dm(y).$$

Since

$$X - \Lambda = A^{-1}Y - \Lambda = A^{-1}(Y - A\Lambda),$$
$$(X-\Lambda)^t\Sigma^{-1}(X-\Lambda) = (Y - A\Lambda)^t(A^{-1})^t\Sigma^{-1}A^{-1}(Y - A\Lambda)$$
$$= (Y - A\Lambda)^t(\Sigma')^{-1}(Y - A\Lambda),$$

with $\Sigma' = A\Sigma A^t$, we see that Y has the law $N(A\Lambda, \Sigma')$.

Note that an arbitrary linear combination $\sum_j \lambda_j X_j$ may appear as a component of Y, so we can state:

Proposition 13.2. *If $X = (X_1, X_2, \ldots, X_n)$ has a multivariate normal law, all linear combinations of the X_j have normal law.*

In fact, the converse also holds.

If instead we set $Y = A(X - \Lambda)$, then Y has the law $N(0, \Sigma')$. It is possible to choose A so that Σ' is the identity matrix. Then

$$X = \Lambda + A^{-1}Y,$$

with Y standard, is the general expression of the multivariate normal law. To see that the choice of A is possible, recall from Theorem 1.4 that there exists an orthogonal matrix M, $M^{-1} = M^t$, such that $M\Sigma M^t = D$ is a diagonal matrix with positive entries. We can obviously extract a square root of D, also diagonal, $D = E^2$, so that

$$M\Sigma M^t = E^2, \quad E^{-1}M\Sigma M^t E^{-1} = I,$$

and $A = E^{-1}M$ satisfies the desired equation.

Chapter 14

Integration on Sub-Manifolds

The next topic addressed in measure and integration theory is how to measure length of sets included in curves, areas of sets included in surfaces, etc., and generally to define an integral for functions defined in sub-manifolds. This can be done in a number of ways. For curves, length can be defined in terms of inscribed polygonals, but area on surfaces cannot be satisfactorily defined in terms of inscribed triangles. Instead, we have chosen to use the change of variable, using local charts, as a very definition for the volume element of a sub-manifold. The surfaces in Euclidean space are particularly studied, but just at first order, that is, we do not consider *curvature*. Still, we study *isothermal* local coordinates in surfaces and show their connection with complex analysis and the Beltrami–Laplace equation.

14.1 Length and Integration on Arcs

14.1.1 First, we define the length covered by a (continuous) arc $\gamma(t), a \leq t \leq b$. In the cinematic interpretation, the distance covered by the moving object. We consider partitions $P : a = t_0 < t_1 < \cdots < t_N = b$ and the corresponding points $p_i = \gamma(t_i)$, which determine a polygonal inscribed in γ^*. Its length is

$$L(P) = \sum_{k=0}^{N} |p_{k+1} - p_k| = \sum_{k=0}^{N} |\gamma(t_{k+1}) - \gamma(t_k)|.$$

The *length* of γ is then defined as

$$L(\gamma) = \sup_P L(P).$$

It is evident that $L(P)$ increases if more points are added, so we may think that $L(\gamma)$ is the limit of $L(P)$ as the partition gets finer. It may be infinite; for instance, this is the case for the graphic of $y = x \sin \frac{1}{x}, 0 \leq x \leq 1$ between the origin ($x = 0$) and any other point. This is because for $x_k = 1/\frac{\pi}{2} + k\pi, p_k = (x_k, y(x_k))$, one has

$$d(p_k, p_{k+1}) \geq c\frac{1}{k},$$

and since $\sum_k \frac{1}{k}$ is divergent, the polygonal given by these points has arbitrarily large length.

Note that in general $L(\gamma)$ depends not only on the range γ^*, as we may visit a point at several times, or cover γ^* several times, for instance $\gamma(t) = (\cos t, \sin t), 0 \leq t \leq 4\pi$, covers twice the unit circle and $L(\gamma) = 4\pi$. When γ is simple, we have a one-to-one correspondence between partitions of $[a, b]$ and polygonals in γ^*, so that the definition only depends on γ^* and can be denoted by $L(\gamma^*)$. The simple arcs for which $L(\gamma^*)$ is finite are called *rectifiable*. An open simple arc $\gamma : (a, b) \to \mathbb{R}^n$ is called *locally rectifiable* if its restriction to every $[c, d] \subset (a, b)$ is rectifiable.

14.1.2 Arc-length parametrization. Suppose $\gamma^* = \gamma(a, b), 0 \in (a, b)$ is simple and locally rectifiable, and fix $p = \gamma(0)$. We consider the length from p to $\gamma(t), t > 0$,

$$s(t) = \sup L(P),$$

where P ranges over all polygonals on γ^* joining p to $\gamma(t)$. If $t < 0$, we define $s(t)$ as the opposite of this length.

Proposition 14.1. *The function s is a continuous strictly increasing function, whence a homeomorphism from (a, b) to its image (α, β).*

Proof. Although rather intuitive, we provide an ε/δ proof. Evidently, if $t < t'$, then $s(t') - s(t) \geq |\gamma(t) - \gamma(t')| > 0$, so $s(t)$ is strictly increasing. Fix $t \in (a, b)$ and $t < d < b$. Given $\varepsilon > 0$ by continuity of γ, there is $\delta_1 > 0$ such that $|\gamma(t) - \gamma(t')| < \varepsilon$ if $|t - t'| < \delta_1$; also, since the length from $\gamma(t)$ to $\gamma(d)$ is finite, there is a partition $P : t = t_0 < t_1 < \cdots < t_N = d$ such that

$$s(d) - s(t) - \varepsilon < L(P).$$

Let $\delta_2 = t_1 - t_0$ and $\delta = \min(\delta_1, \delta_2)$. If $t' - t < \delta$ and adding it to P, it follows that

$$s(d) - s(t) - \varepsilon < |\gamma(t') - \gamma(t)| + s(d) - s(t') < \varepsilon + s(d) - s(t'),$$

showing that $s(t') - s(t) < 2\varepsilon$, and so s is continuous from the right. Continuity from the left is proved analogously. □

Of course, $\beta - \alpha$ equals the total length of γ, finite or not. If $t = t(s)$ denotes the inverse transformation and $\hat{\gamma}(s) = \gamma(t(s))$, we say that $\hat{\gamma}(s)$ is the *arc-length parametrization* and s the *arc-length parameter.*

14.1.3 Every simple arc of class C^1 is rectifiable. This is because $L(P)$ is essentially a Riemann sum of a continuous function. Indeed, if $P : a = t_0 < t_1 < \cdots < t_N = b$,

$$L(P) = \sum_{k=0}^{N} \sqrt{\sum_{j=1}^{n} (x_j(t_{k+1}) - x_j(t_k))^2}.$$

By the mean-value theorem $x_j(t_{k+1}) - x_j(t_k) = x_j'(\xi_{kj})(t_{k+1} - t_k)$, with ξ_{kj} between t_k, t_{k+1} and hence

$$L(P) = \sum_{k=0}^{N} (t_{k+1} - t_k) \sqrt{\sum_{j=1}^{n} |x_j'(\xi_{kj})|^2}.$$

By the uniform continuity of x_j',

$$\left| \sqrt{\sum_{j=1}^{n} |x_j'(\xi_{kj})|^2} - \sqrt{\sum_{j=1}^{n} |x_j'(t_k)|^2} \right|,$$

is arbitrarily small, uniformly in k, if the partition is fine enough. This means that

$$L(P) = \sum_{k=0}^{N} (t_{k+1} - t_k) \sqrt{\sum_{j=1}^{n} |x_j'(t_k)|^2} + o(1).$$

Since the last sum is a Riemann sum of the continuous function $|\gamma'|$, it follows that

$$L = \int_a^b |\gamma'(t)|\, dt.$$

Of course, the cinematic interpretation is clear: as $|\gamma'(t)|$ is the instant speed, $|\gamma'(t)|\, dt$ is the infinitesimal distance covered between time t and time $t + dt$, and so their sum (the integral) is the total distance covered. Note that for a graph $\gamma(x) = (x, f(x))$, we find

$$L = \int_a^b \sqrt{1 + f'(x)^2}\, dx.$$

Note too that if $t = \phi(s)$ is a change of parameter, with ϕ of class C^1, $\hat{\gamma}(s) = \gamma(\phi(s))$, then

$$|\hat{\gamma}'(s)| = |\phi'(s)||\gamma'(\phi(s))|,$$

and by the change of variable formula for integrals

$$\int_c^d |\hat{\gamma}'(s)|\, ds = \int_c^d |\phi'(s)||\gamma'(\phi(s))|\, ds = \int_a^b |\gamma'(t)|\, dt.$$

This formula for the length extends to the so-called *piece-wise C^1 arcs*, meaning that they are of class C^1 except at a finite number of points (vertices) where the derivative has a jump discontinuity. For C^1 simple arcs, we have, with the notations above,

$$s(t) = \int_0^t |\gamma'(\tau)|\, d\tau,$$

implying that $s(t)$ is C^1 with $s'(t) = |\gamma'(t)|$. We refer to $ds = |\gamma'(t)|\, dt$ as the *length element*; it is to be thought as an infinitesimal quantity, the length between $\gamma(t)$ and $\gamma(t+dt)$, approximated by the length of the vector tangent $\gamma'(t)\, dt$. To obtain the length of a subarc $A \subset \gamma^*$, we must add all the infinitesimal lengths

$$l(A) = \int_A ds.$$

In practice, we consider some parametrization $\gamma(t)$, identify $I = \{t : \gamma(t) \in A\}$, and then

$$l(A) = \int_I |\gamma'(t)|\, dt.$$

Of course, $\gamma'(t) \neq 0$ does not imply that the arc is simple, but, as shown in Proposition 3.2, that it is locally simple.

Example 14.1. We compute the length of the arc in $x, y > 0$ given by

$$x^{\frac{2}{3}} + y^{\frac{2}{3}} = a^{\frac{2}{3}}, \quad a > 0.$$

A parametrization is

$$x(t) = a\cos^3 t, \quad y(t) = a\sin^3 t, \quad 0 \leq t \leq \frac{\pi}{2}.$$

Then $\gamma'(t) = 3a(-\sin t\cos^2 t, \cos t\sin^2 t), ds = 3a\sin t\cos t$ and

$$L = 3a\int_0^{\frac{\pi}{2}} \sin t\cos t\, dt = \frac{3a}{2}.$$

14.1.4 Now we ask ourselves how to compute lengths when the arc is given in another coordinate system. We will consider this with full generality later on, now we just consider one specific example, assuming that the arc is given in polar coordinates by $r = r(t), \theta = \theta(t)$. In canonical coordinates

$$x(t) = r(t)\cos\theta(t), \quad y(t) = r(t)\sin\theta(t),$$

and so

$$\begin{aligned} ds &= \sqrt{x'^2 + y'^2}\, dt = \sqrt{(r'\cos\theta - r\theta'\sin\theta)^2 + (r'\sin\theta + r\theta'\cos\theta)^2}\, dt \\ &= \sqrt{r'^2 + r^2(\theta')^2}\, dt. \end{aligned}$$

The interpretation is clear: when traveling from $(r(t), \theta(t))$ to $(r(t+dt), \theta(t+dt))$ (in polar coordinates), we may think that we go through $(r(t+dt), \theta(t))$, covering radially an infinitesimal distance $|r'(t)|\, dt$, and then from $(r(t+dt), \theta(t))$ to $(r(t+dt), \theta(t+dt))$ in the tangential direction covering an arc of $|\theta(t+dt) - \theta(t)| \approx |\theta'(t)|\, dt$ radians, whence of infinitesimal length $r(t)|\theta'(t)|\, dt$, by the very definition of radian. Since these two trajectories are orthogonal, one radial and the other tangential, that is why $ds = \sqrt{r'^2 + r^2(\theta')^2}\, dt$.

Example 14.2. The *cardioid* is given in polar coordinates by $r = 1 + \cos\theta$, that is, $\theta = t$, $r = 1 + \cos t$, $0 \leq t \leq 2\pi$. Then $ds = \sqrt{\sin^2 t + 1 + \cos^2 t + 2\cos t} = \sqrt{2}\sqrt{1 + \cos t} = 2|\cos\frac{t}{2}|, L = 8$.

14.1.5 Integration on regular curves. By Exercise 3.3, a regular curve Γ in a domain $U \subset \mathbb{R}^n$ (a regular sub-manifold of dimension one), if connected, is either homeomorphic to an open interval (a, b) or else homeomorphic to a circle. In both cases, there exists a global parametrization $\Phi(t), \Phi'(t) \neq 0$, which can be turned in the arc-length parametrization $\gamma(s)$.

We can now define integrals of continuous functions on regular curves. If $f : \Gamma \to \mathbb{R}$ is continuous, we consider *Riemann sums* associated to polygonals inscribed on Γ, that written in terms of a parametrization take the form

$$\sum_i f(\gamma(\xi_i))|\gamma(t_{i+1}) - \gamma(t_i)|, \quad t_i \leq \xi_i \leq t_{i+1}.$$

Using the uniform continuity of f on γ^*, it is easily seen that

$$\int_{\gamma^*} f\, ds = \lim_P \sum_i f(\gamma(\xi_i))|\gamma(t_{i+1}) - \gamma(t_i)|$$

exists. In case γ is C^1, we see as before that this sum differs in a $o(1)$ term from a Riemann sum of the continuous function $f(\gamma(t))|\gamma'(t)|$, and therefore,

$$\int_{\gamma^*} f\, ds = \int_a^b f(\gamma(t))|\gamma'(t)|\, dt.$$

This integral has different interpretations; for instance, if $f \geq 0$, it represents the area of the surface in $\mathbb{R}^{n+1}$ with parametrization $(\gamma(t), sf(\gamma(t)), a \leq t \leq b, 0 \leq s \leq 1$. Or else we may think that f represents a mass density, then $f\, ds$ represents the infinitesimal mass, $M = \int_{\gamma^*} f\, ds$ the total mass of γ^*, and the point with coordinates

$$\frac{1}{M} \int_a^b x_i(t) f(\gamma(t))|\gamma'(t)|\, dt,$$

is the *mass center.*

14.2 The Volume Element of a Regular Sub-Manifold

14.2.1 Now, we would like to define the area of a piece A of a surface or more generally of a k-dimensional regular sub-manifold M. Of course, if M is an affine sub-manifold $M = p + V$, we simply transport the k-dimensional Lebesgue measure to M by choosing an orthonormal basis of V. In the general case, a definition is needed. Considering just surfaces by now, $k = 2$, and by analogy with the case of curves, it seems natural to define the area of a surface S in terms of *triangulations*, that is, families of triangles with vertices on S. More specifically, one would define the area of S as the supremum of the areas of such families. However, the next example, known as *the Schwarz or Chinese lantern*, shows that this definition is not the correct one, because it attributes infinite area to a cylinder, an object that on an intuitive basis should have finite area $2\pi rh$, where r is the radius of the basis and h its height.

Example 14.3. We consider the cylinder S with $r = h = 1$ with parametrization

$$x(t,\theta) = \cos\theta, \quad y(t,\theta) = \sin\theta, \quad z(t,\theta) = t, \quad 0 \leq \theta \leq 2\pi, \quad 0 < t < 1.$$

We will show that there are triangulations of S with arbitrarily large area. We partition the total height in M equal parts, and consider, in each of the $M+1$ circles, N equally distributed points. Assume first that we take these points with the same angular distribution, independently of height, that is

$$p_{i,j} = \left(\cos i\frac{2\pi}{N}, \sin i\frac{2\pi}{N}, \frac{j}{M}\right), \quad i = 0, \ldots, N-1, \quad j = 0, \ldots, M.$$

With the triangles with vertices $p_{i,j}, p_{i+1,j}, p_{i,j+1}$ and $p_{i,j+1}, p_{i+1,j+1}, p_{i+1,j}$, the triangulation covers the regular prism of height 1 and having as a basis the regular polygon of N sides. Obviously, the area has limit 2π as $N \to \infty$. Assume now that the points at two consecutive levels are shifted angularly by $\frac{\pi}{N}$; each couple of consecutive points on the same level is the basis of two triangles, one with vertex at the next level and another with vertex at the previous level. In this way, we obtain a triangulation with NM triangles, all of them similar to the triangle with vertices

$$(1,0,0), \quad \left(\cos\frac{2\pi}{N}, \sin\frac{2\pi}{N}, 0\right), \quad \left(\cos\frac{\pi}{N}, \sin\frac{\pi}{N}, \frac{1}{M}\right).$$

The basis of this triangle has length $2\sin\frac{\pi}{N}$ and height

$$\sqrt{\left(1-\cos\frac{\pi}{N}\right)^2 + \frac{1}{M^2}}.$$

Replacing $\sin x$ by x and $1-\cos x$ by $\frac{x^2}{2}$ it is easily checked that the area of this triangle is of the order of $\frac{1}{N}(\frac{1}{M} + \frac{1}{N^2})$ and so the total area is about $1 + \frac{M}{N^2}$. Therefore, choosing $M = N^3$ the total area of the triangulation is arbitrarily large. See Figure 14.1.

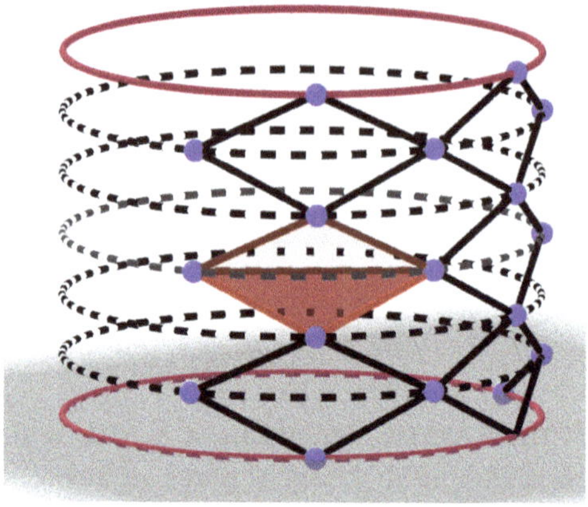

Figure 14.1. The Schwarz's lantern.

14.2.2 There are different, equivalent, possible definitions to be considered. One is in terms of *Haussdorf measures*, see [7]. Here we define it in terms of local charts. Let $\Phi(u_1, u_2, \dots, u_k)$ be a local parametrization; placed at the point $p = \Phi(u)$, we consider the infinitesimal piece of M consisting of the points

$$\Phi(u_1 + \lambda_1 du_1, \dots, u_k + \lambda_k du_k), \quad 0 \leq \lambda_i \leq 1,$$

that is, all points having parameters between u_i and $u_i + du_i, i = 1, \dots, k$. We may think that this piece is a sort of curved k-dimensional rectangle on M. If we replace $\Phi(t)$ by its first-order approximation $\Phi(u) + d\Phi(u)(t-u)$, we get the infinitesimal parallelepiped $p + R$ on $T_p(M)$ where R is spanned by the k linearly independent vectors

$$\frac{\partial \Phi}{\partial u_i}(u)\, du_i, \quad i = 1, \dots, k.$$

Motivated by the proof of the changes of variables formula, we *define* the *k-dimensional volume element* dm_k as the volume of this infinitesimal parallelepiped; as shown in (1.5), this is

$$dm_k(u) = |\det G_\Phi(u)|^{\frac{1}{2}}\, du_1\, du_2 \cdots du_k,$$

where

$$G_\Phi(u) = \left(\langle \frac{\partial \Phi}{\partial u_i}, \frac{\partial \Phi}{\partial u_j} \rangle \right)_{i,j=1,\cdots,k},$$

is the *Gram matrix*. In terms of the $n \times k$ differential matrix $d\Phi(u)$ having these vectors as column vectors, $G_\Phi(u) = (d\Phi(u))^t \times d\Phi(u) = J_k(d\Phi)$ (see Theorem 1.7).

With dm_k we can define the k-dimensional measure $m_k(C)$ of a compact piece $C \subset M$; if C is covered by just one chart Φ and has parameters $D = \Phi^{-1}(C)$, then

$$m_k(C) = \int_D dm_k = \int_D |\det G_\Phi(u)|^{\frac{1}{2}}\, du_1\, du_2 \cdots du_k.$$

In case C is not covered by one chart, we break it into smaller pieces. In most of the examples, a single chart covers the whole of M with the exception of sets of lower dimension, with zero measure, which are then ignored.

14.2.3 An important point is of course that this definition does not depend on the chosen parametrization. Indeed, assume $u = \xi(s)$ is a change of parameters, that is $u_i = u_i(s_1, \ldots, s_k), i = 1, \ldots, k$, and consider the new parametrization $\Psi(s) = \Phi(\xi(s))$ in which C has $D' = (\xi)^{-1}(D), D = \xi(D')$ as domain of parameters. By the chain rule, $d\Psi(s)$ equals the product matrix $=d\Phi(u)d\xi(s)$, so that

$$\begin{aligned} G_\Psi(s) &= (d\Psi(s))^t d\Psi(s) = (d\xi(s))^t (d\Phi(t))^t d\Phi(t) d\xi(s) \\ &= (d\xi(s))^t G_\Phi(\xi(s)) d\xi(s). \end{aligned}$$

Then

$$\int_{D'} |\det G_\Psi(s)|^{\frac{1}{2}}\, ds_1 \cdots ds_k = \int_{D'} |\det G_\Phi(\xi(s))|^{\frac{1}{2}} |\det \xi(s)|\, ds_1 \cdots ds_k,$$

equals by the change of variables formula

$$\int_D |\det G_\Phi(u)|^{\frac{1}{2}}\, du_1 \cdots du_k.$$

14.2.4 An alternative way to define $m_k(C)$ is using the tubular neighborhood Theorem 7.4 in Section 7.3. For δ small enough, the set

$$C_\delta = \{x : d(x, C) = d(x, M) \le \delta\}$$

is the union of (disjoint) balls $N_p(\delta)$ of radius δ in the normal space. It is then quite intuitive that

$$m_k(C) = \lim_{\delta \to 0} \frac{m_n(C_\delta)}{\omega_{n-k}\delta^{n-k}},$$

$\omega_{n-k}\delta^{n-k}$ being the volume of the $(n-k)$-dimensional ball of radius δ.

A formal proof for surfaces in space is as follows. If $\Phi(s,t)$ is a local chart, then

$$N(s,t) = \frac{\Phi_s \times \Phi_t}{|\Phi_s \times \Phi_t|}$$

is a unit normal. In the proof of Theorem 7.4 it was seen that

$$\Psi(s,t,\lambda) = \Phi(s,t) + \lambda N(s,t)$$

is a one-to-one map from $D \times (-\delta, \delta), D = \Phi^{-1}(C)$ onto C_δ. By the change of variables formula,

$$m_n(C_\delta) = \int_{D\times(-\delta,\delta)} |J\Psi(s,t,\lambda)|\, ds\, dt\, d\lambda.$$

Since $\Psi_s = \Phi_s + \lambda N_s, \Psi_t = \Phi_t + \lambda N_t, \Psi_\lambda = N(s,t)$,

$$|J\Psi(s,t,\lambda)| = |\Psi_s \times \Psi_t| = |\Phi_s \times \Phi_t| + O(\lambda^2),$$

which implies

$$m_n(C_\delta) = 2\delta \int_{D\times(-\delta,\delta)} |\Phi_s \times \Phi_t|\, ds\, dt) + O(\delta^3).$$

For example, the $(n-1)$-dimensional volume of the sphere of radius r in $\mathbb{R}^n$ is

$$c_n = \frac{d}{dr}(\omega_n r^n) = n\omega_n r^{n-1}. \tag{14.1}$$

14.2.5 With dm_k we can define integral of functions defined on M. If f is a continuous function on a compact subset $C \subset M$, we define the integral as limit of Riemann sums

$$\int_C f\, dm_k = \sum_i f(p_i) m_k(C_i),$$

where the C_i constitute a partition of C, $p_i \in C_i$, and the limit is as the diameter of C_i go to zero. In case C is covered by a chart $\Phi, C = \Phi(D)$, then it is easily seen that

$$\int_C f\, dm_k = \int_D f(\Phi(u))\, |\det G_\Phi(u)|^{\frac{1}{2}}\, du_1\, du_2 \cdots du_k.$$

If say $f \geq 0$ is a mass or probability density on M, then $\mu = \int_M f dm_k$ is the total mass, and its center is the point $(\overline{x_i})$ (not necessarily on M) whose coordinates are given by

$$\overline{x_i} = \frac{1}{\mu} \int_M x_i f(x)\, dm_k(x).$$

Example 14.4. Assume the surface A of Example 14.5, with $h > 0$ having density z. Its mass is

$$M = 2\pi \int_0^{\arccos h} \cos\phi \sin\phi \, d\phi = 2\pi \int_h^1 x\, dx = \pi(1 - h^2).$$

The center of mass is $(0, 0, \overline{z})$, with

$$\overline{z} = \frac{1}{\pi(1-h^2)} 2\pi \int_0^{\arccos h} \cos^2\phi \sin\phi \, d\phi = \frac{2}{1-h^2} \int_h^1 x^2\, dx = \frac{2}{3}\frac{1-h^3}{1-h^2}.$$

14.2.6 We consider in more detail the case of surfaces, $k = 2$, for which we use the notation dA for the *area element*. Denoting s, t the two parameters, recall that the cross product of the two tangent vectors $\Phi_s = \frac{\partial \Phi}{\partial s}, \Phi_t = \frac{\partial \Phi}{\partial t}$ (also denoted ∂_s, ∂_t) is perpendicular to S at $\Phi(s, t)$ and its length equals the area of the parallelogram they span, whence $dA = |\Phi_s \times \Phi_t|\, ds\, dt$.

Example 14.5. We compute the area of the part A of the unit sphere $x^2 + y^2 + z^2 = 1$ given by $z \geq h, -1 \leq h \leq 1$. Using the parametrization in spherical coordinates

$$\Phi(\phi, \theta) = (\sin\phi \cos\theta, \sin\phi \sin\theta, \cos\phi),$$

we find $dA = \sin\phi\, d\phi\, d\theta$, while A is described by $0 \leq \phi \leq \pi, \cos\phi \geq h, 0 \leq \theta \leq 2\pi$. Therefore, the area equals

$$2\pi \int_0^{\arccos h} \sin\phi\, d\phi = 2(1-h)\pi,$$

the area of the whole sphere being 4π.

It is interesting to repeat the problem using the parametrization in cylindrical coordinates

$$\Phi(\theta, z) = \left(\sqrt{1-z^2}\cos\theta, \sqrt{1-z^2}\sin\theta, z\right),$$

for which we find that $dA = d\theta\, dz$. Since A is given by $0 \leq \theta \leq 2\pi, h \leq z \leq 1$, its area is $2\pi(1-h)$.

In the sphere of radius R, the part $z \geq h$ has area $2\pi R^2(1 - \frac{h}{R}) = 2\pi R(R - h)$. Note that $z \geq h$ means $d(p, q) \leq \sqrt{2R(R-h)}$ where $p = (x, y, z), q = (0, 0, 1)$. So, for an arbitrary point q in a sphere S, the area of the part of S inside the ball $B(q, r)$ is πr^2, independently of the radius of S.

Exercise 14.1. Using the parametrization (3.1) prove that the area of the torus is $2\pi R \times 2\pi r = 4\pi^2 Rr$, as seen intuitively. Can you guess what is the volume enclosed? Repeat the exercise replacing the inner circle by an arbitrary plane simple curve Γ of length L: if C is a ball of small radius r whose center moves along Γ, what is the area of the limiting surface and volume enclosed?

Next, we point out some particular cases. If the surface is the graph $z = f(x, y)$ for f defined on D, the parametrization is $\Phi(x, y) = (x, y, f(x, y))$, the two tangent vectors are $(1, 0, f_x), (0, 1, f_y)$, the normal vector is $(-f_x, -f_y, 1)$ and

$$dA = \sqrt{1 + f_x^2 + f_y^2} dx\, dy. \tag{14.2}$$

Consider now a surface obtained by rotating a graph $\Gamma : z = f(y), 0 \leq a \leq y \leq b$ around the z-axis. A parametrization $\Phi(\theta, r)$ is given by

$$x = r\cos\theta, \quad r\sin\theta, \quad z = f(r), \quad 0 \leq \theta \leq 2\pi, \quad a \leq r \leq b.$$

The two tangent vectors are

$$\Phi_\theta = (-r\sin\theta, r\cos\theta, 0), \quad \Phi_r = (\cos\theta, \sin\theta, f'(r)),$$

and their cross product is

$$\Phi_\theta \times \Phi_r = r(\cos\theta f'(r), \sin\theta f'(r), -1),$$

so that $dA = r\sqrt{1 + f'(r)^2}\, d\theta\, dr$ and the area is given by

$$2\pi \int_a^b r\sqrt{1 + f'(r)^2}\, dr.$$

Analogously, we find

$$2\pi \int_a^b f(r)\sqrt{1 + f'(r)^2}\, dr,$$

for the area in case the graph rotates around the y-axis. More generally, assume that Γ is a curve $z = z(t), y = y(t) \geq 0, a \leq t \leq b$ in the half-space

$y \geq 0$. When rotating say around the z-axis, the resulting surface has parametrization

$$x = y(r)\cos\theta, \quad y = y(r)\sin\theta, \quad z = z(r), \quad a \leq r \leq b, \quad 0 \leq \theta \leq 2\pi.$$

The tangent vectors are

$$\partial_\theta = (-y(r)\sin\theta, y(r)\cos\theta, 0), \quad \partial_r = (y'(r)\cos\theta, y'(r)\sin\theta, z'(r)),$$

the normal vector is

$$(y(r)z'(r)\cos\theta, -y(r)z'(r)\sin\theta, -y(r)y'(r)),$$

the area element is

$$dA = y(r)\sqrt{z'(r)^2 + y'(r)^2}\,dr\,d\theta,$$

and the total area is

$$2\pi\int_a^b y(r)\sqrt{z'(r)^2 + y'(r)^2}\,dr.$$

If we think that the graph is a wire with constant density 1, length $L = \int_a^b \sqrt{z'(t)^2 + y'(t)^2}\,dt$, and center of mass $P = (\overline{x}, \overline{y})$, the above equals $2\pi\overline{y}L$, the product of L with the length of the circle described by P (Pappus' theorem).

Example 14.6. We compute the area of the surface $z = f(r) = 2\sqrt{r}, r = \sqrt{x^2 + y^2}, 0 \leq r \leq 1$, presenting a cusp at the origin

$$A = 2\pi\int_0^1 r\sqrt{1 + f'(r)^2}\,dr = 2\pi\int_0^1 \sqrt{r}\sqrt{1 + r}\,dr.$$

Change variable $r = \tan t$,

$$A = 2\pi\int_0^{\frac{\pi}{4}} \frac{\sin t}{\cos^4 t}\,dt = \frac{2\pi}{3}[\cos^{-3} t]_{t=0}^{t=\frac{\pi}{4}} = \frac{2\pi}{3}(2\sqrt{2} - 1).$$

Example 14.7. Consider a graph $G : y = f(x) \geq 0, 0 < x \leq 1$, assuming f' continuous in $(0, 1]$. Its length is

$$L = \int_0^1 \sqrt{1 + f'(x)^2}\,dx = \lim_{\varepsilon\to 0}\int_\varepsilon^1 \sqrt{1 + f'(x)^2}\,dx,$$

finite or infinite. Let A_1, A_2 be the area of the surfaces obtained by rotating G around the x, y axis, respectively. Let A be the area under the graph, and V_1, V_2, the volume of the body obtained rotating A around the x, y axis. As mentioned above, these quantities are given by

$$A_1 = 2\pi \int_0^1 f(x)\sqrt{1+f'(x)^2}\,dx, \quad A_2 = 2\pi \int_0^1 x\sqrt{1+f'(x)^2}\,dx,$$

$$A = \int_0^1 f(x)\,dx, \quad V_1 = \pi \int_0^1 f(x)^2\,dx, \quad V_2 = 2\pi \int_0^1 xf(x)\,dx.$$

If L is finite, that is, $\int_0^1 |f'|\,dx$ is finite, then

$$\lim_{\varepsilon\to 0} \int_{\varepsilon\to 0}^1 f'\,dx = f(1) - f(\varepsilon)$$

exists, too, whence f is continuous in $[0,1]$. Since f is bounded, all other five quantities are finite, too. But L can be infinite and A_1, A_2 finite, choosing f such that

$$\int_0^1 |f'(x)|\,dx = +\infty, \int_0^1 (x + f(x))|f'(x)|\,dx < +\infty.$$

Choosing f with $f'(0) = \infty$ and such that the last integral is infinite, one has an example of a surface with a singularity with infinite area.

14.3 Area and Metric on Surfaces

14.3.1 We continue here with surfaces. Denoting here u, v the two parameters, by definition,

$$dA = |\det G(\Phi_u, \Phi_v)|^{\frac{1}{2}}\,du\,dv,$$

where $G(\Phi_u, \Phi_v)$ is the Gram matrix of the two tangent vectors. The classical notation is

$$G = \begin{pmatrix} E & F \\ F & G \end{pmatrix}, \quad E = \langle \Phi_u, \Phi_u \rangle, \quad F = \langle \Phi_u, \Phi_v \rangle, \quad G = \langle \Phi_v, \Phi_v \rangle.$$

This matrix is called *the first fundamental form* of S. A curve γ on S given in these coordinates by $u = u(t), v = v(t)$ has the tangent

$$u'(t)\Phi_u(u(t), v(t)) + v'(t)\Phi_v(u(t), v(t)),$$

and therefore,

$$|\gamma'(t)|^2 = Eu'^2 + 2Fu'v' + Gv'^2.$$

This fact is stated by saying that the metric is

$$ds^2 = E\,du^2 + 2F du\,dv + G\,dv^2. \tag{14.3}$$

Here, du^2 means the bilinear map $du \otimes du$, $du\,dv$ is short for $du \otimes dv$, etc. For two tangent vectors $a\Phi_u + b\Phi_v, c\Phi_u + d\Phi_v$, their scalar product is given by

$$acE + (ad + bc)F + bdG.$$

An advantage of this notation is that under a change of coordinates $u = u(\rho, \tau)$, $v = v(\rho, \tau)$, the metric is written in the coordinates ρ, τ by setting $du = u_\rho\, d\rho + u_\tau\, d\tau$, etc.

14.3.2 For curves, we have seen that there is a canonical parametrization by arc-length. It is a natural question whether for a surface a local parametrization $\Phi(s, t)$ exists for which $dA = ds\,dt$, that is Φ preserves the area between sets B in $\mathbb{R}^2$ and sets $\Phi(B)$ in S. This is indeed possible, and quite easy to see. We have seen that under a re-parametrization $(u, v) = \xi(s, t), \Psi(s, t) = \Phi(\xi(s, t))$, the element of area transforms as

$$dA = |\det G_\Phi|^{\frac{1}{2}}\, |\det d\xi|\, ds\, dt.$$

Assuming without loss of generality that the Gramian is positive, we are led to the equation

$$\det d\xi = \det G_\Phi^{-\frac{1}{2}}(\xi(s, t)).$$

So we are given $F(u, v) = |\det G_\Phi|^{-\frac{1}{2}}(u, v)$, and seek for $(A(s, t), B(s, t)) = \xi(s, t)$, such that

$$A_s B_t - A_t B_s = F(A(s, t), B(s, t)).$$

Taking $A = s$, this reduces to

$$B_t = F(s, B(s, t)).$$

This can be regarded as an e.d.o depending on one parameter s, and by the results in Section 8.3 it has a regular solution $B(s, t)$. Example 14.5 shows that cylindrical coordinates are of this type on the sphere, the fact that $dA = d\theta dz$ means that the area of a piece C on the sphere equals that of its projection on the circumscribed cylinder $0 \leq z \leq 1$.

14.3.3 A more interesting question is whether there exist local coordinates u, v such that the first fundamental form is the identity, that is

$$ds^2 = du^2 + dv^2.$$

In other words, Φ preserves distances and S is locally isometric to the plane. This is not always possible. For instance, intuitively we see that no part of the sphere can be folded in a plane surface keeping all distances. There is a very important invariant related to this problem, the *Gauss curvature.* This is defined in terms of second-order derivatives and non treated in this course; intuitively, the curvature K measures how the normal to the surface changes along curves on S. For instance, the sphere has positive non-zero curvature and the plane has zero curvature. The famous *Gauss' Egregium theorem* establishes that the curvature is invariant by isometries, whence pieces of surfaces with different curvatures cannot be isometric. The converse of the Egregium theorem (Minding's theorem) holds for constant curvature; in particular, a surface with zero curvature is isometric to the plane. All these questions pertain to the beautiful classical area of the theory of curves and surfaces (for example, see [6]).

14.3.4 A third question is about the existence of local coordinates u, v for which

$$ds^2 = H(u,v)(du^2 + dv^2).$$

As in paragraph 10.1.3, these are called *isothermal* or *conformal* coordinates. For those, Φ preserves the angles between curves. They locally exist for all surfaces. In the following, we will see which is the partial differential equation involved.

With a slight change of notation with respect to (14.3), we are given a metric say around the origin in $\mathbb{R}^2$

$$ds^2 = E\,dx^2 + 2F\,dx\,dy + G\,dy^2,$$

and want to find a diffeomorphism $x = x(u,v), y = y(u,v)$ such that $ds^2 = H(u,v)(du^2 + dv^2)$.

First, we study the (point-wise) algebraic problem of converting a positive definite polynomial

$$Ex^2 + 2Fxy + Gy^2, \quad E, G > 0, \quad EG - F^2 = W > 0,$$

into $H(u,v)(u^2 + v^2)$ by an invertible linear map $x = x(u,v), y = y(u,v)$. We use complex notation $z = x + iy, w = u + iv, w = az + b\overline{z}, a, b \in \mathbb{C}$ for the linear map $(x, y) \to (u, v)$, as in paragraph 9.5.2. Then the equation becomes, with $\alpha = E - G - 2iF, \beta = 2(E + G)$,

$$\frac{1}{4}(\alpha\, z^2 + \overline{\alpha}\, \overline{z}^2 + \beta\, |z|^2) = H|w|^2 = H|az + b\overline{z}|^2,$$

which we rewrite as

$$\alpha\, z^2 + \overline{\alpha}\, \overline{z}^2 + \beta\, |z|^2 = r|z + \lambda \overline{z}|^2 = r((1 + |\lambda|^2)|z|^2 + \lambda \overline{z}^2 + \overline{\lambda} z^2).$$

Thus, we search for $r > 0, \lambda \in \mathbb{C}$, such that

$$\alpha = r\overline{\lambda}, \quad \beta = r(1 + |\lambda|^2).$$

These imply $\alpha\lambda^2 - \beta\lambda + \overline{\alpha} = 0$, whence for $\alpha \neq 0$ we have two solutions $\lambda_i, i = 1, 2$

$$\lambda = \frac{E + G \pm 2W}{\alpha},$$

satisfying $\alpha\lambda_1\lambda_2 = \overline{\alpha}$. Since now we will make E, F, G vary with the point, we need a smooth choice and take the solution $|\lambda| < 1$, which is

$$\lambda = \frac{E + G - 2W}{\alpha} = \frac{\overline{\alpha}}{E + G + 2W},$$

and for which $r = E + G + 2W > 0$. We may think that λ is a complex number, $|\lambda| < 1$, associated to every metric.

Thus, we have

$$ds^2 = r(x, y)\ |dz + \lambda(x, y) d\overline{z}|^2, \quad r > 0,$$

where the right-hand side means $r(dz + \lambda d\overline{z}) \otimes (d\overline{z} + \overline{\lambda} dz)$.

Now we write the diffeomorphism we are looking for in complex notation $w = f(z), f = u + iv$ and assume without loss of generality that f has positive Jacobian. Since this Jacobian is (paragraph 9.5.2) $|\partial f|^2 - |\overline{\partial} f|^2$, we have $|\partial f| > 0$ and we may write, with $\mu(f) = \frac{\overline{\partial} f}{\partial f}$,

$$df \otimes d\overline{f} = |df|^2 = |\partial f|^2 |dz + \mu(f) d\overline{z}|^2.$$

Comparing, this shows that f defines isothermal coordinates if and only if

$$\frac{\partial f}{\partial \overline{z}} = \lambda \frac{\partial f}{\partial z}. \tag{14.4}$$

Equation (14.4) is called the *Beltrami–Laplace equation.* Their solutions are called *quasi-conformal* maps. It can be proved that for λ smooth enough it has a solution f of class C^1. In fact, Gauss proved the existence of a solution when λ is real-analytic. Note that a C^1 solution is automatically a local difeomorphism because the Jacobian is

$$|\partial f|^2 - |\overline{\partial} f|^2 = (1-\lambda)|\partial f|^2.$$

Once some isothermal coordinates u, v have been found, all other local isothermal coordinates are obtained applying to u, v a conformal map in the plane, that is, a holomorphic or anti-holomorphic transformation.

It is possible to write the Beltrami–Laplace equation in terms of u, v, assuming f of class C^2. Identifying real and imaginary parts, one obtains

$$v_x = \frac{Fu_x - Eu_y}{W}, \quad v_y = \frac{Gu_x - Fu_y}{W},$$

and analogously

$$u_x = \frac{Ev_y - Fv_x}{W}, \quad u_y = \frac{Fv_y - Gv_x}{W},$$

and hence, both u, v satisfy the equation

$$\left(\frac{Fw_x - Ew_y}{W}\right)_y + \left(\frac{Fw_y - Gw_x}{W}\right)_x = 0.$$

which is also called the *Beltrami–Laplace equation.*

Reviewing this procedure, we have written

$$ds^2 = |H|^2 df \otimes d\overline{f},$$

with some complex H. Decomposing ds^2, similarly as in paragraph 9.5.1,

$$ds^2 = \left(\sqrt{E}\, dx + \frac{F+iW}{\sqrt{E}}\, dy\right) \otimes \left(\sqrt{E}\, dx + \frac{F-iW}{\sqrt{E}}\, dy\right),$$

this shows that $\frac{1}{H}$ is an integrating factor,

$$\sqrt{E}\, dx + \frac{F+iW}{\sqrt{E}}\, dy = H df.$$

Conversely, if H satisfies this, then

$$\sqrt{E} = H(u_x + iu_y), \quad \frac{F+iW}{\sqrt{E}} = H(u_y + iv_y),$$

implying

$$E(u_y + iv_y) = (F+iW)(u_x + iv_x).$$

Separating real and imaginary parts and solving in v_x, v_y in terms of u_x, u_y leads to the same equations above. Thus, there is a one-to-one correspondence between solutions of the Beltrami–Laplace equation and integrating factors for the (complex) differential form above.

14.3.5 We can exhibit explicit isothermal coordinates in spheres, relevant in cartography and the so-called *Mercator projection* on the Earth's globe. We start with the spherical coordinates θ, ϕ so that in a sphere S of radius R,

$$x = R\cos\phi\cos\theta,\ y = R\cos\phi\sin\theta,\ z = R\sin\phi,\ 0 \le \theta \le 2\pi,\ -\frac{\pi}{2} \le \phi \le \frac{\pi}{2}.$$

Note that now we are using a different latitude angle, the usual in cartography. In these coordinates,

$$ds^2 = R^2\,cos^2\phi\,d\theta^2 + R^2\,d\phi^2.$$

Since this is already diagonal, it is enough to replace ϕ by $\omega = \omega(\phi)$ to be found. We want

$$R^2\,d\phi^2 = R^2\,cos^2\phi\,d\omega^2, \omega'(\phi) = \frac{1}{\cos\phi}.$$

This means

$$\omega = \int \frac{d\phi}{\cos\phi} = \log\frac{1+\sin\phi}{\cos\phi} + c.$$

We choose $c = 0, \omega(0) = 0$. The map assigning to each point in S_R the point $(R\theta, R\omega(\phi))$ is called the *Mercator projection.* It preserves angles,

$$ds^2 = R^2\cos^2\phi(d\theta^2 + d\omega^2),$$

and we see that the distortion factor for distances is $R\cos\phi$.

This projection must be adapted to an ellipsoid to take into account the real shape of the Earth. The Mercator projection has the inconvenience that the larger the real scale of the map, the larger the distance from the equator. A correction of the Mercator projection was proposed by Gauss, it is the basis of the *Universal Transversal Mercator UTM* projection. The Earth surface between latitudes $-80\%, +80\%$ is divided into 60 parts, with each longitude of 6%. Each of these parts is then represented on its own map in the xy plane, according to the following rules:

(a) The central meridian is mapped to the y axis keeping distances, the $y > 0$ corresponding to the north hemisphere and the $y < 0$ to the south hemisphere.
(b) The intersection of the central meridian with the equator is mapped to $(0, 0)$.
(c) The map is conformal.

Constructing a projection with these properties starting with the Mercator projection requires using techniques of holomorphic functions. Actually, the UTM projection at use is still a slight modification of the above one (x, y), namely $(kx+a, ky)$ where k, a are chosen so that only positive values appear and the distortion of distances away from the central meridian has mean zero.

14.4 Invariant Measures

In this section, we pay attention to an important concept in Analysis, that of *invariant measures*, of which we have seen a number of examples. The Lebesgue measure in $\mathbb{R}^n$ is invariant by translations and rigid motions T,

$$m(T(A)) = m(A), \quad \int f(Tx)\, dm(x) = \int f(x)\, dm(x).$$

Analogously, the Lebesgue measure m_2 in the sphere S of $\mathbb{R}^3$ is invariant by rigid motions

$$m_2(T(A)) = m_2(A), \quad A \subset S.$$

The unit circles in $\mathbb{R}^2$ and $\mathbb{R}^n$ have a group structure, and the Lebesgue measure on those is invariant by the group action. An important group is $SO(3)$, the group of rotations, and now we proceed to define on it an invariant measure $d\mu(T)$, allowing us to integrate functions defined on $SO(3)$,

$$\int_{SO(3)} f(T)\, d\mu(T),$$

with the invariant property

$$\begin{aligned}\int_{SO(3)} f(TT')\, d\mu(T) &= \int_{SO(3)} f(T)\, d\mu(T) \\ &= \int_{SO(3)} f(T'T)\, d\mu(T), \quad T' \in SO(3).\end{aligned}$$

We follow basically [13]. In paragraph 1.2.3, two parametrizations of $SO(3)$ have been indicated. We use (1.8) by interpreting the Euler parameters in a slightly different way. It follows from (1.9) that $\pi - \gamma, \beta$ are the spherical coordinates of $p = T^{-1}(0, 0, 1)$. The point $q = T^{-1}(1, 0, 0)$ is then in the meridian $p^{\perp}$, and an increment $d\alpha$ corresponds to a displacement $d\alpha$ in

$p^\perp$. The points $p, q \in p^\perp$ describe $SO(3)$, too, so it is natural to define $d\mu(T) = dp\, ds(q)$, where ds means arc-length in $p^\perp$. Since $dp = \sin\beta d\gamma\, d\beta$, $ds = d\alpha$,

$$d\mu(T) = \sin\beta\, d\gamma\, d\beta\, d\alpha.$$

If we replace T by TT' or $T'T$, the new points P, Q will be obtained from p, q by a rigid motion of the sphere, whence the invariant property follows.

The total measure of $SO(3)$ is thus $8\pi^2$.

Exercise 14.2. Investigate whether $d\mu(T)$ coincides with the following alternatives

(a) Using the representation (1.7) $T = T(v, \theta)$,

$$d\mu(T) = dv\, d\theta.$$

(b) Considering $SO(3)$ as a sub-manifold of $\mathbb{R}^9$ using either the Euler parametrization or the Olinde Rodrigues parametrization.

Exercise 14.3. Try to parametrize the whole unitary group $O(3)$ and define an invariant measure on it whose total measure should be twice that of $SO(3)$.

Invariant measures, named *Haar measures* can be defined in quite general abstract groups.

Chapter 15

Geometric Measure Theory and Integral Geometry

This chapter is a short introduction to two branches of measure theory. Geometric measure theory studies the interplay between measure and geometric properties of sets. The main reference is [7]. Here, we just deal with the *area and co-area* formulas, in the context of sub-manifolds. Integral geometry, also known as *geometric probability*, is an area pioneered by Blaschke that applies probability ideas to geometric objects, and that later found interesting applications in imaging. The main reference is [15].

15.1 Area and Co-Area Formulas*

This section deals with the so-called *area* and *co-area* formulas. The general versions of these formulas pertain to the field of *Geometric Measure Theory* in the context of Haussdorf measures. Here, we study versions for sub-manifolds in the setting of the constant rank theorem in paragraph 7.2.2.

15.1.1 Theorem 13.6 states how integrals transform under a C^1 map $g : U \subset \mathbb{R}^n \to \mathbb{R}^n$. In this section, we will study how integrals transform under C^1 maps $g : U \subset \mathbb{R}^n \to \mathbb{R}^m$ for $m \neq n$.

We assume first that $n < m, g : U \to \mathbb{R}^m, g = (g_1, \dots, g_m)$ is a one-to-one map of class C^1 in a domain $U \subset \mathbb{R}^n$ and that the rank of $dg(x)$ is $k = n$ for all $x \in U$. We know by Theorem 7.3 that $M = g(U)$ is a regular sub-manifold M of dimension k of R^m. By our definition of dm_k on M, we have for h defined on U,

$$\int_U h(x)J_k g(x)\, dm_k(x) = \int_{g(U)} h(g^{-1}(y))dm_k(y), \tag{15.1}$$

where

$$J_k g(x) = |\det(dg(x)^t dg(x))|^{\frac{1}{2}}.$$

Recall that the term $J_k g(x)$ is the distortion factor of dg when acting on the k-dimensional parallelepiped of area $dm_k(x)$, that is, its image has k-dimensional area $J_k g(x)\, dm_k(x)$.

This formula is called *the area formula* and $J_k g$ is called the *k-th Jacobian of g*. This formula is valid for general g of class C^1, namely

$$\int_U h(x)J_k g(x)\, dm_k(x) = \int_{g(U)} \left(\sum_{g(x)=y} h(x) \right) dm_k(y)$$

holds, but the right-hand side must be interpreted in terms of Haussdorf k-dimensional measure (see [7]).

15.1.2 In this paragraph, we assume that the rank of dg is constant $k = m < n$. In this case, $g(U)$ is an open set in $\mathbb{R}^m$ and for each $y \in g(U)$, the fiber $g^{-1}(y)$ is an $(n-k)$-dimensional sub-manifold. We would like to establish a relationship like (15.1). It is natural to replace $h(g^{-1}(y))$ by

$$\hat{h}(y) = \int_{g^{-1}(y)} h(x)\, dm_{n-k}(x).$$

So, we look for a formula

$$\int_U h(x)\lambda(x)\, dm_n(x) = \int_{g(U)} \hat{h}(y)\, dm_k(y),$$

and ask what should λ be in this case.

We first note that in case g is a projection $g(x) = x_i$ this is just Fubini's theorem. For a linear map given by a $k \times n$ matrix A of rank k, we show that we can choose λ constant. Let N be its kernel, $N^\perp$ the orthogonal complement, and take $h = 1_P$, the characteristic function of a parallelepiped, $P = P_1 \times P_2$, P_1, P_2 parallelepipeds in $N, N^\perp$, respectively. Then $\hat{h} = m_{n-k}(P_1)1_{A(P_2)}$, so if λ is constant, we should have

$$\lambda m_{n-k}(P_1)m_k(P_2) = \lambda m_n(P) = m_k(A(P_2))m_{n-k}(P_1),$$

that is $\lambda m_k(P_2) = m_k(A(P_2))$. By Theorem 1.7 this is precisely the kth Jacobian of A. Then this holds for simple functions and therefore for all functions.

In the spirit of the change of variables formula, this indicates that we should take as $\lambda(x)$ the kth jacobian of $dg(x)$, abbreviated again $J_k g(x)$. Recall that now

$$J_k g(x) = |\det dg(x)(dg(x))^t|^{\frac{1}{2}},$$

that equals $|G(\nabla g_1(x), \dots, \nabla g_k(x))|^{\frac{1}{2}}$.

Theorem 15.1. *In the situation just described,*

$$\int_U h(x) J_k g(x)\, dm_n(x) = \int_{g(U)} \left(\int_{g^{-1}(y)} h(x)\, dm_{n-k}(x) \right) dm_k(y). \quad (15.2)$$

Proof. By a partition of unity argument, it is sufficient to prove the formula locally. We look first to the planar situation, $g = u$ a scalar function. Let u, v be a local bi-orthogonal coordinate system as in paragraph 10.1.2. The system being bi-orthogonal, the Jacobian of the map $(u, v) \to (x, y)$ is

$$|\partial_u||\partial_v| = \frac{1}{|\nabla u|}|\partial_v|,$$

so $|\nabla u|\, dx\, dy = |\partial_v|\, du\, dv$, and the result follows from Fubini's theorem because $|\partial_v|\, dv$ is ds on the curves $u = t$, parametrized by v.

In the general situation, we can consider a coordinate system $u_1, \dots, u_n$ with $u_i = g_i = y_i$, $i = 1, \dots, k$, so that $g^{-1}(y)$ is coordinated by $u_{k+1}, \dots, u_n$. The base of local fields

$$\partial_i = \frac{\partial}{\partial u_i} = \left(\frac{\partial x_1}{\partial u_i}, \dots, \frac{\partial x_n}{\partial u_i} \right), \quad i = 1, \dots, n$$

is dual to du_i,

$$\sum_k \frac{\partial x_k}{\partial u_i} \frac{\partial u_j}{\partial x_k} = \delta_{ij}.$$

We have $dm_{n-k} = G(\partial_{k+1}, \dots, \partial_n)^{\frac{1}{2}}\, du_{k+1} \cdots du_n$, so the right-hand side in (15.2) equals, by Fubini's theorem

$$\int h\, G(\partial_{k+1}, \dots, \partial_n)^{\frac{1}{2}}\, du_1 \cdots du_n,$$

while by the change of variables formula, the left-hand side equals

$$\int h\, G(\nabla u_1(x), \dots, \nabla u_k(x))^{\frac{1}{2}} G(\partial_1, \dots, \partial_n)^{\frac{1}{2}} du_1 \cdots du_n.$$

The later Gramian is the volume of the parallelepiped determined by $\partial_1, \dots, \partial_n$; it equals the $n-k$-dimensional volume of its projection on the tangent space, $G(\partial_{k+1}, \ldots, \partial_n)^{\frac{1}{2}}$, times the kth dimensional volume of its projection P_2 on the normal space N spanned by $\nabla u_1, \ldots, \nabla u_k$. Therefore, we must show that

$$G(\nabla u_1(x), \ldots, \nabla u_k(x))^{\frac{1}{2}} m_k(P_2) = 1.$$

The parallelepiped P_2 is spanned by the projections w_i of ∂_i onto N. Since

$$du_i(w_j) = du_i(\partial_j) = \delta_{ij},$$

P_2 is mapped onto the unit cube of $\mathbb{R}^k$ by $dg(x)$, and so the result follows from Theorem 1.7. □

If $\varphi(y)$ is given and take $h(x) = \varphi(g(x))$ in (15.2), then we have

$$\int_U \varphi(g(x)) J_k g(x)\, dm_n(x) = \int_{g(U)} \varphi(y) m_{n-k}(g^{-1}(y))\, dm_k(y). \qquad (15.3)$$

Formulas (15.2) and (15.3) are called *the co-area formulas*. By Theorem 13.5, the condition that g has constant rank $k = m$ can be dropped if g is smooth enough, because the set S of points where the rank of dg is less than k has $J_k g = 0$ and $g(S)$ has zero measure, so both sides of the formulas are unaffected replacing U by $U \setminus S$.

A comment analogous to that in paragraph 13.3.9 applies as well. Namely, for the coordinate system $u_1, \ldots, u_n$, the last integral

$$\int \hat{h}(u_1, \ldots, u_k)\, du_1 \cdots du_k,$$

is not an integral on a k-dimensional sub-manifold parametrized by u_1, $\ldots, u_k$.

15.1.3 We briefly comment on the case where g is one-to-one with constant rank $k < \min(n, m)$. The natural versions of (15.2) and (15.3) are

$$\int_U h(x) J_k g(x)\, dm_n(x) = \int_M \left(\int_{g^{-1}(y)} h(x)\, dm_{n-k}(x) \right) dm_k(y),$$

$$\int_U \varphi(g(x)) J_k g(x)\, dm_n(x) = \int_M \varphi(y) m_{n-k}(g^{-1}(y))\, dm_k(y),$$

where now $g(U) = M$ is a k-sub-manifold in $\mathbb{R}^m$ and $J_k g(x)$ must be defined as follows. We look at $dg(x)$ intrinsically as a linear map from $\mathbb{R}^n$ to the tangent space $T_{g(x)}(M)$. It is a linear isomorphism from the orthogonal $N^\perp$ of its kernel to the tangent space. Then $J_k g(x)$ is its distortion factor for volumes of k-parallelepipeds lying in $N^\perp$, as explained in paragraph 1.3.3.

15.1.4 It is worth stating the particular case $k = 1$: for a smooth function u with $\nabla u \neq 0$ and $S_t = \{u = t\}$,

$$\int \left(\int_{S_t} h \, dm_{n-1} \right) dt = \int h(x) |\nabla u(x)| \, dm_n(x). \tag{15.4}$$

Of course, for $u(x) = |x|$ this is integration in spherical coordinates,

$$\int \left(\int_{|x|=t} h \, dm_{n-1} \right) dt = \int h(x) \, dm_n(x).$$

Corollary 15.1. *The following formulas hold:*

$$\int_{u \leq T} h(x) |\nabla u(x)| \, dm_n = \int_0^T \left(\int_{S_t} h \, dm_{n-1} \right) dt,$$

$$\frac{d}{dt} \int_{u \leq t} h(x) |\nabla u(x)| \, dm_n = \int_{S_t} h \, dm_{n-1}.$$

The natural question to ask is which functions u satisfy that $|\nabla u|$ is constant.

Proposition 15.1. *The only smooth functions u in $\mathbb{R}^n$ such that $|\nabla u|$ is constant are the linear ones.*

Proof. We assume $|\nabla u| = 1$. Let $\gamma_x(t)$ be the integral curve of ∇u through x, the solution of

$$\frac{d}{dt} \gamma_x(t) = \nabla u(\gamma_x(t)), \quad \gamma_x(0) = x.$$

By hypothesis, t is the arc-length parameter. Then $\frac{d}{dt}(u(\gamma_x(t)) = 1$ and therefore

$$u(\gamma_x(t)) = u(x) + t.$$

Now, by the mean-value Theorem 4.4,

$$|t| = |u(\gamma_x(t)) - u(x)| \leq |\gamma_x(t) - x|.$$

Since the distance between $\gamma_x(t)$ and x is t along the curve γ_x, this implies that γ_x is defined for all t and is a line, $\gamma_x(t) = x + t\nabla u(x)$. So, we have proven

$$u(x + t\nabla u(x)) = u(x) + t. \tag{15.5}$$

Take a point P with $u(P) = 0$ and choose Cartesian coordinates such that $\nabla u(P) = (1, 0, \ldots, 0)$. We will show that $u(0, x') = 0$. Indeed, by (15.5),

$u(t,0) = t$. Using $|u(t,0) - u(0,x')| \leq \sqrt{t^2 + |x'|^2}$, we see that for all t

$$t - \sqrt{t^2 + |x'|^2} \leq u(0,x') \leq (t + \sqrt{t^2 + |x'|^2}),$$

and so, letting $t \to \pm\infty$ we conclude that $u(0,x') = 0$. Then necessarily, $\nabla u(0,x') = (1,0,\ldots,0)$, and so by (15.5) again, $u(x) = x_1$. □

An obvious example of a function with unit gradient, with just one singularity P, that is u is smooth outside P, is $u(x) = |x - P| + c$. These are the only ones; indeed, if u is smooth outside the origin with unit gradient and it is not globally smooth, the proof above shows that all orbits are half-lines collapsing at the origin. The level curves of u are then spheres, so $u(r) = \phi(r)$ for some ϕ. Imposing that $|\nabla u| = 1$ gives $|\phi'| = 1$, so $u = \pm r + c$.

In dimension $n = 3$, other examples are

$$u(x,y,z) = a\sqrt{x^2 + y^2} + bz, \quad a^2 + b^2 = 1.$$

15.1.5 Next, we consider the co-area formula for sub-manifolds. The question is whether for a k-dimensional sub-manifold it is possible to evaluate an integral $\int_M f\, dm_k$ in terms of lower dimensional integrals as we just did with domains. One might ask for Cavalieri's principle for sub-manifolds, in particular whether it is possible to compute k-dimensional measures by adding $k-1$-dimensional ones.

We start with a surface $S = \{g_1 = y_1\}$ in $\mathbb{R}^3$ fibered by a family of curves $\{g_1 = y_1, g_2 = y_2\}$, with $\nabla g_1, \nabla g_2$ linearly independent at every point. We look for a formula

$$\int_S h(p)\lambda(p)\, dA(p) = \int \left(\int_{g_1 = y_1, g_2 = y_2} h(p)\, ds(p) \right) dy_2.$$

Assuming this with λ to be determined, we use (15.2),

$$\begin{aligned} \int_U h(p)|\nabla g_1(p)| dm_3 &= \int \left(\int_{g_1 = y_1} h(p) dA(p) \right) dy_1 \\ &= \int \int \int_{g_1 = y_1, g_2 = y_2} \frac{h(p)}{\lambda(p)}\, ds\, dy_2\, dy_1, \end{aligned}$$

which again by (15.2), equals

$$\int_U \frac{h(p)}{\lambda(p)} J_2(g_1, g_2)\, dm_3.$$

Therefore,

$$\lambda = \frac{J_2(g_1, g_2)}{|\nabla g_1|} = \frac{\sqrt{|\nabla g_1|^2 |\nabla g_2|^2 - (\langle \nabla g_1, \nabla g_2 \rangle)^2}}{|\nabla g_1|}.$$

Theorem 15.2. *For a surface S fibered by curves $\Gamma_t = S \cap \{u = t\}$ (with ∇u not normal to S at any point), one has*

$$\int_S h(p) |\nabla_T u(p)| \, dA(p) = \int \left(\int_{\Gamma_t} h(p) \, ds(p) \right) dt, \qquad (15.6)$$

where $\nabla_T u$ denotes the tangential component of ∇u.

Proof. With $S = \{g_1 = c\}, u = g_2$, it is enough to note that

$$\lambda = |\nabla g_2 - \frac{\langle \nabla g_1, \nabla g_2 \rangle}{|\nabla g_1|^2} \nabla g_1|,$$

and that $\frac{\langle \nabla g_1, \nabla g_2 \rangle}{|\nabla g_1|^2} \nabla g_1$ is the normal component. □

Thus, the area of S is

$$A(S) = \int \left(\int_{\Gamma_t} \frac{1}{|\nabla_T u(p)|} \, ds(p) \right) dt. \qquad (15.7)$$

For $u = z$, and assuming $|\nabla g_1| = 1$ on S,

$$\int_S h(p) \sqrt{1 - (g_1)_z^2} \, dA(p) = \int \left(\int_{\Gamma_t} h \, ds \right) dt,$$

$$\int_S h(p) \, dA(p) = \int \left(\int_{\Gamma_t} \frac{h}{\sqrt{1 - (g_1)_z^2}} \, ds \right) dt.$$

Thus, for instance, the area 4π of the unit sphere is not the sum of the lengths of its sections with $z = t$,

$$\int_{-1}^{1} 2\pi \sqrt{1 - t^2} \, dt = \pi^2,$$

but

$$\int_{-1}^{1} 2\pi \sqrt{1 - t^2} \, dt = \pi^2 = \int_S \sqrt{1 - z^2} dA.$$

15.1.6 We analyze now the analogue of Theorem 15.2 for a plane compact smooth curve Γ. Let $u \in C^1$ such that $\nabla u(p)$ is not normal to Γ when $p \in \Gamma$, that is, each level curve $u = t$ meets Γ transversally. By the inverse function theorem, each $p \in \Gamma \cap \{u = t\}$ is isolated, whence $\Gamma \cap \{u = t\}$ is a finite set.

Theorem 15.3. *In the situation just described, one has*

$$\int_\Gamma h(p)|\nabla_T u(p)|\, ds(p) = \int \left(\sum_{p\in\Gamma\cap\{u=t\}} h(p) \right) dt, \tag{15.8}$$

where $\nabla_T u$ denotes the tangential component of ∇u.

Proof. First note that in case Γ is a graph $y = \varphi(x)$ and $u = x$, the tangential component of $\nabla u = (1, 0)$ is $\frac{1}{\sqrt{1+\varphi'(x)^2}}$ and the result follows. To prove it in general, we can work locally in a coordinate system u, v as in paragraph 10.1.2, where Γ is given by a graph $\gamma(u) = (u, \varphi(u))$, the metric is given by

$$\frac{1}{|\nabla u|^2}\left(du^2 + \frac{1}{\lambda^2} dv^2 \right),$$

and $\partial_u = \frac{\nabla u}{|\nabla u|^2}$. Then $ds = |\gamma'(u)|\, du$, with $\gamma'(u) = \partial_u + \varphi'(u)\partial_v$, $\nabla u = |\nabla u|^2 \partial_u$, the tangential component of ∇u is $\frac{1}{|\gamma'(u)|}$ so that $|\nabla_T u(p)|\, ds(p) = du$ and the left-hand side equals

$$\int h(u, \varphi(u))\, du.$$

□

In particular, one has for the length $l(\Gamma)$ of Γ,

$$l(\Gamma) = \int \left(\sum_{p\in\Gamma, u(p)=t} \frac{1}{|\nabla_T u(p)|} \right) dt, \tag{15.9}$$

which is the version of (15.7) for curves.

15.2 A Hint to Integral Geometry*

In this section, we give an elementary introduction to a nice classical area, *integral geometry*, nowadays revitalized by its applications to stereography.

15.2.1 Taking a linear u in (15.9), one gets a formula for the length of a plane curve Γ in terms of the intersection points with a system of parallel lines. We will see first that if we consider *all lines L*, then it is possible to compute the length just counting intersection points. This is the *Cauchy–Crofton* formula. To add in L we need a parametrization of the set of all lines and a measure on it. A line L is given by an angle $\theta, 0 \leq \theta \leq 2\pi$ and its distance $\tau > 0$ to the origin:

$$L : (x, y) = \tau(\cos\theta, \sin\theta) + \lambda(-\sin\theta, \cos\theta), \quad \lambda \in \mathbb{R}.$$

Alternatively, $L = L(\theta, \tau)$ is given by

$$u_\theta(x, y) = x\cos\theta + y\sin\theta = \tau.$$

Then $dL = d\theta\, d\tau$ is a measure for lines. It is the right measure to consider because of the following fact:

Exercise 15.1. Check that $d\mu$ is invariant by rigid motions, that is, if B is a set of lines, T is a rigid motion, and $T(B)$ denotes the set of lines $T(L), L \in B$, then $\mu(T(B)) = \mu(B)$.

Alternatively, if we let τ range in $\mathbb{R}$, we count every line twice.

Theorem 15.4. *For h defined on Γ,*

$$2\int_\Gamma h\, ds = \int_L \left(\sum_{p \in L \cap \Gamma} h(p) \right) dL.$$

In particular, if $N(\Gamma \cap L)$ denotes the number of points in $\Gamma \cap L$, one has for the length of Γ

$$\int_L N(\Gamma \cap L)\, dL = 2l(\Gamma).$$

Proof. For fixed θ, we apply (15.8) to u_θ to find

$$\int_\Gamma h(p)|\langle \nabla u_\theta, T\rangle(p)|\, ds(p) = \int \left(\sum_{p \in \Gamma \cap L(\theta,\tau)} h(p) \right) d\tau.$$

Now, it is enough to integrate in θ and observe that for a unit vector $T = (a, b)$

$$\int_0^{2\pi} |a\cos\theta + b\sin\theta|\, d\theta = \int_0^{2\pi} |\cos\theta|\, d\theta = 4. \qquad \square$$

In case Γ bounds a convex domain, then for almost all lines $N(\Gamma, L) = 0$ or 2, and we may state that the perimeter of a convex domain equals the measure of all lines meeting it.

Assume that U is a plane bounded domain with smooth boundary $\Gamma = bU$. In the same way, for fixed θ, by Fubini's theorem

$$\int_{-\infty}^{+\infty} \left(\int_{U \cap L(\theta,\tau)} h\, ds \right) d\tau = \int_U h dA,$$

whence

$$\int \left(\int_{U \cap L} h\, ds \right) dL = \pi \int_U h\, dA. \tag{15.10}$$

In particular,

$$\int l(U \cap L)\, dL = \pi A(U).$$

If instead of the length we set $m(U, L) = 1$ whenever L meets U and zero otherwise, then for fixed θ,

$$\int m(U, L)\, d\tau$$

is the width of U in the direction given by θ, so

$$\frac{1}{2\pi} \int m(U, L)\, dL$$

is the *mean width* of U.

15.2.2 Analogously, we can consider the family of all planes Π in $\mathbb{R}^3$ as doubly parametrized by $n \in S^2, \tau \in \mathbb{R}$,

$$\Pi(n, \tau) : \langle p, n \rangle = \tau.$$

Similarly as before, we consider $d\Pi = \frac{1}{2} d\sigma(n)\, d\tau$, where $d\sigma$ is area measure on S^2, as a measure for planes. Using spherical coordinates for n,

$$n = (\cos\phi \cos\theta, \cos\phi \sin\theta, \sin\phi),$$

one has $d\Pi = \frac{1}{2} \sin\phi\, d\phi\, d\theta\, d\tau$.

Exercise 15.2. Check that $d\Pi$ is invariant by rigid motions.

Theorem 15.5. *For h defined on a surface S,*

$$\pi^2 \int_S h\, dA = 2 \int \left(\int_{\Pi \cap S} h\, ds \right) d\Pi. \tag{15.11}$$

Proof. Formula (15.6) for $u(p) = \langle p, n \rangle$ gives for fixed n

$$\int_S h(p)|n_T(p)|dA(p) = \int_{-\infty}^{+\infty} \left(\int_{\Pi(n,\tau)\cap S} h\, ds \right) d\tau.$$

Here, n_T is the tangential component of n, the projection of n onto $T_p(S)$. Integrating in n and since using spherical coordinates,

$$\int_{n \in S^2} |n_T(p)|\, d\sigma(n) = \int_0^{\pi} \int_0^{2\pi} \sin^2 \phi\, d\phi\, d\theta = \pi^2,$$

the result follows. □

In particular, we get for the area of S,

$$\pi^2 A(S) = \int_{n\in S^2} \int_{-\infty}^{+\infty} l(\Pi(n,\tau) \cap S)\, d\tau\, d\sigma(n) = 2 \int l(\Pi \cap S)\, d\Pi. \tag{15.12}$$

For a bounded domain, by Fubini's theorem, for fixed n

$$\int_U h\, dm = \int_{-\infty}^{+\infty} \left(\int_{\Pi(n,\tau)\cap U} h\, dA \right) d\tau,$$

whence integrating in n

$$2\pi \int_U h\, dm = \int \left(\int_{\Pi \cap U} h\, dA \right) d\Pi. \tag{15.13}$$

In particular,

$$2\pi m(U) = \int A(\Pi \cap U)\, d\Pi.$$

As before, if $m(U, \Pi) = 1$ whenever Π meets U and zero otherwise,

$$M(U) = \frac{1}{2\pi} \int m(U, \Pi)\, d\Pi, \tag{15.14}$$

is the *mean width* of U. This quantity has also another interpretation in terms of curvature.

15.2.3 In this paragraph, we consider, instead of all lines in the plane and all planes in space, just those through the origin.

In the plane the lines L through the origin are parametrized by an angle $\theta, -\frac{\pi}{2} \leq \theta \leq +\frac{\pi}{2}$,

$$L_\theta = \left\{u(p) = \arctan \frac{y}{x} = \theta\right\}, \quad p = (x, y).$$

We use $dL = d\theta$. Since $|\nabla u| = \frac{1}{|p|}$, (15.4) gives

$$\int_{0\in L} \left(\int_L hds \right) dL = \int_{\mathbb{R}^2} \frac{h(p)}{|p|} \, dm(p).$$

Note that this follows too from integration in polar coordinates r, θ, because $dL = d\theta$ and $ds = dr$. In particular, for a plane domain U, if $l(L\cap U)$ denotes length,

$$\int_{0\in L} l(L \cap U) \, dL = \int_U \frac{1}{|p|} \, dm(p), \quad \int_{0\in L} \left(\int_{L\cap U} |p| \, ds \right) dL = m(U).$$

For a plane curve Γ, applying (15.8), and noting that

$$\nabla u(x, y) = \frac{1}{|p|^2}(-y, x) = \frac{Jp}{|p|^2}, \quad |\langle T, Jp\rangle| = |\langle p, N\rangle|,$$

where N is the normal to Γ, we get

$$\int_\Gamma h(p) \frac{|\langle p, N\rangle|}{|p|^2} \, ds(p) = \int_{0\in L} \left(\sum_{p\in\Gamma\cap L} h(p) \right) dL. \tag{15.15}$$

In space the planes through the origin are doubly parametrized by $n \in S^2$

$$\Pi(n) = n^\perp : \langle p, n\rangle = 0.$$

We use spherical coordinates for n,

$$n = (\sin\phi \cos\theta, \sin\phi \sin\theta, \cos\phi), \quad d\sigma(n) = \sin\phi \, d\phi \, d\theta.$$

We want to relate

$$\int_{S^2} \left(\int_{\Pi(n)} h \, dA \right) d\sigma(n) = \int_0^{2\pi} \int_0^\pi \int_{\Pi(\phi,\theta)} h \, dA \, \sin\phi \, d\phi \, d\theta,$$

with $\int hW dm$ for some weight W to be found. The equation of $\Pi(n)$ is

$$x \, \sin\phi \cos\theta + y \, \sin\phi \sin\theta + z \, \cos\phi = 0,$$

which we write in the form

$$(\tan\phi)\,(x\cos\theta + y\sin\theta) = -z.$$

For fixed θ, we consider

$$u_\theta(x,y,z) = \frac{z}{x\cos\theta + y\sin\theta},$$

so that the plane has equation $u_\theta = -\tan\phi$. Writing $I(\theta)$ for the integral in ϕ, we thus change variables $t = -\tan\phi$ to get

$$I(\theta) = \int_{-\infty}^{\infty}\left(\int_{u_\theta = t} h\,dA\right)\frac{|t|}{(1+t^2)^{\frac{3}{2}}}\,dt = \int_{-\infty}^{\infty}\left(\int_{u_\theta = t} h\,\frac{|u_\theta|}{(1+u_\theta^2)^{\frac{3}{2}}}\,dA\right)dt.$$

Now, we apply the co-area formula (15.4) to get

$$I(\theta) = \int_{\mathbb{R}^3} h\,\frac{|u_\theta|}{(1+u_\theta^2)^{\frac{3}{2}}}\,|\nabla u_\theta|\,dm(p).$$

Computation shows that

$$\frac{|u_\theta|}{(1+u_\theta^2)^{\frac{3}{2}}}\,|\nabla u_\theta| = \frac{|z|}{|z|^2 + |x\cos\theta + y\sin\theta|^2}.$$

Now, we must integrate $I(\theta)$. We claim that, for $p = (x,y,z)$,

$$J = \int_0^{2\pi}\frac{|z|}{|z|^2 + |x\cos\theta + y\sin\theta|^2}\,d\theta = \frac{2\pi}{|p|}.$$

Clearly, we may assume by homogeneity that $|p| = 1$, and also $y = 0$. Then

$$J = 2\int_{-\frac{\pi}{2}}^{\frac{\pi}{2}}\frac{|z|}{|z|^2\sin^2\theta + \cos^2\theta}\,d\theta.$$

With the change of variable $t = \tan\theta$,

$$J = 2\int_{-\infty}^{+\infty}\frac{|z|}{|z|^2 + t^2}\,dt = 2\pi,$$

establishing the claim. Altogether, we find

$$\int_{0\in\Pi}\left(\int_{\Pi} h\,dA\right)d\Pi = \pi\int_{\mathbb{R}^3}\frac{h(p)}{|p|}\,dm(p),$$

similarly as in the plane. In particular, for a domain U in space

$$\int_{0\in\Pi} A(U\cap\Pi)\,d\Pi = \pi\int_U \frac{1}{|p|}\,dm(p), \pi m(U) = \int_{0\in\Pi}\left(\int_{\Pi\cap U} |p|\,dA(p)\right)d\Pi.$$

Next, we analyze the area of a surface S in terms of lengths of $S\cap\Pi, 0\in\Pi$. Repeating the computation above and using (15.6) instead of (15.4), we find that

$$\int_{0\in\Pi}\left(\int_{S\cap\Pi} h\,ds\right)d\Pi = \int_S h(p)\,W_S(p)\,dA(p),$$

with

$$W_S(p) = \frac{1}{2}\int_0^{2\pi}\frac{|u_\theta|}{(1+u_\theta^2)^{\frac{3}{2}}}\,|\nabla_T u_\theta|\,d\theta,$$

$\nabla_T u_\theta$ denoting the tangential gradient of u_θ on S. If S is a sphere centered at the origin, say the unit sphere S^2, then ∇u_θ is already tangential and the computation is the same as before, leading to

$$\int_{0\in\Pi}\left(\int_{S^2\cap\Pi} h\,ds\right)d\Pi = \pi\int_{S^2} h\,d\sigma.$$

This may be written as

$$2\pi\int_{S^2} h\,d\sigma = \int_{n\in S^2}\left(\int_{S^2\cap n^\perp} h\,ds\right)d\sigma(n), \tag{15.16}$$

which we call *integration through meridians*.

Exercise 15.3. Try to compute W_S for a general surface. You will find an expression in terms of hypergeometric functions and the normal component of the vector position, as in (15.15).

Finally, we consider the family of lines L in space through the origin, doubly parametrized by $n\in S^2$. Using spherical coordinates ρ,ϕ,θ we see that

$$\int_{0\in L}\left(\int_L h\,ds\right)dL = \int_{\mathbb{R}^3}\frac{h(p)}{|p|^2}\,dm(p).$$

The following is the analogue of (15.15) for surfaces in space

$$\int_S h(p)\frac{|\langle p,N\rangle|}{|p|^3}\,dA(p) = \int_{0\in L}\left(\sum_{p\in S\cap L} h(p)\right)dL, \tag{15.17}$$

where N denotes the unit normal to S. To prove it, as usual we may work locally and assume that $S=\{u=t\}$ for $u\in C^1, \nabla u\neq 0$. Then $N=\frac{\nabla u}{|\nabla u|}$

and we use Corollary 15.1 to write the left-hand side as

$$\frac{d}{dt}\int_{u\le t} h(p)\frac{|\langle p, \nabla u\rangle|}{|p|^3}\,dm(p).$$

The last integral equals

$$\int_{0\in L}\left(\int_{L\cap\{u\le t\}} h(p)\frac{|\langle p, \nabla u\rangle|}{|p|}\,ds(p)\right)dL.$$

In $L, p = \rho n$; if $\varphi(\rho) = u(\rho n)$, then

$$\frac{|\langle p, \nabla u\rangle|}{|p|} = |\varphi'(\rho)|,$$

so again by Corollary 15.1,

$$\frac{d}{dt}\int_{L\cap\{u\le t\}} h(p)\frac{|\langle p, \nabla u\rangle|}{|p|}\,ds = \sum_{p\in S\cap L} h(p),$$

and the result follows.

15.2.4 Now, we will combine the results of the previous paragraphs to express $A(S), m(U)$, in $\mathbb{R}^3$ in terms of intersection with lines.

We need a parametrization of the set of lines and a measure to count them. The set of lines L in $\mathbb{R}^3$ is doubly parametrized by a unit direction vector $v(L)$ and the point $p(L) \in L$ closest to the origin, so that

$$L : q = p(L) + \lambda v(L), \quad \langle p(L), v(L)\rangle = 0.$$

We consider the measure given by

$$\mu(B) = \frac{1}{2}\int_S A(\{p_L : L \in B, v(L) = n\})\,d\sigma(n),$$

where B is a set of lines. With a slight abuse of notation, we write $dL = dA_v\,d\sigma(v)$, where dA_v means area measure on $v^\perp$. We may express this parametrization using spherical coordinates for v

$$v = (\cos\phi\cos\theta, \cos\phi\sin\theta, \sin\phi),$$

and

$$v_1 = v_\phi = (-\sin\phi\cos\theta, -\sin\phi\sin\theta, \cos\phi),$$

$$v_2 = \frac{1}{\cos\phi}v_\theta = (-\sin\theta, \cos\theta, 0), \tag{15.18}$$

as orthogonal basis of $v^\perp$, so together with

$$p(L) = \rho v_1(\phi, \theta) + \tau v_2(\phi, \theta),$$

we have a double parametrization in the set of all lines, except a zero measure set. In this parametrization,

$$dL = \frac{1}{2}\, d\rho\, d\tau\, d\sigma(v) = \frac{1}{2} \cos\phi\, d\rho\, d\tau\, d\theta d\phi.$$

Exercise 15.4. Prove that dL is invariant by rigid motions.

Theorem 15.6. *For h defined on a surface S,*

$$\pi \int_S h\, dA = \int \left(\sum_{p \in S \cap L} h(p) \right) dL.$$

In particular, if $N(S \cap L)$ denotes the number of points in $L \cap S$,

$$\pi A(S) = \int N(S \cap L)\, dL.$$

Proof. We consider formula (15.11) and use Theorem 15.4

$$\pi^2 \int_S h\, dA = 2 \int \left(\int_{\Pi \cap S} h\, ds \right) d\Pi = \int \left(\int_{L \subset \Pi} \left(\sum_{p \in L \cap S} h(p) \right) d\mu_\Pi(L) \right) d\Pi.$$

Now, one would like to apply Fubini's theorem and interchange the order of integration. Strictly speaking, we cannot do that because the inner integral is with respect to a measure depending on Π. However, if we proceed, we find

$$\pi^2 \int_S h\, dA = I = \int \left(\int_{L \subset \Pi} d\Pi \right) \left(\sum_{p \in L \cap S} h(p) \right) dL.$$

Intuitively, the inner integral, the measure of the set of all planes containing a fixed line L, is π and we would get the desired result.

The reasoning is not correct, but the result is. To formalize the argument, we make explicit $d\mu_\pi(L)$ as follows. We doubly parametrize the set of planes as in paragraph 15.2.2, $\Pi = \Pi(n, \tau)$; if n_1, n_2 is an orthonormal basis of $n^\perp$ (as in (15.18) replacing v by n), the lines $L \subset \Pi$ are doubly parametrized by $\rho, 0 \le \psi \le 2\pi$,

$$L : q = \tau n + \rho(\cos\psi\, n_1 + \sin\psi\, n_2) + \lambda(-\sin\psi n_1 + \cos\psi n_2), \quad \lambda \in \mathbb{R}.$$

The direction vector of L is $v = -\sin\psi\, n_1 + \cos\psi\, n_2$ and in this parametrization, $d\mu_\Pi(L) = \frac{1}{2}d\rho\, d\psi$. Then

$$I = \int \left(\int_{L\subset\Pi} N(U,L)\, d\mu_\Pi(L)\right) d\Pi = \frac{1}{4}\int\int\int\int N(\rho,\tau,v) d\rho\, d\psi\, d\tau\, d\sigma(n).$$

For fixed ρ, τ, we look at N as a function of $v \in S^2$. The integral in ψ is along $n^\perp$, so by (15.16)

$$\int\int N d\psi d\sigma(n) = 2\pi \int N(U,\rho,\tau,v)\, d\sigma(v).$$

Then

$$I = \frac{\pi}{2}\int\int N(U,\rho,\tau,v)\, d\rho\, d\tau\, d\sigma(v) = \pi\int N(U\cap L)\, dL,$$

because ρ, τ parametrize $v^\perp$. □

Theorem 15.7. *For a bounded domain U in space and h on U,*

$$2\pi\int_U h\, dm = \int\left(\int_{L\cap U} h\, ds\right) dL.$$

In particular,

$$2\pi m(U) = \int l(L\cap U)\, dL.$$

Proof. If we combine (15.13) and (15.10),

$$2\pi^2\int_U h\, dm = \pi\int\left(\int_{U\cap\Pi} h\, dA\right) d\Pi$$

$$= \int\left(\int_{L\subset\Pi}\left(\int_{L\cap U} h\, ds\right), d\mu_\Pi(L)\right) d\Pi = \pi\int\left(\int_{L\cap U} h\, ds\right) dL.$$ □

15.3 A Hint to Minimal Surfaces*

The purpose of this section is to give a hint to another interesting area of mathematics, *calculus of variations*, by looking at minimal surfaces. Calculus of variations deals with *functionals*, functions defined on spaces of functions, and will serve us to introduce some of the concepts in this course in infinite dimension.

Let Γ be a closed curve in space given as the graph of a function $\phi(x, y)$ on the boundary of a plane domain

$$\Gamma = \{(x, y, \phi(x, y)) : (x, y) \in bD\}.$$

If u is defined on $\overline{D}$ and $u = \phi$ on bD, the graph S of u is a surface supported on Γ. Its area is given by (14.2),

$$A(u) = \int_D \sqrt{1 + u_x^2 + u_y^2}\, dm.$$

It is quite natural to consider

$$\inf_u A(u),$$

and ask whether this value is attained at some (unique or not) u. In the plane, the analogous problem has the trivial solution u linear, the segment being the shortest path between two points.

If so, and $\Psi(x, y)$ is compactly supported in D, then $A(u + t\Psi) \geq A(u)$ for small t, and so

$$\frac{d}{dt}_{t=0} A(u + t\Psi) = 0.$$

Note that the left-hand side can be seen as the directional derivative of A at u in the direction Ψ. It equals

$$\int_D \frac{u_x \Psi_x + u_y \Psi_y}{\sqrt{1 + u_x^2 + u_y^2}}\, dm,$$

and so, by (13.2), assuming u of class C^2,

$$\int_D \Psi \left[\left(\frac{u_x}{\sqrt{1 + u_x^2 + u_y^2}} \right)_x + \left(\frac{u_y}{\sqrt{1 + u_x^2 + u_y^2}} \right)_y \right] dm = 0.$$

Since this holds for all Ψ, the other function must be zero; computation shows then, that

$$(1 + u_y^2) u_{xx} + (1 + u_x^2) u_{yy} - 2 u_x u_y u_{xy} = 0.$$

This is the *Euler–Lagrange* equation for the critical points of the functional A. A surface S in space is called *minimal* if it is a critical point of the area functional under compactly supported deformations. When represented locally by a graph $z = u(x, y)$ in Cartesian coordinates, u must satisfy the above equation.

The easiest examples are linear functions and planes S, catenoids and helicoids. It is not easy to find explicit equations of minimal surfaces, but it is easy to visualize many examples. Indeed, soap films attached to rings are minimal surfaces.

The above is a sample of a broad and important area of Mathematics, the *calculus of variations*, very much related through Euler's equations with partial differential equations.

Chapter 16

Line Integrals and Flux

This is the starting chapter of Vector Analysis in this text. Vector Analysis is often quoted as the differential calculus of vector fields. In our presentation, we try to justify this assertion by stressing the analogy with the differential calculus of functions. Functions act on points, and the differential or gradient is a linear approximation, or density, when comparing this action on boundary points of small intervals. In space, vector fields have two possible actions, both motivated by physical considerations. First, they act on *oriented curves* by *circulation*, and, secondly, they act on *oriented surfaces* by *flux*. In this chapter we describe these concepts.

16.1 Regular Sub-Manifolds with Border

In this section, we define new global objects, adding a *border* to the k-dimensional sub-manifolds of R^n defined in Theorem 7.1. We start with the case $k = 1$. We are mainly interested in $n = 2, 3$, but keep the definitions for general n.

16.1.1 A regular curve Γ in a domain $U \subset \mathbb{R}^n$ (a regular sub-manifold of dimension one), if connected, is either homeomorphic to an open interval (a, b) and then there exists a global parametrization $\Phi : (a, b) \to \Gamma$ or else is diffeomorphic to a circle and there is a global parametrization $\Phi : [a, b] \to \Gamma$, $\Phi'(t) \neq 0$, with $\Phi(a) = \Phi(b)$, $\Phi'(a) = \Phi'(b)$. In the first case, if the limits

$$\lim_{t\to a} \Phi(t) = p, \quad \lim_{t\to b} \Phi(t) = q,$$

exist and are different, we add p, q to Γ, keep using $\Phi : [a, b] \to \Gamma$. If moreover,

$$T_p = \lim_{t \to a} \frac{\Phi'(t)}{|\Phi'(t)|}, \quad T_q = \lim_{t \to b} \frac{\Phi'(t)}{|\Phi'(t)|}$$

exist, we say that Γ is a *regular curve with border* $\partial\Gamma = \{p, q\}$. We also say that Γ joints p, q.

If $p = q$ and $T_p = T_q$, we are in the second case and say that Γ is a *regular curve with no border*. If $p = q, T_p \neq T_q$, we say that Γ has no border and one *corner* at $p = q$.

The term *boundary* is often used instead for $\partial\Gamma$; to avoid confusion with the topological boundary, we keep using the term *border*.

If Φ is a homeomorphism but Φ' has a finite number of jump discontinuities at $t = t_i$, that is the tangent changes direction at vertices $\Phi(t_i)$, we use the term *piece-wise regular curve*, with *corners* at points $\Phi(t_i)$, that can be closed or else have a border. There can exist as well cusps at the discontinuities of the tangent, that is, the tangent T changes to $-T$. The curves $\Gamma_i = \Phi([t_i, t_{i+1}])$ are called the *regular pieces* of Γ. See Figure 16.1.

16.1.2 Now, we consider $k = n$, remember that the n-dimensional sub-manifolds are the open sets.

Definition 16.1. We say that a bounded domain U in $\mathbb{R}^n$ has *regular boundary* if for all $p \in bU$ there is a C^1- function u in some ball $B(p, r)$, with $\nabla u \neq 0$, such that $bU \cap B(p, r)$ is given by $u = 0$ and $U \cap B(p, r)$ by $u < 0$.

Thus, bU is a regular $(n - 1)$-dimensional regular sub-manifold, the converse being true if U is locally connected around each point in bU. Using the implicit function theorem, an equivalent definition is that locally around

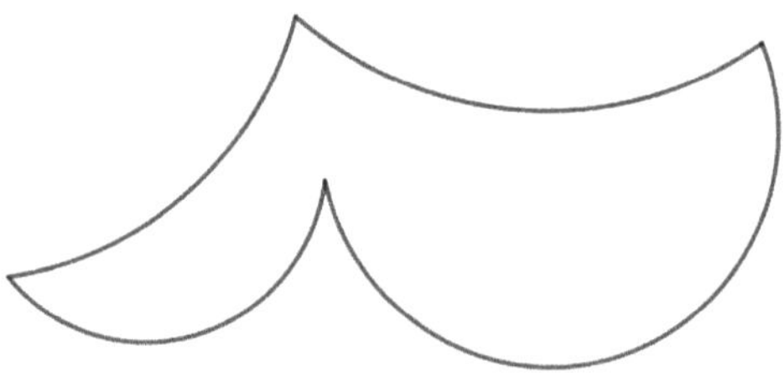

Figure 16.1. A piece-wise regular curve.

each $p \in bU$ and after permutation of the variables, U, bU, are respectively, the sub-graph $x_n > \phi(x')$ and graph $x_n = \phi(x')$ of a function $x_n = \phi(x')$.

Theorem 16.1. *If U is a bounded domain with regular boundary, there is $u \in C^1(\mathbb{R}^n)$ such that $U = \{u < 0\}, bU = \{u = 0\}$ and $\nabla u \neq 0$ on bU.*

Proof. The compact bU is covered by a finite number of balls $B_1, \dots, B_N$, each B_i equipped with a C^1 function u_i as above. Consider the compact $K = \overline{U} \setminus (\cup_i B_i)$ and using theorem 5.4, let u_0 be a C^1 function with compact support in U equal to -1 on K. With $B_0 = \{u < -\frac{1}{2}\}$, the collection $B_i, i = 0, \dots, N$, is an open covering of $\overline{U}$. Using Theorem 5.5, let $\phi_i, i = 0, \dots, N$ be a partition of unity subordinated to this covering. Then $v = \sum_{i=0}^N \phi_i u_i < 0$ in $U, v = 0$ on $bU, v(x) \geq 0, x \notin U$, and $v(x) > 0$ if $x \notin U$ is close enough to bU, say for $d(x, bU) < \alpha$. Moreover, for $x \in bU$,

$$\nabla v = \sum_{i=0}^{N} (\nabla \phi_i) u_i(x) + \sum_{i=0}^{N} \phi_i(x) \nabla u_i(x) = \sum_{i=1}^{N} \phi_i(x) \nabla u_i(x).$$

If $N(x)$ is the outward normal to bU at x, one has $\nabla u_i(x) = \lambda_i N(x)$ for a positive λ_i, because $u_i < 0$ inside U and $u_i > 0$ outside U. Thus,

$$\nabla v(x) = N(x) \sum_{i=1}^{N} \phi_i(x) \lambda_i \neq 0.$$

Adding to v a C^1 function equal to 0 for $d(x, U) < \frac{\alpha}{3}$ and equal to 1 for $d(x, U) > \frac{2\alpha}{3}$ gives the desired u. □

The prototype of a smooth domain is a disk or corona in the plane or a ball in space.

16.1.3 Now, we consider $0 < k < n$.

Definition 16.2. We say that a connected compact set M in $\mathbb{R}^n$ is a *regular k-dimensional sub-manifold with regular border ∂M* if ∂M is compact, the points $p \in M \setminus \partial M$ satisfy the conditions in Theorem 7.1 for k and the points $p \in \partial M$ satisfy the following variant: there is a coordinate system $u_1, \dots, u_n$ in a neighborhood V of $p, u_i(p) = 0$, such that

$$\begin{aligned} M \cap V &= \{u_k \geq 0, u_{k+1} = \cdots = u_n = 0\}, \\ \partial M \cap V &= \{u_k = \cdots = u_n = 0\}, \qquad (16.1) \\ (M \setminus \partial M) \cap V &= \{p \in V : u_k > 0, u_{k+1} = \cdots = u_n = 0\}. \end{aligned}$$

Note that ∂M is then a sub-manifold of dimension $k - 1$.

The intuitive meaning, for $n = 3, k = 2$, is that locally, in suitable coordinates, M is a half-plane and ∂M a line.

Analogously as in Theorem 7.1, it can be seen that the condition on points $p \in \partial M$ is equivalent to the existence of a special local chart: there is an open ball B centered at p, an open set $U \subset \mathbb{R}^k$, and $\Phi : U \to \mathbb{R}^n$ of class $C^1, \Phi(0) = p$, such that $d\Phi(u)$ has rank $k, u = (u_1, \ldots, u_k) \in U$ and Φ is a homeomorphism from $U \cap \{u_k < 0\}$ onto $(M \setminus \partial M) \cap B$ and from $U \cap \{u_k = 0\}$ onto $\partial M \cap B$.

The tangent space $T_p(M)$ makes sense at all points $p \in M$, even those of ∂M, namely the image of $d\Phi'(0)$.

In particular, if M is globally parametrized by a domain $U \subset \mathbb{R}^k$ with regular boundary, that is, $\Phi : \overline{U} \to M$ is a homeomorphism with $d\Phi$ of rank k at all points, M is a regular sub-manifold with regular border $\Phi(bU)$. Also, as a consequence of the definition, if Φ is a diffeomorphism defined in the neighborhood of a sub-manifold M with regular border, so is $\Phi(M)$. Other examples are obtained applying the following proposition:

Proposition 16.1. *Let N be a regular k-dimensional sub-manifold, u a real C^1 function on N and consider the sub-level set*

$$M = \{x \in N : u(x) \leq r\}.$$

Then if the tangential gradient $\nabla_T u(x)$ is not zero when $u(x) = r$, M is a k-dimensional sub-manifold with border

$$\partial M = \{x \in N : u(x) = r\}.$$

Proof. If N is defined by $u_{k+1} = \cdots = u_n = 0$ around p, M is defined by $u_{k+1} = \cdots = u_n = 0, u \leq r$. Moreover, $\nabla_T u(p) \neq 0$ means that $\nabla u(p), \nabla u_i(p)$ are not proportional, whence by Proposition 6.2 they can be enlarged to a local coordinate system and we are done. □

A specific example is, for $q \in N$, taking $u(x) = |x - q|$,

$$M = \{x \in N : |q - x| \leq r\},$$

for r small enough. Indeed, the gradient of $u(x) = |x - q|$ is proportional to $x - q$, which obviously has a non-zero tangential component if r is small enough.

In Figure 16.2, the lateral surface of the cylinder is a regular surface with regular border consisting of two circles.

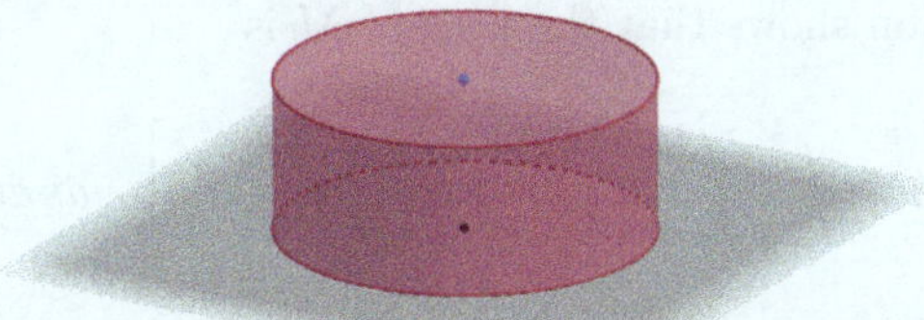

Figure 16.2. The lateral surface has two circles as border.

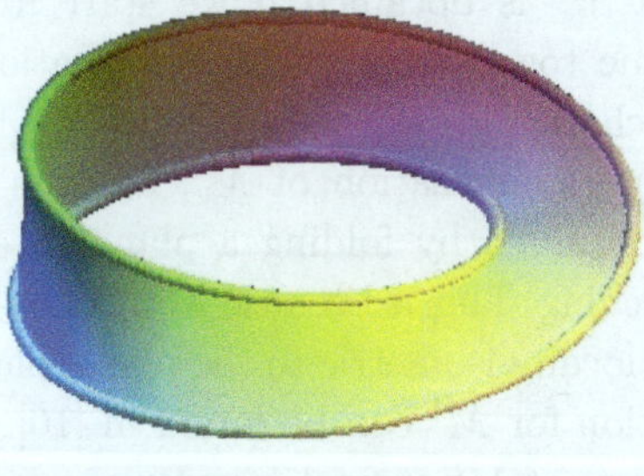

Figure 16.3. A Möbius band.

16.1.4 We may consider the volume element dm_k on a k-dimensional submanifold M with border, defined in the same way in terms of local charts, and dm_{k-1} on ∂M.

Example 16.1. The *Möbius band* M, Figure 16.3, is the surface with border in $\mathbb{R}^3$ described by the parametrization

$$\Phi(s,t) = \left(\left(1 + s\cos\frac{t}{2}\right)\cos t, \left(1 + s\cos\frac{t}{2}\right)\sin t, s\sin\frac{t}{2}\right), \tag{16.2}$$

with $-\frac{1}{2} \leq s \leq \frac{1}{2}, t \in \mathbb{R}$. Φ is 4π-periodic in t and $\Phi(s,t) = \Phi(-s, t+2\pi)$. Fixed t, s parametrizes a segment of length one in the direction

$$v(t) = \Phi_s = \left(\cos\frac{t}{2}\cos t, \cos\frac{t}{2}\sin t, \sin\frac{t}{2}\right).$$

The mid-point $P(t) = (\cos t, \sin t, 0)$ of this segment, corresponding to $s = 0$, describes the unit circle in the xy-plane as $0 \leq t \leq 2\pi$. The vector $v(t)$ rotates in the plane perpendicular to the unit circle at $P(t)$ and changes direction at $t = 0, t = 2\pi$. It is easily seen that Φ_t, Φ_s are orthogonal, whence M is indeed a bordered surface. The boundary ∂M is the curve corresponding to $s = \pm\frac{1}{2}$, parametrized by $\Phi(\frac{1}{2}, t), 0 \leq t \leq 4\pi$.

A computation shows that the area of M is

$$A = \int_{-\frac{1}{2}}^{\frac{1}{2}} \int_0^{2\pi} \left[\left(1 + s\cos\frac{t}{2}\right)^2 + \frac{s^2}{4} \right]^{\frac{1}{2}} ds\, dt,$$

which is greater than

$$\int_{-\frac{1}{2}}^{\frac{1}{2}} \int_0^{2\pi} \left(1 + s\cos\frac{t}{2}\right) ds\, dt = 2\pi.$$

A *different* band M' is obtained if we start from a sheet of paper of sizes $2\pi, 1$ and glue together, without deformations, the two opposite sides of length one changing orientation. Since M has greater area, the above cannot be a parametrization of M', as it is a different band. In fact, M cannot be obtained by folding a plane sheet of area A without deformations or stretching. This is because M has non-zero curvature, while bands like M', being locally isometric to the plane, have zero curvature. An explicit parametrization for M' can be found in [16]. However, the known parametrizations $\Phi(s,t)$ of M' are not isometric.

16.1.5 In the following sections, we will be integrating functions on a k-dimensional sub-manifold M with border ∂M in $\mathbb{R}^n$, with dm_k, and relating these integrals with other integrals on ∂M, with dm_{k-1}.

Strictly speaking, a parallelepiped or tetrahedron has no regular boundary, as it has edges and vertices on its boundary, but still we will be interested in integrating on its boundary. For $k = 1$, we already considered piece-wise regular curves. In the same way, we can allow M or ∂M, or both, to be piece-wise regular. In this paragraph we deal with these generalizations. For simplicity, we consider just $n = 2, 3$.

- First we define plane domains U with piece-wise regular boundary, slightly modifying Definition 16.1. We require that for $p \in bU$ there are local coordinates u_1, u_2 around $p, u_i(p) = 0$ such that either
  - (a) Locally bU is given by $u_2 = 0$ and U by $u_2 > 0$, and then p is a regular point of bU.
  - (b) In the coordinates u_1, u_2, U is a sector of opening $\alpha \neq \pi$. Then p is a *corner*.
  - (c) In the coordinates u_1, u_2, U is described around $(0,0)$ by $0 < u_1, 0 < u_2 < \phi(u_1)$ with ϕ of class C^1, $\phi(0) = 0, \phi'(0) = 0$ and ϕ increasing. Then p is a *cusp*.

In Figure 16.1, the region bounded by the curve is an example. Thus, bU is a piece-wise regular curve with a finite set C of corners and cusps, with an important additional approximation property: for $\varepsilon > 0$, there exists a regular domain U_ε with regular boundary, obtained smoothing the points in C, such that

$$U_\varepsilon \subset U, \quad C \subset U \setminus U_\varepsilon,$$

and such that both the area of $U \setminus U_\varepsilon$ and the length of $bU \setminus U_\varepsilon$ are less than ε. Thus, for continuous f,

$$\int_{U_\varepsilon} f\,dA \to \int_U f\,dA, \quad \int_{bU_\varepsilon} f\,ds \to \int_{bU} f\,ds.$$

- We define now a *regular surface S in $\mathbb{R}^3$ with piece-wise regular border ∂S* in a similar way. We replace the condition on points $p \in \partial S$ in Definition 16.2 by the following: there are local coordinates $u_1, u_2, u_3, u_i(p) = 0$ such that $U \subset \{u_3 = 0\}$ is described in the u_1, u_2 coordinates in one of the three ways described above. Then ∂S is a piece-wise regular curve in $\mathbb{R}^3$. Again, smoothing the corners and cusps, there are surfaces S_ε with regular border such that

$$\int_{S_\varepsilon} f\,dA \to \int_S f\,dA, \quad \int_{bS_\varepsilon} f\,ds \to \int_{\partial S} f\,ds.$$

The tangent plane $T_p(S)$ is defined at all points $p \in S$, including corners and cusps. In Figure 16.4, the part of the sphere in $x, y, z > 0$ is a surface

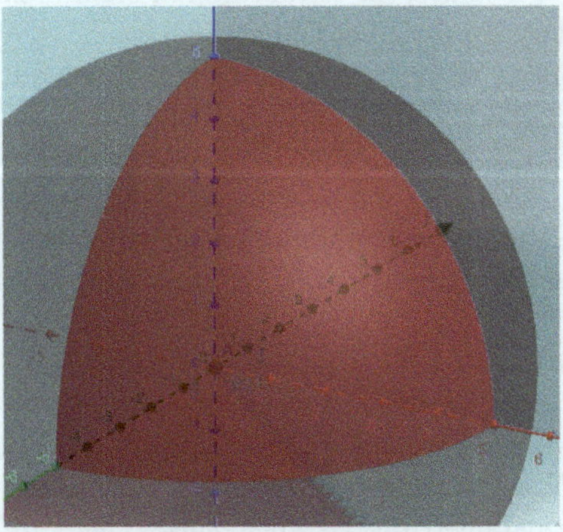

Figure 16.4. A surface with piece-wise regular border.

with piece-wise regular border consisting of three main arcs meeting orthogonally.

- The next concept is that of *piece-wise regular surface with regular or piece-wise regular border*. This is a connected compact set S, which is the union $S = \cup_j S_j$ of a finite number of regular surfaces with piece-wise regular border, and such that $S_i \cap S_j \subset \partial S_i$. The tangent planes $T_p(S_i), T_p(S_j)$ might be equal or not at points on the *edge* $S_i \cap S_j$, and a cusp might exist along this edge. The border ∂S is in this case the closure of

$$\cup_i \partial S_i \setminus (\cup_{i \neq j} \partial S_i \cap \partial S_i).$$

A *vertex* is a point where three or more S_i meet. Obviously, dA is defined on S and ds on ∂S. Figure 16.7 shows an example. If we add the top cover to the cylinder in Figure 16.2, we have a piece-wise regular surface with regular border consisting in one circle.

- Finally, we generalize the concept of bounded domain with regular boundary in $\mathbb{R}^3$. We say that a bounded domain $U \subset \mathbb{R}^3$ is *admissible* if bU is a piece-wise regular surface with no border and the following approximation property holds, by smoothing edges and cusps: for each $\varepsilon > 0$, there is a domain U_ε with regular boundary such that

$$\int_{U_\varepsilon} f\, dm \to \int_U f\, dm, \quad \int_{bU_\varepsilon} f\, dA \to \int_{bU} f\, dA.$$

The prototype of such a domain is a parallelepiped or a general polyhedron. For instance, the tetrahedron in Figure 16.5 has a piece-wise regular boundary consisting of four triangles.

A basic property to be stressed is the fact that in all cases $\partial\partial M = \emptyset$.

16.2 Orientations

We discuss the subject of orientation of k-dimensional sub-manifolds with border just in case $k = 1$ and $k = n - 1$. This will suffice for $n = 2, 3$.

16.2.1 We start with $k = 1$, curves Γ. A regular or piece-wise regular curve Γ, closed or with border, has a global parametrization $\Phi : [a, b] \to \Gamma$. Changing the parametrization means composing with a homeomorphism τ from $[a, b]$ to $[a, b]$, which is either increasing, or decreasing. If τ is increasing, we say that the parametrizations are *related*. There are two equivalence classes, each called an *orientation* of Γ. Choosing one means orienting Γ.

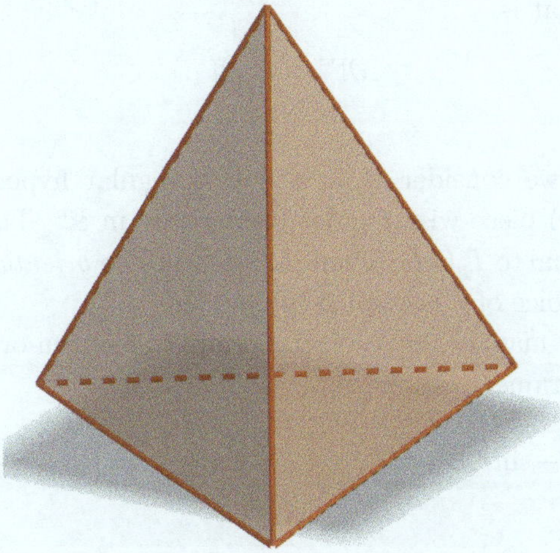

Figure 16.5. The tetrahedron, an admissible domain.

Intuitively it means choosing one of the two possible ways of traveling along Γ. If Γ is regular, there are two unit tangents at every point, $\pm\frac{\Phi'(t)}{|\Phi'(t)|}$, and orienting means choosing one sign or the other. Thus, a curve is always orientable.

From now on we consider that curves are oriented and keep the notation Γ. The same curve equipped with the opposite orientation is denoted $-\Gamma$. Finite linear combinations

$$\sum_i c_i\Gamma_i, \quad c_i \in \mathbb{R},$$

are called 1-*chains*. Chains are added in the obvious way

$$\sum_i c_i\Gamma_i + \sum_i d_i\Gamma_i = \sum_i (c_i + d_i)\Gamma_i.$$

We declare $0 \cdot \Gamma = 0$. Analogously, we consider *oriented points* $\pm p$, that is a point p equipped with one sign, $+$ or $-$, and define 0-*chains* analogously.

If Γ is an oriented curve with border $\partial\Gamma = \{p, q\}$ and the orientation is from p to q, we define the *induced orientation* on $\partial\Gamma$ by attributing to p, q the signs $-, +$, respectively. That is, we define

$$\partial\Gamma = q - p.$$

Note that if Γ is piece-wise regular with regular pieces Γ_i, and p is a corner, $p \in \partial\Gamma_i \cap \partial\Gamma_j$, an orientation of Γ attributes to p opposite induced orientations. That is,

$$\partial\Gamma = \sum_i \partial\Gamma_i.$$

16.2.2 Next, we consider $k = n - 1$, a regular hypersurface M with (possibly empty) piece-wise regular border ∂M in $\mathbb{R}^n$. There are two unit vectors orthogonal to $T_p(M)$. Then, M is said to be *orientable* if there exists a *continuous* choice of a normal $N(p), p \in M$.

The Möbius map is the typical example of a non-orientable surface. Consider the parametrization (16.2).

$$\Phi_t(0,t) = (-\sin t, \cos t, 0), \quad \Phi_s = \left(\cos\frac{t}{2}\cos t, \cos\frac{t}{2}\sin t, \sin\frac{t}{2}\right).$$

Then

$$\Phi_s \times \Phi_t = \left(\cos t \sin\frac{t}{2}, \sin t \sin\frac{t}{2}, -\cos\frac{t}{2}\right).$$

Suppose that $N(p)$ is a continuous unit normal field on M and that $\Phi(0,0,0) = (0,0,1)$. Then necessarily

$$N(\Phi(0,t)) = \left(-\cos t \sin\frac{t}{2}, -\sin t \sin\frac{t}{2}, \cos\frac{t}{2}\right), \quad t > 0.$$

Since this has limit $(0,0,-1)$ as t increases to 2π, we reach a contradiction.

We point out that orientability of M does not imply the existence of $n-1$ vector fields $X_1, \ldots, X_{n-1}$ tangent to M such that the orientation at each p is $[X_1(p), \ldots, X_{n-1}(p)]$; for instance, the unit sphere, by Poincaré's theorem quoted in paragraph 8.1.1, does not have non-vanishing tangent vector fields.

If U is a bounded domain with regular boundary in $\mathbb{R}^n$, $M = bU$ is orientable. Indeed, we define $N(p)$ as the unit normal pointing outward. Thus, if U is defined by $\{u < 0\}$ as in Proposition 16.1, then

$$N = \frac{1}{|\nabla u|}\nabla u.$$

16.2.3 In this paragraph, we consider just surfaces S in $\mathbb{R}^3$. If S is a bordered surface oriented with N, we define an *induced orientation* on the border curve ∂S as follows. For $p \in \partial S$ as in (16.1), we choose the sign in the tangent $T = \pm\partial_1$ to bS that satisfies

$$\det(N(p), T, \partial_2) > 0.$$

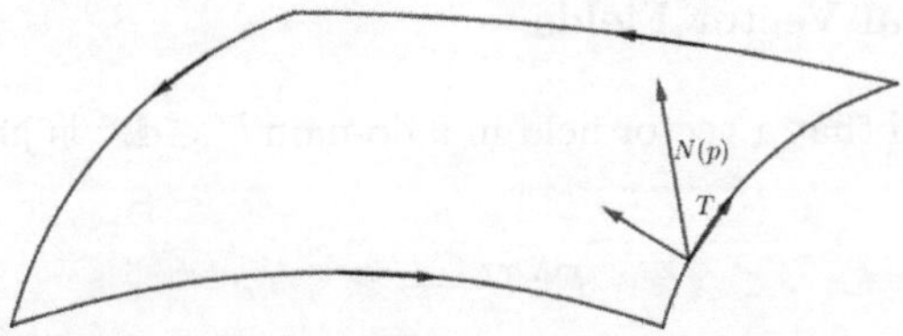

Figure 16.6. The right-hand side rule.

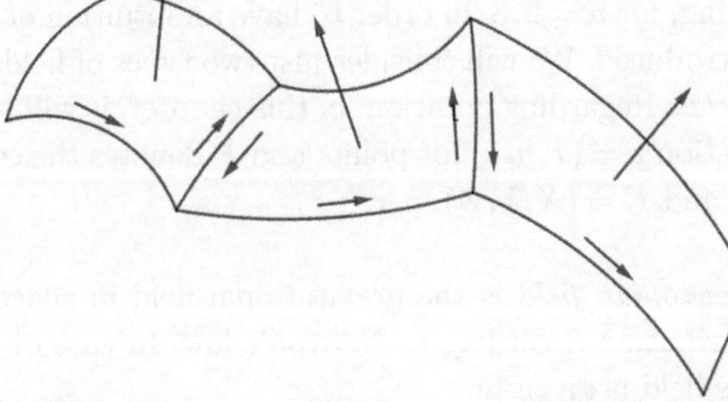

Figure 16.7. Orientation in a piece-wise regular surface.

Equivalently, N has the direction of $T \times \partial_2$. The intuitive meaning is given by the *right-hand side rule*: if the index points in direction T and middle finger in direction to S, the thumb points in the direction of $N(p)$. Alternatively, if the fingers in the right-hand side curl along the orientation of ∂S, the thumb points in the direction of N. See Figure 16.6.

16.2.4 Finally, we define orientation on a piece-wise regular surface S in $\mathbb{R}^3$ with regular pieces S_i. We say that S is orientable if each S_i can be equipped with an orientation such that

$$\sum_i \partial S_i$$

contains no edges, only terms in ∂S. That is, the induced orientations on an edge $S_i \cap S_j$ by S_i, S_j are opposite. The above sum is then taken as a definition of the induced orientation ∂S. See Figure 16.7.

From now on all surfaces are supposed oriented, we keep the notation S and ∂S denotes the border equipped with the induced orientation.

16.3 Physical Vector Fields

16.3.1 Recall that a vector field in a domain $U \subset \mathbb{R}^n$ is just a continuous map

$$F : U \longrightarrow \mathbb{R}^n.$$

Instead of looking at F as a transformation from U to $F(U)$, we look at F as assigning to each $p \in U$ the free vector $F(p)$ with origin at p. We may consider as well fields depending in time $F(p,t)$.

In this section, it will be convenient to have in mind vector fields with a physical meaning, for $n = 2, 3$, in order to have an intuition of the concepts that will be introduced. We will consider just two types of fields: *force fields* and *velocity fields*. Regarding notation, in this chapter, it will be convenient to use the notation $p = (x, y, z)$ for points and P denotes the column vector $P = (x, y, z)^t$, and $F = (X, Y, Z)$.

16.3.2 A *Newtonian field* is the gravitational field in space determined by a mass distribution according to Newton's law. In case of a mass m at the origin, this field is given by

$$F(p) = -km\frac{p}{|p|^3}, \quad p \neq 0,$$

and MF is the force acting on a mass M. For a mass distribution given by a density ρ on a body C, the forces due to $m = \rho(q)\, dm(q)$ add up to

$$F(p) = -k\int_C \frac{p-q}{|p-q|^3}\rho(q)\, dm(q), \quad p \notin C.$$

The mass distribution might be on a surface or a curve, in which case we replace the volume element $dm(q)$ by dA or ds. The electrostatic fields created by charge distributions look the same, since mathematically Coulomb's law is the same as Newton's.

The total force acting on a body A exterior to C with mass density τ will be the sum of all forces acting on the $M = \tau(p)\, dm(p)$, that is, by Fubini's theorem

$$F = -k\int_{p\in A}\int_{q\in C} \frac{p-q}{|p-q|^3}\rho(q)\,\tau(p)\, dm(p)dm(q).$$

For further use, we note that Newtonian fields are gradient fields. Indeed, a computation shows that

$$-\frac{p}{|p|^3} = \nabla\frac{1}{|p|},$$

whence the function

$$u(p) = kM \int_C \frac{1}{|p-q|} \rho(q)\, dm(q), \quad p \notin C,$$

satisfies $\nabla u = F$. This function is called a *Riesz potential.*

In dimension $n = 2$, there is an analogue of the Newtonian field that can be motivated as follows. Writing $\mathbb{R}^3 = \mathbb{R}^2 \times \mathbb{R}$, $p = (p', p'')$, $p'' \in \mathbb{R}$, assume that $C = C' \times I$ where $I = [-1, 1]$ and that $\rho(q', q'') = \rho(q')$ is independent of p''. At points $p = (p', 0)$, the third component of $F(p)$ is zero, while the p'-component is

$$F(p) = -kM \int_{q' \in C'} \int_{-1}^{1} \frac{p' - q'}{|p-q|^3} \rho(q')\, dq'\, dq'', \quad p' \notin C'.$$

The integral in q''

$$\int_{-1}^{1} \frac{1}{(|p'-q'|^2 + q''^2)^{\frac{3}{2}}}\, dq''$$

is of the order of $|p' - q'|^2$ for say $|p' - q'| \leq 100$. This leads to the *plane Newtonian field* for a mass distribution ρ in the plane (again with standard notation p, q)

$$F(p) = -kM \int_C \frac{p-q}{|p-q|^2} \rho(q)\, dm(q).$$

Since $\nabla \log |p| = \frac{p}{|p|^2}$, this is again a gradient field $\nabla \gamma$ with the *logarithmic potential*

$$u(p) = -\int_C \log |p-q| \rho(q)\, dm(q).$$

16.3.3 Another example is that of the *velocity fields of fluids in the plane or in the space.* To describe the motion of a fluid we need first the velocity field $F(p, t)$, the velocity of the particle at point p at time t, and secondly the fluid density $\rho(p, t)$, so that

$$\int_C \rho(p, t)\, dm(p)$$

is the total mass inside C at time t. If both F, ρ are independent of t, we talk about a *stationary fluid.*

Together with $F(p,t)$ we consider its *trajectories* $\varphi(t,s,p)$, the position at time t of the particle that at time s is at p. Recall (Theorem 8.5) that these are the solutions of the Cauchy problem

$$\frac{d\varphi}{dt} = F(\varphi(t),t), \quad \varphi(s) = p.$$

The trajectories are not to be confused with the integral curves of the field $F(p,t_0)$ for a fixed t_0, which are the solutions of the autonomous system

$$\frac{d\varphi}{dt} = F(\varphi(t),t_0).$$

16.3.4 Assume that F is a linear field, in matrix notation, with $P = p^t$ and a 3×3 matrix M,

$$F(p) = MP.$$

We decompose M into its symmetric and anti-symmetric parts

$$M^s = \frac{1}{2}(M + M^t), \quad M^a = \frac{1}{2}(M - M^t).$$

Since M^s is symmetric, it diagonalizes in some orthonormal basis. On the other hand,

$$M^a P = \begin{pmatrix} 0 & -\gamma & \beta \\ \gamma & 0 & -\alpha \\ -\beta & \alpha & 0 \end{pmatrix} \begin{pmatrix} x \\ y \\ z \end{pmatrix} = \omega \times p,$$

with $\omega = (\alpha,\beta,\gamma)$. Thus, every linear mapping given by a matrix M decomposes as a sum of a diagonal one in some orthonormal basis and the operation of cross product with a fixed vector. The interpretation of these two components is better understood looking at them as vector fields and analyzing the trajectories.

To solve say

$$\frac{d\varphi(t,p)}{dt} = M^s(\varphi(t,p)), \quad \varphi(0,p) = p,$$

working in the orthonormal basis in which M^s diagonalizes we are led to three decoupled equations for x, y, z like

$$\frac{dx}{dt} = \lambda x, \quad x(0) = a,$$

where λ is a eigenvalue of M^s and $p = (a,b,c)$. The solution is $x(t) = ae^{\lambda t}$. The qualitative behavior of the trajectories depends on the signs of the eigenvalues. Such linear field is called a *deformation field.*

To solve

$$\frac{d\varphi(t,p)}{dt} = \omega \times (\varphi(t,p)), \quad \phi(0,p) = p,$$

we may assume, working in a suitable coordinate system, that $\omega = (0,0,\alpha)$. Then the equations become

$$\frac{dx}{dt} = -\alpha y(t), \quad \frac{dy}{dt} = \alpha x(t), \quad \frac{dz}{dt} = 0,$$

with initial conditions $(x(0), y(0), z(0)) = p = (a,b,c)$. The solutions are

$$x(t) = a\cos\alpha t - b\sin\alpha t, \quad y(t) = a\sin\alpha t + b\cos\alpha t, \quad z(t) = c,$$

representing a rotation in the xy-plane of angle αt.

If F is a smooth field, we can consider the Taylor expansion around (p_0, t_0)

$$F(p,t) = F(p_0,t_0) + d_pF(p_0,t_0)(p-p_0) + O(|p-p_0|^2) + O(|t-t_0|).$$

By the smooth dependence on initial conditions and parameters in Theorem 8.5, the trajectories $\varphi(p,t)$ of $F(p,t)$ will not differ much, for p close to p_0 and $t - t_0$ small of those of the field

$$F(p_0,t_0) + d_pF(p_0,t_0)(p-p_0).$$

The linear field $d_pF(p_0,t_0)(p-p_0)$ decomposes as before as sum of a symmetric and an anti-symmetric one. Thus, infinitesimally, every vector field is at first order the sum of a translation, a deformation field and a rotation around an axis.

The axis $\omega(p)$ corresponding to the anti-symmetric part of d_pF is, if $F = (X,Y,Z)$,

$$\omega(p) = \frac{1}{2}(Z_y - Y_z, X_z - Z_x, Y_x - X_y).$$

So, we encounter as in paragraph 9.3.3, the field

$$\nabla \times F = (Z_y - Y_z, X_z - Z_x, Y_x - X_y),$$

called the *rotational or curl* of F, also denoted $\operatorname{rot} F$.

Note that $\operatorname{rot} F = 2\omega$ if $F(p) = \omega \times p$.

16.4 Line Integrals, Work and Circulation

16.4.1 Let Γ be an oriented regular curve in $\mathbb{R}^n$ and T, the unit tangent provided by the orientation. If F is a vector field defined on Γ, we define its *line integral or circulation* along Γ, $\mathcal{C} = \mathcal{C}(F, \Gamma)$ as the sum of all its tangential components

$$\mathcal{C} = \int_\Gamma \langle F, T\rangle \, ds,$$

where ds denotes arc-length on Γ, whenever this integral makes sense (for instance if F is continuous). Of course, this definition extends to piece-wise regular oriented curves.

In terms of a parametrization $\gamma : [a, b] \to \Gamma$ compatible with the orientation,

$$T = \frac{\gamma'(t)}{|\gamma'(t)|}, \quad ds = |\gamma'(t)| \, dt,$$

and therefore,

$$\mathcal{C} = \int_a^b \langle F(\gamma(t)), \gamma'(t)\rangle \, dt.$$

If $F = (F_1, \ldots, F_n), \gamma(t) = (x_1(t), \ldots, x_n(t))$, one has

$$\mathcal{C} = \int_a^b (F_1(\gamma(t))x_1'(t) + \cdots + F_n(\gamma(t))x_n'(t)) \, dt, \tag{16.3}$$

an expression that is often written in terms of a 1-form

$$\mathcal{C} = \int_\Gamma \sum_i F_i dx_i.$$

At this point, we note that this definition makes sense for a general piece-wise regular path γ, even though if $\gamma'(t) = 0$ at some points and is not one-to-one. We use the notation $\mathcal{C}(F, \gamma)$. See Example 17.2.

If $|F| \leq M$ on Γ and Γ has length L, it follows from the definition that

$$|\mathcal{C}| \leq ML(\gamma). \tag{16.4}$$

Example 16.2. Let Γ be the half circle defined by $x^2 + y^2 + z^2 = 1, x + y + z = 0, z \geq 0$ oriented from $p = (-\frac{1}{\sqrt{2}}, +\frac{1}{\sqrt{2}}, 0)$ to $q = (+\frac{1}{\sqrt{2}}, -\frac{1}{\sqrt{2}}, 0)$ and $F = (y, x, z)$. First, we parametrize Γ; the unit vectors

$v_1 = (-\frac{1}{\sqrt{2}}, \frac{1}{\sqrt{2}}, 0), v_2 = (-\frac{1}{\sqrt{6}}, -\frac{1}{\sqrt{6}}, \frac{2}{\sqrt{6}})$ are orthogonal vectors in the plane $x+y+z=0$. The parametrization

$$\gamma(t) = (\cos t)\, v_1 + (\sin t)\, v_2, \quad 0 \le t \le \pi,$$

goes from p to q. Then

$$\begin{aligned}
\mathcal{C} &= \int_\Gamma (y\,dx + x\,dy + z\,dz) \\
&= \int_0^\pi \left(\frac{1}{\sqrt{2}}\cos t - \frac{1}{\sqrt{6}}\sin t\right)\left(\frac{1}{\sqrt{2}}\sin t - \frac{1}{\sqrt{6}}\cos t\right) dt \\
&\quad + \int_0^\pi \left(-\frac{1}{\sqrt{2}}\cos t - \frac{1}{\sqrt{6}}\sin t\right)\left(-\frac{1}{\sqrt{2}}\sin t - \frac{1}{\sqrt{6}}\cos t\right) dt \\
&\quad + \int_0^\pi \frac{2}{\sqrt{6}}\sin t \frac{2}{\sqrt{6}}\cos t\, dt.
\end{aligned}$$

All integrals including $\sin t, \cos t$ are zero, and those in $\sin^2 t, \cos^2 t$ are equal, so $\mathcal{C} = 0$.

16.4.2 For a first physical interpretation in space, assume F is a force field and under its action a particle with mass M moves along Γ, say from p to q following the path $\gamma(t)$, that is, $\gamma(t)$ denotes position at time t. Then only the tangential component matters, $F_T = \langle F, T\rangle$ and

$$\mathcal{C} = \int F_T\, ds.$$

Since $F_T\, ds$ represents the work done to move the particle the amount ds, $\mathcal{C}$ represents the total work done. On the other hand, by Newton's law, $F_T = m\frac{dv}{dt} = m|\gamma'(t)|$, so

$$\mathcal{C} = M\int_a^b \frac{dv}{dt} v\, dt = \frac{1}{2}M\int_a^b \frac{dv^2}{dt} dt = \frac{1}{2}Mv(q)^2 - \frac{1}{2}Mv(p)^2,$$

therefore, the total work done equals the variation of the kinetic energy.

A second interpretation occurs for closed regular curves with no boundary. Assume first that $n = 2$ and that Γ is a small circle or square centered at p, oriented counterclockwise. If X is a force field and we think of Γ as a rigid wire that can rotate around a fixed p, $\mathcal{C}$, as sum of the tangential components, represents the total force acting on Γ that will make it rotate counterclockwise if $\mathcal{C}$ is positive or clockwise if negative. As a particular case, the circulation of a constant field will be zero along Γ.

For instance, for the unit circle Γ and $F = (|y|, 0)$, $\langle F, T\rangle$ is odd and so the circulation is zero. Intuitively, the tangential force on the upper part is the opposite of the one on the lower part. For $X = (y, 0)$, we find, since $T = (\sin t, -\cos t)$,

$$\mathcal{C} = \int_0^{2\pi} \sin^2 t\, dt = \pi.$$

In case X is velocity field of a fluid, we may consider that by friction a force field proportional to F acts on the rigid wire and, eventually, makes it turn.

We point out a common misunderstanding. The fact that $\mathcal{C} \neq 0$ for a closed curve does not mean that the trajectories of the fluid turn around the curve, as shown by the previous examples.

16.4.3 The following examples are relevant for the proof of Stokes' theorem in Section 17.4.

Example 16.3. Assume $n = 3, F(p) = \omega \times (p - q)$ and that Γ is the border of a parallelogram R

$$R = \{q' + tv_1 + sv_2, 0 \leq s, t \leq 1\},$$

containing q, oriented by $N = \frac{v_1 \times v_2}{|v_1 \times v_2|}$. We compute $\mathcal{C}$. Since $F(p) = \omega \times (p - q') + \omega \times (q' - q)$ and the latter is orthogonal to v_1, v_2, we may replace q by q' and assume $q' = 0$. On the first regular piece of Γ, $p = tv_1, 0 \leq t \leq 1, T = \frac{v_1}{|v_1|}$, and $\langle F, T\rangle = 0$, so it contributes with zero circulation; similarly, the contribution of the last piece $p = sv_2, 0 \leq s \leq 1$ is zero. In the second piece, $p = v_1 + sv_2, 0 \leq s \leq 1, T = \frac{v_2}{|v_2|}$ and since $\omega \times v_2$ is orthogonal to v_2, the scalar product

$$\langle F, T\rangle = \left\langle \omega \times v_1, \frac{v_2}{|v_2|}\right\rangle,$$

is constant. Therefore, the contribution to circulation of this piece of length $|v_2|$ is

$$\langle \omega \times v_1, v_2\rangle.$$

Now, we observe that

$$\langle \omega \times v_1, v_2\rangle = \langle v_2 \times \omega, v_1\rangle = \langle v_1 \times v_2, \omega\rangle. \tag{16.5}$$

For the contribution of the third piece, we must replace v_1 by v_2 and v_2 by $-v_1$ so we get

$$\mathcal{C} = 2\langle v_1 \times v_2, \omega\rangle = 2A(R)\langle N, \omega\rangle,$$

$A(R)$ denoting the area of S. In particular, if the direction of rotation is orthogonal to the plane

$$\omega = \rho N,$$

we find

$$\mathcal{C} = 2\rho A(S).$$

Example 16.4. Again with $F(p) = \omega \times (p - q)$, let now v_1, v_2 be unit orthogonal vectors and consider the ellipse Γ given by the parametrization

$$\gamma(t) = q' + a(\cos t)v_1 + b(\sin t)v_2, \quad 0 \leq t \leq 2\pi,$$

containing q. Again, we may replace q by q' and assume $q' = 0$. Then $\gamma'(t) = -a \sin t v_1 + b \cos t v_2$, and

$$\langle F, \gamma'(t)\rangle = -ab \sin^2 t\langle \omega \times v_2, v_1\rangle + ab \cos^2 t\langle \omega \times v_1, v_2\rangle,$$

which making use again of (16.5) turns out to be constant equal to $ab\langle v_1 \times v_2, \omega\rangle$, whence we get

$$\mathcal{C} = 2\pi ab\langle v_1 \times v_2, \omega\rangle.$$

Again, if the direction of rotation is perpendicular to the plane spanned by v_1, v_2, we encounter

$$\mathcal{C} = 2\rho\pi ab,$$

twice the speed of rotation times the area bounded by the ellipse.

16.4.4 A continuous field F is determined by all its line integrals and can be identified with the map

$$\Gamma \longrightarrow \mathcal{C}(F, \Gamma).$$

This is obvious, for the line integral along the segment from p to $p + \varepsilon v$ equals

$$I(\varepsilon) = \int_0^\varepsilon \langle F(p + \varepsilon v), v\rangle dt,$$

and so

$$\langle F, v\rangle = \lim_{\varepsilon \to 0} \frac{I(\varepsilon)}{\varepsilon}.$$

However, as it will become clear later, the line integrals along *closed curves* do not determine F.

16.5 Surface Integrals and Flux

16.5.1 Let M be an oriented hyper-surface in $\mathbb{R}^n$ and N the unit normal provided by the orientation. If F is a vector field defined on M, we define its *flux* across M, $\mathcal{F} = \mathcal{F}(F, M)$ as the sum of all its normal components

$$\mathcal{F} = \int_M \langle F, N\rangle \, dm_{n-1},$$

where dm_{n-1} denotes the volume element on M, whenever this integral makes sense. The definition extends to oriented piece-wise regular hyper-surfaces in $\mathbb{R}^n$ in the obvious way.

Let $\Phi(u_1, \ldots, u_{n-1}), u \in U$, be a local chart. Then, with $\partial_i = \frac{\partial \Phi}{\partial u_i}$,

$$v = \partial_1 \times \cdots \times \partial_{n-1}$$

is orthogonal to M. If $N = \frac{v}{|v|}$, we call (U, Φ) a chart *coherent with the orientation*, or that the basis ∂_i is *positively oriented*. Then

$$dm_{n-1} = |v| \, du_1 \cdots du_{n-1},$$

therefore for F supported on the range of Φ,

$$\mathcal{F} = \int_U \langle F, \partial_1 \times \cdots \times \partial_{n-1}\rangle \, du_1 \cdots du_{n-1}.$$

That is, the flux is obtained adding up the signed infinitesimal volume of the parallelepiped spanned by F and the tangent vectors ∂_i.

For instance, in $n = 3$, expanding the determinant one finds, if $F = (X, Y, Z)$ and $\Phi(s,t) = (x(s,t), y(s,t), z(s,t))$

$$\mathcal{F} = \int_U (X(y_s z_t - z_s y_t) + Y(z_s x_t - x_s z_t) + Z(x_s y_t - y_s z_t)) \, ds \, dt, \quad (16.6)$$

usually expressed (see paragraph 8.1.5)

$$\int_M X \, dy \wedge dz + Y \, dz \wedge dx + Z \, dx \wedge dy.$$

Again, this can be taken as a definition even if $d\Phi$ has not full rank and is not one-to-one.

In case S is a graphic $z = g(x,y), (x,y) \in U$, oriented with the unit normal with positive z-component, with parametrization $\Phi(x,y) = (x, y, g(x,y))$, then

$$\mathcal{F} = \int_U (Z - g_x X - g_y Y) dx \, dy.$$

If $|F| \leq C$ on M and M has measure A, it follows from the definition that

$$|\mathcal{F}| \leq CA.$$

16.5.2 For an intuitive interpretation in space, consider a fluid in motion with velocity field v and density ρ. The flux of v measures the rate of change of the volume of fluid crossing M, at time t, while the flux of ρv

$$\int_M \rho(p,t)\,\langle v(p,t), N(p)\rangle\, dA(p),$$

measures the flux of the mass of fluid. Indeed, a particle that at time t is at $p \in dA$ is at $p + v(p, t + dt)$ at time $t + dt$. In this time interval, dA covers a sort of parallelepiped with base dA and height $\langle v(p,t), N(p)\rangle\, dt$ whence $\langle v, N\rangle\, dA\, dt$ represents the infinitesimal amount of volume crossing dA during this time interval. Then the sum

$$dt \int_M \langle v, N\rangle\, dA$$

is the total volume crossing M in the time interval dt.

Example 16.5. If F is the Coulombian field due to a positive charge C at the origin, $F = kC\frac{p}{|p|^3}$, and M is a sphere of radius R centered at the origin, oriented with the exterior normal, $N = \frac{p}{|p|}$, then $\langle F, N\rangle = \frac{kq}{R^2}$ and therefore the flux is $4\pi kC$.

Example 16.6. If F is a constant field and ∂R is the boundary of a parallelepiped oriented with the exterior normal N, $\langle F, N\rangle$ is an odd function on ∂R, whence the flux is zero. The same applies to spheres or a general symmetric domain.

The definition of flux makes sense whenever M is orientable, there is a well-defined volume element on M and continuous functions have finite integrals. In particular, M might have singular points around which it has finite area:

Example 16.7. Let M be the regular surface with boundary defined by

$$z^2 = 4\sqrt{x^2 + y^2}, \quad 0 \leq z \leq 1.$$

It is the surface of revolution spanned by the curve $z = 2\sqrt{y}, 0 \leq y \leq 1$ on the zy-plane when it rotates around the z-axis. It has a singular point at the origin, a cusp, but still the area is finite (see Example 14.6). We orient M with the unit normal N with negative z component. For $F = (x, y, x^2+y^2)$,

$$\mathcal{F} = \int_{x^2+y^2 \le 1} (Z - g_x X - g_y Y)\, dA = 2\pi \int_0^1 (\sqrt{r} - r^2) r\, dr = \frac{3\pi}{10}.$$

16.5.3 The bi-dimensional version of flux is equivalent to circulation. If Γ is a regular plane curve oriented by $T = (\alpha, \beta)$, we consider the normal vector $N = (\beta, -\alpha)$. If $F = (X, Y)$ is a continuous field, then

$$\langle F, N \rangle = \beta X - \alpha Y = \langle JF, T \rangle,$$

where $JF = (-Y, X)$ (multiplication by i in the complex plane), so the flux of F is the circulation of JF.

16.5.4 The following examples will be relevant for the proof of Gauss' theorem.

Example 16.8. We compute the flux of a linear field $F(p) = M(P - Q)$, $M = (m_{ij})$ being a constant matrix, across the boundary ∂R of a parallelepiped in space spanned by 3 vectors v_1, v_2, v_3

$$R = \{q' + u_1 v_1 + u_2 v_2 + u_3 v_3, 0 \le u_i \le 1\},$$

containing q, oriented by the outward normal. F differs from $M(P - Q')$ by a constant field, so by previous example we can replace q by q' and assume $q' = 0$. On the face $u_3 = 1$, the basis v_1, v_2 is positively oriented and the flux is

$$\int_0^1 \int_0^1 \det(M(u_1 v_1 + u_2 v_2 + v_3), v_1, v_2)\, du_1\, du_2,$$

while on the face $u_3 = 0$, it is

$$-\int_0^1 \int_0^1 \det(M(u_1 v_1 + u_2 v_2), v_1, v_2)\, du_1\, du_2.$$

Therefore, they add up to

$$\int_0^1 \int_0^1 \det(M(v_3), v_1, v_2)\, du_1\, du_2.$$

If $M(v_3) = \sum_i \lambda_i v_i$, this equals $\lambda_3 \det(v_3, v_1, v_2)$. The same applies to the other two couples of opposite sides, whence the flux is exactly

$$\operatorname{trace}(M) \det(v_1, v_2, v_3),$$

the trace of M times the volume of R. The same result holds in general dimension n.

Example 16.9. With F as in the previous example and B a ball of radius r in $\mathbb{R}^n$ centered at q' containing q and oriented by the outward normal, again we may assume $q = q' = 0$, $N = \frac{1}{r}P$,

$$\langle F, N\rangle = \frac{1}{r}\langle MP, P\rangle.$$

All terms but those corresponding to the diagonal of M are odd. Therefore the flux is

$$\operatorname{trace}(M)\frac{1}{nr}\int_{\partial B} r^2\, dm_{n-1} = \operatorname{trace}(M)\omega_n r^n,$$

again the trace times the volume.

16.5.5 A continuous vector field F is completely determined by its flux across hyper-surfaces and can be identified with the map

$$M \longrightarrow \mathcal{F}(F, M).$$

For instance, if $n = 3, F = (X, Y, Z)$, if $D(p, \varepsilon)$ denotes the disk in the xy-plane centered at p of radius ε oriented with $N = (0, 0, 1)$, the flux equals $I(\varepsilon) = \int_{D(p,\varepsilon)} Z\, dA$, whence

$$Z(p) = \lim_{\varepsilon\to 0}\frac{I(\varepsilon)}{\pi\varepsilon^2}.$$

Analogously to circulation, the flux of F across surfaces with no border does not determine F.

Chapter 17

The Basic Theorems of Vector Analysis

The two actions of fields, on curves through circulation and on surfaces through flux, have their differential counterpart, *rotational* and *divergence*, introduced in this chapter as densities for these actions. The corresponding analogues of the fundamental theorem of calculus, in a sense different than that in Section 12.8, are *Stokes'* and *Gauss'* theorems. In order to stress the analogy with the setting of functions and the gradient, we include in the list a version of the fundamental theorem of calculus along curves. As a particular case of Stokes' theorem, we describe Green's formula. Our proofs of these theorems rely on the multidimensional version of the fundamental theorem of calculus just mentioned and use minimal regularity assumptions, an important fact regarding the *Cauchy's formula*, that we present as a particular case of Green's formula.

17.1 The Fundamental Theorem of Calculus for Curves

The theorems in this and next sections have a common feature with the fundamental theorem of calculus. If M is an oriented k-dimensional piecewise regular sub-manifold and ∂M is its border with the induced orientation, the theorems establish a relationship between integration on ∂M of a certain object Ω and integration on M of another object $D\Omega$, defined locally from Ω by a differentiation process. All theorems are in the spirit of Theorem 12.13 about densities.

The case $k = 1$ is the following, an a straightforward consequence of Theorem 12.10:

Theorem 17.1. *Let $\Gamma \subset \mathbb{R}^n$ be a piece-wise regular curve oriented by T and $\partial\Gamma = q - p$.*

(a) *If u has a directional derivative $D_T u$ at all points of Γ and $D_T u$ is Lebesgue integrable on Γ, then*

$$\int_\Gamma D_T u \, ds = u(q) - u(p).$$

(b) *If u is differentiable, then $D_T u = \langle \nabla u, T \rangle$ and therefore,*

$$\mathcal{C} = \int_\Gamma \langle \nabla u, T \rangle \, ds = u(q) - u(p).$$

Corollary 17.1. *A continuous gradient field ∇f has zero circulation along a closed oriented curve. In particular, constant fields have zero circulation.*

Example 17.1. Assume that F is a linear field $F(p) = MP$ given by symmetric matrix $M, M^t = M$. Then F is a gradient field, $F = \nabla u$ with $u(p) = \frac{1}{2}\nabla P^t MP$, whence it has zero circulation along closed oriented curves.

We saw before that if F is a force field and under its action a unit mass particle moves along Γ, say from q to p, the total work done equals the variation of the kinetic energy $\frac{1}{2}v(p)^2 - \frac{1}{2}v(q)^2$. On the other hand, a Newtonian field is a gradient field. Using the normalization $F = -\nabla U$, the work is also $U(q) - U(p)$, so that the sum $U + K$ remains constant, the law of conservation of mechanical energy. The function U is called the *potential energy*.

For the next example, we adopt the more general point of view of fields acting on general paths.

Example 17.2. In the plane, an *argument* of a point $(x, y) \neq (0, 0)$ is a real number θ such that $x = r\cos\theta, y = r\sin\theta$, with $r^2 = x^2 + y^2$. The set of its arguments is a coset in $\mathbb{R}/2\pi\mathbb{Z}$. Among those, the unique argument θ such that $0 \leq \theta < 2\pi$ is called the *principal argument.* A *determination of the argument* in a set A not containing the origin is a continuous function ϕ assigning an argument to each point in A. Obviously, this is not possible for the whole of $\mathbb{R}^2 \setminus (0, 0)$. For instance, the principal argument is a determination of the argument in $A = \mathbb{R}^2 \setminus x \geq 0$ and is not in

$A \setminus (0,0)$ because it is discontinuous at all points in the positive semi-axis. Analogously, there is a determination ϕ_L in the complement of a semi-axis L. If ϕ_1, ϕ_2 are determinations of the argument, $\phi_1 - \phi_2$ is continuous and takes integer values, so it is constant if A is connected. In particular, a determination differs locally from one ϕ_L by an integer constant, whence it is differentiable with

$$\nabla\phi = \nabla\phi_L = F = \left(-\frac{y}{x^2+y^2}, \frac{x}{x^2+y^2}\right).$$

If the range γ^* does not meet a semi-axis L, it follows from these considerations that if γ goes from p to q, then

$$\mathcal{C} = \phi_L(q) - \phi_L(p)$$

is the variation of the argument along γ. Breaking it into pieces, the same holds for a general path not passing through $(0,0)$. Thus, if γ is closed, $p = q$, it follows that

$$\frac{1}{2\pi}\mathcal{C} = I(\gamma)$$

is an integer counting the number of turns of γ around the origin. Replacing γ by $\gamma - p$ gives the number of turns around p, called the *index of* γ with respect to p. It can be proved that $I(\gamma, p)$ is constant in every connected component in the complement of γ^*. In Figure 17.1, the index with respect to the points A, B, C, D, E is, respectively, $3, -1, 2, -1, 1$.

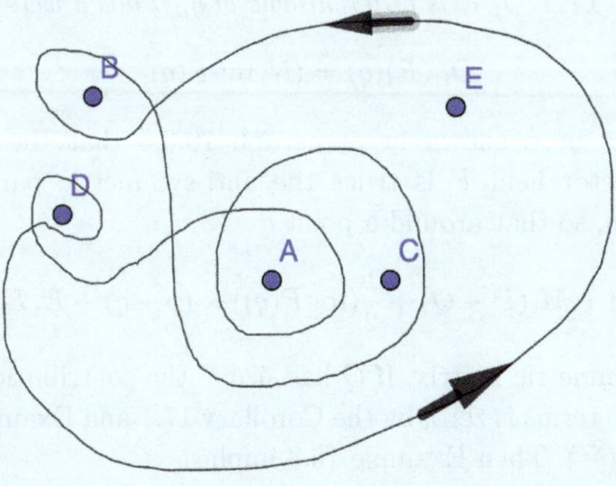

Figure 17.1. The index is constant in connected components.

17.2 Green's Formula

17.2.1 The directional derivative of a continuous function f along a curve Γ is a density for the set function

$$\Phi(I) = f(q) - f(p),$$

where I is an arc in $\Gamma, \partial I = q - p$. Now we want to define an analogous concept for vector fields instead of functions, oriented surfaces instead of oriented curves. In this paragraph, we consider the case of planes, for which fewer technicalities are needed.

Let Π be a plane in $\mathbb{R}^3$ oriented by a unit normal $N, q \in \Pi$ and let Q denote a generic square in Π containing q.

Definition 17.1. We say that a continuous vector field F defined on Π has a *circulation density* $D_\Pi(F)$ at $q \in \Pi$ if

$$D_\Pi(F)(q) = \lim_{Q \to q} \frac{\mathcal{C}(F, \partial Q)}{A(Q)}$$

exists, where $A(Q)$ denotes the area of Q and the limit is understood as Q shrinks to q.

For instance, for $F(p) = \omega \times p$, Example 16.3 shows that the ratio above is constant equal to $2\langle N, \omega\rangle$. This field has constant circulation density on every plane. Also, Corollary 17.1 implies that gradient fields have zero density on every plane.

Proposition 17.1. *If F is differentiable at q, it has a well-defined density*

$$D_\Pi(F)(q) = \langle N, \operatorname{rot} F(q)\rangle.$$

Proof. It has been shown in paragraph 16.3.4 that the curl of a differentiable vector field F is twice the anti-symmetric part of its linear approximation, so that around a point q

$$F(p) = F(q) + M_s(P - Q) + \frac{1}{2}(\operatorname{rot} F(q)) \times (p - q) + E, E = o(|p - q|),$$

with M_s a symmetric matrix. If Q has size δ, the contribution to $\mathcal{C}(F, \partial Q)$ of the first two terms is zero, by the Corollary 17.1 and Example 17.1, while that of E is $o(\delta^2)$. Then Example 16.3 implies

$$\mathcal{C}(F, \partial Q) = A(Q)\langle \operatorname{rot} F(q), N\rangle + o(A(Q)). \qquad \square$$

17.2.2 The following result is a version of the fundamental theorem of calculus 12.10 in this setting.

Theorem 17.2. *Let $U \subset \Pi$ be an admissible domain, oriented by N and let ∂U have the induced orientation. Then,*

(a) *If F is a continuous vector field on Π having a density $D_\Pi(F)$ at all points $q \in \Pi$, and $D_\Pi(F)$ is Lebesgue integrable on U, one has*

$$\int_{\partial U} \langle F, T\rangle \, ds = \int_U D_\Pi(F)\, dA.$$

(b) *In particular, if F is differentiable and $\langle N, \operatorname{rot} F\rangle$ is Lebesgue integrable in U,*

$$\int_{\partial U} \langle F, T\rangle \, ds = \int_U \langle N, \operatorname{rot} F\rangle dA.$$

When $U \subset \mathbb{R}^2$ is a plane domain and $F = (X, Y)$, then $N = (0, 0, 1)$ and the last formula is the classical *Green's formula*

$$\int_{\partial U} (X dx + Y dy) = \int_U (Y_x - X_y)\, dA,$$

whenever X, Y are differentiable and $(Y_x - X_u)$ is Lebesgue integrable on U, and bU is oriented counter clockwise.

Proof. We assume $\Pi = \mathbb{R}^2$ and by approximation that U has regular boundary. Since $\Phi(Q) = \mathcal{C}(F, \partial Q)$ is a set function, Theorem 12.13 proves that $\Phi(Q) = \int_Q D_\Pi(F)\, dA$. Then this holds as well for unions of cubes B

$$\int_{\partial B} \langle F, T\rangle \, ds = \int_B D_\Pi(F)\, dA.$$

Now, by Proposition 11.1, U is exhausted by elementary sets B, so that

$$\int_B D_\Pi(F)\, dA \to \int_U D_\Pi(F)\, dA.$$

Therefore, it is enough to show that $\mathcal{C}(F, \partial B) \to \mathcal{C}(F, \partial U)$ as B exhausts U. In turn, using a partition of unity, it is enough to prove this locally, when U, bU are the subgraph and graph of $y = \phi(x), a \leq x \leq b$ and F is

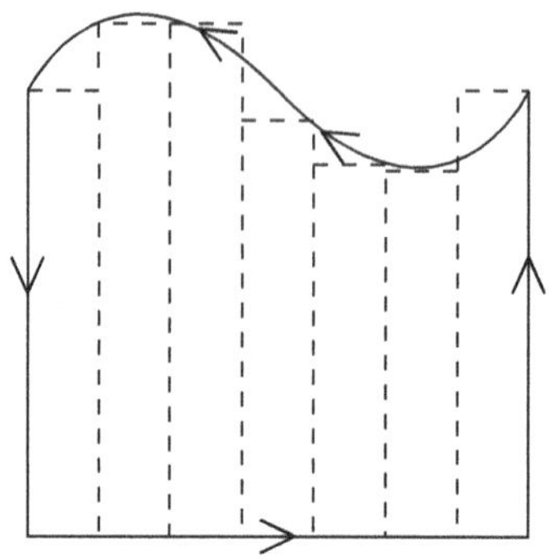

Figure 17.2. Green's theorem for a subgraph.

supported there (Figure 17.2). If $a = x_0 < x_1 < \cdots < x_N = b$ is a partition of $[a, b]$ and $\phi(\xi_i) = \min\{\phi(x), x_i \le x \le x_{i+1}\}$, the upper part of ∂B is given by the horizontal segments $x_i \le x \le x_{i+1}$ and the vertical segments determined by $\phi(\xi_i)$. Then if $F = (X, Y), \mathcal{C}(F, \partial U)$ is given by

$$\int_a^b [X(x, \phi(x)) + Y(x, \phi(x))\phi'(x)]\, dx,$$

while $\mathcal{C}(F, \partial B)$ is given by

$$\sum_{i=0}^{N-1} \int_{x_i}^{x_{i+1}} X(x, \phi(\xi_i))\, dx + \sum_{i=0}^{N-1} \int_{\phi(\xi_i)}^{\phi(\xi_{i+1})} Y(x_{i+1}, y)\, dy.$$

Changing variables we write the last term as

$$\sum_{i=0}^{N-1} \int_{\xi_i}^{\xi_{i+1}} Y(x_{i+1}, \phi(x))\phi'(x)\, dx.$$

Their difference has two terms. The first is

$$\begin{aligned} I &= \int_a^b X(x, \phi(x))\, dx - \sum_{i=0}^{N-1} \int_{x_i}^{x_{i+1}} X(x, \phi(\xi_i))\, dx \\ &= \sum_{i=0}^{N-1} \int_{x_i}^{x_{i+1}} [X(x, \phi(x)) - X(x, \phi(\xi_i)]\, dx, \end{aligned}$$

which tends to zero by the uniform continuity of $X(x, \phi(x))$. The second term in the difference is

$$\int_a^b Y(x, \phi(x))\phi'(x)\,dx - \sum_{i=0}^{N-1} \int_{\xi_i}^{\xi_{i+1}} Y(x_{i+1}, \phi(x))\phi'(x)\,dx$$

$$= \sum_{i=0}^{N-1} \int_{\xi_i}^{\xi_{i+1}} [Y(x, \phi(x)) - Y(x_{i+1}, \phi(x))\phi'(x)\,dx,$$

and tends to zero by the uniform continuity of $Y(x, \phi(x))$. □

Note that while $\mathcal{C}(F, \partial B) \to \mathcal{C}(F, \partial U)$ holds, obviously

$$\lim \int_{\partial B} f\,ds \neq \int_{\partial U} f\,ds,$$

for instance the length of ∂B does not converge to the length of ∂U.

17.2.3 If U is as in Green's formula, one can compute the area of U in terms of its boundary:

Corollary 17.2.

$$A(U) = \int_{\partial U} x\,dy = \frac{1}{2} \int_{\partial U} (-y\,dx + x\,dy).$$

Example 17.3. The *Descartes folium* is the curve defined by the equation

$$x^3 + y^3 - 3axy = 0, \quad x, y \geq 0.$$

The curve intersects $y = tx, t > 0$ at

$$x = \frac{3at}{1+t^3}, \quad y = \frac{3at^2}{1+t^3},$$

so we can consider this as a parametrization, $0 \leq t \leq +\infty$. The area is then

$$A = \frac{1}{2} \int_0^{+\infty} (x\,dy - y\,dx) = 9a^2 \int_0^{+\infty} \frac{t^2}{(1+t^3)^2}\,dt = 3a^2.$$

As mentioned in paragraph 16.1.5, Green's formula is also valid if ∂U has cusps around which U has finite area and ∂U has finite length, because approximation by regular domains is also possible.

Example 17.4. We consider the area of the plane region enclosed by the curve

$$|x|^\alpha + |y|^\alpha = 1,$$

presenting cusps at the points $(\pm 1, 0), (0, \pm 1)$ if $\alpha < 1$. We use the parametrization

$$x = (\cos t)^{\frac{2}{\alpha}}, \quad y = (\sin t)^{\frac{2}{\alpha}}, \quad 0 \le t \le \frac{\pi}{2},$$

in the first quadrant. The area is

$$2\int (x\,dy - y\,dx) = \frac{4}{\alpha}\int_0^{\frac{\pi}{2}} (\sin t \cos t)^{\frac{2}{\alpha}-1}\,dt = \frac{2^{2-\frac{2}{\alpha}}}{\alpha}\int_0^{\frac{\pi}{2}} (\sin 2t)^{\frac{2}{\alpha}-1}\,dt.$$

This can be computed explicitly for $\alpha = 2/(1+k), k = 0, 1, \ldots$. For $\alpha = \frac{2}{3}$,

$$A = \frac{3}{2}\int_0^{\frac{\pi}{2}} (\sin 2t)^2\,dt = \frac{3\pi}{8}.$$

Exercise 17.1. Assume that u is a C^2 function satisfying the equation of the minimal surfaces

$$(1 + u_y^2)u_{xx} + (1 + u_x^2)u_{yy} - 2u_x u_y u_{xy} = 0.$$

For a regular domain D in the plane, the graph $S : z = u(x, y), (x, y) \in D$ is the minimal surface with border, the curve $\Gamma : z = u(x, y), (x, y) \in bD$. Show that

$$A(S) = \frac{1}{2}\int_{bD} \frac{1}{\sqrt{1 + u_x^2 + u_y^2}}[(x + uu_x)dy - (y + uu_y)dx + (xu_y - yu_x)dz].$$

17.2.4 The following example deals again with the variation of the argument from a slightly different point of view.

Example 17.5. Assume that Γ is a piece-wise regular curve not passing through the origin such that each ray meets Γ in at most one point. We apply Green's formula to the domain U in Figure 17.3 and

$$F = \left(-\frac{y}{x^2 + y^2}, \frac{x}{x^2 + y^2}\right),$$

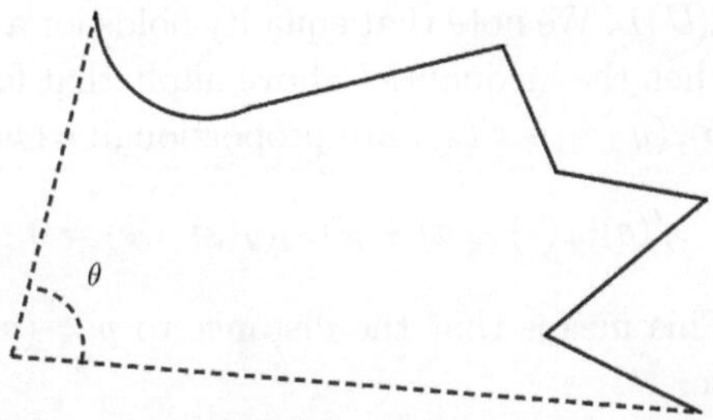

Figure 17.3. Angle of vision.

as in Example 17.2, for which $X_y - Y_x = 0$. On the rays of bU, the line integral is zero (because the tangent has direction (x, y) and F is proportional to $(-y, x)$) so we obtain, with Γ' the arc of bU on $r = \varepsilon$

$$|\mathcal{C}(F, \Gamma)| = |\mathcal{C}(F, \Gamma')| = |\mathcal{C}(F, \Gamma'')| = L(\Gamma''),$$

where Γ'' is the projection of Γ on the unit circle and $L(\Gamma'')$, its length. So $|\mathcal{C}(F, \Gamma)|$ is the size of the angle θ of vision of Γ from the origin.

17.2.5 In this paragraph, we consider *iso-perimetric inequalities.* A first one is a direct consequence of Green's formula. With the notations above, clearly for an arbitrary point (a, b)

$$A(U) = \frac{1}{2} \int_{bU} (x - a)\, dy - (y - b)\, dx.$$

Incidentally, this formula is the basis of the classical *planimeters*, devices used to measure areas enclosed by contours. In Section 15.2, other methods are explained. We choose $p = (a, b)$ so that $\max_{q \in bU} d(p, q)$ attains a minimum $R(U)$ at p; that is, the disc $B(p, R(U))$ is the smallest disc containing U. Then, if L is the length of bU, parametrizing bU by arc-length

$$\begin{aligned} 2A(U) &= \int_0^L [(x(s) - a)y'(s) - (y(s) - b)x'(s)]\, ds \\ &\le \int_0^L |(x(s) - a)y'(s) - (y(s) - b)x'(s)|\, ds \\ &\le \int_0^L \sqrt{(x(s) - a)^2 + (y(s) - b)^2} \sqrt{x'^2 + y'^2}\, ds \le R(U)L. \end{aligned}$$

Thus, $2A(U) \le R(U)L$. We note that equality holds for a circle. Conversely, if equality holds, then the inequalities above imply that for all s, the vectors $(x(s)-a, y(s)-b), (y'(s), -x'(s))$ are proportional, whence

$$x'(s)(x(s)-a) + y'(s)(y(s)-b) = 0,$$

identically in s. This means that the distance to $p = (a,b)$ is constant on bU and U is a disc.

Since $R \le 2L$, one has $A \le L^2$. The classical iso-perimetric inequality is a better result:

Theorem 17.3. *For a plane bounded domain U with smooth boundary, one has*

$$4\pi A \le L^2,$$

and equality holds if and only if U is a disc.

Proof. We reproduce the proof from [8]. We may assume by homogeneity that $L = 2\pi$. Let $(x(s), y(s))$ be a parametrization of bU by arc-length, $x'(s)^2 + y'(s)^2 = 1$. By a rigid motion, we may assume too that $x(0) = x(\pi) = 0$. By Green's theorem,

$$A = \int_{bU} x\,dy = \int_0^{2\pi} x(s)y'(s)\,ds,$$

and it is enough to prove that the integrals I, J on $[0,\pi], [\pi, 2\pi]$ are both bounded by $\frac{\pi}{2}$. One has

$$I \le I' = \frac{1}{2}\int_0^{\pi} x(s)^2 + y'(s)^2\,ds = \frac{1}{2}\int_0^{\pi} (x(s)^2 + 1 - x'(s)^2)\,ds.$$

To compare with the disc situation, we write everything in terms of $\varphi(s) = \frac{x(s)}{\sin s}$. By the assumption $x(0) = x(\pi) = 0$, φ is continuous on $[0,\pi]$ with boundary values $x'(0), x'(\pi)$. Also,

$$\varphi'(s)\sin s = x'(s) - \cos s\,\frac{x(s)}{\sin s}$$

extends continuously to $[0,\pi]$. Since

$$\begin{aligned} x(s)^2 + 1 - x'(s)^2 &= \varphi^2(s)\sin^2 s + 1 - (\varphi'(s)\sin s + \varphi(s)\cos s)^2 \\ &= 1 - \varphi'(s)^2\sin^2 s - (\varphi^2(s)\sin s\cos s)', \end{aligned}$$

one has

$$I \le I' = \frac{1}{2}\int_0^{\pi} (1 - \varphi'(s)^2 \sin^2 s)\, ds \le \frac{\pi}{2},$$

as claimed. If $A = \pi$, then $I' = \frac{\pi}{2}$ whence $\varphi'(s) = 0$, φ is constant and $x(s) = c \sin s$ and similarly for $y(s)$ working with J, J'. □

17.3 Cauchy's Formula

17.3.1 We point out that the proof of Green's formula we have presented is not the standard one, which uses the fundamental theorem of calculus in one variable to prove first Green's formula for intervals. Our proof has the added value that it requires the integrability hypothesis only on $Y_x - X_y$, and not on the components of $\operatorname{rot} F$ or Y_x, X_y separately. As an application of this fact, in this section we deal with Cauchy's theorem, a fundamental result in complex function theory.

First, we must consider the complex version of line integrals. Let Γ be an oriented regular curve in the complex plane with parametrization $z(t) = x(t) + iy(t), a \le t \le b$, and $f = u + iv$, a complex-valued continuous function on Γ. The line integral

$$\begin{aligned}\int_\Gamma f(z)\, dz &= \int_\Gamma (f(z)\, dx + i f(z)\, dy) = \int_\Gamma (u\, dx - v\, dy) + i \int_\Gamma (v\, dx + u\, dy) \\ &= \int_\Gamma (u + iv) dx + (-v + iu) dy,\end{aligned}$$

(with real part the circulation of $(u, -v)$ and imaginary part the circulation of (v, u)), in parametrized form

$$\int_a^b f(z(t)) z'(t)\, dt,$$

can be thought as the limit of the Riemann sums $\sum_i f(z_i)(z_{i+1} - z_i)$. We note that

$$(-v + iu)_x - (u + iv)_y = -(v_x + u_y) + i(u_x - v_y) = 2i\bar{\partial} f.$$

Therefore, Green's formula implies the first part of the next theorem.

Theorem 17.4. *Let U be a domain in the complex plane with piecewise regular border oriented counterclockwise. If f is differentiable in a neighborhood of $\overline{U}$ and $\overline{\partial} f$ is Lebesgue integrable in U, one has:*

(a) $\int_{bU} f(z)\, dz = 2i \int_U \overline{\partial} f dA$.
(b) *For $w \in U$,*

$$f(w) = \frac{1}{2\pi i} \int_{bU} \frac{f(z)}{z-w}\, dz - \frac{1}{\pi} \int_U \frac{\overline{\partial} f(z)}{z-w}\, dA(w).$$

Proof. To prove the second part, we use the first one for $g(z) = \frac{f(z)}{z-w}$ in the domain consisting in U with a disc $D(w,\varepsilon)$ removed. Using that $\overline{\partial} g(z) = \frac{\overline{\partial} f(z)}{z-w}$, we get

$$\int_{bU} \frac{f(z)}{z-w}\, dz = \int_{|w-z|=\varepsilon} \frac{f(z)}{z-w}\, dz + 2i \int_U \frac{\overline{\partial} f(z)}{z-w}\, dA(z).$$

Parametrizing the circle $|w-z| = \varepsilon, z = w + \varepsilon e^{it}, dz = ie^{it} dt$, the second integral equals $i \int_0^{2\pi} f(w + \varepsilon e^{it})\, dt$ and so the result follows making $\varepsilon \to 0$. $\square$

17.3.2 Another way of looking at the previous result is that $\overline{\partial} f$ is the density for complex line integrals. If f is differentiable in $U, p \in U$ and D_ε is the disk or a square centered at p of diameter ε with boundary oriented counter clockwise,

$$\lim_{\varepsilon \to 0} \frac{\int_{\partial D_\varepsilon} f(z)\, dz}{m(D_\varepsilon)} = 2i \frac{\partial f}{\partial \overline{z}}(p),$$

Theorem 17.5. *A differentiable function $f = u + iv$ in a plane domain U is holomorphic if and only if $\int_{\partial D} f(z)\, dz = 0$ for every disc $\overline{D} \subset U$. The same statement holds for squares or triangles instead of discs.*

Later on, in Theorem 18.2, we will show that a continuous function satisfying the hypothesis is automatically differentiable and whence holomorphic.

17.3.3 Next corollary is Cauchy's representation formula, a cornerstone in complex analysis:

Theorem 17.6. *If U as above, f is holomorphic in a neighborhood of $\overline{U}$ and $w \in U$, one has*

$$f(w) = \frac{1}{2\pi i} \int_{bU} \frac{f(z)}{z-w}\, dz, \quad w \in U.$$

As a consequence, holomorphic functions in a plane domain have derivatives of all orders, they are infinitely holomorphic, *given in a subdomain* $V \subset U$ *by*

$$f^{(k)}(w) = \frac{k!}{2\pi i} \int_{bV} \frac{f(z)}{(z-w)^{k+1}}\, dz, \quad w \in V.$$

In fact, they are *complex-analytic*, meaning that they are locally the sum of complex power series, see paragraph 9.5.3. Indeed, assume f is holomorphic in a disc $D(a, R)$ and apply the formula with $U = D(a, r)$, $r < R$. Using the development

$$\frac{1}{z-w} = \frac{1}{z-a} \sum_{k=0}^{\infty} \left(\frac{w-a}{z-a} \right)^k = \sum_{k=0}^{\infty} \frac{(w-a)^k}{(z-a)^{k+1}},$$

and previous formula for $f^{(k)}(a)$, we find

$$f(w) = \sum_{k=0}^{\infty} c_K (w-a)^k, \quad c_k = \frac{f^{(k)}(a)}{k!}, \quad |w-a| < r. \tag{17.1}$$

Together with the results in paragraph 9.5.3, this shows that a function is complex-analytic iff it is holomorphic.

Exercise 17.2. This exercise complements Exercise 5.2. Prove that for f holomorphic in U and $K \subset U$ compact, there are positive constants C, r such that

$$|f^{(k)}(z)| \leq C r^k k!, \quad z \in K.$$

These are called the *Cauchy inequalities.*

17.4 Stokes' Theorem

17.4.1 Now we would like to generalize Green's formula to an oriented surface S in $\mathbb{R}^3$. A first step is to define a circulation density on S as

$$D_S(F)(q) = \lim_{S' \to q} \frac{\mathcal{C}(F, \partial S')}{A(S')},$$

for pieces $S' \subset S$ shrinking to q. If S is oriented by N, here $\partial S'$ has the induced orientation, that is $\partial S', N$ satisfy the right-hand rule. To be precise, we consider pieces $S' = \Phi(Q)$ that are squares Q in a local chart $(U, \Phi(s,t)), 0 = (0,0) \in U, q = \Phi(0)$. Since

$$A(\Phi(Q)) = \int_Q |\Phi_s \times \Phi_t| \, ds \, dt,$$

$$D_S(F)(q)|\Phi_s(0) \times \Phi_t(0)| = \lim_{Q \to (0,0)} \frac{\mathcal{C}(F, \Phi(\partial Q))}{m(Q)}.$$

If this limit exists, we say that F has a *circulation density* on S. Obviously, gradient fields have zero circulation density on every surface.

Theorem 17.7. *If F is differentiable at q, it has a circulation density given by*

$$D_S(F)(q) = \langle N, \nabla \times F(q) \rangle.$$

Proof. Assume $Q \subset U$ has side $\delta, Q = [a, a+\delta] \times [b, b+\delta], 0 = (0,0) \in Q$ and let us evaluate the circulation along $\partial\Phi(Q)$. We consider again the linear approximation of F around q:

$$F(p) = F(q) + dF(q)(P - Q) + E, \quad E = o(|p - q|).$$

As before, by Corollary 17.1, the constant field $F(q)$ contributes with zero circulation, as it does the symmetric part of $dF(q)$. The field E contributes, by (16.4), with $o(\delta^2)$. It remains to estimate the circulation of $G(p) = \frac{1}{2}(\nabla \times F(q)) \times (p - q)$.

We consider the linear approximation L of Φ around $(0,0)$,

$$\Phi(s,t) = L(s,t) + o(\delta), \quad L(s,t) = q + \Phi_s(0)u_1 + \Phi_t(0)u_2.$$

By Example 16.3,

$$\mathcal{C}(G, \partial L(Q)) = \delta^2 \langle (\nabla \times F(q)), \Phi_s(0) \times \Phi_t(0) \rangle.$$

We will see that $\mathcal{C}(G, \partial L(Q))$ and $\mathcal{C}(G, \partial\Phi(Q))$ differ in $o(\delta^2)$. Both are parametrized by ∂Q; in the first piece of $\partial\Phi(Q), a \le s \le a + \delta, t = b$, their difference is

$$\Delta = \int_a^{a+\delta} (\langle G(\Phi(s,b)), \Phi_s(s,b) \rangle - \langle G(L(s,b)), \Phi_s(0,0) \rangle) ds.$$

Making use of

$$G(\Phi(s,b)) - G(L(s,b)) = G(\Phi(s,b) - L(s,b)) = G(o(\delta)) = o(\delta),$$

$$\Phi_1(s,b) - \Phi_1(0,0) = o(1), \quad G(L(s,b)) = O(\delta),$$

we see that

$$\Delta = \int_a^{a+\delta} o(\delta)ds = o(\delta^2).$$

The other sides are dealt similarly. Altogether,

$$\mathcal{C}(F, \partial\Phi(Q)) = \delta^2 \langle \Phi_s(0) \times \Phi_t(0), \nabla \times F(q) \rangle + o(\delta^2).$$

Therefore, if Q shrinks to $(0,0)$,

$$\lim \frac{\mathcal{C}(F, \partial\Phi(Q))}{m(Q)} = \langle \Phi_s(0) \times \Phi_t(0), \nabla \times F(q) \rangle, \quad q = \Phi(0,0), \tag{17.2}$$

and so, exactly as with planes,

$$D_S(F)(q) = \langle N, \nabla \times F(q) \rangle. \qquad \square$$

17.4.2 Stokes' theorem is in the other direction, from density to circulation.

Theorem 17.8 (Stokes' theorem). Let S be an admissible surface in $\mathbb{R}^3$ with border oriented by N and let ∂S have the induced orientation. If F has a Lebesgue integrable circulation density $D_S(F)$, then

$$\int_{\partial S} \langle F, T \rangle \, ds = \int_S D_S(F) \, dA.$$

In particular, if F is differentiable around S, and $\langle N, \nabla \times F \rangle$ is Lebesgue integrable on S, one has

$$\int_{\partial S} \langle F, T \rangle \, ds = \int_S \langle \nabla \times F, N \rangle \, dA.$$

Proof. By approximation, we may assume that S is a regular surface with regular border. The first part is a consequence of a surface version of Theorem 12.13, for S can be tiled with arbitrarily small pieces of type $\Phi(R)$. Instead of checking this point in detail, we give a direct proof of the second part, using a partition of unity argument. By compactness, S is covered by a finite number of local charts (U_i, Φ_i). Let ϕ_i be a partition of unity subordinated to the covering U_i as in Theorem 5.5 and set $F_i = \phi_i F$.

Then of course the circulation of F is the sum of circulations of F_i while a computation shows that

$$\nabla \times F_i = \phi_i \nabla \times F + \nabla \phi_i \times F.$$

Therefore, since $\sum_i \phi_i = 1$ on a neighborhood of S,

$$\sum_i \nabla \times F_i = \nabla \times F + \nabla \left(\sum \phi_i \right) \times F = \nabla \times F.$$

Thus, it is enough to prove the theorem when F is supported in a local chart (U, Φ). With this chart fixed, we consider the set function

$$\Psi(Q) = \int_{\partial \Phi(Q)} \langle F, T \rangle \, ds,$$

for squares $Q \subset U$. The hypothesis means that it has a density given by (17.2). By Theorem 12.13,

$$\Psi(Q) = \int_Q \langle \Phi_s(0) \times \Phi_t(0), \nabla \times F(q) \rangle \, ds \, dt = \int_{\Phi(Q)} \langle N, \nabla \times F \rangle \, dA.$$

This proves the theorem for $\Phi(Q)$, whence for $\Phi(B)$ as well for B a union of squares.

Now, there are two types of charts. If there are no border points, $\Phi(U) \cap \partial S = \emptyset$, we may choose B such that the support of F lies in $\Phi(\mathring{B})$,

$$\int_S \langle \nabla \times F, N \rangle \, dA = \int_{\Phi(B)} \langle \nabla \times F, N \rangle \, dA = \int_{\partial \Phi(B)} \langle F, T \rangle \, ds = 0$$
$$= \int_{\partial S} \langle F, T \rangle \, ds.$$

If in the coordinates s, t, S is like the half-plane, we may choose B such that (Figure 17.4)

$$\int_S \langle \nabla \times F, N \rangle \, dA = \int_{\Phi(B)} \langle \nabla \times F, N \rangle \, dA$$
$$= \int_{\partial \Phi(B)} \langle F, T \rangle \, ds = \int_{\partial S} \langle F, T \rangle \, ds. \qquad \square$$

Corollary 17.3. *A continuous rotational field $\nabla \times F$ (in particular, a constant field) has zero flux across oriented surfaces with no border.*

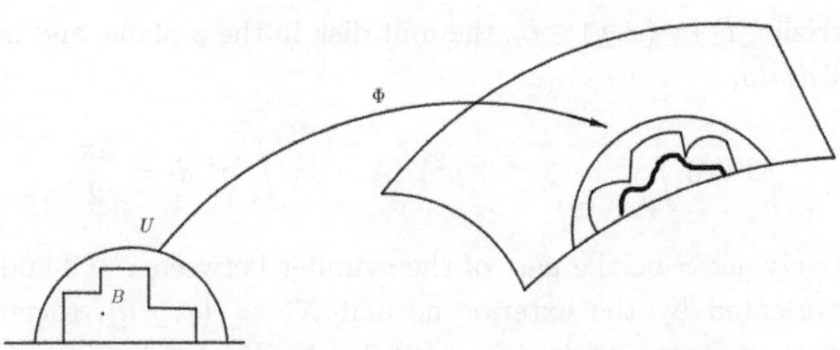

Figure 17.4. Second type of chart.

Equivalently, if S_1, S_2 are oriented surfaces with common border, so that $S_1 \cup S_2$ has no border, the flux of $\nabla \times F$ across S_1, S_2 is the same. In particular, constant fields have the same flux, as the intuition dictates.

Example 17.6. Let S be the piece of the unit sphere defined by $x + y + z \geq 1$, oriented with $N_1 = (x, y, z)$, and $F = (a, b, c)$; the flux across S equals the flux across the disk D in the plane $x + y + z = 1$ limited by S, oriented by $N_2 = \frac{1}{\sqrt{3}}(1, 1, 1)$, whence it equals

$$\frac{a + b + c}{\sqrt{3}} A(D),$$

where $A(D)$ is the area of D. Its center is $(\frac{1}{3}, \frac{1}{3}, \frac{1}{3})$, its radius $\frac{\sqrt{4}}{3}$, so the flux is $(4\pi(a + b + c))/9\sqrt{3}$.

17.4.3 For fields $F = (X(x, y), Y(x, y))$ in dimension $n = 2$ and U a plane domain, we may consider F as a field in $\mathbb{R}^3$ parallel to $\mathbb{R}^3, F = (X(x, y), Y(x, y), 0)$ and U as a surface in $\mathbb{R}^3$. If bU is oriented counter clockwise, then $N = (0, 0, 1)$ and Stokes' theorem becomes Green's formula. We may say that in dimension $n = 2$, $\operatorname{rot} F$ is a scalar quantity; for differentiable fields

$$\operatorname{rot} F = Y_x - X_y.$$

Example 17.7. Let D be the disc in the plane $x + y + z = 1$ interior to the cylinder $x^2 + y^2 = 1$, oriented with $N = \frac{1}{\sqrt{3}}(1, 1, 1)$ and $F = (-y^3, x^3, -z^3)$. We wish to compute the circulation $\mathcal{C}$ of F along ∂D with the induced orientation. Since $\operatorname{rot} F = (0, 0, 3x^2 + 3y^2)$, it equals

$$\frac{1}{\sqrt{3}} \int_D 3(x^2 + y^2)\, dA.$$

Parametrizing D by $(x, y) \in U$, the unit disk in the z-plane, and using that $dA = \sqrt{3}\, dx\, dy$,

$$\mathcal{C} = 3 \int_{x^2+y^2 \leq 1} (x^2 + y^2)\, dA = 6\pi \int_0^1 r^3\, dr = \frac{3\pi}{2}.$$

Alternatively, let S be the part of the cylinder between $z = 0$ and $x + y + z = 1$, oriented by the exterior normal $N_1 = (x, y, 0)$, together with the base U, oriented with $N_2 = (0, 0, -1)$. Its border is $-\partial D$. Since $\langle \nabla \times F, N_1 \rangle = 0$,

$$\mathcal{C} = -\int_U \langle \nabla \times F, N_2 \rangle\, dA = 3 \int_{x^2+y^2 \leq 1} (x^2 + y^2)\, dA.$$

17.5 Gauss' Theorem

17.5.1 Now we look at vector fields as acting on oriented hyper surfaces in $\mathbb{R}^n$ by flux and would like to consider the corresponding notion of derivative.

Definition 17.2. We say that a continuous vector field F has a *flux density* $\operatorname{div}(F)$ at q if

$$\operatorname{div}(F)(q) = \lim_{Q \to q} \frac{\mathcal{F}(F, \partial Q)}{m(Q)}$$

exists, where $m(Q)$ denotes the volume of a cube $Q, q \in Q$ shrinking to q. This function is called the *divergence* of F.

By Corollary 17.3, a rotational field in $\mathbb{R}^3$ has zero flux density.

Theorem 17.9. *If F is differentiable at q, it has a flux density given by*

$$\operatorname{div}(F) = \langle \nabla, F \rangle = \sum_i D_i F_i.$$

Proof. Let Q be a cube of size δ containing q and F a differentiable field at q, that we expand again around q,

$$F(p) = F(q) + dF(q)(P - Q) + E, \quad E = o(|p - q|),$$

The contribution to $\mathcal{F}(F, \partial Q)$ of the constant field $F(q)$ is zero, while that of E is $o(\delta^n)$. Example 16.8 implies if $F = (F_1, \ldots, F_n)$,

$$\mathcal{F}(F, \partial Q) = (D_1 F_1 + \cdots + D_n F_n)(q) m(Q) + o(m(Q)). \qquad \square$$

17.5.2 Next is Gauss' theorem, from densities to flux:

Theorem 17.10. *Let U be a bounded admissible domain in $\mathbb{R}^n$ oriented by the outward normal and F, a continuous field defined in a neighborhood of $\overline{U}$.*

(a) *If F has a flux density* $\operatorname{div}(F)$ *at all points of U and* $\operatorname{div}(F)$ *is Lebesgue integrable in U, one has*

$$\mathcal{F} = \int_{bU} \langle F, N \rangle \, dm_{n-1} = \int_U \operatorname{div}(F) \, dm_n.$$

(b) *In particular, if F is differentiable and* $\operatorname{div} F$ *is Lebesgue integrable in U,*

$$\mathcal{F} = \int_{bU} \langle F, N \rangle \, dm_{n-1} = \int_U \langle \nabla, F \rangle \, dm_n.$$

Proof. The proof of the first part goes along the same lines as in Green's formula. We consider the set function

$$\Psi(Q) = \int_{\partial Q} \langle F, N \rangle \, dm_{n-1},$$

which is obviously additive, whose density is $\operatorname{div}(F)$. Theorem 12.13 proves the result for cubes, whence for union of cubes B as well. Assuming as we may that bU has regular boundary, and exhausting U by sets B, it is enough to show that the flux across bB approaches the flux across bU. We show this in detail for $n = 3$. Again, we may work locally, assuming that U is a sub-graph $0 < z < \phi(x, y), a \leq x \leq b, c \leq y \leq d$. We consider a partition of $[a, b] \times [c, d]$ in squares $R_{i,j}$ of size δ centered at points $(x_i, y_j), i, j = 0, \ldots, N$ and define B_δ as the union of the $R_{i,j} \times [0, z_{i,j}]$, with $z_{i,j} = \phi(x_i, y_j) - \omega(\delta)$. Clearly, we can choose $\omega(\delta) \to 0$ as $\delta \to 0$ such that B_δ lies below the graph and the top part of bR approaches bU as $\delta \to 0$. The flux across bU is, if $F = (X, Y, Z)$,

$$\int_a^b \left(\int_c^d [Z(x, y, \phi(x, y)\right) - \phi_x(x, y) X(x, y, \phi(x, y))$$

$$- \phi_y(x, y) Y(x, y, \phi(x, y)) \, dy) \, dx = I + II + III.$$

The flux across the top part of bR has also three terms. The horizontal squares contribute with

$$I' = \sum_{i,j} \int_{R_{ij}} Z(x, y, z_{i,j})\, dx\, dy,$$

that has limit I as $\delta \to 0$ by the uniform continuity of $Z(x, y, \phi(x, y))$. The vertical faces with $x = x_j$ on which $N = (\pm 1, 0, 0)$ contribute with

$$II' = -\sum_{i,j} \int_{y_j}^{y_{j+1}} \int_{z_{i,j}}^{z_{i+1,j}} X(x_{i+1}, y, z)\, dy\, dz.$$

The integral in z equals, using the change of variable $z = \phi(x, y_j)$,

$$\begin{aligned}
&\int_{\phi(x_i,y_j)}^{\phi(x_{i+1},y_j)} X(x_{i+1}, y, z - \omega(\delta))\, dz \\
&\quad = \int_{x_i}^{x_{i+1}} X(x_{i+1}, y, \phi(x, y_j) - \omega(\delta))\phi_x(x, y_j)\, dx,
\end{aligned}$$

whence

$$II' = -\sum_{i,j} \int_{y_j}^{y_{j+1}} \int_{x_i}^{x_{i+1}} X(x_{i+1}, y, \phi(x, y_j) - \omega(\delta))\phi_x(x, y_j)\, dx\, dy.$$

This term has limit II as $\delta \to 0$ by the uniform continuity of the function $X(x, y, \phi(x, y))\phi_x(x, y)$. The flux across the vertical faces with $y = y_j$ is seen to have limit III in the same way. □

Example 17.8. We compute the flux of $F = (x^3, z^3, y^2)$ through the upper half M of the unit sphere, oriented by the outward normal, applying Gauss' theorem to the half-ball U. On the one hand, if B is the whole ball,

$$\begin{aligned}
\int_U \langle \nabla, F\rangle\, dm &= \int_U (3x^2)\, dm = \frac{3}{2}\int_B x^2\, dm \\
&= \frac{1}{2}\int_B (x^2 + y^2 + z^2)\, dm = 2\pi \int_0^1 \rho^4\, d\rho = \frac{2\pi}{5}.
\end{aligned}$$

On the other hand, $bU = M - D$, where D is the unit disc in the xy plane oriented by $(0, 0, 1)$. Therefore, the flux through M is

$$\frac{2\pi}{5} + \int_D y^2 dA = \frac{2\pi}{5} + \frac{\pi}{4} = \frac{13}{20}\pi.$$

We saw before that the Coulombian field F created by a charge C placed at q has flux $4\pi kC$ across a sphere centered at q. Now, a straightforward calculation shows that $\operatorname{div} F = 0$ off q. If U is a domain containing q in its interior and B is a ball centered at $q, B \subset U$, an application of Gauss' theorem shows that the flux across bU equals that across bB a, $4\pi kC$. As a consequence, if F is a Coulombian field created by a charge density,

$$F(p) = -k \int_C \frac{p-q}{|p-q|^3} \rho(q)\, dm(q), \quad p \notin C,$$

and U is a domain containing C, then the flux across bU is

$$4\pi k \int_C \rho(q)\, dm(q),$$

proportional to the total charge contained in U (Gauss' law of electrostatics).

Example 17.9. This example is parallel to Example 17.5. Consider again $F(p) = \frac{p}{|p|^3}$ in space and let S be an oriented surface not containing the origin such that each ray from the origin meets S in at most one point. Let U be the cone-shaped domain consisting of the union of such rays, with a small ball B centered at the origin removed. Since F is radial, its flux across the rays is zero; using $\operatorname{div} F = 0$ and Gauss' theorem, we find that

$$\left| \int_S \langle F, N \rangle\, dA \right| = \left| \int_{\partial B \cap \partial U} \langle F, N \rangle\, dA \right| = A(S'),$$

the area of the projection of S on the unit sphere. This value is called the *solid angle* determined by S, it measures the angle of vision of an observer looking at S from the origin. The unit in this context is called a *steredian.*

The following exercise is related to paragraph 17.2.5.

Exercise 17.3.

(a) For a bounded domain $U \subset \mathbb{R}^3$ with smooth boundary, prove that

$$3m(U) = \int_{bU} \langle F, N \rangle\, dA,$$

with $F = (x-a, y-b, z-c)$.

(b) Prove that if $R = R(U)$ is the radius of the smallest ball containing U and A is the area of bU, then

$$3m(U) \leq RA,$$

and that equality holds if and only if U is a ball.

The iso-perimetric inequality in $\mathbb{R}^3$ is the statement that

$$36\pi m(U)^2 \leq A^3,$$

with equality if and only if U is a ball. Its proof requires tools not covered in this course. There are other less known iso-perimetric inequalities in space relating $m(U), A$ and the mean width M defined in (15.14) (see [2]):

$$48\pi^2 m(U) \leq M^3, \quad 3m(U)M \leq A^2.$$

17.5.3 By the remark in paragraph 16.5.3, in dimension $n = 2$, Gauss' and Stokes' theorems and Green's formula are equivalent formulations: if $F = (X, Y)$ and $JF = (-Y, X)$, then

$$\operatorname{rot} F = Y_x - X_y = -\operatorname{div} JF,$$

so Gauss' theorem for JF is Stokes' theorem, Green's formula for F.

17.5.4 The divergence is thus a density of flux per unit volume. In case of fluids, it is possible to reach the same interpretation in terms of the trajectories $\varphi(t, s, p)$ of the fluid. Consider a portion of the fluid that at time s fills a domain U. At time t, the particle at $p \in U$ has moved to $\varphi(t, s, p)$, so U has shifted to $\varphi(t, s, U)$ and fills a volume

$$V(t) = m(\varphi(t, s, U)) = \int_U \det J(\varphi(t, s, U))\, dm(p).$$

Therefore,

$$V'(t) = \int_U \frac{d}{dt} \det J(\varphi(t, s, p))\, dm(p).$$

Now we claim that

$$\frac{d}{dt}|_{t=s} \det J(\varphi(t, s, p)) = \operatorname{div} F(s, p). \tag{17.3}$$

From Theorem 8.6, we know in fact the system satisfied the derivatives of φ at all times. Since we are just interested in $t = s$, we proceed directly, as it is simpler. We have, with $p = (x, y, z)$,

$$J(\varphi(t, s, p)) = \begin{pmatrix} \frac{\partial \varphi(t,s,p)}{\partial x} \\ \frac{\partial \varphi(t,s,p)}{\partial y} \\ \frac{\partial \varphi(t,s,p)}{\partial z} \end{pmatrix}.$$

To differentiate a determinant we must differentiate each row successively. Assume we differentiate the first one, to get

$$I = \begin{pmatrix} \frac{\partial^2 \varphi(t,s,p)}{\partial t \partial x} \\ \frac{\partial \varphi(t,s,p)}{\partial y} \\ \frac{\partial \varphi(t,s,p)}{\partial z} \end{pmatrix}$$

and now set $t = s$. Since $\varphi(s,s,p) = p$, the second and third rows become $(0,1,0)$ and $(0,0,1)$, respectively. The first one equals

$$\frac{\partial^2 \varphi(t,s,p)}{\partial x \partial t} = \frac{\partial}{\partial x} F(t,p).$$

Therefore, at $t = s$ one has $I = \frac{\partial X}{\partial x}$. Similarly, the other rows contribute with $\frac{\partial Y}{\partial y}$ and $\frac{\partial Z}{\partial z}$ and (17.3) is proved. Thus,

$$V'(s) = \int_U \operatorname{div} F \, dm(p), \quad \frac{V'(s)}{V(s)} = \frac{1}{m(U)} \int_U \operatorname{div} F(s,p) \, dm(p).$$

If U is an infinitesimal ball around p, this means that $\operatorname{div} F(s,p)$ can be thought as the rate of expansion per unit volume at p.

17.5.5 It is worth stressing why Stokes' and Gauss' theorems can be viewed as multidimensional versions of the fundamental theorem of calculus. The derivative f' of a function is a density for the action of f on oriented intervals $[a,b]$ by $f(b) - f(a)$, $\nabla \times F$ is a density for the action of F on closed oriented curves by circulation, and $\langle \nabla, F \rangle$ is a density for the action of F on oriented surfaces by flux. The three actions have the common feature

$$f(b) - f(a) = \sum_i (f(b_i) - f(a_i)), \int_{\partial S} \langle F, T \rangle \, ds = \sum_i \int_{\partial S_i} \langle F, T \rangle \, ds,$$

$$\int_{bU} \langle F, N \rangle \, dA = \sum_i \int_{bU_i} \langle F, N \rangle \, dA,$$

if $[a,b], S$ or U are split in infinitesimal pieces, so that Theorem 12.13 applies.

Chapter 18

Conservative and Solenoidal Fields

From this chapter on we work with continuous functions/vector fields on a domain U. In order to define gradient, rotational and flux in this category, we use a weak formulation, taking Stokes' and Gauss' theorems as definitions. Then we introduce two basic classes of vector fields, the gradient fields or *conservative* and the rotational fields or *solenoidal*. These are global notions implying the local conditions, *rotational-free* and *divergence-free*, respectively. We show that they are equivalent in star-shaped domains via the Poincaré lemma, for which we give a proof valid in the continuous category. Finally, we explain the language of *forms* in dimension n as objects acting on oriented k-sub-manifolds, and discuss how the basic theorems generalize to this setting.

18.1 The Weak Formulation

18.1.1 The following definitions are motivated by Stokes' and Gauss' theorems.

Definition 18.1. Given a continuous function u and a continuous vector field F in U, we say that $\nabla u = F$ *in Stokes' sense* if

$$\int_\Gamma \langle F, T\rangle \, ds = u(q) - u(p),$$

where Γ is a path within U from p to q. As shown soon, this is in fact equivalent to requiring that u is a C^1-function and $\nabla u = F$ in the classical sense. We are considering this apparently wider sense only to stress the analogy with the next definitions.

Definition 18.2. Given a continuous function u and a continuous vector field F in U, we say that $\operatorname{div} F = u$ *in Stokes' sense* if

$$\int_{\partial V} \langle F, N \rangle \, dm_{n-1} = \int_V u \, dm_n,$$

for all bounded domains with smooth boundary $V \subset U$. The condition amounts to saying that F has flow density u. If F is differentiable, $\operatorname{div} F = u$ in Stokes' sense amounts to $\langle \nabla, F \rangle = u$ point-wise.

One may have $\operatorname{div} F = u$ without F being C^1. For instance, a field $F = \operatorname{rot} G$ with G of class C^1 in $\mathbb{R}^3$, G not C^2, has zero divergence, combining Stokes' and Gauss' theorems. This notion may be seen as a weaker form of differentiability; for instance, if $F = (X, Y, Z)$, it implies

$$\begin{aligned} &\frac{\int_0^h \int_0^h [Z(x+t, y+s, z+h) - Z(x+t, y+s, h)] \, dt \, ds}{h^3} \\ &\quad + \frac{\int_0^h \int_0^h [Y(x+t, y+h, z+s) - Y(x+t, y, z+s)] \, dt \, ds}{h^3} \\ &\quad + \frac{\int_0^h \int_0^h [X(x+h, y+t, z+s) - X(x, y+t, z+s)] \, dt \, ds}{h^3} \\ &\quad \to u(x, y, z). \end{aligned}$$

The following definition is specific for $\mathbb{R}^3$:

Definition 18.3. We say that a continuous field F in $U \subset \mathbb{R}^3$ has *continuous rotational G in Stokes' sense* if

$$\int_{\partial S} \langle F, T \rangle \, ds = \int_S \langle G, N \rangle \, dA,$$

for all oriented bordered surfaces $S \subset U$. Equivalently, F has circulation density on S equal to $\langle G, N \rangle$. We write $\operatorname{rot} F = G$. If F is differentiable, $\operatorname{rot} F = G$ in Stokes' sense amounts to $\nabla \times F = G$ point-wise.

Again, one may have $\operatorname{rot} F = G$ without F being C^1. For instance, an arbitrary gradient field $F = \nabla u$ with u a C^1 function, not C^2, has zero rotational because

$$\int_{\partial S} \langle F, T \rangle \, ds = 0.$$

The condition implies weaker differentiability properties, for instance, with $F = (X, Y, Z), G = (A, B, C)$

$$\frac{\int_0^h (X(x+t,y,z) - X(x+t,y+h,z) + Y(x+h,y+t,z) - Y(x,y+t,z))\,dt}{h^2}$$

has limit $C(x, y, z)$ as $h \to 0$.

By paragraphs 16.5.3 and 17.5.3, in dimension $n = 2$, $\operatorname{rot} F$ is a scalar quantity and the statement $\operatorname{rot} F = g$ is equivalent to $\operatorname{div} JF = -g$, so Definition 18.3 reduces to Definition 18.2. Whenever we refer to $\operatorname{rot} F$ or $F \times G$, the context is $\mathbb{R}^3$.

18.1.2 We will be using the following identities. If F has continuous divergence in Stokes' sense (continuous rotational in $\mathbb{R}^3$, respectively) and u is of class C^1, then so does uF and

$$\operatorname{div}(uF) = u \operatorname{div} F + \langle F, \nabla u\rangle, \quad \operatorname{rot}(uF) = \nabla u \times F + u \operatorname{rot} F, \tag{18.1}$$

If F, G have continuous rotational and divergence in $\mathbb{R}^3$ in Stokes' sense, then so does $F \times G$ and

$$\operatorname{div}(F \times G) = \langle G, \operatorname{rot} F\rangle - \langle F, \operatorname{rot} G\rangle,$$

$$\operatorname{rot}(F \times G) = (\operatorname{div} G)F - \langle F, \nabla\rangle G - (\operatorname{div} F)G + \langle G, \nabla\rangle F. \tag{18.2}$$

Here, $\langle F, \nabla\rangle = \sum_i F_i D_i$ acts component-wise on G. A straightforward computation shows this for differentiable fields. For continuous fields, we use the approximation as follows.

For a continuous function, u we consider $u_\varepsilon = u * \phi_\varepsilon$ as in paragraph 13.2.2. For a field F, we define F_ε component-wise. Then $u_\varepsilon, F_\varepsilon$ are smooth in $U_\varepsilon = \{p : d(p, U^c) > \varepsilon\}$ and $u_\varepsilon \to u, F_\varepsilon \to F$ uniformly on compact sets.

Now, if $\nabla u = F$ or $\operatorname{div} F = u$ or $\operatorname{rot} F = G$ in Stokes' sense in U, we claim that $\nabla u_\varepsilon = F_\varepsilon, \operatorname{div} F_\varepsilon = u_\varepsilon, \operatorname{rot} F_\varepsilon = G_\varepsilon$ in U_ε. Indeed, we may write

$$F_\varepsilon(p) = \int \phi_\varepsilon(q) F(p-q)\, dm_n(q)$$

as a linear combination of translates of F. Then, if Γ joints P to Q,

$$\begin{aligned}\int_\Gamma \langle F_\varepsilon(p), T\rangle\, ds(p) &= \int \phi_\varepsilon(q) \left(\int_\Gamma \langle F(p-q), T\rangle\, ds(p)\right) dm_n(q)\\ &= \int \phi_\varepsilon(q) \left(\int_{\Gamma - q} \langle F(p), T\rangle\, ds(p)\right) dm_n(q)\\ &= \int \phi_\varepsilon(q)(u(Q-q) - u(P-q))\, dm_n(q) = u_\varepsilon(Q) - u_\varepsilon(P).\end{aligned}$$

Similarly, for $S \subset U$,

$$\int_{\partial S} \langle F_\varepsilon, T\rangle \, ds = \int \phi_\varepsilon(q) \left(\int_{\partial S} \langle F(p-q), T\rangle ds(p) \right) dm_n(q)$$

$$= \int \phi_\varepsilon(q) \left(\int_{\partial(S-q)} \langle F, T\rangle \, ds \right) dm_n(q) = \int \phi_\varepsilon(q) \left(\int_{S-q} \langle G, N\rangle \, dA \right) dm_n(q)$$

$$= \int \phi_\varepsilon(q) \left(\int_S \langle G(p-q), N\rangle \, dA \right) dm_n(q) = \int_S \langle G_\varepsilon, N\rangle \, dA.$$

This holds for all S, therefore, $\operatorname{rot} F_\varepsilon = G_\varepsilon$. In the same way, we prove that $\operatorname{div} F_\varepsilon = u_\varepsilon$.

Then the proofs of (18.2) and (18.1) proceed by approximation. For instance, $\operatorname{rot} F_\varepsilon = G_\varepsilon \to G = \operatorname{rot} F$ uniformly on compact sets. Now, if $V \subset U$, taking limits in

$$\int_{\partial V} \langle F_\varepsilon \times G_\varepsilon, N\rangle \, dA = \int_V (\langle G_\varepsilon, \operatorname{rot} F_\varepsilon\rangle - \langle F_\varepsilon, \operatorname{rot} G_\varepsilon\rangle) \, dm_n,$$

we find

$$\int_{\partial V} \langle F \times G, N\rangle \, dA = \int_V (\langle G, \operatorname{rot} F\rangle - \langle F, \operatorname{rot} G\rangle) \, dm_n,$$

proving the first in (18.2) for continuous fields. The proof of the other one and (18.1) is similar. It would be interesting to find a direct proof of these facts.

As a consequence, for all sub-domains $V \subset U$,

$$\int_{bV} u\langle F, N\rangle \, dm_{n-1} = \int_V (u \operatorname{div} F + \langle F, \nabla u\rangle) \, dm_n,$$

$$\int_{bV} \langle F \times G, N\rangle \, dA = \int_V (\langle G, \operatorname{rot} F\rangle - \langle F, \operatorname{rot} G\rangle) \, dm_n, \qquad (18.3)$$

If in these we take F, G constants, respectively,

$$\int_{bV} u\langle F, N\rangle \, dm_{n-1} = \int_V \langle \nabla u, F\rangle \, dm_n,$$

$$\int_{bV} \langle F \times G, N\rangle \, dA = \int_V \langle G, \operatorname{rot} F\rangle \, dm_n.$$

Using in the second one that $\langle F \times G, N\rangle = \langle N \times F, G\rangle$, we get

$$\int_{bV} \langle N \times F, G\rangle \, dA = \int_V \langle \operatorname{rot} F, G\rangle \, dm_n,$$

and since this holds for all constant G, we obtain the vector equality

$$\int_{bV} (N \times F) \, dA = \int_V \operatorname{rot} F \, dm_n. \tag{18.4}$$

Similarly, since the first one holds for all constant F, we get the vector equation

$$\int_{bV} uN \, dm_{n-1} = \int_V \nabla u \, dm_n.$$

Applying this to uv, we obtain

$$\int_V u\nabla v \, dm_n + \int_V v\nabla u \, dm_n = \int_{bV} uvN \, dm_{n-1},$$

$$\int_V uD_i v \, dm_n + \int_V vD_i u \, dm_n = \int_{bV} uvN_i \, dm_{n-1}, \tag{18.5}$$

which is the general version of (13.1).

18.1.3 Next, we will reformulate these concepts in terms of weak derivatives. We use the coupling between continuous vector fields F, G given by

$$(F, G) = \int_U \langle F, G\rangle \, dm_n,$$

whenever it makes sense, for instance if G has compact support. We use the same notation for the coupling between continuous functions

$$(u, \phi) = \int_U u\phi \, dm_n,$$

or between functions and fields

$$(\phi, F) = \int_U \phi \, F \, dm_n.$$

The important point is that the action of F, u on compactly supported G, ϕ, respectively, determines F, u uniquely, too. We look at these as being scalar products.

These couplings have two basic properties, besides (18.5). By (18.3), if F has continuous rotational and divergence, and ϕ, L are C^1 and compactly supported,

$$(F, \operatorname{rot} L) = (\operatorname{rot} F, L), \quad (F, \nabla\phi) = -(\operatorname{div} F, \phi). \tag{18.6}$$

We may state that rot is self-adjoint and ∇ and div are anti-adjoint operators. This motivates the following definition, analogous to Definition 13.1.

Definition 18.4. For continuous F, u in $U \subset \mathbb{R}^n$, we say that $\nabla u = F$ in the weak sense if

$$-(u, \nabla\phi) = (\phi, F)$$

for all ϕ compactly supported and smooth, and say that $\operatorname{div} F = u$ in the weak sense if

$$(F, \nabla\phi) = -(u, \phi)$$

for all ϕ compactly supported and smooth. For continuous F, G in $U \subset \mathbb{R}^3$, we say that $\operatorname{rot} F = G$ in the weak sense if

$$(G, L) = (F, \operatorname{rot} L)$$

for all L compactly supported smooth fields.

For u of class $C^1, \nabla u = F$ in the weak sense iff $\nabla u = F$ in the classical sense. For fields F of class C^1, $\operatorname{rot} F = G, \operatorname{div} F = u$ in the weak sense iff they hold in the classical sense.

The relationship with Definitions 18.1, 18.2 and 18.3 is given by the next propositions. We prove the first and third.

Proposition 18.1. *The following are equivalent for continuous u, F:*

(a) $\nabla u = F$ *in Stokes' sense.*
(b) $\nabla u = F$ *in the weak sense.*
(c) u *is of class* C^1 *and* $\nabla u = F$ *in the classical sense.*

Proof. We have already seen that (c) implies (a) in Theorem 17.1. Conversely, if (a) holds, for $p \in U$ and a unit direction v, with Γ the segment from p to $p + tv$,

$$u(p+tv) - u(p) = \int_\Gamma \langle F, T\rangle\, ds = \int_0^t \langle F(p+sv), v\rangle\, ds, \tag{18.7}$$

whence $D_v u(p)$ exists and equals $\langle F(p), v\rangle$. Assuming (b), let $u_\varepsilon = u * \phi_\varepsilon$, which is a smooth function in U_ε. By hypothesis,

$$\nabla u_\varepsilon(x) = \int u(y)\nabla_x \phi_\varepsilon(y-x)\, dm_n(y) = -\int u(y)\nabla_y \phi_\varepsilon(y-x)\, dm_n(y)$$

$$= \int F(y)\phi_\varepsilon(y-x)\, dm_n(y) = F_\varepsilon(x).$$

Now, since $u_\varepsilon \to u, \nabla u_\varepsilon = F_\varepsilon \to F$ uniformly on compacts, it follows that u is C^1 and $\nabla u = F$. □

Proposition 18.2. *The following properties are equivalent:*

(a) $\operatorname{div} F = u$ *in Stokes' sense, that is, for all sub-domains* $V \subset U$ *with regular boundary*

$$\int_{\partial V} \langle F, N\rangle\, dm_{n-1} = \int_V u\, dm_n.$$

(b) $\operatorname{div} F = u$ *in the weak sense, that is, for all smooth functions* ϕ *with compact support in* U

$$\int_U \langle F, \nabla\phi\rangle\, dm_n = -\int u\,\phi\, dm_n.$$

(c) *There exists smooth fields* F_ε *such that* $F_\varepsilon \to F, \operatorname{div} F_\varepsilon \to u$ *uniformly on compact sets.*

Proposition 18.3. *The following properties are equivalent in a domain* $U \subset \mathbb{R}^3$:

(a) $\operatorname{rot} F = G$ *in Stokes' sense, that is, for all surfaces* $S \subset U$

$$\int_{\partial S} \langle F, T\rangle\, ds = \int_S \langle G, N\rangle\, dA.$$

(b) $\operatorname{rot} F = G$ *in the weak sense, that is, for all smooth vector fields* L *with compact support in* U

$$\int_U \langle F, \operatorname{rot} L\rangle\, dm_n = \int_U \langle G, L\rangle\, dm_n.$$

(c) *There exist smooth fields F_ε such that $F_\varepsilon \to F, \operatorname{rot} F_\varepsilon \to G$ uniformly on compact sets.*

Proof. We know already that (a), (c) are equivalent and trivially (c) implies (b). It is enough to show that (b) also implies $\operatorname{rot} F_\varepsilon = G_\varepsilon$. The third component of $\operatorname{rot} F_\varepsilon$ is with $p = (x, y, z), q = (x', y', z')$,

$$(X_\varepsilon)_y(p) - (Y_\varepsilon)_x(p) = \int [X(q)D_y\phi_\varepsilon(p-q) - Y(q)D_x\phi_\varepsilon(p-q)]\, dm_n(q)$$
$$= -\int [X(q)D_{y'}\phi_\varepsilon(p-q) - Y(q)D_{x'}\phi_\varepsilon(p-q)]\, dm_n(q).$$

This is

$$-\int \langle F, \operatorname{rot} L_\varepsilon(p-q)\rangle\, dm_n(q),$$

with $L_\varepsilon = (0, 0, \phi_\varepsilon)$, so it equals

$$\int \langle G, L_\varepsilon(p-q)\rangle\, dm_n(q),$$

the third component of G_ε. □

18.2 Conservative and Solenoidal Fields

In this paragraph, our aim is the description of fields appearing globally as gradient or rotational in Stokes' sense in U.

Theorem 18.1. *The following are equivalent for a continuous vector field F:*

(a) *There is a continuous function u such that $\nabla u = F$ in Stokes' sense or weak sense.*
(b) *There is a C^1 function u such that $\nabla u = F$ in the classical sense.*
(c) *It has zero circulation along all closed curves in U. Equivalently, the circulation along an oriented curve joining p to q just depends on p, q.*

Proof. We need to just prove that (c) implies (a). We fix $p \in U$ and define

$$u(q) = \int_\Gamma \langle F, T\rangle\, ds, \tag{18.8}$$

where Γ is a path within U joining p to q. The hypothesis means that this definition is independent of the choice of Γ, so u is well defined. It is obvious that $\nabla u = F$ in Stokes' sense. □

The fields F satisfying the conditions in Theorem 18.1 are called *conservative*, and the function u with $\nabla u = F$ is called a *potential function.* In paragraph 16.3.2, we have seen that Newtonian fields are conservative and have potential functions. In terms of 1-forms $\omega = \sum_i F_i\, dx_i$, these forms are called *exact.*

Proposition 18.4. *If U is convex, a continuous field is conservative if and only if it has zero circulation along all triangles in U. A 1-form ω is exact if and only if it has line integral zero along all triangles in U.*

Proof. We fix $p \in U$ and define u as in (18.8) choosing Γ the segment from p to q. Then the argument in (18.7) works the same and $du = \omega$. □

Theorem 18.2 (Morera's theorem). A continuous function f in a plane domain is holomorphic if and only if

$$\int_{b\Delta} f(z)\, dz = 0,$$

for all closed triangles $\Delta \subset U$.

Proof. We already know from Theorem 17.5 that the statement holds for differentiable functions, so we need to prove just the sufficiency in a disc. If u is differentiable with $du = f\, dz$, and since $du = \partial u\, dz + \overline{\partial} u\, d\overline{z}$, it follows that $\overline{\partial} u = 0$. Therefore, u is holomorphic and $u' = f$. By Theorem 17.6, u is infinitely holomorphic and so is f. □

Theorem 18.3. *The following are equivalent for a continuous vector field F in a domain $U \subset \mathbb{R}^3$:*

(a) *It is the rotational in Stokes' or weak sense of a continuous vector field G.*
(b) *It has zero flux across all oriented surfaces with no border in U. Equivalently, if $S_1, S_2 \subset U$ have the same border, F has the same flux across S_1, S_2.*

Proof. If F is a rotational, $F = \operatorname{rot} G$ in Stokes' sense, and $S \subset U$ is a closed surface with no border in U

$$\int_S \langle F, N\rangle\, dA = \int_{\partial S} \langle G, T\rangle\, ds = 0.$$

The converse is a difficult result, in fact, it is a weak version of de Rham's theorem, which we do not prove, see paragraph 18.5.6. The author does not know a constructive proof of this theorem in a general domain. For convex or star-shaped domains, a proof is provided in Section 18.4. □

Using the remark in paragraph 16.5.3, the version of this theorem in dimension $n = 2$ is:

Theorem 18.4. *The following are equivalent for a continuous vector field F in a domain $U \subset \mathbb{R}^2$:*

(a) *There exists a function of class C^1 in U such that $F = J\nabla u$.*
(b) *It has zero flux across all closed oriented curves oriented in U. Equivalently, if $\Gamma_1, \Gamma_2 \subset U$ have the same end-points, F has the same flux across Γ_1, Γ_2.*

The vector fields F fulfilling condition (b) in the two previous theorems are called *solenoidal* or *incompressible.* In $\mathbb{R}^3$, a vector field G such that $\operatorname{rot} G = F$ is called a *potential vector.* In dimension $n = 2$, a field F is solenoidal if and only if JF is conservative.

18.3 Rotational-Free and Divergence-Free Vector Fields

18.3.1 Obviously, a necessary condition for a continuous field F in $\mathbb{R}^2$ or $\mathbb{R}^3$ to be conservative is that $\operatorname{rot} F = 0$ in Stokes' sense or in the weak sense: for compactly supported L

$$\int_U \langle F, \operatorname{rot} L\rangle \, dm_n = 0.$$

We call F *rotational-free* or *irrotational.* If $F = (X, Y, Z), L = (\phi, 0, 0)$, this means

$$\int_U (Y D_z\phi - X D_y\phi) \, dm_n = 0,$$

that is $D_zY = D_yX$ in the weak sense. Analogously, we find that $D_zX = D_xZ, D_zY = D_yZ$ in the weak sense.

Along the same lines, a necessary condition for F to be solenoidal in $\mathbb{R}^3$ is that $\operatorname{div} F = 0$ in Stokes' sense: for compactly supported smooth ϕ

$$\int_U \langle F, \nabla\phi\rangle \, dm_n = 0.$$

We call it *divergence-free.* If $F = \nabla \times G$ with G twice differentiable, this is again Schwarz's rule

$$\langle \nabla, \nabla \times G\rangle = 0.$$

These two conditions, expressing essentially that some density is zero, are of *local character*. For differentiable fields they are, respectively, $\nabla \times F = 0$ and $\langle \nabla, F \rangle = 0$. In dimension two, $F = (X, Y)$, respectively,

$$X_y = Y_x, \quad X_x + Y_y = 0.$$

18.3.2 In general, these two necessary conditions, of local character, do not imply that F is conservative or solenoidal, respectively. For instance, in $U = \mathbb{R}^2 \setminus (0,0)$, the field in Example 17.2

$$F = \left(-\frac{y}{x^2+y^2}, \frac{x}{x^2+y^2} \right),$$

is locally a gradient, the differential of a local determination of the argument, but it is not globally a gradient in U, since no global determination exists. Along the same lines, the field F in Example 17.9, $F(p) = \frac{p}{|p|^3}$ is divergence-free in $\mathbb{R}^3 \setminus (0,0,0)$, but it is not solenoidal because $\int_{\partial B} \langle F, N \rangle \, dA \neq 0$ for a ball centered at the origin.

If U satisfies certain topological conditions, the local conditions imply the global ones. These conditions are expressed in terms of the vanishing of certain *cohomology groups* (see Section 18.5). Here we will describe them intuitively.

A domain $U \subset \mathbb{R}^n$ is called *simply connected* if every closed curve in U can be continuously contracted within U to a point. Intuitively, it means that every closed curve in U is the border of some surface in U. For plane domains it means that U has no holes. It is then clear, by Stokes' theorem and Green's formula, that a continuous field with $\operatorname{rot} F = 0$ in Stokes' sense is conservative and so it is a gradient of a C^1-function. Thus, in plane simply connected domains, conservative fields are the gradients of C^1 functions, and solenoidal fields are those of the form $J\nabla u$.

Similarly, if every closed surface $S \subset U \subset \mathbb{R}^3$ is the border of some sub-domain, a continuous field with zero divergence in Stokes' sense is solenoidal, and by de Rham's Theorem 18.3 it is a rotational in Stokes' sense. In this case, we say that *U has no cavities.*

18.4 Poincaré's Lemma

18.4.1 In this paragraph, our aim is to prove constructively, without appealing to de Rham's Theorem 18.3, that in balls, or more generally in a star-shaped domain, the conditions $\operatorname{rot} F = 0$ ($\operatorname{div} F = 0$) in Stokes' sense imply that F is conservative (resp., solenoidal). A domain U is called *star-shaped* with respect to a point $p \in U$ if for all $q \in U$ the segment from p to q is within U. Our aim is to find explicit operators I_1, I_2 acting on continuous fields such that $I_1 F$ is a C^1-function with $\nabla I_1 F = F$ if $\operatorname{rot} F = 0$ in Stokes' sense, and $\operatorname{rot} I_2 F = F$ in Stokes' sense if $\operatorname{div} F = 0$ in Stokes' sense. We will see that a choice of I_1 leads automatically to a choice of I_2. We work in $\mathbb{R}^3$. Recall that in $\mathbb{R}^2$, since $\operatorname{rot} F = -\operatorname{div} JF$, we need to just consider conservative fields.

Assuming $U \subset \mathbb{R}^3$ star-shaped with respect to the origin, for instance a ball, the simplest choice of I_1 is to use the segment from the origin to p as in Theorem 18.1. If $F = (X, Y, Z), p = (x, y, z)$,

$$I_1 F(p) = \int_0^1 \langle F(tp), p \rangle dt = \int_0^1 (xX(tp) + yY(tp) + zZ(tp))\, dt.$$

From Theorem 18.1, we already know that I_1 works. Now, assume F is a smooth field. We know that $\nabla I_1 F = F$ if $\operatorname{rot} F = 0$. This indicates that $F - \nabla I_1 F$ should depend only on $\operatorname{rot} F$. Computing,

$$D_x I_1 F(p) = \int_0^1 [X(tp) + txX_x(tp) + tyY_x(tp) + tzZ_x(tp)]\, dt,$$

and inserting $\frac{d}{dt}(tX(tp)) = X(tp) + txX_x(tp) + tyX_y(tp) + tzX_z(tp)$, we get

$$D_x I_1 F(p) = X(p) + \int_0^1 [ty(Y_x - X_y)(tp) + tz(Z_x - X_z)(tp)]\, dt.$$

A similar computation with the other components leads to

$$(F - \nabla I_1 F)(p) = -\int_0^1 tp \times (\operatorname{rot} F)(tp)\, dt.$$

Thus, with

$$I_2 G(p) = -\int_0^1 tp \times G(tp)\, dt,$$

one has

$$F = \nabla(I_1 F) + I_2(\operatorname{rot} F). \tag{18.9}$$

Applying rot to both sides, we get

$$\operatorname{rot} F = \operatorname{rot} I_2(\operatorname{rot} F).$$

Thus, for rotational fields $G = \operatorname{rot} F$ one has $G = \operatorname{rot} I_2 G$. This indicates that for a general smooth field F the difference $F - \operatorname{rot} I_2 F$ should depend only on $\operatorname{div} F$. To compute $\operatorname{rot} I_2F$, we note that

$$I_2F(p) = -p \times B, \quad B(p) = \int_0^1 tF(tp)\,dt,$$

and use the second in (18.2)

$$\operatorname{rot}(A \times B) = (\operatorname{div} B)A - (\operatorname{div} A)B + \langle B, \nabla\rangle A - \langle A, \nabla\rangle B,$$

with $A = p$. Recall that here $\langle B, \nabla\rangle = B_1D_x + B_2D_y + B_3D_z$ is applied component-wise and similarly $\langle A, \nabla\rangle$, so that $\langle B, \nabla\rangle A = B$. On the other hand,

$$\begin{aligned}\langle A, \nabla\rangle B &= \frac{d}{ds}_{s=1} B(sp)\\ &= \frac{d}{ds}_{s=1} \int_0^1 tF(tsp)\,dt = \frac{d}{ds}\Big|_{s=1} \frac{1}{s^2} \int_0^s tF(tp)\,dt = -2B(p) + F(p).\end{aligned}$$

Finally, $\operatorname{div} A = 3$, and

$$\operatorname{div} B(p) = \int_0^1 t^2 \operatorname{div} F(tp)\,dt.$$

Altogether, we find

$$F = \operatorname{rot} I_2F + I_3(\operatorname{div} F), \tag{18.10}$$

where I_3 acts on functions

$$I_3(\phi)(p) = \left(\int_0^1 t^2\phi(tp)\,dt\right)p.$$

For a plane field $F = (X, Y), \operatorname{rot} F = Y_x - X_y, p = (x, y)$. Treating it as $F = (X, Y, 0)$, we find

$$F(p) = \nabla\left(\int_0^1 \langle F(tp), p\rangle dt\right) - \left(\int_0^1 t \operatorname{rot} F(tx, ty)\,dt\right)(y, -x),$$

that is, $I_2(\operatorname{rot} F)$, the obstruction to be conservative, is tangential to circles through the origin depending on $\operatorname{rot} F$. If we apply it to $JF = (-Y, X)$,

$$JF(p) = \nabla\left(\int_0^1 \langle JF(tp), p\rangle dt\right) - \left(\int_0^1 \operatorname{div} X(tx, ty)\,dt\right)(y, -x),$$

$$F(p) = J\nabla\left(\int_0^1 \langle F(tp), Jp\rangle dt\right) + \left(\int_0^1 \operatorname{div} X(tx, ty)\,dt\right)(x, y),$$

and we see that the obstruction to be solenoidal is radial depending on $\operatorname{div} F$.

Formulas (18.9) and (18.10) are known together as the *Poincaré lemma.* These computations show that $I_1(F), I_2(F)$ are of class C^1 if F is of class C^1 and

$$\nabla I_1 F = F \quad \text{if } \operatorname{rot} F = 0, \quad \operatorname{rot} I_2 F = F \quad \text{if } \operatorname{div} F = 0.$$

Now, we note that I_2 makes sense on a continuous field, too. If F has zero divergence in Stokes' sense, by Proposition 18.2 we can choose smooth approximation vector fields F_ε with $\operatorname{div} F_\varepsilon = 0$ convergent to F on compacts. Then $I_2(F_\varepsilon)$ converges to $I_2(F)$ uniformly as well, and taking limit as $\varepsilon \to 0$ in

$$\int_{\partial S} \langle I_2 F_\varepsilon, T\rangle \, ds = \int_S \langle F_\varepsilon, N\rangle \, dA,$$

we get

$$\int_{\partial S} \langle I_2 F, T\rangle \, ds = \int_S \langle F, N\rangle \, dA,$$

proving that $F = \operatorname{rot} I_2 F$ in Stokes' sense.

We collect everything in the following statement.

Theorem 18.5.

(a) *The continuous fields F in a domain U such that* $\operatorname{rot} F = 0$ *in Stokes' sense are exactly those which are locally the gradient on a C^1 function u. In a star-shaped domain, this holds globally, an explicit solution being $u_0 = I_1 F$. The general solution is $u = u_0 + c$.*

(b) *The continuous fields in $U \subset \mathbb{R}^3$ with* $\operatorname{div} F = 0$ *in Stokes' sense are exactly those which are locally a rotational in Stokes' sense of a continuous field G. In a star-shaped domain, this holds globally, an explicit solution being $G_0 = I_2 F$. The general solution is $G = G_0 + \nabla u, u \in C^1(U)$.*

(c) *The continuous fields in $U \subset \mathbb{R}^2$ with* $\operatorname{div} F = 0$ *in Stokes' sense are exactly those locally of the form $J\nabla u$. In a star-shaped domain, this holds globally.*

18.4.2 We will now study the field $I_2 F$ in more detail and understand better its structure. Rather to understand it point-wise, let us try to understand its action on an oriented curve Γ by circulation. Assume that Γ is such that each ray from the origin meets Γ in at most one point, as in Example 17.5. Those rays meeting Γ define a surface, that we orient so that the induced orientation on Γ is the given one. See Figure 17.3.

Now, we have already noted that $I_2F(p) = p \times B(p)$ for some field B. Therefore, $I_2F(p)$ is orthogonal to p and so has zero circulation along rays. Since $\operatorname{rot} I_2F = F$ in Stokes' sense, the conclusion is that

$$\int_\Gamma \langle I_2F, T\rangle\, ds = \int_S \langle F, N\rangle\, dA.$$

This is much in the spirit of the definition of I_1. The action of I_1F on points is a circulation of F, and we see here that the action of I_2F on curves is a flux of F.

Alternatively, we can use this as a definition of I_2F. Using this point of view it is intuitively clear that $\operatorname{rot} I_2F = F$ in Stokes' sense. Indeed, if $\partial S' = \Gamma$ and S is as explained, S and S' have a common border, whence using that F has zero divergence in Stokes' sense,

$$\int_{S'} \langle F, N\rangle\, dA = \int_S \langle F, N\rangle\, dA,$$

which by definition equals

$$\int_\Gamma \langle F, T\rangle\, ds.$$

As a final remark, we point out that I_2F is not in general differentiable for continuous F.

18.4.3 For C^1 fields F in a ball or the whole space, to solve $\nabla u = F$ if $\operatorname{rot} F = 0$ or $\operatorname{rot} G = F$ if $\operatorname{div} F = 0$, instead of using the Poincaré operators I_1, I_2, one proceeds by anti-differentiation, as in paragraph 4.7.3.

Example 18.1. The field

$$F = (\sin xyz + xyz \cos xyz), \quad x^2 z \cos xyz + y, x^2 y \cos xyz + z),$$

satisfies $\operatorname{rot} F = 0$. Anti-differentiating $u_x = \sin xyz + xyz \cos xy$, we get $u = x \sin xyz + C(y, z)$ for some function C. Then $u_y = x^2 z \cos xyz + C_y$, $C_y = y, C(y, z) = \frac{y^2}{2} + D(z)$. Then $u_z = x^2 y \cos xyz + D'(z)$, so $D' = z$, $D = \frac{z^2}{2} + c$ and $u = x \sin xyz + \frac{1}{2}(y^2 + z^2) + c$.

To solve $\operatorname{rot} G = F, G = (A, B, C), F = (X, Y, Z)$ that is

$$C_y - B_z = X, \quad A_z - C_x = Y, \quad B_x - A_y = Z,$$

we seek for a solution with $C = 0$. Then the first two equations mean

$$B(x,y,z) = -\int_0^z X(x,y,t)\,dt + B_1(x,y), A(x,y,z) = \int_0^z Y(x,y,t)\,dt + A_1(x,y).$$

Then we must choose A_1, B_1 such that

$$B_x - A_y = -\int_0^z X_x(x,y,t)\,dt + (B_1)_x(x,y) - \int_0^z Y_y(x,y,t)\,dt - (A_1)_y(x,y) = Z.$$

Since $\langle \nabla, F\rangle = 0$, the sum of the integrals is $-\int_0^z Z_z(x,y,t)\,dt = Z(x,y,0) - Z(x,y,t)$, whence the condition on A_1, B_1 becomes

$$(B_1)_x(x,y) - (A_1)_y(x,y) = Z(x,y,0).$$

A solution is $A_1 = 0, B_1(x,y) = \int_0^x Z(t,y,0)\,dt$.

18.5 The Language of Forms and Chains*

In this section, we address how to generalize the concepts of vector analysis in dimension 2, 3 to higher dimensions.

18.5.1 The first point is to precise the domain of integration, *regular sub-manifolds M with border ∂M*. This concept has already been defined, requiring that around $p \in \partial M$, in a suitable coordinate system $u_1, \ldots, u_n, M$ is given by $u_k \leq 0, u_{k+1} = \cdots = u_n = 0$ and ∂M by $u_k = u_{k+1} = \cdots = u_n = 0$.

The subject of orientation, which for $k = 1$, $k = n - 1$ we have defined simply as a continuous choice of a unit tangent or a unit normal, respectively, must be reformulated in general as follows.

In a linear space V of dimension k, two different ordered basis $[v_1, \ldots, v_k]$, $[w_1, \ldots, w_k]$ are said to be equivalent if the transition matrix has positive determinant. Thus, they are classified in two different classes. *Orienting* V means choosing one of them, and the ordered basis in the chosen class are said to be *coherent or have positive orientation*. The orientation we choose for $\mathbb{R}^n$ is the one given by the canonical basis.

A sub-manifold M of dimension k is called *orientable* if it is possible to orient each tangent space $T_p(M)$ (that is, choosing one of the two equivalent classes of ordered basis) in a continuous way. This amounts to the existence of an atlas (U_i, Φ_i) of M such that the transition maps $\Phi_i\Phi_j^{-1}$ have positive determinant. For $k = n$, domains are obviously orientable and we consider them oriented choosing the canonical basis; the positive basis are those with positive determinant.

In case $k = n - 1$, orientability is equivalent to the existence of a continuous normal field N, the positive basis $v_1, \ldots, v_{n-1}$ being those for which $\det(N, v_1, \ldots, v_{n-1}) > 0$, that is, N has the direction of $v_1 \times \cdots \times v_{n-1}$. A bounded domain with regular boundary, $U = \{u < 0\}$, is orientable; the induced orientation on bU is the one defined by the outward normal $N = \frac{\nabla u}{|\nabla u|}$.

If the coordinates $u_1, \ldots, u_k$ of M around p are coherent with the orientation of M (that is $\partial_i, i = 1, \ldots, k$ is a basis coherent with the orientation of M), the orientation of ∂M is the one for which $(-1)^k \partial_i$, $i = 1, \ldots, k-1$ is a positive basis.

18.5.2 Now about the objects to integrate. For $k = 1$, $k = n - 1$, circulation and flux of fields have already been defined.

The need for another kind of object, other than fields, is best understood when looking at circulation density. We can consider closed curves lying on surfaces and define exactly in the same way a circulation density at each point $p \in S$,

$$D_S(F, p) = \lim_{S_i \to p} \frac{\mathcal{C}(\partial S_i)}{A(S_i)}. \tag{18.11}$$

In $\mathbb{R}^3$ and for differentiable fields, all these densities are coded by $\operatorname{rot} F$ and $D_S(F, p) = \langle F, N \rangle$ where N is the normal to S at p. In dimension $n > 3$, what we have, fixed F and letting S vary, is a map that assigns a number to each couple (p, Π) consisting of a point p and an oriented plane Π through it. This cannot be coded by a vector field, just think of the number of degrees of freedom.

So, another concept is needed to deal with circulation density and to extend the concept of circulation and flux to the intermediate cases.

Circulation and flux are obtained adding up the action of a vector field F on tangent fields T and pairs of tangent fields Φ_s, Φ_t defined, respectively, by

$$\langle F, T \rangle, \quad \langle F, \Phi_s \times \Phi_t \rangle = \det(T, \Phi_s, \Phi_t).$$

The first is linear in T, the second is bilinear in Φ_s, Φ_t and antisymmetric, it equals $|F|\cos\theta$ times the area of the parallelogram spanned by Φ_s, Φ_t. So these numbers are signed lengths and signed areas. Thus, in general we want to consider maps B assigning to each family $v_1, \ldots, v_k$ of vectors a number

$$B(v_1, v_2, \ldots, v_k) = \pm\lambda m_k(P(v_1, v_2, \ldots, v_k)), \quad \lambda \in \mathbb{R},$$

with $P(v_1, v_2, \ldots, v_k)$ being the parallelepiped these vectors determine in the subspace V they span. Moreover, we want B to have different signs on $v_1, \ldots, v_k$ and $w_1, \ldots, w_k$ if they are basis of the space V with opposite orientation (that is, the transition matrix has negative determinant). Fixing a basis $w_1, \ldots, w_k$, assume it is orthonormal, we see that necessarily

$$B(v_1, v_2, \ldots, v_k) = (\det A)\, B(w_1, \ldots, w_k),$$

A being the transition matrix between the two basis. This is the general form of a k-*linear alternate* map

$$B : V \times V \times \cdots \times V \to \mathbb{R},$$

that is B is linear in each variable and

$$B(v_{\sigma(1)}, \ldots, v_{\sigma(k)}) = \epsilon(\sigma) B(v_1, v_2, \ldots, v_k),$$

$\epsilon(\sigma)$ denoting the index of the permutation σ. The space of such forms, 1-dimensional, is called $\Lambda^k V$.

Now, we need such a thing for each $p \in M$, so we are lead to the concept of a k-*differential form on* M. We already met the concept of a differential form in paragraph 8.1.2. This is a map ω assigning to each $p \in M$ an $\omega(p) \in \Lambda^k(T_p(M))$. This is usually the restriction to M of the same kind of map defined in the whole of $\mathbb{R}^n$ or a domain $U \subset \mathbb{R}^n$. Thus, a k-differential form in $\mathbb{R}^n$ is a map assigning to each p an $\omega(p) \in \Lambda^k(\mathbb{R}^n)$, a k-linear alternate map acting on $\mathbb{R}^n = T_p(\mathbb{R}^n)$. The space $\Lambda^k(\mathbb{R}^n)$ no longer has dimension 1, of course. Restricting ω to a k-manifold means restricting $\omega(p), p \in M$, to $T_p(M)$.

The k-linear alternate maps in $\mathbb{R}^n$ are manipulated in terms of a basis of 1-forms. For 1- forms $\omega_1, \ldots, \omega_k$, one defines the k-linear map

$$(\omega_1 \otimes \cdots \otimes \omega_k)(v_1, \ldots, v_k) = \omega_1(v_1) \cdots \omega(v_k),$$

and makes it alternate, defining

$$(\omega_1 \wedge \cdots \wedge \omega_k)(v_1, \ldots, v_k) = \sum_{\sigma} \epsilon(\sigma)(\omega_1 \otimes \cdots \otimes \omega_k)(v_{\sigma(1)}, \ldots, v_{\sigma(k)}).$$

This is called the *exterior or wedge product.* It is straightforward to prove that if $\omega_1, \dots, \omega_n$ is a basis of 1-forms, the k-forms

$$\omega_{i_1} \wedge \cdots \wedge \omega_{i_k}, \quad i_1 < i_2 < \cdots < i_k, \quad i_j = 1, \dots, n,$$

constitute a basis of $\Lambda^k(\mathbb{R}^n)$, whence this space has dimension $\binom{n}{k}$.

Choosing $dx_1, \dots, dx_n$ as basis of 1-forms, the general expression of a k-differential form in a domain U is

$$\omega = \sum_I f_I(x)\, dx^I, \quad dx^I = dx_{i_1} \wedge \cdots \wedge dx_{i_k}, \quad I = i_1 < i_2 < \cdots < i_k,$$

with f_I functions. It is said to be continuous, C^1, etc. if the f_I are. For $k = 1$ and $k = n-1$, there is an identification with vector fields $F = (F_1, \dots, F_n)$ through

$$\sum_i F_i dx_i, \sum_i F_i \cdots \wedge \widehat{dx_i} \wedge \cdots . \tag{18.12}$$

For $k = n$, n-differential forms are identified with functions,

$$\omega = f dx_1 \wedge \cdots \wedge dx_n. \tag{18.13}$$

The exterior product of $\eta = \sum_I f_I dx^I, \mu = \sum_J g_J dx^J$ is defined in the obvious way

$$\eta \wedge \mu = \sum_{I \cap J = \emptyset} f_I g_J dx^I \wedge dx^J.$$

18.5.3 Now we want to define $\int_M \omega$ if ω is a k-differential form and M is an oriented k-sub-manifold. Now, on M there is a distinguished k-form η on $T_p(M)$, namely the one defined setting

$$\eta(v_1, \dots, v_k) = m_k(P(v_1, \dots, v_k)),$$

for a positively oriented basis of $T_p(M)$. The form η is called the *volume form* of M. It follows from these considerations that the restriction of a k-form ω to M equals $f\eta$ for some function f on S. Then, by definition,

$$\int_M \omega = \int_M f\, dm_k.$$

Thus, orientability allows to assign a function to a k-form; we cannot integrate k-forms on a non-orientable k-sub-manifold, but we can indeed integrate functions.

In terms of a chart (U, Φ) coherent with the orientation (meaning that $d\Phi$ transforms the canonical basis in the positively oriented basis $D_1\Phi, \ldots, D_k\Phi$),

$$\omega(D_1\Phi, \ldots, D_k\Phi) = f m_k(P(D_1\Phi, \ldots, D_k\Phi)),$$

so for f supported in $\Phi(U)$,

$$\int_M \omega = \int_M f\, dm_k = \int_U f m_k(P(D_1\Phi, \ldots, D_k\Phi))\, du_1 \cdots du_k$$
$$= \int_U \omega(D_1\Phi, \ldots, D_k\Phi)\, du_1 \cdots du_k. \tag{18.14}$$

This is the general version of (16.3) for $k = 1$ and (16.6) for $k = 2$.

The right-hand side of the last formula is the integral of the k-form

$$\Phi^*(\omega)(u)(v_1, \ldots, v_k) = \omega(\Phi(u))(d\Phi(u)(v_1), \ldots, d\Phi(u)(v_k)),$$

called the *pull-back* of ω. Thus, we have the functor property

$$\int_{\Phi(U)} \omega = \int_U \Phi^*(\omega).$$

In coordinates, if $\omega = \sum_I f_I(x)\, dx_{i_1} \wedge \cdots \wedge dx_{i_k}$,

$$\Phi^*(\omega) = \sum_I f_I(\Phi(u))\, d\Phi_{i_1} \wedge \cdots \wedge d\Phi_{i_k}.$$

Note that in case $k = n$, this is coherent with the change of variables formula, since if $x = \Phi(u)$ is a diffeomorphism preserving orientations, its Jacobian $J\Phi$ is positive and

$$\Phi^*(f(x)dx_1 \wedge \cdots \wedge dx_n) = f(\Phi(u))J\Phi(u)du_1 \wedge \cdots \wedge du_n.$$

18.5.4 We comment now on the version of Stokes' and Gauss' theorems in higher dimensions. We already stated Gauss' theorem for hyper-surfaces in $\mathbb{R}^n$, this corresponds to integration of $n-1$-forms through the identification in (18.12).

Remember that now we integrate differential forms on oriented sub-manifolds with border. We are given a k-differential form and a $k+1$-dimensional sub-manifold M with boundary and we want to find out which is the $k+1$-differential form Ω such that

$$\int_{\partial M} \omega = \int_M \Omega.$$

We may assume that $\omega = f(x)\, \omega_1 \wedge \cdots \wedge \omega_k$, with ω_j 1-forms among $dx_1, \ldots, dx_n$. Assume for simplicity that M is the $k+1$-dimensional parallelepiped in $\mathbb{R}^n$

$$M = \left\{ x = \sum_{j=1}^{k+1} u_j v_j, 0 \leq u_j \leq 1 \right\},$$

oriented declaring $v_1, \ldots, v_{k+1}$ a positive basis. In the face $u_{k+1} = 0$ of ∂M, the induced orientation is $(-1)^{k+1}$ times that of $v_1, \ldots, v_k$ while at the face $u_{k+1} = 1$, is the opposite. Thus, the contributions of these two faces in ∂M is, using Definition 18.14,

$$(-1)^k \int_{0 \leq u_j \leq 1} \left[f\left(\sum_{j=1}^{k} u_j v_j + v_{k+1} \right) - f\left(\sum_{j=1}^{k} u_j v_j \right) \right]$$

$$(\omega_1 \wedge \cdots \wedge \omega_k)(v_1, \ldots, v_k)\, du_1 \cdots du_k$$

$$= (-1)^k \int_{0 \leq u_j \leq 1} df\left(\sum_{i=1}^{k+1} u_i v_i \right) (v_{k+1}) (\omega_1 \wedge \cdots \wedge \omega_k)(v_1, \ldots, v_k) du_1 \cdots du_k\, du_{k+1}.$$

Now, in

$$(df(q) \wedge \omega_1 \wedge \cdots \wedge \omega_k)(v_1, v_2, \ldots, v_{k+1})$$

$$= \sum_\sigma \epsilon(\sigma)(df(q) \otimes \omega_1 \otimes \cdots \otimes \omega_k)(v_{\sigma(1)}, \ldots, v_{\sigma(k+1)}),$$

the terms in the last integral are those for which $\sigma(1) = k+1$. Similarly, the faces $u_i = 0, 1$ contribute with the terms for which $\sigma(1) = i$. Therefore,

$$\int_{\partial M} \omega = \int_{0 \leq u_j \leq 1} \left[df\left(\sum_{i=1}^{k+1} u_i v_i \right) \wedge \omega_1 \wedge \cdots \wedge \omega_k \right]$$

$$(v_1, \ldots, v_{k+1}) du_1 \cdots du_k\, du_{k+1},$$

which by Definition 18.14 is $\int_M df \wedge \omega_1 \wedge \cdots \wedge \omega_k$.

This indicates that for $\omega = \sum_I f_I(x)\, dx_{i_1} \wedge \cdots \wedge dx_{i_k}$, we must choose

$$\Omega = \sum_I df_I(x) \wedge dx_{i_1} \wedge \cdots \wedge dx_{i_k}.$$

This $k+1$-form is called the *exterior derivative* of ω, denoted $d\omega$.

Exercise 18.1. Check the following properties:

(a) $d^2\omega = 0$ for a twice differentiable ω.
(b) $d(\omega \wedge \eta) = d\omega \wedge \eta + (-1)^l \omega \wedge d\eta$, η an l-differential form.

Thus, for a 1-differential form ω or field through the identification (18.12), its circulation density as defined by (18.11) is

$$d\omega(p)(v_1, v_2),$$

with v_1, v_2 an orthonormal basis of $T_p(S)$. If $n = 3$,

$$d\omega(p)(v_1, v_2) = \langle \operatorname{rot} F, v_1 \times v_2 \rangle.$$

With the identifications (18.12), (18.13), for $\omega = \sum_i F_i \cdots \wedge \widehat{dx_i} \wedge \cdots$

$$d\omega = \sum_i D_i F_i dx_1 \wedge \cdots \wedge dx_n,$$

and we recover the divergence. Altogether, the exterior derivative $d\omega$ is the unifying concept for the rotational and the divergence.

Theorem 18.6 (General version of Stokes' theorem). Let M be an oriented $k+1$-dimensional sub-manifold with border ∂M with the induced orientation and ω a k-differential form of class C^1 defined around M. Then

$$\int_{\partial M} \omega = \int_M d\omega.$$

We have proved it for a parallelepiped, but in fact this implies the general case. Indeed, using again a partition of unity argument, it is enough to prove the theorem for $\Phi(Q)$ where (U, Φ) is a local chart coherent with the orientation and Q is an interval in $\mathbb{R}^{k+1}$. But

$$\int_{\Phi(Q)} d\omega = \int_Q \Phi^*(d\omega), \quad \int_{\partial\Phi(Q)} \omega = \int_{\Phi(\partial Q)} \omega = \int_{\partial Q} \Phi^*(\omega).$$

The result then follows from the fact that exterior differentiation and pull-back commute, $\Phi^*(d\omega) = d(\Phi^*(\omega))$, routinely checked applying the definitions.

Thus, Stokes' theorem is equivalent to the statement

$$\int_{\partial Q} \Phi^*(\omega) = \int_Q \Phi^*(d\omega),$$

for a chart Φ. Now, it is plain that this holds for a general C^1-map, without Φ being one-to-one or having full rank, or the target $\Phi(Q)$ being orientable. This is called *Stokes' theorem for chains.*

Stokes' theorem says that in the pairing between forms and submanifolds, the operators d and ∂ are formally adjoint. As a consequence of $\partial^2 = 0$, one has $d^2 = 0$ for

$$\int_M d^2\omega = \int_{\partial M} d\omega = \int_{\partial^2 M} \omega = 0.$$

Of course, this can be easily checked too in coordinates. In terms of fields in $\mathbb{R}^3$,

$$\operatorname{rot} \nabla u = 0, \quad \operatorname{div}(\operatorname{rot} F) = 0.$$

Example 18.2. Consider a bounded domain U with smooth boundary, oriented by the outward normal. Let $v_1, \ldots, v_{n-1}$ be unit orthonormal vectors, $v = v_1 \times v_2 \times \cdots \times v_{n-1}$, positively oriented, $N = \frac{v}{|v|}$. Then

$$(-1)^{i-1}\widehat{dx_i}(v_1, \ldots, v_n) = N_i|v| = N_i m(P(v_1, \ldots, v_{n-1}).$$

Therefore,

$$\int_{bU} X_i(x)(-1)^{i-1}\widehat{dx_i} = \int_{bU} X_i N_i \, dm_{n-1},$$

$$\sum (-1)^{i-1} N_i \widehat{dx_i} = dm_{n-1} \quad \text{on } bU.$$

Example 18.3. Stokes' theorem does not apply to the Möbius band M, which is non-orientable. However, a valid version can be obtained if we look at it as a two-chain. Let us cut M along one position L of the moving segment, say between points p, q, and let us move ε distance apart the two parts. We obtain a surface M_ε, orientable, whose border consists in two copies L_1, L_2 of L at distance ε and two curves Γ_1, Γ_2 whose limit is ∂M as $\varepsilon \to 0$. If we orient M_ε, the induced orientation on L_1, L_2 do not cancel when $\varepsilon \to 0$. If the induced orientation on L is from p to q, the induced orientation on both Γ_1, Γ_2 is from q to p. Making $\varepsilon \to 0$ we obtain a version of Stokes' theorem for M: if $M \setminus L$ is oriented by N, then

$$\int_M \langle \operatorname{rot} F, N \rangle \, dA = \int_{\Gamma_1} \langle F, T \rangle \, ds + \int_{\Gamma_2} \langle F, T \rangle \, ds + 2 \int_L \langle F, T \rangle \, ds.$$

Using the parametrization Φ in (16.2), this is Stokes' theorem for the chain Φ.

Exercise 18.2. Check this formula for the model M in (16.2) and $F = (x, 0, y)$.

18.5.5 As mentioned above, $d\omega$ is a unifying concept for rot and div; it serves as well to unify the concept of conservative and solenoidal fields. If ω is a $k-1$-differential form and η a k-differential form in a domain U, we say that $d\omega = \eta$ in Stokes' sense if

$$\int_{bM} \omega = \int_M \eta,$$

whenever $M \subset U$ is an oriented k-sub-manifold with border. Forms of type $d\omega$ are called *exact*, and forms with $d\omega = 0$ are called *closed*. Since $d^2 = 0$, every exact form is closed.

If U is star-shaped with respect to the origin, the Poincaré operator I is defined on a k-form

$$\omega = \sum_I f_I(x) \, dx_{i_1} \wedge \cdots \wedge dx_{i_k},$$

as

$$I(\omega)(x) = \sum_I \sum_{j=1}^{k} (-1)^{j-1} \left(\int_0^1 t^{k-1} f_I(tx) dt \right) x_{i_j} \, dx_{i_1} \wedge \cdots \widehat{dx_{i_j}} \wedge \cdots \wedge dx_{i_k}.$$

Exercise 18.3. (Poincaré lemma). Prove that $\omega = I(d\omega) + d(I\omega)$.

Poincaré lemma implies that in a star-shaped domain every closed form is exact.

18.5.6 In this last paragraph, we give an idea of the de Rham's theorem in the Euclidean context. A main point is to interpret k-forms ω as objects acting on k-dimensional oriented sub-manifolds M by integration, $(\omega, M) = \int_M \omega$. We also think of oriented k-sub-manifolds as objects acting on k-forms, so we can consider linear combinations $C = \sum \lambda_i M_i$ called k-chains in U. $-M$ denotes M with the reverse orientation and

$$\int_C \omega = \sum_i \lambda_i \int_{M_i} \omega.$$

The k-chains form a group $\mathcal{C}_k(U)$. Of course, one defines

$$\partial C = \sum_i \lambda_i \partial M_i.$$

Thus, $\partial : \mathcal{C}_{k+1}(U) \to \mathcal{C}_k(U)$. Let us call a chain C with ∂C *closed* if $\partial C = 0$ and *exact* if it is of the border of a chain. The quotient space of closed chains modulo exact ones is called the kth *singular homology group* $H_k(U)$.

Two closed k-forms ω_1, ω_2 are said to be *related* if $\omega_1 - \omega_2$ is exact. The set of equivalence classes, that is the quotient space of closed forms modulo exact ones is called *the k-th de Rham cohomology group* $H^k_{dr}(U)$.

Stokes' theorem

$$\int_{\partial C} \omega = \int_C d\omega, \ (\omega, \partial C) = (d\omega, C),$$

states that d, ∂ are adjoint. If $d\omega = 0$, then $\int_{\partial C} \omega = 0$, whence

$$\int_D \omega = \int_{D'} \omega,$$

if $D - D'$ is the border of some chain. This means that every closed form ω defines an element $I(\omega)$ in the dual of $H_k(U)$, named the kth *singular cohomology group*

$$(I(\omega), [C]) = \int_C \omega.$$

If ω is exact, then by Stokes' theorem, $I(\omega) = 0$, so in fact we have a well-defined map

$$I : H^k_{dr}(U) \to (H_k(U))^*.$$

The de Rham's theorem is the statement that I is in fact an isomorphism.

Let us rephrase the statement that I is one-to-one. Assume that ω is a k-form and $\int_M \omega = 0$ whenever $\partial M = 0$. Taking M the boundary of a small $k+1$-dimensional ball, we see that $d\omega = 0$ and $I(\omega) = 0$. By de Rham's theorem, the class of ω in $H^k_{dr}(U)$ is zero, so ω is exact. In Theorem 18.3, we used this theorem in the language of solenoidal fields instead of 2-forms in $\mathbb{R}^3$. A reference for de Rham's theorem is [9].

Chapter 19

Harmonic Functions

In this chapter, we deal with *harmonic fields*, those being simultaneously conservative and solenoidal, and exactly gradients of *harmonic functions*. We study the properties of harmonic functions in arbitrary dimension, in particular their mean value properties, both continuous and discrete. In the last section, we reach Poisson's equation $\Delta u = f$ in relation to the decomposition of vector fields as sum of a conservative and a solenoidal component.

19.1 Harmonic Fields

19.1.1 Recall from paragraph 9.5.2 that holomorphic functions $f = u+iv$ in a domain $U \subset \mathbb{R}^2$ are the differentiable ones satisfying the Cauchy–Riemann equations

$$u_y = -v_x, \quad u_x = v_y.$$

This means that the field $F = (u, -v)$ is both locally conservative and locally solenoidal and justifies the following definition:

Definition 19.1. A continuous vector field F in a domain $U \subset \mathbb{R}^n, \quad n = 2, 3$, is called a *harmonic field* if $\operatorname{rot} F = 0, \operatorname{div} F = 0$ in Stokes' sense:

$$\int_{\partial S} \langle F, T\rangle \, ds = 0, \quad \int_{\partial V} \langle F, N\rangle \, dA = 0,$$

for every oriented regular surface $S \subset U$ and for every sub-domain $V \subset U$ with smooth boundary. Said otherwise, F is both locally conservative and locally solenoidal. In a star-shaped domain, F is globally conservative and solenoidal.

We denote by $\mathcal{C}, \mathcal{S}$, respectively, the space of continuous fields which are locally conservative and locally solenoidal:

$$\mathcal{C} = \{F : \operatorname{rot} F = 0 \text{ in Stokes' sense}\},$$

$$\mathcal{S} = \{F : \operatorname{div} F = 0 \text{ in Stokes' sense}\},$$

and their subspaces

$$\mathcal{C}_0 = \{F = \nabla\phi, \phi \in C_c^1(U)\} \subset \mathcal{C}, \quad \mathcal{S}_0 = \{F = \operatorname{rot} G, G \in C_c^1(U)\} \subset \mathcal{S}.$$

We have seen in Propositions 18.3 and 18.2 that

$$\mathcal{C} = (\mathcal{S}_0)^\perp = \{F : (F, G) = 0, G \in \mathcal{S}_0\},$$

$$\mathcal{S} = (\mathcal{C}_0)^\perp = \{F : (F, G) = 0, G \in \mathcal{C}_0\}.$$

We denote by $\mathcal{H}$ the space of harmonic fields. Then,

$$\mathcal{H} = \mathcal{C} \cap \mathcal{S} = (\mathcal{C}_0 + \mathcal{S}_0)^\perp = \{F : (F, \operatorname{rot} G) = 0, G \in C_c^1, (F, \nabla\phi) = 0, \phi \in C_c^1\}.$$

For a differentiable field $F = (X, Y, Z)$ in $\mathbb{R}^3$, harmonicity means $\operatorname{rot} F = 0, \langle \nabla, F \rangle = 0$:

$$X_y = Y_x, X_z = Z_x, \quad Y_z = Z_y, \quad X_x + Y_y + Z_z = 0.$$

The main example is the Newtonian field $\frac{x}{|x|^3}$ outside the origin.

If F is harmonic, locally $F = \nabla u$ for a C^1-function and also

$$\int_U \langle \nabla u, \nabla \phi \rangle \, dm_n = 0$$

for every C^1-function with compact support. Using (18.6) and since

$$\operatorname{div} \nabla\phi = \Delta\phi, \quad \Delta = \sum_i D_{ii},$$

we find that

$$\int_U u \Delta\phi \, dm_n = 0,$$

that is, $\Delta u = 0$ in the weak sense, see Definition 13.1. For instance, the field $F = (x, -y)$ is harmonic in $\mathbb{R}^2$ and $F = \nabla u$ with $u = \frac{1}{2}(x^2 - y^2)$.

We thus encounter again the Laplacian operator Δ. We will see soon that these functions are in fact smooth and $\Delta u = 0$ in the classical sense. By now we consider the following definition, which we state in general dimension.

Definition 19.2. A twice differentiable (real-valued) function u in a domain $U \subset \mathbb{R}^n$ is called harmonic if $\Delta u = 0, \Delta = \sum_i D_{ii}$.

19.1.2 In the following paragraphs, we study the properties of harmonic functions in general dimension. Note that if $n = 1$, they are the linear functions. For $n = 2$, we saw in Theorem 9.5 that the real harmonic functions are locally the real parts of holomorphic functions. Using the power series expansion (17.1) of f we conclude that a real harmonic function in the plane has a local expansion, say in $D(0, R)$,

$$u(re^{it}) = \sum_{j \in \mathbb{Z}} c_j r^{|j|} e^{ijt}, \quad c_{-j} = \overline{c_j}. \tag{19.1}$$

For fixed r, this is of course the Fourier series expansion of $u(re^{it})$.

Example 19.1. Let us check when a polynomial $P(x) = \sum_\alpha c_\alpha x^\alpha$ is harmonic. We write $P = \sum_k P_k$, with P_k homogeneous of degree k. Note than ΔP_k is homogeneous of degree $k-2$, and so P is harmonic iff every P_k is. Obviously, the linear term P_1 is harmonic. A homogeneous polynomial of degree 2

$$P_2(x) = \sum_{i=1}^{n} a_i x_i^2 + \cdots,$$

is harmonic iff $a_1 + a_2 + \cdots + a_n = 0$. For $n = 2, k = 3$ $P(x, y) = ax^3 + by^3 + cx^2y + dxy^2$ is harmonic iff $3a + d = 0, c + 3b = 0$, that is, it is a linear combination of $x^3 - 3xy^2, y^3 - 3yx^2$. If $n = 3$, the space of harmonic homogeneous polynomials of degree 3 is 4.

19.1.3 The following result shows why Δ is a particularly important operator from a mathematical viewpoint. Assume

$$L = \sum_{|\alpha| \leq N} a_\alpha(x) D^\alpha,$$

is a partial differential operator and let $T(x) = y$ be a rigid motion, in matrix notation

$$Y = A + UX,$$

where A is a column translation vector and U is a unitary matrix. The operator L is called *invariant by rigid motions* if for all smooth functions

$$(Lu) \circ T = L(u \circ T).$$

Theorem 19.1. *The differential operators invariant by rigid motions are exactly those of type $L = P(\Delta)$ where P is a polynomial in one variable.*

Proof. It is straightforward to check that Δ, whence $P(\Delta)$, is invariant. Conversely, if L is invariant, obviously the a_α must be constant. Next, we check with $u_\xi(x) = e^{\langle \xi, x\rangle}$. Since $u_\xi(Ux) = e^{\langle \xi, Ux\rangle} = e^{\langle U^t\xi, x\rangle}$, we obtain

$$L(u_\xi \circ U)(x) = \sum_\alpha a_\alpha D^\alpha[e^{\langle U^t\xi, x\rangle}] = \left[\sum_\alpha a_\alpha (U^t\xi)_1^{\alpha_1} \dots (U^t\xi)_n^{\alpha_n}\right] e^{\langle U^t\xi, x\rangle},$$

which we write as

$$\left[\sum_\alpha a_\alpha (U^t\xi)^\alpha\right] e^{\langle U^t\xi, x\rangle}.$$

On the other hand,

$$Lu_\xi(x) = \left(\sum_\alpha a_\alpha \xi^\alpha\right) e^{\langle \xi, x\rangle},$$

$$Lu_\xi(Ux) = \left(\sum_\alpha a_\alpha \xi^\alpha\right) e^{\langle \xi, Ux\rangle} = \left(\sum_\alpha a_\alpha \xi^\alpha\right) e^{\langle U^t\xi, x\rangle}.$$

Therefore,

$$\sum_\alpha a_\alpha \xi^\alpha = \sum_\alpha a_\alpha (U^t\xi)^\alpha$$

for all U. This amounts to saying that the polynomial $\sum_\alpha a_\alpha \xi^\alpha$ is radial, that is, it has the desired form. □

19.1.4 In this context, it is natural to look at harmonic radial functions $u(x) = \varphi(r), r = |x|$. Then $D_i u = \varphi'(r) D_i r = \varphi'(r)\frac{x_i}{r}$,

$$D_i^2 u = \varphi''(r)\frac{x_i^2}{r^2} + \varphi'(r)\frac{1}{r} - \varphi'(r) x_i \frac{x_i}{r^3},$$

$$\Delta u = \varphi''(r) + n\varphi'(r)\frac{1}{r} - \varphi'(r)\frac{1}{r} = \varphi''(r) + \frac{n-1}{r}\varphi'(r).$$

Then $\Delta u = 0$ means $\varphi'(r) = Cr^{1-n}$, whence

$$\varphi(r) = k_1 r^{2-n} + k_2, \quad n > 2, \quad \varphi(r) = k_1 \log r + k_2, \quad n = 2.$$

In all cases there is a singularity at $r = 0$ unless $k_1 = 0$, that is, constants are the only radial harmonic functions in a ball.

The function

$$G(x) = d_n|x|^{2-n} \quad \text{if } n > 2, \quad d_2 \log|x| \quad \text{if } n = 2, \tag{19.2}$$

where d_n is a normalization constant to be chosen later on, is called *Green's function.* It is harmonic outside the origin, whence a finite linear combination $u(x) = \sum_{i=1}^{N} c_i G(x - a_i)$ is an example of harmonic function. More generally, we can consider infinite linear combinations

$$u(x) = \int_K \phi(y) G(x - y)\, dm_n(y).$$

where K is compact and ϕ is integrable on K. This is a harmonic function in the complement of K.

19.1.5 We will be using the so-called *Green's identities*, a consequence of Gauss' divergence theorem. Let U be a bounded admissible domain oriented by the outward normal N as in Gauss' theorem and assume that u, v are twice differentiable functions defined in an open set containing $\overline{U}$. We apply Gauss' theorem to $F = u\nabla v$. On the one hand,

$$\langle u\nabla v, N\rangle = u\langle \nabla v, N\rangle = uNv.$$

On the other hand, computing

$$\operatorname{div} F = (uv_x)_x + (uv_y)_y + (uv_z)_z = \langle \nabla u, \nabla v\rangle + u\Delta v.$$

So assuming this is integrable one gets the *first Green's identity*

$$\int_{bU} u\, Nv\, dm_{n-1} = \int_U (\langle \nabla u, \nabla v\rangle + u\Delta v)\, dm_n.$$

Permuting u, v and subtracting leads to the *second Green's identity*

$$\int_{bU} (u\, Nv - v\, Nu)\, dm_{n-1} = \int_U (u\Delta v - v\Delta u)\, dm_n, \tag{19.3}$$

whenever the right-hand side makes sense. We note as special cases, for u harmonic of class C^1 up to bU

$$\int_{bU} Nu\, dm_{n-1} = 0, \quad \int_{bU} u\, Nu\, dm_{n-1} = \int_U |\nabla u|^2 dm_n. \tag{19.4}$$

In particular, if $u = 0$ on bU, then $u = 0$ in U, and if $Nu = 0$ on bU, then u is constant.

19.1.6 To have an intuition of the meaning of the equation $\Delta u = 0$, we consider the Taylor expansion around $a \in U$:

$$u(x) = u(a) + \langle \nabla u(a), x - a \rangle + \frac{1}{2} Hu(a)(x - a, x - a) + o(|x - a|^2).$$

Recall that $Hu(a)$ denotes the Hessian

$$\frac{1}{2} Hu(a)(x - a, x - a) = \frac{1}{2} \sum_{i,j=1}^{n} D_{ij}u(a)(x_i - a_i)(x_j - a_j).$$

To single out the terms with $i = j$, we evaluate the mean $M(a, r)$ of u on the sphere $S(a, r)$. All linear terms, and all terms with $i \neq j$, have zero mean. Therefore, if $c_n = m_{n-1}(S^{n-1})$ (computed in (14.1)),

$$\begin{aligned} M(a, r) &= u(a) + \frac{1}{2} \sum_{i=1}^{n} D_{ii}u(a) \frac{1}{c_n r^{n-1}} \int_{S(a,r)} (x_i - a_i)^2 \, dm_{n-1}(x) + o(r^2) \\ &= u(a) + \frac{1}{2n} r^2 \Delta u(a) + o(r^2), \end{aligned}$$

proving that

$$\frac{1}{2n} \Delta u(a) = \lim_{r \to 0} \frac{M(a, r) - u(a)}{r^2}.$$

This motivates the following definition:

Definition 19.3. A continuous function u in U is said to satisfy the *mean value property* if

$$u(a) = \frac{1}{c_n r^{n-1}} \int_{|x-a|=r} u(x) \, dm_{n-1}(x) = \frac{1}{c_n} \int_{|w|=1} u(x + rw) \, dm_{n-1}(w),$$

whenever $S(a, r) \subset U$.

We have seen that a twice differentiable function u enjoying the mean value property is harmonic. We will see that the converse holds. If u is harmonic, by (19.4)

$$\begin{aligned}\frac{d}{dr}M(a,r) &= \frac{d}{dr}\frac{1}{c_n}\int_{|w|=1} u(a+rw)\,dm_{n-1}(w)\\ &= \frac{1}{c_n}\int_{|w|=1}\sum_i D_i u(a+rw)w_i\,dm_{n-1}(w)\\ &= \frac{1}{c_n}\int_{|w|=1}\frac{\partial u}{\partial n}(a+rw)\,dm_{n-1}(w)\\ &= \frac{1}{c_n r^{n-1}}\int_{|x-a|=r}\frac{\partial u}{\partial n}(x)\,dm_{n-1}(x)=0.\end{aligned}$$

Therefore, $M(a,r)$ is constant; since its limit as $r \to 0$ is $u(a)$, it follows that $M(u,a,r) = u(a)$. Therefore, a twice differentiable function is harmonic if and only if it has the mean value property.

In dimension $n = 1$, the mean value property is

$$u\left(\frac{a+b}{2}\right) = \frac{1}{2}(u(a)+u(b)), \quad a,b \in \mathbb{R}.$$

This alone implies that u is linear, because u equals $\frac{x-a}{b-a}u(a) + \frac{b-x}{b-a}u(b)$ on all dyadic points in between. The same holds in general dimension. To see this, it is enough to show that a function with the mean value property is smooth. Indeed, let χ be a smooth radial function, $\chi(x) = \Psi(|x|)$, with compact support in $B(0,1)$, normalized by $\int \chi(x)\,dm_n(x) = 1$. If $\chi_\varepsilon(x) = \varepsilon^{-n}\chi(x/\varepsilon)$, then $u_\varepsilon = u * \chi_\varepsilon$ is smooth in $\{d(x,U^c) > \varepsilon\}$. But

$$\begin{aligned}u_\varepsilon(x) &= \int u(y)\chi_\varepsilon(x-y)\,dm_n(y) = \int u(x-y)\chi_\varepsilon(y)\,dm_n(y)\\ &= \int_{|y|<\varepsilon} u(x-y)\varepsilon^{-n}\chi(y/\varepsilon)\,dm_n(y) = \int_{|y|<1} u(x-\varepsilon y)\chi(y)\,dm_n(y)\\ &= \int_0^1\int_{|y|=1} u(x-\varepsilon ry)\Psi(r)r^{n-1}\,dm_{n-1}(y)\,dr = c_n u(x)\int_0^1 \Psi(r)r^{n-1}\,dr\\ &= u(x)\int \chi(y)\,dm_n(y) = u(x).\end{aligned}$$

Therefore, u is smooth in U. Summarizing:

Theorem 19.2. *The following are equivalent for a continuous function u:*

(a) *u is harmonic.*
(b) *It has the mean value property.*
(c) *$\Delta u = 0$ in the weak sense.*

Harmonic functions are smooth.

Proof. It remains to prove that $\Delta u = 0$ in the weak sense implies that u is harmonic (this is known as *Weyl's lemma*). But

$$\Delta u_\varepsilon = \int u(y)\Delta_x \chi_\varepsilon(x-y)\,dm_n(y) = \int u(y)\Delta_y \chi_\varepsilon(x-y)\,dm_n(y) = 0.$$

Therefore, u_ε is harmonic and has the mean value property. Since $u_\varepsilon \to u$ uniformly on compact sets, u has the mean value property, too. □

Corollary 19.1. *Harmonic vector fields are exactly local gradients of harmonic functions, their components are harmonic functions and are smooth.*

19.1.7 We have seen in paragraph 19.1.3 that the Laplacian is invariant by rigid motions and therefore harmonic functions are as well. But there exist transformations that while not preserving the Laplacian do preserve harmonic functions. For instance, in dimension $n = 2$ any holomorphic transformation Ψ between domains U, V preserves harmonic functions, meaning that $u(\Psi(x))$ is harmonic whenever u is. This is because harmonic functions are locally real parts of holomorphic functions and composing holomorphic functions is holomorphic.

Theorem 19.3. *The homeomorphisms between two domains that preserve harmonic functions are exactly:*

(a) *Holomorphic or anti-holomorphic transformations in dimension $n = 2$.*
(b) *Rigid motions if $n > 2$.*

Proof. Let $\Psi = (\Psi_1, \ldots, \Psi_n) : U \to V, y = \Psi(x)$ be a continuous transformation between two domains U, V and assume that $v \circ \Psi$ is harmonic whenever v is. Then the components Ψ_i are harmonic, and let us compute $\Delta(v \circ \Psi)$. One finds

$$\Delta(v\circ\Psi)(x) = \sum_{i,k,l=1}^{n} D_{lk}v(\Psi(x))D_i\Psi_l(x)D_i\Psi_k(x) + \sum_{k=1}^{n} D_k v(\Psi(x))\Delta\Psi_k(x).$$

Taking $v = y_l^2 - y_k^2$, we get

$$\sum_{i=1}^{n} (D_i \Psi_l)^2 = \sum_{i=1}^{n} (D_i \Psi_k)^2, \quad k, l = 1, \ldots, n,$$

and taking $v = y_k y_l, l \neq k$, we get

$$\sum_{i=1}^{n} D_i \Psi_l D_i \Psi_k = 0, \quad k \neq l.$$

This means that the Jacobian matrix J_Ψ is a scalar multiple of an orthogonal matrix, that is, Ψ is conformal. Thus, in dimension two, they are either holomorphic or anti-holomorphic. If $n > 2$ by Liouville's Theorem 10.1, they are composition of rigid motions with inversions. But a computation shows that inversions are harmonic only if $n = 2$. $\square$

19.1.8 When considering mean value properties, one might ask whether spheres can be replaced by more general sets. First, we point out that the mean value property for spheres is equivalent to the ball mean value property. If $MB(a, R)$ is the mean on $B(a, R)$,

$$\begin{aligned} MB(a, R) &= \frac{n}{c_n R^n} \int_{|x-a| \leq R} u(x)\, dm_n(x) \\ &= \frac{n}{c_n R^n} \int_0^R r^{n-1} \left(\int_{|w|=1} u(a + rw)\, dm_{n-1}(w) \right) dr \\ &= \frac{n}{R^n} \int_0^R r^{n-1} M(a, r)\, dr. \end{aligned}$$

If $M(a, r) = u(a)$, it follows that $MB(a, R) = u(a)$. Since

$$\frac{d}{dR} MB(a, R) = -\frac{n}{R} (MB(a, R) - M(a, R)),$$

the converse holds, too.

A consequence of the mean value property for balls is the next result, known as *Liouville's theorem.*

Corollary 19.2. *A bounded harmonic function in the whole space $\mathbb{R}^n$ is constant. In case $n = 2$, a harmonic function u such that $u(z) = c \log |z| + O(1)$ as $|z| \to +\infty$ is constant.*

Proof. Assume $|u| \le C$. By the ball mean value property, for each $r > 0$ big enough,

$$u(a) - u(0) = \frac{n}{c_n r^n}\left\{\int_{B(a,r)} u(x)\,dm_n(x) - \int_{B(0,r)} u(x)\,dm_n(x)\right\},$$

$$\begin{aligned}|u(a) - u(0)| &\le C\frac{n}{c_n r^n}\left\{\int_{|x-a|<r,|x|>r} dm_n(x) + \int_{|x-a|>r,|x|<r} dm_n(x)\right\}\\ &\le C\frac{n}{c_n r^n} m(\{x : r - |a| < |x| < r + |a|\})\\ &= O(r^{-n}((r+|a|)^n - (r-|a|)^n) = O(r^{-1}).\end{aligned}$$

In case $n = 2$, if $u = \operatorname{Re} f$ with f entire, it follows that $|e^f| = e^u = O(|z|^N)$. By the Taylor expansion of e^f, it follows that e^f is a polynomial with no zeros, whence it is constant. □

19.1.9 One may consider general polyhedra instead of balls and spheres. Analyzing this situation will lead us to some curious, not widely known results. Assume for simplicity that $n = 2$, let $P_k, k > 2$ denote the regular polygon of k-sides determined by the k-roots of unity $\omega_j = e^{2\pi i \frac{j}{k}}, j = 1, \ldots, k$, L_k its length, and set $P_k(a,r) = a + rP_k$. For a continuous u, we consider the means

$$M_k(a,r) = \frac{1}{rL_k}\int_{\partial P_k(a,r)} u(z)\,ds(z), \quad N_k(a,r) = \frac{1}{k}\sum_{j=1}^{k} u(a + r\omega_j),$$

so formally $M_\infty(a,r) = N_\infty(a,r) = M(a,r)$.

The following computations are preparatory for the theorem stated in what follows. We compute $\frac{d}{dr}M_k(a,r)$ for a smooth u:

$$M_k(a,r) = \frac{1}{L_k r}\int_{\partial P_k(a,r)} u(z)\,ds(z) = \frac{1}{L_k}\int_{\partial P_k} u(a+rw)\,ds(w).$$

In each edge L of ∂P_k, with p_L its mid-point, $w = p_L + tv_L, -1 \le t \le 1$

$$\int_L u(a+rw)\,ds(w) = \int_{-1}^{1} u(a + rp_L + trv_L)|v_L|\,dt,$$

$$\frac{d}{dr}\int_L u(a+rw)\,ds(w) = |v_L|\int_{-1}^{1} \langle \nabla u(a + rp_L + trv_L), p_L + tv_L\rangle\,dt$$

$$= |v_L| \int_{-1}^{1} \langle \nabla u(a + rp_L + trv_L), p_L \rangle \, dt$$

$$+ |v_L| \int_{-1}^{1} t \langle \nabla u(a + rp_L + trv_L), v_L \rangle \, dt = I + II.$$

If $\phi(t) = u(a+rp_L+trv_L), \phi'(t) = \langle \nabla u(a+rp_L+trv_L), rv_L \rangle$, so the second term II is

$$|v_L| \frac{1}{r} \int_{-1}^{1} t\phi'(t) \, dt = |v_L| \frac{1}{r} (\phi(1) + \phi(-1) - \int_{-1}^{1} \phi(t) \, dt)$$

$$= |v_L| \frac{1}{r} (u(a_L) + u(b_L)) - \frac{1}{r} |v_L| \int_{-1}^{1} u(a + rp_L + trv_L) \, dt,$$

where a_L, b_L denote the vertices of $a + rL$. As $|v_L| = \frac{d_k}{2n}, |p_L| = b$ do not depend on L and p_L is normal to L, we obtain adding on L

$$\frac{d}{dr} M_k(a, r) = \frac{b}{L_k} \int_{\partial P_k(a,r)} Nu \, ds(w) + \frac{1}{r} N_k(a, r) - \frac{1}{r} M_k(a, r).$$

Theorem 19.4. *For a real continuous function in a plane domain, the following are equivalent,* $k > 2$:

(a) $u(a) = N_k(a, r)$ *whenever* $P_k(a, r) \subset D$.
(b) $u(a) = M_k(a, r)$ *whenever* $P_k(a, r) \subset D$.
(c) u *is a harmonic polynomial of degree at most* $k - 1$, *that is,*

$$u(z) = \operatorname{Re} \sum_{l=0}^{k-1} c_l z^l.$$

Proof. If u is harmonic, by (19.4) the first integral above is zero, and therefore

$$\frac{d}{dr}(rM_k(a, r)) = N_k(a, r).$$

It follows from this that the first two statements are equivalent for a harmonic function.

For linear u, obviously $M_k(0,r) = N_k(0,r) = 0$. For $u = 2xy, v = x^2 - y^2$, their discrete means in P_k are the real and imaginary parts of the discrete mean of z^2, which is zero. Therefore, for $u = xy$, one has

$$M_k(0,r) = N_k(0,r) = 0,$$

while $u = x^2, u = y^2$ have the same means $M_k(0,r) = \alpha_k r^2, N_k(0,r) = \beta_k r^2$ for some positive constants α_k, β_k.

Assume first that u is smooth. The Taylor expansion implies

$$M_k(a,r) = u(a)+\frac{1}{2}\Delta u(a)\alpha_k r^2+o(r^2), N_k(a,r) = u(a)+\frac{1}{2}\Delta u(a)\beta_k r^2+o(r^2).$$

Then, both (a), (b) imply that u is harmonic and therefore they are equivalent. To show (c), we use the expansion (19.1)

$$u = \operatorname{Re} f, \quad f = \sum_{l=0}^{\infty} c_l z^l.$$

We compute $N_k(a,r)$ for $a \in D(0,R)$ and r small for z^l

$$\begin{aligned} N_k(a,r) &= \frac{1}{k}\sum_{j=1}^{k}(a + r\omega_j)^l = \frac{1}{k}\sum_{j=1}^{k}\sum_{m=0}^{l}\binom{l}{m}a^{l-m}r^m\omega_j^m \\ &= \sum_{m=0}^{l}\binom{l}{m}a^{l-m}r^m\frac{1}{k}\sum_{j=1}^{k}\omega_j^m. \end{aligned}$$

The mean $\frac{1}{k}\sum_{j=1}^{k}\omega_j^m$ equals 1 if m is a multiple of k and zero otherwise. Therefore, for $l \leq k-1$, the discrete mean is a^l, while for $l \geq k$ it equals

$$a^l + \binom{l}{k}a^{l-k}r^k + \binom{l}{2k}a^{l-2k}r^{2k} + \cdots + .$$

Thus, the discrete mean of $u = \operatorname{Re}\sum_{l=0}^{k-1} c_l z^l$ is $u(a)$. For higher degrees, say between k and $2k-1$, we get extra terms

$$u(a) + \operatorname{Re}\sum_{l=k}^{2k-1} c_l\binom{l}{k}a^{l-k}r^k,$$

that are zero for all a, r if and only if $c_l = 0, l \geq k$. This proves the theorem for u smooth. It remains to prove that a continuous u satisfying (a) or (b) is of type (c). As before, we consider the approximation u_ε. Since convolution is translation invariant, it is also invariant by means of translations. This means that u_ε satisfies (a), (b), and it is therefore a harmonic polynomial of degree at most $k-1$. But $u_\varepsilon \to u$ uniformly on compact sets, so u itself is such a polynomial. $\square$

Note that for $k = 2$, a continuous function $u(x, y)$ such that

$$\frac{1}{2}(u(x-r, y) + u(x+r, y)) = u(x, y)$$

has the general form $u(x, y) = \phi_1(y)x + \phi_2(y)$.

19.1.10 It follows from (19.4) that if u is harmonic in a bounded admissible domain U, smooth enough up to bU and $u = 0$ on bU, then $u = 0$. This can be generalized with the following *maximum modulus principle.* Note that in the real line this is an obvious statement.

Theorem 19.5. *If u is continuous on $\overline{U}$ and harmonic in U,*

$$\max_{x \in bU} u(x) = M = \max_{x \in \overline{U}} u(x), \quad \min_{x \in bU} u(x) = m = \min_{x \in \overline{U}} u(x).$$

If m or M are attained at some point $x \in U$, then u is constant.

Proof. Let $M = \max_{x \in \overline{U}} u(x)$. It is enough to prove that if $u(x_0) = M$ at some $x_0 \in U$, then $u = M$ in U. Define $V = \{x \in U : u(x) = M\}$, a non-empty closed set in U. We claim that it is also open. If $a \in V$ and $\overline{B}(a, r) \subset U$, by the ball mean value property,

$$M = u(a) = \frac{1}{m(B(a, r)} \int_{|x-a| \leq r} u(x)\, dm_n(x),$$

which we write as

$$\frac{1}{m(B(a, r))} \int_{B(a,r)} (u(x) - u(a))\, dm_n(x) = 0.$$

Since $u(x) - u(a) = u(x) - M \leq 0$ it follows that $u = u(a) = M$ in $B(a, r)$ and so $B(a, r) \subset V$. So V is open and $V = U$. $\square$

Note from the proof that the result holds for $u \in C(\overline{U})$ satisfying the *weak mean value property*: for each $a \in U$, there is a ball $B(a, r) \subset U$ such that $u(a)$ is the mean of u on $B(a, r)$ (in particular if for all $t < r, u(a)$ is the mean of u on the sphere $S(a, t)$). This condition is only apparently weaker, as it turns out that these functions are in fact harmonic, as shown in Corollary 21.1.

Example 19.2. We consider in $\mathbb{R}^3$ the harmonic polynomial $P(x,y,z) = x^3 - 3xy^2 + x^2 + y^2 - 2z^2$. The absolute maximum and minimum in the closed unit ball are the same as those in the sphere, and we can find them with Lagrange Method. We obtain the points $(\pm 1, 0, 0), (\pm\frac{1}{2}, \pm\frac{\sqrt{3}}{2}, 0), (0, 0, \pm 1)$, the maximun is 2 and the minimum is -2.

19.2 Poisson's Equation

Assume that U is the whole space or a bounded star-shaped domain in $\mathbb{R}^n, n = 2, 3$. Then, with the notations in Section 19.1, $\mathcal{C} = \mathcal{S}_0^\perp$ is the space of conservative fields, $\mathcal{S} = \mathcal{C}_0^\perp$ is the space of solenoidal fields and $\mathcal{H} = \mathcal{C} \cap \mathcal{S} = (\mathcal{C}_0 + \mathcal{S}_0)^\perp$ is the space of harmonic fields, equal to the space of gradients of harmonic functions. We think of $\mathcal{C}_0, \mathcal{S}_0$ as the subspaces of $\mathcal{C}, \mathcal{S}$ consisting of fields vanishing at bU.

Without this being a proof, this indicates that a continuous field F in a suitable class $\mathcal{V}$ should have a unique decomposition

$$F = F_1 + F_2 + F_3,$$

with F_1, F_2 conservative and solenoidal fields, respectively, vanishing on bU in some sense and F_3 a harmonic field. If F itself vanishes on bU, then F_3 is a harmonic field vanishing on bU, and by the maximum modulus principle, $F_3 = 0$.

This is the situation we will be considering in the next chapters, starting with the unbounded case. Assume in the whole space that F vanishes at infinity, has a continuous divergence $\operatorname{div} F$ in Stokes' sense and we want to decompose $F = F_1 + F_2$ with F_1 conservative and F_2 solenoidal, both vanishing at infinity. Then $\operatorname{div} F_1 = \operatorname{div} F$, so we need to solve the equation $\operatorname{div} F_1 = \operatorname{div} F$ in the weak sense with F_1 conservative. We point out that solving the equation

$$\operatorname{div} F_1 = f,$$

with a given continuous function f, is accomplished along the lines of Poincaré lemma. Applying div to (18.10), we get

$$\operatorname{div} F = \operatorname{div} I_3(\operatorname{div} F).$$

Thus, $f = \operatorname{div} I_3(f)$ for $f = \operatorname{div} F$, so a fortiori a solution is $I_3 f$.

Exercise 19.1. Prove that in a star-shaped domain U a solution of $\operatorname{div} F_1 = f$ is given by the radial field $F_1 = I_3(f)$

$$F_1(x) = \left(\int_0^1 t^{n-1} f(tx)\, dt\right) x.$$

The important point is that we want the solution F_1 to be conservative. If $F_1 = \nabla u$, then $\operatorname{div} F_1 = \Delta u$, and we must solve the equation $\Delta u = \operatorname{div} F$ in the weak sense with ∇u vanishing at infinity. This is how we arrive at the equation

$$\Delta u = f,$$

for a given f. This is called *Poisson's equation.*

If u_1, u_2 are two such solutions, then $u = u_1 - u_2$ is harmonic and ∇u (whose components are also harmonic) vanishes at infinity, whence it is zero by Liouville's theorem. Thus there is at most one solution to this problem, showing that F_1 must depend only on $\operatorname{div} F$.

We can proceed the other way around and find F_2. Assuming that F has a rotational in Stokes' sense, then $\operatorname{rot} F_2 = \operatorname{rot} F$, so we need to solve this equation in the weak sense with F_2 solenoidal and vanishing at infinity. If F_2 is solenoidal, $F_2 = \operatorname{rot} H$; the potential vector H is not unique, $H = H_0 + \nabla\psi$ being the general form. A way to normalize H is to require it to be H solenoidal, too. We will see later that this is indeed possible. A computation shows that for a general smooth field

$$\operatorname{rot}\operatorname{rot} H = \nabla(\operatorname{div} H) - \Delta H,$$

where Δ acts component-wise. In particular, $\operatorname{rot}\operatorname{rot} H = -\Delta H$ for a smooth solenoidal field. By the usual approximation procedure, the same holds in the weak sense for a continuous field for which $\operatorname{rot}\operatorname{rot} H$ makes sense. So we are lead to solve

$$\Delta H = -\operatorname{rot} F$$

in the weak sense, with H solenoidal, and $\operatorname{rot} H$ vanishing at infinity, and then we will take $F_2 = \operatorname{rot} H$. As before, this problem has at most one solution, showing that F_2 must depend only on $\operatorname{rot} F$.

Thus, we look for a linear operator R acting on functions and fields such that

$$F = R(\operatorname{div} F) + R(\operatorname{rot} F),$$

whenever F has a continuous divergence and rotational and vanishes at infinity. For a continuous function f, Rf should be the unique conservative field F_1 vanishing at infinity such that $\operatorname{div} F_1 = f$ in the weak sense. For a solenoidal continuous field L, $R(L)$ should be the unique solenoidal field F_2 vanishing at infinity such that $\operatorname{rot} F_2 = L$ in the weak sense. We have seen that both equations are reduced to Poisson's equation and then apply ∇ or rot.

In the plane, $\operatorname{rot} F = -\operatorname{div} JF$. To solve $\operatorname{rot} F_2 = g$, we solve $\operatorname{div} F_1 = -g$ and take $F_2 = -JF_1$. So when $n = 2$, we are just interested in solving the divergence equation.

Chapter 20

The Divergence and Rotational Equations, Poisson's Equation

The purpose of this chapter is to establish the Helmholtz decomposition of continuous vector fields in the full space. For this, we first study *potentials* as solutions of the rotational equation $\operatorname{rot} F = G$, the divergence equation $\operatorname{div} F = u$ and Poisson's equation $\Delta u = f$ in the whole space.

20.1 Potentials*

20.1.1 We saw before that Green's function $G(x)$ is up to constants the unique non-constant radial harmonic function. The constant d_n in the definition of G is chosen so that

$$R(x) = \nabla G(x) = \frac{1}{c_n} \frac{x}{|x|^n},$$

that is, $d_n = \frac{1}{(2-n)c_n}$ if $n > 2$ and $d_2 = \frac{1}{c_2} = \frac{1}{2\pi}$. The field R is radial and harmonic. Actually, it is up to constants the only radial harmonic field, because if $F = \nabla u$ is such a field, necessarily u is radial. Indeed, if $|p| = |q| = r$ and γ is a path on the sphere $|x| = r$ joining p, q, then $\langle F, T \rangle = 0$ and therefore

$$u(p) - u(q) = \int_\Gamma \langle F, T \rangle \, ds = 0.$$

Then u is a radial harmonic function, whence $F = cR$ for some constant. The field R is called the *Newtonian field.* The following proposition summarizes the main properties of G, R.

Proposition 20.1.

(a) *Both G, R are locally integrable.*

(b) *If U is an admissible domain and u is a C^1 function in a neighborhood of $\overline{U}$, then for $x \in U$,*

$$u(x) = \int_{bU} u(y)\langle R(y-x), N\rangle \, dm_{n-1}(y) - \int_U \langle \nabla u(y), R(y-x)\rangle \, dm_n(y).$$

Therefore, for a compactly supported $\phi \in C^1_c(U)$

$$\phi(x) = -\int_U \langle \nabla \phi(y), R(y-x)\rangle \, dm_n(y).$$

(c) *If u is of class C^2,*

$$u(x) = \int_{bU} (u(y) N_y G(y-x) - G(y-x) N u(y)) \, dm_{n-1}(y) + \int_U G(y-x)\Delta u(y) \, dm_n(y). \tag{20.1}$$

Therefore, for a compactly supported ϕ of class C^2,

$$\phi(x) = \int_U G(y-x)\Delta\phi(y) \, dm_n(y).$$

(d) *If $n = 3$, for a compactly supported field L of class C^1,*

$$\int_U \langle \operatorname{rot} L(y), R(y-x)\rangle \, dm_n(y) = 0.$$

Proof. Using spherical coordinates it is clear that both G, R are integrable near zero. To prove the second part we proceed as in the proof of Cauchy's formula. Given $x \in U$, we apply Gauss' theorem to $uR(y-x)$ in $U \setminus B(x, \varepsilon)$. If both $bU, bB(x, \varepsilon)$ are oriented by the exterior unit normal, and using

$$\begin{aligned} \operatorname{div}_y(uR(y-x)) &= \langle \nabla u(y), R(y-x)\rangle + u \operatorname{div}_y R(y-x) \\ &= \langle \nabla u(y), R(y-x)\rangle, \end{aligned}$$

we get

$$\begin{aligned} \int_{bU} u(y)\langle R(y-x), N(y)\rangle \, dm_{n-1}(y) - \int_{|x-y|=\varepsilon} u(y)\langle R(y-x), N(y)\rangle \, dm_{n-1}(y) \\ = \textstyle\int_U \langle \nabla u(y), R(y-x)\rangle \, dm_n(y). \end{aligned}$$

In $|y-x| = \varepsilon, R(y-x) = \frac{1}{c_n \varepsilon^{n-1}} N(y)$, whence the integral is the mean of u. Making $\varepsilon \to 0$, we obtain (b).

For (c), we apply again Gauss' theorem to $G(y-x)\nabla u(y)$, observing that

$$\operatorname{div}_y(G(y-x)\nabla u(y)) = \langle \nabla u(y), R(y-x)\rangle + G(y-x)\Delta u(y),$$

$$\int_{bU} G(y-x)\langle \nabla u(y), N(y)\rangle \, dm_{n-1}(y) - \int_{|x-y|=\varepsilon} G(y-x)\langle \nabla u(y), N(y)\rangle \, dm_{n-1}(y)$$

$$= \int_U \langle \nabla u(y), R(y-x)\rangle \, dm_n(y) + \int_U G(x-y)\Delta u(y) \, dm_n(y).$$

Now the integral over the sphere is $O(\varepsilon)$, and (c) follows subtracting both equations. For the last part, we apply again Gauss' theorem to the same domain $U \setminus B(x,\varepsilon)$ and the field $R(y-x) \times L(y)$. On the one hand,

$$\begin{aligned}\operatorname{div}(R(y-x) \times L(y)) &= \langle L(y), \operatorname{rot} R(y-x)\rangle - \langle R(y-x), \operatorname{rot} L(y)\rangle \\ &= -\langle R(y-x), \operatorname{rot} L(y)\rangle.\end{aligned}$$

The flux on bU is zero because $L = 0$ and is also zero on $|y-x| = \varepsilon$ because $R(y-x) \times L(y)$ is orthogonal to $R(y-x)$ and $N(y), R(y-x)$ are proportional there. This proves the last part. □

The decomposition formula in part (c) is called the *Riesz decomposition.* The three integrals involving G receive different names. The integrals

$$G\phi(x) = \int_A G(y-x)\phi(y)\, dm_n(y),$$

are called *Riesz's potentials*, also *logarithmic potential* in case $n = 2$. An integral like

$$\int f(y)R(y-x)\, dm_n(y)$$

is sometimes called a *Riesz potential of the first order*. An integral like

$$\int_{\partial U} \phi(y)G(y-x)\, dm_{n-1}(y)$$

is called a *single layer potential.* The third type of integral,

$$\int_{\partial U} \phi(y) N_y G(y-x)\, dm_{n-1}(y),$$

is called a *double layer potential,* because it can be interpreted as the superposition of two single layer potentials. All these potentials are harmonic functions outside the domain of integration.

There is as well a Green's function in dimension $n=1$, namely $G(x) = \frac{1}{2}|x|$. The version of the formula in part (c) is for u of class C^2 in $[a,b]$

$$2u(x) = u(a) + u(b) - (|x-b|u'(b) - |x-a|u'(a)) + \int_a^b |x-y|u''(y)\, dy,$$

obtained integrating by parts.

20.1.2 Proposition 20.1 means that in the weak sense

$$\Delta_y G(y-x) = \operatorname{div}_y R(y-x) = \delta(x-y), \quad \operatorname{rot}_y R(y-x) = 0.$$

For this reason, G is also called the *fundamental solution* of Δ and R the fundamental solution of div. The kernel R also satisfies the cancellation property

$$\int_{|y-x|=\varepsilon} R(x-y)\, dm_{n-1}(y) = 0.$$

From Section 19.2, we are interested in solving Poisson's equation $\Delta u = f$ for a given f. Formally, if

$$Gf(x) = \int f(y) G(x-y)\, dm_n(y),$$

then

$$\begin{aligned} \Delta Gf(x) &= \int \Delta_x (f(y) G(x-y))\, dm_n(y) \\ &= \int f(y) \Delta_x G(x-y)\, dm_n(y) = \int f(y) \delta(x-y) dm_n(y) = f(x). \end{aligned}$$

So $u = Gf$ is formally a solution.

Also, from Section 19.2, we are interested in $F_1 = \nabla u$ when $f = \operatorname{div} F$. Also, formally,

$$F_1(x) = \nabla Gf(x) = \int \nabla_x(f(y)G(x-y))\,dm_n(y)$$

$$= \int f(y)\nabla_x G(x-y)\,dm_n(y) = \int f(y)R(x-y)\,dm_n(y).$$

We were interested as well in $F_2 = \operatorname{rot} H$ where H is solenoidal solving $\Delta H = -\operatorname{rot} F$. The latter is formally accomplished with

$$H(x) = -\int G(x-y)\operatorname{rot} F(y)\,dm_n(y).$$

The field H is indeed formally solenoidal, for using (18.1) first in x and then in y,

$$\operatorname{div} H(x) = -\int \langle \nabla_x G(x-y), \operatorname{rot} F(y)\rangle\,dm_n(y)$$

$$= \int \langle \nabla_y G(y-x), \operatorname{rot} F(y)\rangle\,dm_n(y) = 0.$$

Now, also formally,

$$F_2(x) = \operatorname{rot} H(x) = -\int \operatorname{rot}_x(\operatorname{rot} F(y)G(x-y))\,dm_n(y)$$

$$= \int \operatorname{rot} F(y) \times \operatorname{rot}_x G(x-y)dm_n(y) = \int \operatorname{rot} F(y) \times R(x-y)dm_n(y).$$

Thus, we are led to the study of four operators (note we change notation $\operatorname{rot} F$ to F)

$$Gf(x) = \int f(y)G(x-y)\,dm_n(y), Rf(x) = \int f(y)R(x-y)\,dm_n(y),$$

$$GF(x) = \int G(x-y)F(y)\,dm_n(y), RF(x) = \int \operatorname{rot} F(y) \times R(x-y)\,dm_n(y).$$

By the comments in paragraph 16.5.3, in dimension $n = 2$ we need to just consider Rf, Gf.

Our next goal is to fully justify these formal developments by a careful study of these operators. Now, the convergence issues in the integrals are different, Gf, GF might be undefined while Rf, RF are. This is why we shall investigate first Rf, RF directly without relying on Gf, GF.

20.2 The Divergence and Rotational Equations in Space*

In this section, we study the operators Rf, RF. We are interested in $\mathbb{R}^2, \mathbb{R}^3$ in relation to decomposition of fields, but for Rf, everything goes through to general dimension n.

20.2.1 Recall that

$$Rf(x) = \int f(y)R(x-y)\,dm_n(y) = \frac{1}{c_n}\int f(y)\frac{x-y}{|x-y|^n}\,dm_n(y).$$

RF acts component-wise on F so we study Rf. We assume f continuous.

First, note that R is locally integrable but not integrable at infinity. If $|f|$ is locally bounded and satisfies

$$\int_{\mathbb{R}^n} |f(y)||y|^{1-n}\,dm_n(y) < +\infty, \tag{20.2}$$

then the integral defining $Rf(x)$ is absolutely convergent: near x because $R(x-y)$ is locally integrable and at infinity because of condition (20.2). For instance, a decay $|f(y)| = O((1-|y|)^{-1-\varepsilon})$ will do.

Exercise 20.1. If f is radial, $f(y) = h(|y|)$ and (20.2) holds, that is

$$\int_0^{+\infty} |h(t)|\,dt < +\infty,$$

check that Rf is a radial field given by

$$Rf(x) = \left(\int_0^s r^{n-1}h(r)\,dr\right)\frac{x}{|x|^n}, \qquad s = |x|.$$

Then

$$|Rf(x)| = s^{1-n}\int_0^s r^{n-1}h(r)\,dr = \int_0^s \left(\frac{r}{s}\right)^{n-1} h(r)\,dr,$$

vanishes at infinity, by the dominated convergence theorem.

Under (20.2), Rf is locally integrable, in fact continuous. For instance, to prove continuity of Rf at a, we split $f = f_1 + f_2$, where f_1 is continuous supported in $K = B(a, 2r)$ and f_2 is zero in $B(a,r)$ satisfying (20.2). For $x \in B(a, \frac{r}{2})$ and $|a-y| > r$, $|x-y|$ is comparable to $|a-y|$, whence

$$R(f_2)(x) = \int f_2(y)R(x-y)\,dm_n(y)$$

is continuous in $B(a, \frac{r}{2})$ because $f_2(y)R(x-y)$ is continuous in x and uniformly dominated by $f(y)|a-y|^{1-n}$ (Proposition 12.3). In fact, Rf_2 is smooth in $B(a, \frac{r}{2})$ because $D_x^\beta f(y)R(x-y)$ is uniformly dominated by $|f(y)||y|^{1-n-\beta}$ there and we can apply again the criteria in Proposition 12.3. In particular, we note

$$\operatorname{div} Rf(a) = \sum_i D_i(Rf)_i = \int f_2(y) \operatorname{div}_x R(x-y)\, dm_n(y) = 0.$$

The term Rf_1 is continuous using continuity of translations in L^1, because

$$Rf_1(x) - Rf_1(z) = \int_K f_1(y)(R(x-y) - R(x-z))\, dm_n(y),$$

$$|Rf_1(x) - Rf_1(z)| \le C\|\tau_x g - \tau_z g\|_1,$$

where C is a bound of f near x, with $g = 1_K R$ integrable.

In fact, it is possible to prove that Rf is almost of class C^1:

Exercise 20.2. Prove that in fact one has on compacts K

$$|Rf(x) - Rf(z)| \le C(K)|x-z||\log|x-z||, \quad x, z \in K.$$

Theorem 20.1. *Assume that f, F are continuous and* (20.2) *is satisfied by the least radial decreasing majorant of $\psi = |f|, |F|$, that is*

$$\int_0^{+\infty} |h(t)|\, dt < +\infty, \quad h(t) = \sup_{|y|>t} |\psi(y)|, \tag{20.3}$$

Then:

(a) *Rf is the unique continuous conservative field F_1 vanishing at infinity such that $\operatorname{div} F_1 = f$ in Stokes' sense, that is*

$$\int_\Gamma \langle Rf, T\rangle\, ds = 0, \quad \int_{bU} \langle Rf, N\rangle\, dm_{n-1} = \int_U f\, dm_n,$$

for all closed curves Γ and all bounded admissible domains U.

(b) *For $n = 3$, if F is solenoidal, RF is the unique continuous solenoidal field F_2 vanishing at infinity such that $\operatorname{rot} F_2 = F$ in Stokes' sense,* *that is*

$$\int_{\partial S} \langle RF, T\rangle \, ds = \int_S \langle F, N\rangle \, dA, \quad \int_{bU} \langle RF, N\rangle \, dA = 0,$$

for all closed surfaces S and all bounded admissible domains U.

Proof. To show that Rf is conservative, we must show that

$$\int \langle Rf(x), \operatorname{rot} L(x)\rangle \, dm_n(x) = 0,$$

for all compactly supported smooth L. The left-hand side equals by Fubini's theorem

$$\int f(y)(\int \langle R(x-y), \operatorname{rot} L(x)\rangle \, dm_n(x)) \, dm_n(y).$$

The integral in x is zero, by part (d) in Proposition 20.1.

To show that $\operatorname{div} R = f$ in the weak sense, we must show

$$\int \langle Rf(x), \nabla\phi(x)\rangle \, dm_n(x) = -\int f\phi \, dm_n.$$

The left-hand side equals by Fubini's theorem

$$\int f(y) \int \langle R(x-y), \phi(x)\rangle \, dm_n(x) \, dm_n(y).$$

The integral in x equals, by Proposition 20.1, $-\phi(x)$ and we are done.

To show that RF is solenoidal, we must show that for compactly supported ϕ,

$$\int \langle RF(x), \nabla\phi(x)\rangle \, dm_n(x) = 0.$$

This equals

$$\begin{aligned} &\int \int \langle F(y) \times R(x-y), \nabla\phi(x)\rangle \, dm_n(y) \, dm_n(x) \\ &= \int \int \langle R(x-y), \nabla\phi(x) \times F(y)\rangle \, dm_n(y) \, dm_n(x). \end{aligned}$$

But $\nabla\phi(x) \times F(y) = \operatorname{rot}_x(\phi(x)F(y))$, whence by part (d) of Proposition 20.1, the integral in x is zero.

To show that $\operatorname{rot} RF = F$ in the weak sense, we must show

$$\int \langle RF(x), \operatorname{rot} L(x)\rangle \, dm_n(x) = -\int \langle F, L\rangle \, dm_n,$$

for compactly supported L. The right-hand side is, by Fubini's theorem

$$\int\int \langle F(y) \times R(x-y), \operatorname{rot} L(x)\rangle \, dm_n(x)\, dm_n(y).$$

The integral in x is

$$\int \langle R(x-y), \operatorname{rot} L(x) \times F(y)\rangle \, dm_n(x).$$

Consider the compactly supported $\phi(x) = \langle F(y), L(x)\rangle$. A computation shows that

$$\nabla\phi(x) = (F(y)\cdot\nabla_x)L(x) + F(y) \times \operatorname{rot} L(x),$$

where $F(y)\cdot\nabla_x$ is understood as a linear differential operator acting component-wise on L. Therefore, by property (b) in Proposition 20.1, the integral in x equals

$$-\phi(y) - \int \langle R(x-y), (F(y)\cdot\nabla_x)L(x)\rangle \, dm_n(x),$$

and so it is enough to prove that

$$\int\int \langle R(x-y), (F(y)\cdot\nabla_x)L(x)\rangle \, dm_n(x)\, dm_n(y) = 0,$$

for all L. If $L = (\psi, 0, 0)$, then $(F(y)\cdot\nabla_x)L(x) = (F(y)\cdot\nabla_x)(\psi, 0, 0)$ and

$$\langle R(x-y), (F(y)\cdot\nabla_x)L(x)\rangle = h(x-y)\langle F(y), \nabla_x\psi\rangle,$$

where h denotes the first component of R. So, it is enough to show that

$$\int h(x-y)\langle F(y), \nabla\psi(x)\rangle \, dm_n(x)\, dm_n(y) = 0$$

for compactly supported ψ. We write it as

$$I = \int \langle F(y), \int h(x-y)\nabla\psi(x)\, dm_n(x)\rangle \, dm_n(y).$$

Consider

$$\hat{\psi}(y) = \int h(x-y)\psi(x)\, dm_n(x) = \int h(x)\psi(x+y)\, dm_n(x).$$

Then

$$\nabla\hat{\psi}(y) = \int h(x)\psi(x+y)\,dm_n(x) = \int h(x-y)\nabla\psi(x)\,dm_n(x).$$

Therefore,

$$I = \int \langle F(y), \nabla\hat{\psi}(y)\rangle\,dm_n(y).$$

If $\hat{\psi}$ had compact support, and since $\operatorname{div} F = 0$, this would imply that $I = 0$. One has instead, as $|y| \to +\infty$,

$$\hat{\psi}(y) = O(|y|^{1-n}).$$

We consider the field $\hat{\psi}F$, whose divergence is, by (18.1), $\hat{\psi}\operatorname{div} F + \langle\nabla\hat{\psi}, F\rangle$, and apply Gauss' theorem to the ball $B(0, R)$ to obtain

$$I = \lim_R \int_{|y|=R} \langle\hat{\psi}(y)F, N\rangle dm_{n-1}.$$

This is $O(h(R))$ and so has limit zero.

Now we show that Rf, RF are zero at infinity. It is enough to prove it for Rf. One has

$$\begin{aligned}|Rf(x)| &\leq c\int h(|y|)|x-y|^{1-n}\,dm_n(y)\\ &= c\int_0^{+\infty} h(t)\Big(\int_{|y|=t} |x-y|^{1-n}\,dm_{n-1}(y)\Big)\,dt.\end{aligned}$$

The integral on $|y| = t$ just depends on $|x| = s$. It is $O(1)$ if $s < \frac{t}{2}$, $O(|\log(1-\frac{t}{s})|)$ if $\frac{t}{2} < s < 2t$ and $O((\frac{t}{s})^{n-1})$ if $s > 2t$. Therefore,

$$|Rf(x)| \leq \int_0^{\frac{s}{2}} h(t)(\frac{t}{s})^{n-1} + \int_{\frac{s}{2}}^{2s} h(t)\left|\log\left(1-\frac{t}{s}\right)\right|\,dt + \int_{2s}^{+\infty} h(t)\,dt.$$

It is clear that the third term tends to zero as $s \to +\infty$ and so does the first one by dominated convergence. The third one equals

$$s\int_{\frac{1}{2}}^{2} h(xs)|\log(1-x)|\,dx,$$

which is bounded by $csh(\frac{s}{2})$. Since h is integrable and decreasing, this term has limit zero, too.

Finally, the unicity is clear, for if F_1 is conservative, vanishes at infinity and $\operatorname{div} F_1 = f$, then $F_1 - Rf$ is a harmonic field vanishing at infinity, whence it is identically zero by Liouville's theorem. Similarly for RF. □

20.2.2 In case Rf, RF are of class C^1, they are classical solutions of $\operatorname{div} F_1 = f, \operatorname{rot} F_2 = F$. A sufficient condition is as follows:

Theorem 20.2. *If f satisfies a local Lipschitz condition for some $\alpha > 0$*

$$|f(x) - f(y)| \leq C|x-y|^\alpha, \quad x, y \in B(a, R),$$

then Rf is of class C^1 around a.

Proof. Splitting $f = f_1 + f_2$ as in paragraph 20.2.1, it is enough to deal with Rf_1 near a. We cannot differentiate under the integral sign because

$$D_j \frac{x_i - y_i}{|x-y|^n} = \frac{\delta_{ij}}{|x-y|^n} - n\frac{(x_i - y_i)(x_j - y_j)}{|x-y|^{n+2}}$$

behaves like $|x-y|^{-n}$ and is not locally integrable. However, we can exploit again the fact that they have mean zero on spheres

$$\int_{|x-y|=\varepsilon} D_j R(y-x)\, dm_{n-1}(y) = 0, \tag{20.4}$$

(if $i \neq j$, the integrand is odd with respect the ith axis, if $i = j$, the two terms have the same integral). As a consequence,

$$\int_{R \geq |x-y| \geq \varepsilon} D_{j,x} R(x-y)\, dm_n(y) = 0, \quad 0 < \varepsilon < R.$$

Consider $\chi \in C^\infty(\mathbb{R})$ such that $\chi(t) = 0$ in $(-1,1), \chi(t) = 1$ if $|t| \geq 2$ and consider the truncated version

$$v_\varepsilon(x) = \int f_1(y) R(x-y) \chi\left(\frac{|x-y|}{\varepsilon}\right) dm_n(y).$$

Then $v_\varepsilon(x) \to Rf(x)$ uniformly on compacts. Since the singularity is killed, we can differentiate under the integral sign to obtain

$$\begin{aligned} D_j v_\varepsilon(x) &= \int f_1(y) D_{j,x} R(x-y) \chi\left(\frac{|x-y|}{\varepsilon}\right) dm_n(y) \\ &\quad + \int f_1(y) R(x-y) \frac{1}{\varepsilon}\frac{x_j - y_j}{|x-y|} \chi'\left(\frac{|x-y|}{\varepsilon}\right) dm_n(y) \\ &= A^j_\varepsilon(x) + B^j_\varepsilon(x). \end{aligned}$$

Using the cancellation above, we may write

$$A^j_\varepsilon(x) = \int (f_1(y) - f_1(x)) D_{j,x} R(x-y) \chi\left(\frac{|x-y|}{\varepsilon}\right) dm_n(y).$$

In spherical coordinates, $y = x + \rho w$, $|w| = 1$, the ith component of B^j_ε is

$$\begin{aligned}B^{ij}_\varepsilon(x) &= \int_{|w|=1}\int_\varepsilon^{2\varepsilon} f_1(x+\rho w)\frac{1}{\varepsilon}\frac{w_iw_j}{\rho^{n-1}}\,\rho^{n-1}\chi'\left(\frac{\rho}{\varepsilon}\right)d\rho\,dm_{n-1}(w)\\ &= \int_{|w|=1}\int_1^2 f_1(x+t\varepsilon w)w_iw_j\chi'(t)dt\,dm_{n-1}(w).\end{aligned}$$

If $i \neq j$, $\int_S w_iw_j dm_{n-1}(w) = 0$, so

$$B^{ij}_\varepsilon = \int_S (f_1(x+t\varepsilon w) - f_1(x))w_iw_j\chi'(t)dt\,dm_{n-1}(w)$$

is $0(\varepsilon^\alpha)$, uniformly on compacts. If $i = j$, $\int_S w_i^2 dm_{n-1}(w) = c_n/n$, so

$$\begin{aligned}B^{ij}_\varepsilon &= \frac{1}{c_n}\int_1^2\int_S (f_1(x+t\varepsilon w) - f_1(x))w_i^2\chi'(t)dt\,dm_{n-1}(w)\\ &\quad+\frac{1}{c_n}f_1(x)\left(\int_1^2\chi'(t)dt\right)\left(\int_S w_i^2dm_{n-1}(w)\right) = 0(\varepsilon^\alpha) + \frac{1}{n}f_1(x).\end{aligned}$$

Therefore, $B^{ij}_\varepsilon(x) \to 0$ if $i \neq j$ and $B^{ij}_\varepsilon(x) \to \frac{1}{n}f_1(x)$ if $i = j$, uniformly on compacts. As for A^j_ε, the hypothesis on f implies that the integral is absolutely convergent, and we may set $\varepsilon = 0$. Altogether, $D_jv_\varepsilon(x)$ has uniform limit on compacts

$$\int (f_1(y) - f_1(x))D_jR(x-y)dm_n(y) + B(y),$$

where B is the vector with $B_i = 0$ if $i \neq j$ and $B_j = \frac{1}{n}f_1(x)$ if $i = j$ uniformly on compacts. Since this expression is continuous (this is seen similarly as for Rf because the singularity is integrable), it follows that $Rf \in C^1(\mathbb{R}^n)$. □

We already know that $\operatorname{div} Rf = f$ in the weak sense, so this holds in the classical sense, too. In fact, we can check this directly: for $x \in B(a, \frac{r}{2})$,

$$\begin{aligned}\operatorname{div} R(f_1) &= \lim_\varepsilon \operatorname{div} v_\varepsilon\\ &= f_1(x) + \int (f(y) - f(x))\operatorname{div}_x R(x-y)dm_n(y) = f_1(x) = f(x),\end{aligned}$$

because $\operatorname{div}_x R(x-y) = \Delta G(x-y) = 0$.

20.3 Poisson's Equation in $\mathbb{R}^n$*

20.3.1 Now, we will carry out a similar study for the potential $G(f)$ and the field $G(F)$, for continuous f, F. Even though we are interested in these potentials in relations to fields in $\mathbb{R}^2, \mathbb{R}^3$, the analysis for $G(f)$ holds in general dimension.

They both make sense point-wise if $\psi = f, F$ satisfies

$$\int_{\mathbb{R}^n} \psi(y)|G(y)|dm_n(y) < +\infty. \tag{20.5}$$

Example 20.1. Consider the single layer potential of a unit charge distribution on $|y| = r$,

$$u(x) = \int_{|y|=r} G(x-y)\, dm_{n-1}(y).$$

This is obviously radial in x, and being harmonic near 0 and outside the ball, must be constant in $B(0, r)$ and equal to $kG(x)$ outside. The value of the constant in $|x| < r$, its value at $x = 0$, is

$$\int_{|y|=r} G(y)\, dm_{n-1}(y) = \frac{r}{2-n},$$

if $n > 2$ and $r \log r$ if $n = 2$. Moreover,

$$k = \lim_{|x|\to\infty} \int_{|y|=r} \frac{G(x-y)}{G(x)}\, dm_{n-1}(y) = c_n r^{n-1}.$$

Note that this is a continuous function in the whole space. The computation shows that the electric field created by a uniform distribution of charge on a sphere is zero in its interior.

Example 20.2. Assume $f(y) = h(|y|)$, where h is continuous in $[0, +\infty)$ and such that (20.5) holds for f

$$\int_0^{+\infty} t|h(t)|\, dt < +\infty, n = 3, \quad \int_0^{+\infty} |\log t|t|h(t)|\, dt < +\infty, n = 2. \tag{20.6}$$

Then $G(f)$ is again radial, $G(f)(x) = H(|x|)$. We compute H; with $|x| = s$,

$$\begin{aligned} H(s) &= \int_{\mathbb{R}^n} h(|y|)G(x-y)\, dm_n(y) \\ &= \int_0^{+\infty} h(r) \left(\int_{|y|=r} G(x-y)\, dm_{n-1}(y) \right) dr. \end{aligned}$$

The integral over $|y| = r$ has been computed in the previous example 20.1 depending on whether $s < r$ or $s > r$. We obtain

$$H(s) = \frac{1}{2-n}\left(s^{2-n}\int_0^s r^{n-1}h(r)\,dr + \int_s^{+\infty} rh(r)\,dr\right),$$

if $n = 3$ and

$$H(s) = \log s \int_0^s h(r)r\,dr + \int_s^{+\infty} h(r)r\log r\,dr,$$

if $n = 2$. We can differentiate once

$$H'(s) = s^{1-n}\int_0^s r^{n-1}h(r)\,dr,$$

and once again

$$H''(s) = (1-n)s^{-n}\int_0^s r^{n-1}h(r)\,dr + h(s).$$

As $\Delta G(f)(x) = H''(s) + \frac{n-1}{s}H'(s)$, we find that indeed $\Delta G(f)(x) = f(x)$ in the classical sense. We check that H has limit zero at infinity if $n > 2$. This is clear for the second term, while the first is bounded in size by

$$\int_0^s \left(\frac{r}{s}\right)^{n-2} |h(r)|r\,dr,$$

which has limit zero by dominated convergence. In case $n = 2$, however, the function $H(s)$ is of the order of $\log s$.

Theorem 20.3. *Let f be continuous in $\mathbb{R}^n$ with $\psi = f$ satisfying* (20.5). *Then,*

(a) *$G(f)$ is of class C^1, and $\nabla G(f) = R(f)$ (note that* (20.5) *implies* (20.2)).
(b) *$\Delta G(f) = f$ in the weak sense.*
(c) *If f satisfies a local Lipschitz condition, $G(f)$ is of class C^2 and $\Delta G(f) = f$ in the classical sense.*
(d) *If $h(t) = \sup_{|x|\geq t}|\psi(x)|$ satisfies* (20.6), *then $G(f)$ vanishes at infinity if $n > 2$ (resp., have growth $\log|x|$ as $|x| \to +\infty$ if $n = 2$) and $u = G(f)$ is the unique solution of $\Delta u = f$ vanishing at infinity (resp., with logarithmic growth).*

Proof. To prove (a), given a we split $f = f_1 + f_2$ as was done for Rf before Theorem 20.1. As in there, we see that Gf_2 is smooth near a, in fact harmonic. Since $\nabla G(x-y) = R(x-y)$ and $R(x-y)$ is locally integrable, it makes sense to differentiate under the integral sign. For a formal argument, let us prove that

$$\lim_{h\to 0} \frac{G(f_1)(x+he_i) - G(f_1)(x)}{h} - \int_{\mathbb{R}^n} f_1(y) R_i(x-y)\, dm_n(y)$$

$$= \lim_{h\to 0} \int_K f_1(y) \big(\frac{G(x+he_i-y) - G(x-y) - hD_iG(x-y)}{h} dm_n(y) = 0.$$

Here, $K = B(a, 2r)$. Write for a.e. y, $G(x+he_i-y) - G(x-y) = h\int_0^1 D_iG(x+the_i-y)\, dt$, so it is enough to prove

$$\int_K \int_0^1 |D_iG(x+the_i-y) - D_iG(x-y|\, dt dm_n(y) \to 0, \quad h \to 0.$$

Since the integral on K equals, with $g = 1_K D_iG$

$$\int_0^1 \|\tau_{x+the_i} g - \tau_x g\|_1\, dt,$$

this follows from the continuity of translations. Altogether,

$$D_iG(f)(x) = \int_{\mathbb{R}^n} f(y) D_iG(x-y)\, dm_n(y),$$

and $\nabla G(f) = Rf$.

To see that $\Delta G(f) = f$ in the weak sense, we use Fubini's theorem and Proposition 20.1

$$\int G(f)(x)\Delta\phi(x)\, dm_n(x) = \int\int f(y) G(x-y)\Delta\phi(x)\, dm_n(x)\, dm_n(y)$$

$$= \int f(y)\,\phi(y)\, dm_n(y).$$

Part (c) follows from Theorem 20.2. With the notations of Example 20.2, one has $|Gf(y)| \le H(|y|)$, so (d) follows. The unicity statement follows from Liouville's Theorem 19.2. □

20.3.2 The statement for GF is as follows:

Theorem 20.4. *Let F be a continuous field in $\mathbb{R}^3$ with $\psi = F$ satisfying* (20.5). *Then*

(a) *$G(F)$ is of class C^1, and* $\operatorname{rot} G(F) = R(F)$ *(note that* (20.5) *implies* (20.2)).
(b) *$G(F)$ is solenoidal if F is.*
(c) *If $h(t) = \sup_{|x|\geq t} |\psi(x)|$ satisfies* (20.6), *then $G(F)$ vanishes at infinity.*

Proof. The regularity properties of RF and the decay at infinity are the same as those of Rf because R acts component-wise. Differentiating under the integral sign proves $\operatorname{rot} G(F) = R(F)$. To see that GF is solenoidal if F is, simply write

$$GF(y) = -\int G(y-x)F(x)\,dm_n(x) = -\int G(x)F(y-x)\,dm_n(x),$$

which exhibits GF as a superposition of translates of F. □

20.3.3 All the regularity statements regarding Gf are local. Combining them with Weyl's lemma (part (c) of Theorem 19.2) we get

Proposition 20.2. *Assume that u, f are continuous in U and $\Delta u = f$ in the weak sense. Then u is of class C^1 and ∇u satisfies a local condition*

$$|\nabla u(x) - \nabla u(y)| = O(|x-y|\,|\log|x-y||).$$

If f satisfies locally a Lipschitz condition, u is of class C^2 in U.

20.4 The Helmholtz's Decomposition, Unbounded Case*

20.4.1 Now, we can combine the results of the last sections to obtain the so-called *Helmholtz–Hodge decomposition* of fields in the plane or the full space. This is a decomposition result, but also an existence result. We assume that F is continuous, vanishes at infinity and has a continuous divergence and rotational in Stokes' sense. The next theorem is a restatement of Liouville's theorem for harmonic functions.

Theorem 20.5. *F is uniquely determined by* $\operatorname{div} F, \operatorname{rot} F$.

Let us try to find an explicit expression of F in terms of $\operatorname{div} F, \operatorname{rot} F$. A simple way to obtain one is as follows. We assume by now that F satisfies a local Lipschitz condition; consider a bounded domain U with smooth boundary and the field

$$G(F)(x) = \int_U F(y)G(x-y)\,dm_n(y).$$

By Theorem 20.4, $G(F)$ is a field of class C^2 in U and $\Delta G(F) = F$ in U. On fields in $\mathbb{R}^3$, we already noticed that

$$\Delta H = \nabla(\operatorname{div} H) - \operatorname{rot}\operatorname{rot} H, \tag{20.7}$$

so

$$F = \nabla(\operatorname{div} G(F)) - \operatorname{rot}\operatorname{rot} G(F).$$

In $\operatorname{div} G(F), \operatorname{rot} G(F)$ we can differentiate once because ∇G is integrable. Using (18.2),

$$\begin{aligned}\operatorname{div} G(F)(x) &= \int_U \langle F(y), \nabla_x G(x-y)\rangle\,dm_n(y)\\ &= -\int_U \langle F(y), \nabla_y G(x-y)\rangle\,dm_n(y).\end{aligned}$$

Now, we use again (18.2) in y, $\operatorname{div}_y(F(y)G(x-y)) = \langle F(y), \nabla_y G(x-y)\rangle + G(x-y)\operatorname{div} F$, and use Gauss' theorem as in previous occasions, observing that the boundary term around x has limit zero, to get

$$\begin{aligned}\operatorname{div} G(F)(x) &= \int_U G(x-y)\operatorname{div} F(y)\,dm_n(y)\\ &\quad - \int_{bU} G(x-y)\langle F(y), N\rangle\,dm_{n-1}(y).\end{aligned}$$

Similarly, using (18.1)

$$\begin{aligned}\operatorname{rot} G(F)(x) &= \int_U \nabla_x G(x-y)\times F(y)\,dm_n(y)\\ &= -\int_U \nabla_y G(x-y)\times F(y)\,dm_n(y),\end{aligned}$$

Using again (18.1) in $y, \operatorname{rot}_y(G(x-y)F(y)) = \nabla_y G(x-y)\times F(y) + G(x-y)\operatorname{rot} F(y)$,

$$\begin{aligned}\operatorname{rot} G(F)(x) &= -\int_U \operatorname{rot}_y(G(x-y)F(y))\,dm_n(y)\\ &\quad + \int_U G(x-y)\operatorname{rot} F(y)\,dm_n(y).\end{aligned}$$

By (18.4) (in fact a version of it in $U \setminus x$ and observing again that the boundary term around x has limit zero),

$$\operatorname{rot} G(F)(x) = -\int_{bU} N \times (G(x-y)F(y))\, dm_{n-1}(y) + \int_U G(x-y) \operatorname{rot} F(y)\, dm_n(y).$$

Thus,

$$F(x) = \nabla \left[\int_U G(x-y) \operatorname{div} F(y)\, dm_n(y) - \int_{bU} G(x-y)\langle F(y), N\rangle\, dm_{n-1}\right] + \operatorname{rot}\left[\int_{bU} N(y) \times (G(x-y)F(y))\, dm_{n-1}(y) - \int_U G(x-y) \operatorname{rot} F(y)\, dm_n(y)\right]. \tag{20.8}$$

Applying this to $U = B(0,R)$ and making $R \to \infty$, we see that if $F(y) = o(\frac{1}{|y|})$, then formally, and assuming that the integrals make sense,

$$\begin{aligned} F(x) &= \nabla \int G(x-y) \operatorname{div} F(y)\, dm_n(y) - \operatorname{rot} \int G(x-y) \operatorname{rot} F(y)\, dm_n(y) \\ &= \int \operatorname{div} F(y) R(x-y)\, dm_n(y) + \int \operatorname{rot} F(y) \times R(x-y)\, dm_n(y). \end{aligned}$$

If $n = 2$, (20.7) becomes

$$\Delta H = \nabla(\operatorname{div} H) + J\nabla(\operatorname{rot} H),$$

so

$$F = \nabla(\operatorname{div} G(F)) + J\nabla(\operatorname{rot} G(F)).$$

Since $\operatorname{rot} G(F) = -\operatorname{div} JGF = -\operatorname{div} GJF$, the decomposition takes the form

$$F(x) = \nabla \int G(x-y)(\operatorname{div} F)(y)\, dA(y) + J\nabla \int G(x-y)(\operatorname{rot} F)(y)\, dA(y).$$

20.4.2 The next theorems make these decompositions precise and moreover show that $\operatorname{div} F, \operatorname{rot} F$ can be arbitrarily prefixed.

Theorem 20.6. *Helmholtz decomposition in* $\mathbb{R}^3$.

(a) *If f is a given continuous function, G a given continuous solenoidal field, and $\psi = |f| + |G|$ satisfies*

$$\int_0^\infty h(t)\,dt < +\infty, \quad \text{where } h(t) = \max_{t\leq|y|} \psi(y), \tag{20.9}$$

then $F = R(f) + R(G)$ is a continuous field with $\operatorname{div} F = f, \operatorname{rot} F = G$ in Stokes' sense vanishing at infinity.

(b) *So if F is continuous vanishing at infinity, with continuous divergence and rotational in Stokes' sense and $\psi = |\operatorname{div} F| + |\operatorname{rot} F|$ satisfies (20.9),*

$$F(x) = \int \operatorname{div} F(y) R(x-y)\,dm_n(y) + \int \operatorname{rot} F(y) \times R(x-y)\,dm_n(y),$$

is the unique decomposition $F = F_1 + F_2$ as a sum of a conservative and a solenoidal field vanishing at infinity.

(c) *In case the stronger condition*

$$\int_0^\infty t h(t)\,dt < +\infty \tag{20.10}$$

is satisfied, the potentials can be made explicit, and

$$F(x) = \nabla \int G(x-y) \operatorname{div} F(y)\,dm_n(y) - \operatorname{rot} \int G(x-y) \operatorname{rot} F(y)\,dm_n(y),$$

is the unique decomposition $F = \nabla u + \operatorname{rot} H$ with H solenoidal and u, H of class C^1 vanishing at infinity.

Proof. The first part is a consequence of Theorem 20.1, the second of Theorem 20.5, the third of Theorems 20.3 and 20.4. □

The version for $n = 2$ is:

Theorem 20.7. *Helmholtz decomposition in $\mathbb{R}^2$*

(a) *If f, g are given continuous functions, and $\psi = |f| + |g|$ satisfies*

$$\int_0^\infty h(t)\,dt < +\infty, \quad \text{where } h(t) = \max_{t\leq|y|} \psi(y), \tag{20.11}$$

then $F = R(f) + JR(g)$ *is a continuous field with* $\operatorname{div} F = f, \operatorname{rot} F = g$ *in Stokes' sense vanishing at infinity.*

(b) *So if* F *is continuous vanishing at infinity, with continuous divergence and rotational in Stokes' sense and* $\psi = |\operatorname{div} F| + |\operatorname{rot} F|$ *satisfies* (20.9),

$$F(x) = \int \operatorname{div} F(y) R(x-y)\, dm_n(y) + J \int \operatorname{rot} F(y) R(x-y)\, dm_n(y),$$

is the unique decomposition $F = F_1 + F_2$ *as a sum of a conservative and a solenoidal field vanishing at infinity.*

(c) *In case the stronger condition*

$$\int_0^{+\infty} |t \log t\, h(t)|\, dt < +\infty,$$

is satisfied, the potentials can be made explicit, and

$$F(x) = \nabla \int G(x-y) \operatorname{div} F(y)\, dm_n(y)$$
$$+ J\nabla \operatorname{rot} F(y) \int G(x-y) \operatorname{rot} F(y)\, dm_n(y),$$

is the unique decomposition $F = \nabla u + J\nabla v$ *with* u, v *of class* C^1 *with logarithmic growth at infinity.*

A decay $\psi(x) = O(|x|^{-1-\varepsilon})$ implies (20.9), while $\psi(x) = O(|x|^{-2-\varepsilon})$ is sufficient for (20.10).

20.4.3 It is natural to ask whether the decomposition $F = F_1 + F_2$ is orthogonal with respect to the coupling of fields. Write $F_1 = \nabla u$ as in part (c). Applying Gauss' theorem in $B(0, R)$ to uF_2, for which $\operatorname{div}(uF_2) = \langle \nabla u, F_2 \rangle = \langle F_1, F_2 \rangle$, we find

$$\int_{|y| \le R} \langle F_1, F_2 \rangle\, dm_n = \int_{|y|=R} u \langle F_2, N \rangle\, dm_{n-1},$$

so that this will be the case if uF_2 decays faster than $|y|^{1-n}$. For instance, if $n = 3$ and $h(t)$ decays like $t^{-2-\varepsilon}$, then checking in the proof of Theorem 20.1 and Example 20.2 we find that $F_1(y), F_2(y)$ decay like $|y|^{-1-\varepsilon}$ and $u(y)$ decays as $|y|^{-\varepsilon}$, so if $\varepsilon > \frac{1}{2}$, then $\int_{\mathbb{R}^3} \langle F_1, F_2 \rangle\, dm_n$ is absolutely convergent and equals zero. In case $n = 2, F_1, F_2$ decay like $|y|^{-1-\varepsilon}$ and u has logarithmic growth so $\varepsilon > 0$ suffices.

20.4.4 Since F_1 is unique, it is also a natural question whether it is possible to obtain an expression of F_1 in terms of F alone. This is indeed possible by another application of Gauss' theorem. So we assume that F vanishes at infinity, $\operatorname{div} F$ satisfies (20.2), and want to find

$$F_1(x) = R(\operatorname{div} F)(x) = \int \operatorname{div} F(y) R(x-y)\, dm_n(y),$$

only in terms of F. We use Gauss' theorem to $R_i(x-y)F(y)$ in $B(x,R) \setminus B(x,\varepsilon)$

$$\begin{aligned}
&\int_{|x-y|=R} R_i(x-y)\langle F, N\rangle\, dm_{n-1}(y) - \int_{|x-y|=\varepsilon} R_i(x-y)\langle F, N\rangle\, dm_{n-1}(y) \\
&= \int_{\varepsilon<|x-y|<R} (\operatorname{div} F)(y) R_i(x-y)\, dm_n(y) + \int_{\varepsilon<|x-y|<R} \langle \nabla_y R_i(x-y), F(y)\rangle\, dm_n(y).
\end{aligned}$$

The first term has limit zero as $R \to +\infty$ because F vanishes at infinity. The second term is, if $F = (F_1, F_2, F_3)$,

$$\begin{aligned}
&-\int_{|x-y|=\varepsilon} \frac{x_i - y_i}{|x-y|^n} \langle F, N\rangle\, dm_{n-1} \\
&\quad = \int_{|x-y|=\varepsilon} \frac{x_i - y_i}{|x-y|^n} \sum_j F_j(x) \frac{x_j - y_j}{|y-x|}\, dm_{n-1}(x).
\end{aligned}$$

Its limit as $\varepsilon \to 0$ does not change replacing $F_j(x)$ by $F_j(y)$; then the terms with $j \neq i$ have limit zero while those with $j = i$ have limit $-\frac{1}{n}$. Altogether, its limit is $\frac{1}{n} F_i(x)$. Assuming that at infinity

$$\int |F(x||x|^{-n} dm_n(x) < +\infty,$$

we get

$$F_1(y) = \frac{1}{n} F(x) - \lim_{\varepsilon \to 0} \int_{\varepsilon<|x-y|} \langle \nabla_x R_i(y-x), F(x)\rangle\, dm_n(x).$$

Here, $\langle \nabla_x R_i(y-x), F(x)\rangle$ is to be understood as the matrix $(D_{j,x} R_i(y-x)$ acting on $F(x)$.

The kernel $D_{j,x} R_i(y-x)$ has singularity $O(|x-y|^{-n})$ and is not integrable. The limit above is an example of a *singular integral.* They appear in many areas of analysis and applied mathematics, particularly this one in fluid mechanics, and their properties are well studied. The kernel enjoys a cancellation property. Indeed,

$$D_{j,x}R_i(y-x) = -\frac{\delta_{i,j}}{|x-y|^n} + n\frac{(x_i-y_i)(x_j-y_j)}{|x-y|^{n+2}}$$

has integral zero on every ball centered at y. Therefore, the limit does not change replacing $F(x)$ by $F(x) - F(y)$. If F is locally Lipschitz, then the integral becomes absolutely convergent and we may write

$$F_1(y) = \frac{1}{n}F(x) - TF(y), F_2 = F - F_1 = \frac{n-1}{n}F + T(F),$$

$$TF(y) = \lim_{\varepsilon\to 0}\int_{\varepsilon<|x-y|} \langle \nabla_x R_i(y-x), F(x) - F(y)\rangle \, dm_n(x).$$

It is possible in fact to take this as a definition of F_1 and prove directly that

$$\operatorname{rot} T(F) = \frac{1}{n}\operatorname{rot} F, \quad \operatorname{div} T(F) = \frac{1-n}{n}\operatorname{div} F,$$

whenever F is Lipschitz and has a continuous divergence and rotational.

Chapter 21

The Dirichlet and Neumann Problems

The Dirichlet and Neumann problems are two boundary value problems related to Poisson's equation $\Delta u = f$ ubiquitous in mathematical physics. Here, we consider them in connection with again the Helmholtz decomposition of vector fields, this time in bounded domains. We describe the solutions to these problems in terms of *Green's functions.* Taking the unit ball as model domain, we compute explicitly the Green's functions and solve Dirichlet's and Neumann's problems with careful boundary estimates. These allow obtaining different Helmholtz-type (unique) decompositions of continuous vector fields in terms of a conservative component, a solenoidal component and a harmonic one.

21.1 Green's Functions

Even though we are mainly interested in $n = 2, 3$ in relation to field decompositions, all results in this section are general.

21.1.1 For a bounded domain $U \subset \mathbb{R}^n$ with regular boundary, the formula in (19.4)

$$\int_{bU} uNu\, dm_{n-1} = \int_U |\nabla u|^2\, dm_n,$$

implies that a harmonic function vanishing on bU is zero, and is constant if $Nu = 0$ in bU. Thus, a general function u should be determined by Δu

in U and u in bU, and up to a constant by Δu in U and Nu on bU. This indicates that there is a redundancy in the decomposition formula (20.1).

In this section, we will obtain decomposition formulas where the boundary integrals contain only either u or Nu. These formulas will allow us to study the so-called *Dirichlet and Neumann problems.*

Definition 21.1. Dirichlet's problem consists in finding, given continuous functions φ, f in bU, U respectively, a continuous function u on $\overline{U}$, such that $u = \varphi$ on bU and $\Delta u = f$ in the weak sense.

Definition 21.2. Neumann's problem consists in finding, given continuous functions ψ, f in bU, U, respectively, a function $u \in C(\overline{U})$, with Nu continuous up to the boundary, such that $Nu = \psi$ on bU and $\Delta u = f$ in U in the weak sense. By (19.3) with $v = 1$, the data must satisfy the compatibility condition

$$\int_{bU} \psi \, dm_{n-1} = \int_U f \, dm_n.$$

By Proposition 20.2, in both cases $u \in C^1(U)$. In some texts Neumann's problem is stated with $u \in C^1(\overline{U})$, but in this case something more is needed on ψ to ensure that the tangential derivatives are smooth up to the boundary too, continuity alone does not suffice. When $f = 0$, the problems are said to be *homogeneous.*

By the maximum modulus principle, Dirichlet's problem has at most one solution. By (19.4), Neumann's problem has at most one solution up to constants. Obviously, Dirichlet's problem splits in two problems

$$D_1 : \Delta u = 0, \quad u = \varphi \text{ in } bU; \quad D_2 : \Delta u = f, \quad u = 0 \text{ in } bU,$$

called *homogeneous and inhomogeneous.* Similarly, Neumann's problem splits in two problems

$$N_1 : \Delta u = 0, \quad Nu = \psi \text{ in } bU, \quad \int_{bU} \psi = 0;$$

$$N_2 : \Delta u = f, \quad Nu = 0 \text{ in } bU, \quad \int_U f = 0.$$

Indeed, if v is an arbitrary smooth function such that

$$\int_{bU} Nv = \int_{bU} \psi = \int_U f = \int_U \Delta v,$$

and u_1, u_2 are solutions to N_1, N_2 with data $\psi - Nv, f - \Delta v$, respectively, then $u = u_1 + u_2 + v$ is a solution of N.

A general remark is that D_1 implies D_2 for bounded f. Indeed, assume D_1 can be solved for arbitrary data $\varphi \in C(bU)$ and let f be continuous and bounded in U. Then, by Theorem 20.3

$$G(f)(x) = \int_U f(y)G(x-y)\,dm_n(y),$$

is of class C^1 in $\mathbb{R}^n$ and $\Delta G(f) = f$ in the weak sense. If φ is the restriction of $G(f)$ to bU and u solves D_1 with boundary data φ, then $G(f) - u$ solves D_2 with data f.

The same remark applies to Neumann's problem. Assume N_1 can be solved, let f as above with $\int_U f = 0$, consider again $G(f)$ and let $\psi = NG(f)$ on bU. We claim that ψ satisfies the compatibility condition $\int_{bU} \psi = 0$. First, we note that

$$Gf(x) = \int_U f(y)G(x-y)\,dm_n(y) = \int_U f(y)(G(x-y) - G(x))\,dm_n(y),$$

$$\nabla Gf(x) = \int_U f(y)(\nabla G(x-y) - \nabla G(x))\,dm_n(y).$$

One has $|\nabla G(x-y) - \nabla G(x)| = O(|x-y|^{-n})$ uniformly in $y \in U$ and $|x|$ large, whence $|\nabla Gf(x)| = O(|x|^{-n})$ as $|x| \to +\infty$. Applying (19.4) to Gf,

$$\int_{|x|=r} NGf\,dm_{n-1} = \int_{bU} NGf\,dm_{n-1},$$

which implies that the right-hand side is zero. Now, if u is a solution of N_1 with data $\psi = NGf$, then $Gf - u$ is a solution of N_2.

21.1.2 Our first goal is to eliminate the term Nu in (20.1), for u of class C^2 in $\overline{U}$. Let v be an harmonic function in a neighborhood of $\overline{U}$. By Green's second identity (19.3)

$$\begin{aligned} 0 &= \int_U u\Delta v\,dm_n \\ &= \int_{bU} (u(y)Nv(y) - v(y)Nu(y))\,dm_{n-1}(y) + \int_U v(y)\Delta u(y)\,dm_n(y). \end{aligned}$$

Subtracting it from (20.1), we obtain

$$\begin{aligned} u(x) = &\int_{bU} (u(y)N_y(G(y-x) - v(y)) - (G(y-x) - v(y))Nu(y))\,dm_{n-1}(y) \\ &+ \int_U (G(y-x) - v(y))\Delta u(y)\,dm_n(y). \end{aligned}$$

The idea is to choose $v = v_x$ such that $v_x(y) = G(x-y)$ for $y \in bU$. Then $G_U(x,y) = G(x-y) - v_x(y)$ will satisfy the desired equation

$$u(x) = \int_{bU} u(y) N_y G_U(x,y)\, dm_{n-1}(y) + \int_U G_U(x,y) \Delta u(y)\, dm_n(y). \tag{21.1}$$

Finding v_x consists in solving a particular homogeneous Dirichlet problem. The function G_U is called *the Dirichlet–Green function of U with pole x*. It is harmonic in $y \in U \setminus \{x\}$, and $G_U(x,y) = 0, y \in bU$.

By the maximum principle,

$$\max_{y \in \overline{U}} |v_x(y) - v_z(y)| = \max_{y \in bU} |G(x-y) - G(z-y)|,$$

implying that v_x is jointly continuous in $x, y \in U$, so it is $G_U(x,y)$ outside the diagonal.

Proposition 21.1. *If $G_U(x,y)$ exists, then $G_U \leq 0$ and $G_U(x,y) = G_U(y,x), x \neq y \in U$.*

Proof. By the second identity in (19.3) in $U \setminus (B_\varepsilon(x) \cup B_\varepsilon(y))$ with $u(z) = G_U(x,z), v(z) = G_U(y,z)$,

$$\int_{|z-x|=\varepsilon} (uNv - vNu)\, dm_{n-1} = -\int_{|z-y|=\varepsilon} (uNv - vNu)\, dm_{n-1}.$$

Near x the contribution of uNv tends to zero with ε, and the one of vNu is seen to have the limit $v(x)$, just as in the proof of (20.1). Analogously, near y we get $u(y)$.

$G_U(x,y)$ is harmonic in $y \in U \setminus B_\varepsilon(x)$. If $y \in bU, G_U(x,y) = 0$ and if $|y-x| = \varepsilon, G_U(x,y) = G(x,y) - v_x(y)$ has the limit $-\infty$ as $\varepsilon \to 0$. By the maximum principle, $G_U(x,y) \leq 0$. □

Example 21.1. It is worth looking at the case $n = 1$; we compute the first Green's function of $U = (a,b)$ with pole at $x \in (a,b)$. Recall that $G(x-y) = \frac{1}{2}|x-y|$, so v_x is the linear function in (a,b)

$$\begin{aligned} v_x(y) &= \frac{1}{2}|a-x| + \frac{1}{2}\frac{(|b-x| - |a-x|)}{b-a}(y-a) \\ &= \frac{1}{2}(x-a) + \frac{1}{2}\frac{b-x-(x-a)}{b-a}(y-a) = \frac{1}{2}(x-a) + \frac{1}{2}\frac{b+a-2x}{b-a}(y-a). \end{aligned}$$

From this,

$$G_U(x,y)=\frac{1}{2}|x-y|-v_x(y)=\begin{cases}\dfrac{(a-y)(b-x)}{b-a} & \text{if } y<x,\\ \dfrac{(x-a)(y-b)}{b-a} & \text{if } x<y.\end{cases}$$

The function

$$P_U(x,y)=N_yG_U(x,y), x\in U, y\in bU$$

is called the *Poisson kernel*. Thus, the formula above becomes

$$u(x)=\int_{bU}u(y)P_U(x,y)\,dm_{n-1}(y)+\int_U\Delta u(y)G_U(x,y)\,dm_n(y). \tag{21.2}$$

In particular, $P_U(x,y)$ has integral one in y.

Thus, if the Dirichlet's problem $\Delta u=f, u=\varphi$ on bU has a solution, it must be

$$u(x)=P_U(\varphi)+G_U(f), P(\varphi)(x)=\int_{bU}\varphi(y)P_U(x,y)\,dm_{n-1}(y),$$

$$G_U(f)(x)=\int_U f(y)G_U(x,y)\,dm_n(y),$$

the first integral being the solution of problem D_1 and the second one the solution of problem D_2. Checking that $P(\varphi)$ is harmonic in x is easy, for G_U being harmonic in $x\in U$ for $y\in bU$, P_U is harmonic in $x\in U$, too. Similarly, we can easily check that $\Delta G(f)=f$ in the weak sense. Indeed, the formula above implies that for a compactly supported smooth Ψ,

$$\psi(x)=\int_U\Delta\psi(y)G_U(x,y)\,dm_n(y).$$

Then

$$\int_U G_U(f)(x)\Delta\Psi(x)\,dm_n(x)$$
$$=\int_U\left(\int_U\Delta\Psi(x)G_U(x,y)\,dm_n(x)\right)f(y)\,dm_n(y).$$

Using the symmetry of G_U, this equals

$$\int_U f(y)\Psi(y)\,dm_n(y).$$

The boundary behavior of $P_U(\varphi), G_U(f)$ is the only thing to be checked. Formally, it is clear that $G_U(f)$ vanishes on bU, because $G(x,y) = 0, x \in bU$, but one needs to prove that $G_U(f)$ is continuous up to the boundary. Later we analyze this behavior in the ball. Without knowing explicitly G_U, when f is bounded, a proof is as follows. If $G_U(x,y) = G(x-y) - v_x(y)$, since $G(f)$ is continuous in the whole space, it is enough to show that

$$u_0(x) = \int_U f(y) v_x(y)\, dm_n(y)$$

is continuous on $\overline{U}$. Write

$$\begin{aligned} u_0(x) - u_0(z) &= \int_U f(y)(v_x(y) - v_z(y))\, dm_n(y) \\ &= \int_{|y-x|\leq\delta} + \int_{|y-x|\geq\delta} = I + II. \end{aligned}$$

Since $G_U \leq 0, |v_x(y)| \leq |G(y-x)|$ if $n > 2$ and $|v_x(y)| \leq |G(y-x)| + C(U)$ if $n = 2$, so the term I is estimated

$$|I| \leq C \int_{|y-x|\leq\delta} (|G(x-y)| + |G(z-y)| + 2C(U))\, dm_n(y).$$

The integral of $|G(x-y)|$ on a ball $B(a,\delta)$ is $O(\delta^2)$ uniformly in a, x (in fact, is maximal when $x = a$); therefore $I = O(\delta^2)$ for all x, z: The term II, for $|x-z| < \frac{\delta}{2}$, is arbitrarily small by dominated convergence.

21.1.3 Analogously, we could try to delete the term in u in (20.1) by choosing another harmonic function w_x such that $Nw_x = N_y G(x-y)$ for $y \in bU$. This is inconsistent because by (19.3) one must have

$$\int_{bU} N w_x dm_{n-1} = \int_U \Delta w_y\, dm_n = 0.$$

But

$$\int_{bU} N_y G(x-y)\, dm_{n-1}(y)$$

equals, using that $G(x-y)$ is harmonic in $U \setminus \{x\}$,

$$\int_{|x-y|=\varepsilon} N_y G(x-y)\, dm_{n-1}(y) = 1.$$

We set Neumann's problem, $N_y w_x = N_y G - \frac{1}{m}, \Delta w_x = 0$, with $m = m_{n-1}(bU)$, which is now consistent. If w_x is a solution, with $H_U(x,y) = w_x(y) - G(x-y)$, we obtain

$$u(x) = \frac{1}{m}\int_{bU} u(y)\, dm_{n-1}(y)$$
$$+ \int_{bU} H_U(x,y)Nu(y)\, dm_{n-1}(y) - \int_U H_U(x,y)\Delta u(y)\, dm_n(y). \tag{21.3}$$

The function $H_U(x,y)$ is harmonic in $U \setminus x$ and $N_y H_U(x,y) = -\frac{1}{m}$ for $y \in bU$. This is not unique; w_x is unique up to constants, so we can add to $H_U(x,y)$ an arbitrary function $c(x)$ of x. Let us see which is the condition so that H_U becomes symmetric. We repeat the argument in Proposition 21.1 with $u(z) = H_U(x,z), v(z) = H_U(y,z)$

$$\int_{bU} (uNv - vNu)\, dm_{n-1} = \left(\int_{|z-x|=\varepsilon} + \int_{|z-y|=\varepsilon}\right)(uNv - vNu)\, dm_{n-1}.$$

The limit of the right-hand side is $-v(x) + u(y)$. The integral on bU is

$$\int_{bU} (uNv - vNu)\, dm_{n-1} = \frac{1}{m}\int_{bU} (v-u)\, dm_{n-1},$$

so it should hold

$$\int_{bU} H_U(y,z)\, dm_{n-1}(z) = \int_{bU} H_U(x,z)\, dm_{n-1}(z).$$

This means that replacing H_U bU $H_U + c(x)$ with

$$c(x) = -\int_{bU} H_U(x,z)\, dm_{n-1}(z),$$

we get a symmetric kernel, and all possible symmetric kernels differ by a constant. From now on we assume that H_U is symmetric.

The function H_U is called the *Neumann–Green function of* U *with pole at* x. If Neumann's problem has a solution, the solution up to a constant should be

$$u = Q_U(\psi) + H_U(f), \quad Q_U(\psi)(x) = \int_{bU} \psi(y) H_U(x,y)\, dm_{n-1}(y),$$

$$H_U(f)(x) = -\int_U f(y) H_U(x,y)\, dm_n(y).$$

$Q_U(\psi), H_U(f)$ being, respectively, the solutions of N_1, N_2. Note that changing H_U to $H_U + c(x)$ does not change Q, H due to the compatibility conditions $\int_{bU} \psi = 0 = \int_U f$.

Since (21.3) implies $\phi = H_U(\Delta\phi)$ for a smooth compactly supported ϕ and $H_U(x,y)$ is symmetric, it follows as before that $\Delta Hf = f$ in the weak sense. Formally, it is also clear that $NHf = 0$ on bU; this is because H_U being symmetric, $N_x H(x,y) = N_x H(y,x) = -\frac{1}{m}$ for $x \in bU$, and so

$$NHf(x) = \int_U f(y) N_x H(x,y)\, dm_n(y) = -\frac{1}{m}\int_U f(y)\, dm_n(y) = 0.$$

While for G_U we saw in the previous paragraph that $G_U f$ solves D_2 for f bounded, for H_U a more detailed knowledge of $H_U(x,y)$ is needed.

21.2 Dirichlet and Neumann Problems in the Ball*

21.2.1 It can be proved that Dirichlet and Neumann problems can be solved in a bounded domain with smooth boundary. Consequently, both Green's functions G_U, H_U exist. In this and the next paragraph, we will deal with the unit ball $B = B(0,1)$, that serves as a model for a general domain, and find them by inspection. We denote $S = bB$.

For G_B, for each $x \in B$ we must find a harmonic function v_x such that

$$v_x(y) = G(x-y), \quad |y| = 1.$$

We use the *reflection method.* If $x = 0$, obviously $v_x = d_n$; if $x \neq 0$, let $x' = \frac{x}{|x|^2}$, the image of x by the inversion centered at 0. Then

$$|x'-y|^2 = \left|\frac{x}{|x|^2} - y\right|^2 = \frac{|x|^2}{|x|^4} + |y|^2 - \frac{2\langle x,y\rangle}{|x|^2},$$

$$|x|^2|x'-y|^2 = 1 + |x|^2|y|^2 - 2\langle x,y\rangle = |y|^2|y'-x|^2. \tag{21.4}$$

Then $|x-y| = |x||x'-y|$ if $|y| = 1$, and we choose

$$v_x(y) = |x|^{2-n} G(x'-y) = d_n |x|^{2-n} |x'-y|^{2-n}, \quad n > 2,$$

and

$$v_x(y) = G(x'-y) + d_2 \log|x| = d_2 \log(|x||x'-y|), \quad n = 2.$$

Thus,

$$G_B(x,y) = d_n(|x-y|^{2-n} - |x|^{2-n}|x'-y|^{2-n}), \quad n > 2,$$

$$G_B(x,y) = d_2 \log \frac{|x-y|}{|x||x'-y|}, \quad n = 2.$$

Next we compute Poisson's kernel. If $y = rw, N = \frac{\partial}{\partial r} = \sum y_i D_i$. Then, if $n > 2$

$$N_y G(x-y) = d_n \frac{\partial}{\partial r}|x-y|^{2-n} = d_n \frac{\partial}{\partial r}\left(|x|^2 + r^2 - 2r\sum_i x_i w_i\right)^{1-\frac{n}{2}}$$

$$= d_n(1-\frac{n}{2})(2r - 2\langle x, w\rangle)|x-y|^{-n} = \frac{1}{c_n}\frac{|y|^2 - \langle x, y\rangle}{|y|}|x-y|^{-n}.$$

In case $n = 2$, we obtain the same result

$$N_y G(x-y) = d_2 \frac{\partial}{\partial r}\log|x-y| = \frac{1}{c_2}\frac{|y|^2 - \langle x, y\rangle}{|y||x-y|^2}.$$

For $n = 2$, we get the same expression for $N_y v_x$ replacing x by x'. For $n > 2$,

$$N_y v_x = d_n|x|^{2-n}\frac{\partial}{\partial r}|x'-y|^{2-n} = \frac{1}{c_n}|x|^{2-n}\frac{|y|^2 - \langle x', y\rangle}{|y|}|x'-y|^{-n}.$$

So, for $|y| = 1$, we get in all cases, using again $|x-y| = |x||x'-y|$,

$$P_B(x,y) = \frac{1}{c_n}|x-y|^{-n}\{1 - \langle x, y\rangle - (1 - \langle x', y\rangle)|x|^2\} = \frac{1}{c_n}\frac{1-|x|^2}{|x-y|^n}. \tag{21.5}$$

21.2.2 In paragraph 13.2.2, we introduced approximations of the identity. We write here $\varphi_\varepsilon(x,y) = \frac{1}{\varepsilon^n}\varphi(\frac{y-x}{\varepsilon})$ for $\varphi \in L^1(\mathbb{R}^n)$, $\int \varphi = 1$. They satisfy:

(a) $\int \varphi_\varepsilon(x,y)\, dm_n(y) = 1$.
(b) $\int |\varphi_\varepsilon(x,y)|\, dm_n(y) = \int |\varphi| = 1$.
(c) Given $\delta > 0$, $\int_{|y-x|>\delta} |\varphi_\varepsilon(x,y)|\, dm_n(y) \to 0, \varepsilon \to 0$.

There we showed that $\int f(y)\varphi_\varepsilon(x,y)\, dm_n(y) \to f(x)$ in various senses, formally $\varphi_\varepsilon(x,y)\, dm_n(y) \to \delta(y-x), \varepsilon \to 0$.

The following are analogous properties for a kernel $K(x,y)$, $x \in B$, $y \in S$:

(a) $\int_S K(x,y)dm_{n-1}(y) = 1, x \in B$.
(b) $\int_S |K(x,y)|dm_{n-1}(y) \leq C, x \in B$.
(c) For $w \in S$ fixed, $C(\delta, x) = \int_{|y-w|>\delta} |K(x,y)|\, dm_{n-1}(y) \to 0, x \to w$.

A kernel satisfying these properties in the present context is also called an approximation of the identity. They imply that $K(x,y)\, dm_{n-1}(y) \to \delta(y-w)$ as $x \to w$, that is, for φ continuous on S,

$$\int_S \varphi(y)K(x,y)\, dm_{n-1}(y) \to \varphi(w).$$

Indeed, the size of the difference is

$$\int_S (\varphi(y) - \varphi(w))K(x,y)\, dm_{n-1}(y),$$

and is estimated by

$$\begin{aligned}&\int_{|w-y|<\delta} |\varphi(y) - \varphi(w)||K(x,y)|\, dm_{n-1}(y) + 2MC(\delta, x)\\ &\quad \leq C \sup_{|w-y|<\delta} |\varphi(y) - \varphi(w)| + 2MC(\delta, x).\end{aligned}$$

The first is arbitrarily small for all x if δ is small, by the continuity of φ; the second term is arbitrarily small as $x \to w$.

It is clear that P_B satisfies the three properties. Thus, we have proved:

Theorem 21.1. *The solution of Dirichlet's problem $\Delta u = 0, u = \varphi$ on S is*

$$u(x) = \int_S \varphi(y)P_B(x,y)\, dm_{n-1}(y).$$

Example 21.2. We compute the solution u_i for $\varphi_i(x) = x_i^2$. Since $x_i^2 - x_j^2$ is harmonic, $u_i - u_j = x_i^2 - x_j^2$ in B, that is $u_i - x_i^2$ is independent of i. The sum $\sum_i u_i$ is the harmonic function with boundary values $\sum_i x_i^2 = 1$, whence $\sum_i u_i = 1$ in B so $\sum_i(u_i - x_i^2) = 1 - |x|^2$ and $u_i(x) = x_i^2 + \frac{1}{n}(1-|x|^2)$.

The above is an example of the so-called *spherical harmonics*. If $\mathcal{H}_k$ is the space of harmonic homogeneous polynomials of degree k, one can identify it as a space of functions in S.

Exercise 21.1. The aim of this exercise is to prove that the spaces $\mathcal{H}_k$ are mutually orthogonal with respect to the inner product in $L^2(S)$, that is the pairing

$$(f,g) = \int_S f\, g\, dm_{n-1}.$$

Prove first that $kP(x) = \sum_i x_i D_i P(x)$ for all k-homogeneous polynomials. It follows that $NP = kP$ on S. Using this and the second Green's identity, prove that $\int_S P\,Q\,dm_{n-1} = 0$ for $P \in \mathcal{H}_k, Q \in \mathcal{H}_l, k \neq l$.

When $n = 2$, $\mathcal{H}_k$ has dimension two and is spanned by the real and imaginary parts of z^k. It can be shown that the spaces $\mathcal{H}_k$ span the whole of $L^2(S)$ so that each $f \in L^2(S)$ has a unique expansion,

$$f = \sum_{k=0}^{+\infty} P_k, \quad P_k \in \mathcal{H}_k.$$

When $n = 2$, this is the expansion in (19.1) making $r = 1$.

Next, we show an application of the theorem to still another characterization of harmonic functions.

Corollary 21.1. *Let u be a continuous function in a domain U with the* weak mean value property, *that is, for every $a \in U$ there is* one *ball $B(a,r)$ such that $u(a)$ is the mean of u on $B(a,r)$. Then u is harmonic in U.*

Proof. Let $B \subset U$ be a ball and let v, the solution of Dirichlet's problem $\Delta v = 0, v = u$ on bB. Then $u - v$ has the weak mean value property and equals zero on bB. It was pointed out after its statement that Theorem 19.5 holds as well for these functions. Therefore, $v = u$ in B and u is harmonic. □

The next result deals with isolated singularities of harmonic functions.

Corollary 21.2. *If u is harmonic in $B(a,R)$ and $u(x) = o(G(x-a))$ as $x \to a$, u is harmonic in $B(a,R)$.*

Proof. We assume $a = 0$; we consider the radialized function obtained by averaging over the group $SO(3)$ of rotations (Section 14.4):

$$v(x) = \frac{1}{8\pi^2} \int_{SO(3)} u(T(x))\, d\mu(T).$$

Then v is harmonic and radial, whence $v(x) = aG(x) + b$. The hypothesis implies $a = 0$, so v is constant. Since u, v have the same means on spheres, it follows that the means of u on spheres are constant. Therefore, u is a continuous function with the weak mean value property and so is harmonic. □

21.2.3 We turn now to the solution of problem D_2. Given a continuous function f in B, we are interested in finding $u \in C(\overline{B})$ such that $u = 0$ in S and $\Delta u = f$ in the weak sense. As noted, the solution should be the potential

$$G_B f(x) = \int_B f(y) G_B(x, y)\, dm_n(y).$$

We will carry out a study of G_B analogous to that of Gf in Theorem 20.3. We bound ourselves to $n > 2$, the case $n = 2$ being similar.

Given $x \in B, G_B(x, y)$ vanishes once at $|y| = 1$, because $N_y G_B(x, y) = P_B(x, y) > 0$. Therefore, a required integrability condition on f is that

$$\int_B |f(y)|(1 - |y|)\, dm_n(y) < +\infty. \tag{21.6}$$

This is known as *Blaschke's condition.*

The potential $G_B f$ has the same local regularity properties in B as Gf. Indeed, fixed $a \in B$, let $\tau = \frac{1}{2}(1 - |a|)$; and write $f = f_1 + f_2$ with f_1 supported in $B(a, \tau)$ and $f_2 = 0$ in $B(a, \tau/2)$. Then $G_B f_2$ is smooth near a, say for $x \in B(a, \tau/4)$, because then $D_x^\alpha G_B(x, y)$ is uniformly dominated by $|f(y)|(1 - |y|)$. As for $G_B f_1$, we use $G_B = G - v_x$,

$$G_B f_1(x) = G f_1(x) - d_n |x|^{2-n} \int f_1(y) |x' - y|^{2-n}\, dm_n(y).$$

Clearly, the second integral is smooth near a, whence $G_B f_1$ has the same regularity properties near a than Gf_1. Using the results in Theorem 20.3, we conclude that it is of class C^1 and of class C^2 if f satisfies a local Lipschitz condition.

We already saw in paragraph 21.1.2 that $\Delta G_B f = f$ in the weak sense. It remains to show that $G_B f(x) \to 0$ as $|x| \to 1$. This is clear if f is bounded, because using again formula (21.1) for $u = |x|^2$

$$|G_B f(x)| \le C \int_B |G_B(x, y)|\, dm_n(y) = \frac{C}{2}(1 - |x|^2).$$

The sole integrability condition (21.6) is not enough to ensure $G_B f(x) \to 0$ as $|x| \to 1$. The situation is similar to that of Gf, to ensure that it vanishes at infinity we require a condition stronger than (20.5). We look first at the radial case, $f(y) = h(r), r = |x|$, so the integrability condition is $\int_0^1 (1-r)|h(r)|\,dr < +\infty$. Then $G_B(x) = H(s), s = |x|$ is also radial,

$$H(s) = \int_0^1 h(r)\left(\int_{|y|=r} G_B(x,y)\,dm_{n-1}(y)\right)dr.$$

Now $G_B = G - v_x$; the integral of G was computed in Example 20.1. If $n \geq 3$, it equals $1/(2-n)r$ if $s < r$ and $1/(2-n)r^{n-1}s^{2-n}$ if $s > r$. As v_x is harmonic, its integral on $|y| = r$ equals $c_n r^{n-1} v_x(0) = 1/(2-n)r^{n-1}$. Thus

$$|H(s)| \leq (s^{2-n} - 1)\int_0^s |h(r)|r^{n-1}\,dr + \int_s^1 |h(r)|r(1-r^{n-2})\,dr.$$

In case $n = 2$, the estimate for H is

$$|H(s)| \leq |\log s| \int_0^s r|h(r)|\,dr + \int_s^1 |h(r)|r|\log r|\,dr.$$

In both cases the last term has limit zero as $s \to 1$, but the first behaves like $(1-s)\int_0^s |h(r)|\,dr$. If $|h|$ is non-decreasing, the integrability condition implies $(1-r)h(r) \to 0$ and so H has limit zero. Thus, if the least increasing majorant satisfies the Blaschke condition, $G_B f(x) \to 0$ as $|x| \to 0$.

Later on we will need a criterion on f ensuring that G_B is C^1 up to the boundary. Using the same arguments, this will be the case if

$$\int_B |\nabla_x G_B(x,y) f(y)|\,dm_n(y) < +\infty, |x| = 1.$$

By the same computation that led to Poisson's kernel, permuting x, y, this is

$$\int_B \frac{1-|y|^2}{|x-y|^3}|f(y)|\,dm_n(y) < +\infty.$$

It is easily checked that $|f(y)| = O((1-|y|)^{\varepsilon-1})$ is enough.

Altogether, we may state:

Theorem 21.2. *Let f be a continuous function in B such that*

$$\int_0^1 h(r)(1-r)\,dr < +\infty, \quad h(r) = \max_{|x|\leq r}|f(x)|.$$

Then $G_B f$ is a function of class C^1 in B such that $\Delta G_B f = f$ in the weak sense and $G_B f(x) \to 0$ as $|x| \to 1$. If $|f(y)| = O((1-|y|)^{\varepsilon-1})$, $G_B f \in C^1(\overline{B})$.

21.2.4 The computation of $H_B(x,y)$ is more involved. Now we must find a harmonic function w_x such that $Nw_x = N_y G(x-y) - \frac{1}{c_n}$ for $|y|=1$. If in (21.5) we add $N_y G, N_y v_x$ instead, we get

$$N_y G + N_y v_x = \frac{1}{c_n}, \quad n=2, = \frac{1}{c_n}|x-y|^{2-n}, \quad n>2.$$

Thus, we take $w_x = -v_x$ if $n=2$ to obtain

$$H_B(x,y) = -\frac{1}{2\pi}\log(|x-y||x||x'-y|).$$

By (21.4), this is already symmetric. For $n>2$, one must search for a harmonic function h_x, such that

$$N_y h_x = \frac{1}{c_n}(1-|x-y|^{2-n}).$$

In this case, the same computations show that no function of $|x'-y|$ alone can work. In [19], an explicit function is found in general dimension. For $n=3$, it appears previously in [4] and we reproduce it here. The function is

$$h_x(y) = \frac{1}{4\pi}\log E, \quad E = \frac{\langle x'-y, x\rangle}{|x|} + |x'-y|.$$

Computation shows that h_x is harmonic, and

$$4\pi \nabla h_x = \frac{1}{|y|}\left(1 - \frac{1}{|x||x'-y|}\right).$$

By (21.4), this equals $1 - \frac{1}{|x-y|}$ if $|y|=1$, as desired. Thus, we get

$$\begin{aligned} &-G(x-y) - v_x(y) - h_x(y) \\ &= \frac{1}{4\pi}\left(\frac{1}{|x-y|} + \frac{1}{|x||x'-y|} - \log E\right). \end{aligned}$$

For a symmetric expression, we write $E = \frac{1-\langle x,y\rangle + |y||y'-x|}{|x|}$ and delete the term in $\log|x|$ to finally get

$$H_B(x,y) = \frac{1}{4\pi}\left(\frac{1}{|x-y|} + \frac{1}{|x||x'-y|} - \log(1-\langle x,y\rangle + |y||y'-x|)\right).$$

21.2.5 Now, we would like to solve N_1, N_2 by means of

$$Q_B(\psi)(x) = \int_S \psi(y) H_B(x,y)\, dA(y), H_B f(x) = -\int_B f(y) H_B(x,y)\, dm_n(y).$$

We start studying $H_B f$. Now $H_B(x,y)$ does not vanish for $y \in S$ so we must require that f be continuous and integrable on B. In the same way as in previous paragraph with G_B it is easily seen that $H_B f$ has the same local regularity properties as Gf, namely, it is of class C^1 in B, of class C^2 if f satisfies a local Lipschitz condition. We already know that $\Delta H_B f = f$ in the weak sense. We study the behavior of $\nabla H_B f$ near S. One has, with $F = 1 - \langle x, y\rangle + |y||y' - x|$

$$4\pi \nabla_x H_B(x,y) = \frac{y - x}{|y - x|^3} + \frac{y' - x}{|y||y' - x|^3} - \frac{1}{F}\left(-y + |y| \frac{x - y'}{|y' - x|}\right). \quad (21.7)$$

First, we look at the normal component. With $N = \sum x_i D_i, N_x H_B(x,y) = \langle \nabla_x H_B(x,y), x\rangle$. A computation shows that

$$4\pi N_x H_B(x,y) = \frac{\langle x, y\rangle - |x|^2}{|x - y|^3} + \frac{1 - \langle x, y\rangle}{|x|^3 |y - x'|^3} - 1. \quad (21.8)$$

Incidentally, $N_y H_B(x,y)$ has the same expression permuting x, y and we see again that $N_y H_B(x,y) = -\frac{1}{4\pi}$ for $y \in S$. Then, since f has integral zero,

$$4\pi N H_B f(x) = \int_B f(y) \left(\frac{\langle x, y\rangle - |x|^2}{|x - y|^3} + \frac{1 - \langle x, y\rangle}{|x|^3 |y - x'|^3}\right) dm_n(y).$$

The kernel $K(x,y)$ inside the integral does not vanish for $y \in S$ so we need the same hypothesis of integrability for f as we had for H_B. The kernel $K(x,y)$ is zero if $|x| = 1$. To prove rigorously that NH_B has limit zero, we need, as in the previous paragraphs, a stronger hypothesis than just integrability of f, namely

$$\int_0^1 h(r)\, dr < +\infty, \quad h(r) = \max_{|y| \le r} |f(y)|.$$

For instance, $|f(y)| = O((1-|y|)^{\varepsilon-1})$ will do. We indicate briefly how to estimate $NH_B(x)$; of course, the contribution I near x, say $|y-x| \le \delta = \frac{1}{2}(1-|x|)$, and the contribution II away must be treated separately. In I,

$$\left|\frac{\langle x,y\rangle - |x|^2}{|x-y|^3}\right| \le c|x-y|^{-2}, \quad \left|\frac{1-\langle x,y\rangle}{|x|^3|y-x'|^3}\right| \le c\delta^{-2},$$

and $|f| \le c\delta^{\varepsilon-1}$, obtaining $I = O(\delta^\varepsilon)$. In $|y-x| \ge \delta$,

$$|K(x,y)| \le c\left(\frac{\delta^2(1-|y|^2)}{|x-y|^5} + \frac{\delta(1-|y|^2)}{|x-y|^4}\right),$$

so, up to constants,

$$\begin{aligned}
&\int_{|y-x|\ge\delta} |K(x,y)|f(y)|\,dm_n(y)\\
&\quad\le \int_{|y-x|\ge\delta} \frac{\delta^2(1-|y|^2)^\varepsilon}{|x'-y|^2|x-y|^3}\,dm_n(y) + \int_{|y-x|\ge\delta} \frac{\delta(1-|y|^2)^\varepsilon}{|x'-y|^2|x-y|^2}\,dm_n(y)\\
&\quad\le \int_{|y-x|\ge\delta} \frac{\delta^2}{|x-y|^{5-\varepsilon}}\,dm_n(y) + \int_{|y-x|\ge m} \frac{\delta}{|x-y|^{4-\varepsilon}}\,dm_n(y) = O(\delta^\varepsilon).
\end{aligned}$$

For the full gradient, from (21.7) we see that

$$|\nabla_x H_B(x,y)| = O\left(\frac{1}{|x-y|^2}\right).$$

Since

$$\int \frac{(1-|y|)^{\varepsilon-1}}{|x-y|^2}\,dm_n(y),$$

is still finite for $x \in S$, it is easily seen that H_Bf is of class C^1 up to $\overline{B}$.

Thus, we have proved:

Theorem 21.3. *If f is a continuous function in B and $|f(y)| = O((1-|y|)^{\varepsilon-1})$ for some $\varepsilon > 0$, H_Bf is a C^1-function in $\overline{B}$ such that $\Delta H_Bf = f$ in the weak sense and $NH_Bf(x) = 0$ for $x \in S$.*

21.2.6 Now we deal with

$$Q_B(\psi)(x) = \int_S \psi(y)H_B(x,y)\,dA(y).$$

Clearly, $Q_B(\psi)$ is harmonic in B. By (21.8) and the cancellation condition on ψ,

$$NQ_B(\psi)(x) = \frac{1}{4\pi}\int_S \psi(y)\left(\frac{\langle x,y\rangle - |x|^2}{|x-y|^3} + \frac{1-\langle x,y\rangle}{|x|^3|y-x'|^3} - 1\right) dA(y)$$

$$= \int_S \psi(y)\left(P_B(x,y) - \frac{1}{4\pi}\right) dA(y) = \int_S \Psi(y) P_B(x,y)\, dA(y).$$

By Theorem 21.1, the boundary value of $NQ_B(\psi)$ is ψ.

In fact, problem N_1 can be reduced to D_1 in the ball in any dimension. A computation shows that $Nu = \sum_i x_i D_i u$ is harmonic if u is, whence if u is a solution of N_1,

$$Nu(x) = \int_S \psi(y) P_B(x,y)\, dm_{n-1}(y).$$

In spherical coordinates, $N = \rho\frac{\partial}{\partial\rho}$, thus, if $x = \rho w$, $w \in S$,

$$u(x) = \int_0^\rho Nu(sw)\frac{ds}{s} = \int_0^1 Nu(tx)\frac{dt}{t}.$$

Now we would like to insert the formula above for Nu and apply Fubini's theorem. To make the integral in t convergent near zero, we use the cancellation condition and replace $P_B(x,y)$ by $P_B(x,y) - P_B(0,y) = P_B(x,y) - \frac{1}{c_n}$ to obtain

$$u(x) = \int_S \psi(y) H(x,y)\, dm_{n-1}(y),$$

with

$$H(x,y) = \int_0^\rho \left(P_B(sw,y) - \frac{1}{c_n}\right)\frac{ds}{s}$$

$$= \int_0^1 \left(P_B(tx,y) - \frac{1}{c_n}\right)\frac{dt}{t}, \quad x \in B, \quad y \in S.$$

Then $N_x H(x,y) = P_B(x,y) - \frac{1}{c_n}$ and as before u is the solution of problem N_1.

Incidentally, this shows that $H_B(x,y) = H(x,y)$ for $y \in S$. Since $H_B(x,y) = w_x(y) - G(x-y)$ for a harmonic w_x, w_x is known on S, and so it is possible to write a closed expression for $H_B(x,y)$ in terms of H, P_B and G.

The function $Q_B(\psi)$ is not in general of class C^1 up to the boundary. This is most easily seen when $n = 2$. Then, in polar coordinates $x = re^{i\theta}$

$$\begin{aligned} u(x) &= -\frac{1}{2\pi}\int \psi(y)\log(|x-y|^2)\,ds(y) \\ &= -\frac{1}{2\pi}\int_0^{2\pi} \psi(e^{it})\log(1+r^2-2r\cos(\theta-t))\,dt. \end{aligned}$$

Then

$$\frac{\partial u}{\partial \theta} = -\frac{1}{2\pi}\int_0^{2\pi} \psi(e^{it})\frac{2r\sin(\theta-t)}{1+r^2-2r\cos(\theta-t)}\,dt,$$

becomes a singular integral when $r \to 1$. This is in fact the *conjugate function* or *Hilbert transform* of Ψ. It is known that there are continuous functions with discontinuous conjugate.

Exercise 21.2. Prove that if ψ satisfies a Lipschitz condition on S, then $Q_B(\psi)$ is C^1 up to the boundary.

Altogether, we may state:

Theorem 21.4. *Let f be a continuous function in B such that*

$$|f(y)| = O((1-|y|)^{\varepsilon-1})$$

for some $\varepsilon > 0$, and ψ a Lipschitz function on S satisfying the compatibility condition $\int_S \Psi = \int_B f$. Then the Neumann problem has the unique solution (up to constants) $Q_B(\psi) + H_B(f)$ in $C^1(\overline{B})$.

21.3 Decomposition of Vector Fields on Smooth Domains*

21.3.1 In this section, we study the Helmholtz decomposition of vector fields in the ball B in $\mathbb{R}^n, n = 2, 3$. First we assume $n = 3$. All results hold in a general bounded domain U with smooth boundary for which Dirichlet's and Neumann's problems can be solved. We deal with fields F continuous up to S, and assume they have continuous divergence and rotational in Stokes' sense in B. We denote by $F_n = \langle F, N\rangle N, F_t$, respectively, the normal and tangential component of F at points in S. We say that F is *tangential* if $F_n = 0$ and *normal* if $F_t = 0$.

The main difference with the unbounded case is that now there are non-trivial harmonic fields. Thus, if any, there would exist infinite decompositions of F a sum of a solenoidal and a conservative field. Paragraph 20.4.1

is such a decomposition, and gives F in terms of $\operatorname{div} F, \operatorname{rot} F$ and F on S. However, this decomposition is not the natural one in the bounded case, as shown by the next result, a restatement of the uniqueness in Dirichlet and Neumann problems. It indicates that only the tangential component or the normal component of F are needed.

Theorem 21.5.

(a) *A field is completely determined in B by* $\operatorname{div} F, \operatorname{rot} F$ *and either F_n or F_t.*

(b) *A harmonic field F is completely determined by either F_n, F_t on S, and is zero if either $F_n = 0$ or $F_t = 0$.*

Proof. It is enough to prove the second assertion. Write $F = \nabla u$ with $u \in C^1(\overline{B})$, harmonic. If $F_n = Nu = 0$, then by (19.4) u is constant and $F = 0$. If $F_t = 0$, then u is constant on S, therefore is constant too in B and $F = 0$. □

To express F in terms of F_n on S for harmonic F, write $F = \nabla u$ as above, and use the representation formula (21.3)

$$u(x) = \int_S Nu(y) H_B(x, y)\, dA(y) + c.$$

Therefore,

$$F(x) = \int_S F_n(y) \nabla_x H_B(x, y)\, dA(y).$$

Recall that in the ball $N_x H_B = P_B - 1$ for $y \in S$, P_B being Poisson's kernel. This defines too a conservative field with given normal component. We may write

$$F_t(x) = \int_S F_n(y) \nabla_{t,x} H_B(x, y)\, dA(y),$$

where $\nabla_{t,x}$ denotes tangential gradient in x. The kernel

$$\nabla_{t,x} H_B(x, y) = CP(x, y),$$

which becomes singular as $x \to bU$ (as showed before in the disk) allows to express F_t in terms of F_n on S. That is, the vector-valued operator

$$CP(\psi)(x) = \int_S \psi(x) CP(x, y)\, dA(y),$$

satisfies $F_t = CP(F_n)$ if F is harmonic. In the ball it is called the *conjugate Poisson kernel.* The inverse operator from F_t to F_n is harder to express and also of singular nature. Both CP and its inverse have a simple expression in the upper half-space, in terms of the so-called *Riesz transforms.* As shown before, these singular integrals become absolutely convergent if ψ satisfies a Lipschitz condition. Using CP and its inverse, it is possible to see that there is a conservative field with given preassigned tangential component, whenever the latter satisfies a Lipschitz condition.

21.3.2 As in Section 20.4, our aim is to establish the existence and uniqueness of decompositions $F = F_1 + F_2$, with F_1 conservative and F_2 solenoidal depending only on $\operatorname{div} F, \operatorname{rot} F$ and either F_n, F_t. Moreover, some of these decompositions will be orthogonal with respect to the coupling of fields

$$(F, G) = \int_B \langle F(x), G(x)\rangle \, dm_n(x),$$

and therefore unique, too. We introduce the following spaces of fields. First is the space

$$\mathcal{C}_n = \{F \in C(\overline{B}), \operatorname{rot} F = 0, F_t = 0\},$$

of conservative normal fields. Secondly, is the space

$$\mathcal{S}_t = \{F \in C(\overline{B}), \operatorname{div} F = 0, F_n = 0\},$$

of solenoidal tangential fields. Finally, $\mathcal{H}$ denotes the space of harmonic fields.

Proposition 21.2. *F is conservative (resp., solenoidal) if and only if it is orthogonal to $\mathcal{S}_t$ (resp., $\mathcal{C}_n$).*

Proof. Since $\mathcal{S}_t$ (resp., $\mathcal{C}_n$) contain the rotational and gradient of compactly supported fields and functions, respectively, we already know that a field orthogonal to $\mathcal{S}_t$ (resp., $\mathcal{C}_n$) is conservative (resp., solenoidal). For the converse, let us see under what boundary conditions a conservative F is orthogonal to a solenoidal G. If $F = \nabla u, G = \operatorname{rot} H$, by (18.2), (18.1) and Gauss' theorem

$$\int_S u\langle G, N\rangle \, dA = (F, G) = -\int_S \langle F \times H, N\rangle \, dA.$$

This is zero if either G is tangential or F is normal. $\square$

Corollary 21.3. *The three spaces $\mathcal{C}_n, \mathcal{S}_t, \mathcal{H}$ are pair-wise orthogonal.*

This implies that every possible decomposition into components in those is unique.

21.3.3 Assume F is continuous up to $\overline{B}$ with continuous divergence and we want to decompose $F = F_1 + F_2$ with F_1 conservative and F_2 solenoidal. For technical reasons we assume that F satisfies locally a Lipschitz condition. Then

$$\operatorname{div} F_1 = \operatorname{div} F, \quad \operatorname{rot} F_2 = \operatorname{rot} F.$$

On the other hand, by Theorem 21.5, each of them, say F_1, is completely determined by either its normal or tangential component. There are thus four natural choices each leading to unique decompositions $F = F_1^i + F_2^i$, $i = 1, 2, 3, 4$.

The first is choosing F_1 normal, $(F_1)_t = 0$, then $(F_2)_t = F_t$. We want then $F_1 = \nabla u$ with $\Delta u = \operatorname{div} F_1 = \operatorname{div} F$ and u constant on S, say zero. This is Dirichlet's problem D_2, whose solution is

$$u(x) = G_B(\operatorname{div} F)(x) = \int_B \operatorname{div} F(y) G_B(x, y)\, dm_n(y).$$

If $\operatorname{div} F(x) = O((1 - |x|)^{\varepsilon - 1}), \varepsilon > 0$, by Theorem 21.2 $u \in C^1(\overline{B}), u = 0$ on S. Then

$$F_1^1(x) = \nabla u(x) = \int_B \operatorname{div} F(y) \nabla_x G_B(x, y)\, dm_n(y).$$

The second choice is $\langle F_1, N\rangle = \langle F, N\rangle$, then $(F_2)_n = 0$ and F_2 is tangential. Then $F_1 = \nabla u$ where u is the solution of Neumann's problem $\Delta u = \operatorname{div} F, Nu = \langle F, N\rangle$ on S. The compatibility condition is ensured by Gauss' theorem. By Theorem 21.4, this problem has a unique solution up to constants

$$u = Q_B(\langle F, N\rangle) + H_B(\operatorname{div} F),$$

so

$$F_1^2(x) = F_{11}^2 + F_{12}^2, \quad F_{11}^2(x) = \int_S \langle F(y), N\rangle \nabla_x H_B(x, y)\, dA(y),$$

$$F_{12}^2(x) = \int_B \operatorname{div} F(y) \nabla_x H_B(x, y)\, dm_n(y). \tag{21.9}$$

The first term F_{11}^2 is harmonic with normal component F_n.

A third choice is $(F_1)_n = 0$, then $(F_2)_n = F_n$. The solution is the second term F_{12}^2 in (21.9).

The fourth choice is $(F_1)_t = F_t$, then $(F_2)_t = 0$. This is the more delicate case, as the operator CP and its inverse enter into the picture. If $\varphi = CP^{-1}(F_t)$, a Lipschitz function, then

$$F_1^4 = F_{11}^4 + F_{12}^4, F_{11}^4 = \nabla Q_B(\varphi) = \int_S \varphi(y)\nabla_x H_B(x,y)\,dA(y), F_{12}^4 = F_1^1.$$

The first term is harmonic with tangential component F_t and the second term has zero tangential component.

Thus, we may state the *Helmholtz's decomposition theorem*:

Theorem 21.6. *If F is a Lipschitz field on $\overline{B}$ with continuous divergence*

$$\operatorname{div} F(y) = O((1-|y|)^{\varepsilon-1}),$$

then F can be uniquely decomposed $F = F_1 + F_2$ with F_1 conservative and F_2 solenoidal, both continuous, satisfying one of the following additional properties:

(a) *F_1 is normal.*
(b) *F_2 is tangential.*
(c) *F_1 is tangential.*
(d) *F_2 is normal.*

The first two decompositions are orthogonal with respect to the coupling of fields.

In fact, it can be shown that in all cases the F_i are also Lipschitz.

21.3.4 Using Corollary 21.3, it is possible to rephrase the orthogonal decompositions as arising from a three component decomposition, called *the Hodge decomposition*:

Theorem 21.7. *Under the same assumptions, F has a unique orthogonal decomposition*

$$F = G_1 + H + G_2,$$

with $G_1 = G_1(\operatorname{div} F)$ conservative and normal, $G_2 = G_2(\operatorname{rot} F)$ solenoidal and tangential, and H a harmonic field.

Proof. With the above notations, if $F_1^2 = \nabla u$, we decompose

$$u(x) = \int_S u(y) P_B(x,y)\, dA(y) + \int_B \operatorname{div} F(y) G_B(x,y)\, dm_n(y),$$

so $F_1^2 = H(x) + F_1^1$, with H harmonic. So $G_1 = F_1^1, G_2 = F_2^2$ and

$$H(x) = \nabla \int_S u(y) P_B(x,y)\, dA(y)$$

is the required decomposition. □

Doing similar splittings in the other cases leads to other three types of decompositions, with both G_1, G_2 tangential, or both G_1, G_2 normal or G_1 tangential and G_2 normal, but these are not unique.

21.3.5 It is not easy to find explicit expressions for the solenoidal components in the decompositions. These are solutions to the problem $\operatorname{rot} F_2 = \operatorname{rot} F$ with boundary conditions on $(F_2)_t$ or $(F_2)_n$. As in full space, one way to deal with this problem is to look for F_2 in the form $F_2 = \operatorname{rot} G$ with solenoidal G. Then the problem becomes a vectorial Poisson's equation

$$\Delta G = \nabla \operatorname{div} G = -\operatorname{rot}\operatorname{rot} G = -\operatorname{rot} F_2 = -\operatorname{rot} F.$$

Adding the boundary conditions on the normal or tangential components of F_2 leads to a system of three scalar Poisson's equations with coupled boundary conditions.

However, in dimension $n = 2$, $\operatorname{rot} F$ is a scalar quantity and the situation becomes simpler, as shown in the next paragraph.

21.3.6 In the plane the situation is simpler using

$$\operatorname{rot} F = -\operatorname{div} JF.$$

In particular, JF is harmonic iff F is. We denote $T = JN$ the unit tangent to S, so

$$F = \langle F, N\rangle N + \langle F, T\rangle T,$$
$$J(F) = \langle F, N\rangle T - \langle F, T\rangle N,$$

gives the relationship between tangential and normal components of F, JF. Using this, all problems in which a non-zero tangential component is prefixed can be rephrased in terms of normal components and be solved by means of the operators P_B, G_B, Q_B, H_B. For instance, if F is a harmonic field with prescribed tangential component $\langle F, T\rangle = \varphi$, then JF is also harmonic with normal component $-\varphi$, whence $JF = -\nabla Q_B(\varphi)$ and $F = J(\nabla Q_B(\varphi)$. This is the case for the harmonic term F_{11}^4 with tangential component F_t in the fourth type decomposition.

By the same reason, now the second terms F_2 can be made explicit, too. We want $\operatorname{rot} F_2 = \operatorname{rot} F$ with prescribed conditions on the normal and tangential components of F_2. Then $\operatorname{div} JF_2 = -\operatorname{rot} F$ and we simply translate everything in terms of JF_2.

For instance, in the first choice we want $\langle F_2, T\rangle = \langle F, T\rangle$, so $N(J(F_2)) = -\langle F, T\rangle$. The solution for $J(F_2)$ is

$$
\begin{aligned}
J(F_2) &= -\nabla Q_B(\langle F, T\rangle) - \nabla H_B(\operatorname{rot} F), \\
F_2(x) &= J(\nabla Q_B(\langle F, T\rangle) + J\nabla H_B(\operatorname{rot} F) \\
&= \int_S \langle F, T\rangle J(\nabla_x H_B(x,y))\, ds(y) + \int_B \operatorname{rot} F(y) J(\nabla_x H_B(x,y))\, dA(y).
\end{aligned}
$$

Analogously, in all other cases, we see that all decompositions consist in combining a number of pieces with the J operator. If $\operatorname{div} F = f, \operatorname{rot} F = g$, these pieces are $G_B(f), G_B(g), H_B(f), H_B(g), Q_B(F_t), Q_B(F_n)$.

Chapter 22

Additional Exercises

1. Consider the 2×2 matrix
$$M = \begin{pmatrix} a & b \\ c & d \end{pmatrix}.$$
Compute $|M|$ assuming either $a^2 + c^2 = b^2 + d^2$ or $ab + cd = 0$. Find a formula for $|M|$ in the general case using $M^t M$.
2. Let M be as above, looked as a linear transformation from $\mathbb{R}^2$ to $\mathbb{R}^2$. Define
$$|M|_{p,q} = \sup\left\{\frac{|Mv|_p}{|v|_q}, v \neq 0\right\}.$$
Try to find explicitly $|M|_{p,q}$ in terms of a, b, c, d in the four cases $p, q = 1, \infty$.
3. (a) Prove that if A is an $n \times k$ matrix with rank k, then $A^t A$ is invertible and if it has rank n, then AA^t is invertible.
   (b) Prove that if V is an linear subspace of dimension $k, 0 < k < n$ in $\mathbb{R}^n$, $v_1, \ldots, v_k$ is a basis of V, and A is the matrix $n \times k$ with column vectors v_i and $y \notin V$, the orthogonal projection w of y onto V is $W = AX$, where the column vector X is the solution of
$$A^t AX = A^t Y,$$
that is, $X = (A^t A)^{-1} A^t Y, W = A(A^t A)^{-1} A^t Y$.
   (c) Here is an equivalent statement of the previous item. With A as above, with rank $k < n$, set $V = A(\mathbb{R}^k)$. The system has no solution if $Y \notin V$ and if $Y \in V$ if it has a unique solution. If $Y \notin V$,

we define an approximate solution of $AX = Y$ as the solution X_0 of the least-square problem

$$\min_X \|AX - Y\|^2.$$

That is, AX_0 is the orthogonal projection of Y onto V. By the previous item, $X_0 = (A^tA)^{-1}A^tY$, whence the map $Y \to X_0$ is lineal. This gives an interpretation of $(A^tA)^{-1}A^t$, a left-inverse of A.

(d) Assume V of dimension k is given by a system of $n-k$ independent linear equations $AX = 0$, that is A is a $(n-k)\times n$ matrix of rank $n-k$, and V is the kernel of A). Given $X \notin V$, prove that the projection of X onto $V^\perp$ is $A^t(AA^t)^{-1}AX$.

(e) A rephrasing of the previous item is as follows. The system $AX = Y$ has non-unique solutions for all Y; then, among all solutions we single out the unique solution $X_0 \in V^\perp$. Equivalently, X_0 is the projection onto the orthogonal $V^\perp$ of the kernel V. The previous item shows that $X_0 = A^t(AA^t)^{-1}Y$. This provides an interpretation of $A^t(AA^t)^{-1}$, a right-inverse of d'A.

4. This exercise complements paragraph 4.5.2. Let A be an $n\times m$ matrix, regarded as a linear map from $\mathbb{R}^m$ to $\mathbb{R}^n$. Let N be the kernel of A and R, its range. The restriction of A to $N^\perp$ is an isomorphism B between $N^\perp$ and R. Consider the map $A^+(Y) = B^{-1}(P_RY)$; that is, first we project Y onto R and among all solutions of $AX = PY$ we choose the one in $N^\perp$. Prove that A^+ is the unique $m\times n$ matrix such that

   (a) $AA^+A = A$.
   (b) $A^+AA^+ = A^+$.
   (c) AA^+ i A^+A are symmetric matrices.

   The matrix A^+ is called *the Moore–Penrose pseudo-inverse of* A. Note that A is the pseudoinverse of A^+. We have seen that if the rank $r = m$, then $A^+ = (A^tA)^{-1}A$, and when $r = n$, $A^+ = A^t(AA^t)^{-1}$.

5. Let A be an $n\times m$ matrix with rank m, that is, $y = Ax$ is one-to-one. We know that $|Ax|$ has a maximum and a minimum in $|x| = 1$. Assume that $Ax = y_0$ has the unique solution x_0, allow a small increment Δy_0 and assume that $Ax = y_0+\Delta y_0$ has a unique solution x. We may think that y_0 comes from a measure and an error has been made. Prove that

$$\frac{|x - x_0|}{|x_0|} \le C(A)\frac{|\Delta y_0|}{|y_0|},$$

with

$$C(A) = \frac{\max\{|Ax|, |x| \le 1\}}{\min\{|Ax|, |x| \le 1\}},$$

that is, the relative error in x is controlled by the relative error on y and by $C(A)$. In case $n = m$ (A invertible), prove that $C(A) = \|A\|\|A^{-1}\|$. The number $C(A)$ is called *the condition number* of A.

6. For a symmetric matrix A, compute $J_k(A)$ in terms of the eigenvalues.
7. Prove $\lim_{p\to+\infty} |v|_p = |v|_\infty$.
8. Give an example of a closed set A which is not the closure of its interior.
9. Let (x^k) be a convergent sequence to a point x in $\mathbb{R}^n$. Prove that the set of points in the sequence, together with x, is compact.
10. Prove that a bounded sequence is convergent to p if and only if whenever a subsequence is convergent, its limit is p.
11. The *graph* of a function $f : A \to \mathbb{R}^m, A \subset \mathbb{R}^n$, is the set

$$G(f) = \{(x, f(x)), x \in A\}.$$

Prove that if f is continuous, $G(f)$ is closed. Prove the converse if A is compact.

12. Define the sum $A + B$ of two sets in $\mathbb{R}^n$ as

$$A + B = \{z : z = x + y, x \in A, y \in B\}.$$

(a) Prove that $A + B$ is closed if A is closed and B compact.
(b) Prove that $A + B$ is bounded if both are bounded.
(c) Prove that $A + B$ is compact if both are.
(d) Give an example of two closed sets with non closed sum.

13. The *Haussdorf distance* between sets A, B is defined

$$\delta(A, B) = \max\left(\sup_{x\in A} d(x, B), \sup_{y\in B} d(y, A)\right).$$

Prove that this is indeed a distance in the class X of compact sets in $\mathbb{R}^n$ and that X is complete with this distance.

14. Describe:

(a) In cylindrical coordinates the intersection of $x^2 + y^2 + z^2 \le 1$ and $x^2 + y^2 \le \frac{3}{4}$.
(b) In spherical coordinates the intersection of $x^2 + y^2 + z^2 \le 1$ with $z^2 \le x^2 + y^2$.

(c) In cylindrical and spherical coordinates the intersection of the balls $x^2 + y^2 + z^2 \leq 1, x^2 + y^2 + (z-1)^2 \leq 1$.

15. Find sets A as big as possible where $u = x^3 - 3xy^2, v = 3x^2y - y^3$ is a coordinate system. Hint: use the complex variable $z = x + iy$.
16. Analyze in which plane domains the functions

$$u = x^2 - xy, \quad v = y^2 + xy$$

define a coordinate system and compute x_u, x_v, y_u, y_v in terms of x, y.

17. Give a parametrization of:

(a) The curve intersection of $\frac{x^2}{12} + \frac{y^2}{24} + \frac{z^2}{4} = 1$ and $y = x^2$ from $(2, 4, 0)$ to $(0, 0, 2)$.

(b) The intersection of the same ellipsoid with the plane $x + z = 2$.

18. Find an equation $f(x, y, z) = 0$ for the following surfaces given in parametric form:

(a) $x = s\cos t,\ y = s\sin t,\ z = s$.

(b) $x = 4\cos\phi\cos\theta,\ y = 2\cos\phi\sin\theta,\ z = \sin\phi$.

(c) $x = 3\cos t,\ y = s,\ z = 3\sin t$.

(d) $x = s\cos t,\ y = s\sin t,\ z = s^2$.

(e) $x = s\cosh t,\ y = s\cosh t,\ z = s^2$.

19. Let $E, k > 0$ and consider the curve given in polar coordinates

$$r(1 + E\cos\theta) = k.$$

(a) Transform the equation to Cartesian coordinates.

(b) Prove that if $0 < E < 1$ is an ellipse, a parabola if $E = 1$ and a hyperbola if $E > 1$.

20. Study the continuity, existence of directional derivatives, differentiability, etc. of the following functions:

$$f(x, y) = (x^3 - y^2)\sin\left(\frac{1}{x^2 + y^2}\right),$$

$$f(x, y) = \frac{x\sin y^2}{x^2 + y^2}, \quad f(x, y) = \frac{x^2 y}{x^4 + y^2},$$

$$f(x, y) = x^\alpha y^\beta, \quad f(x, y) = \frac{x^3 - y^3}{x^2 + y^2},$$

$$f_\alpha(x, y) = \frac{xy}{(x^4 + y^4)^\alpha}, \quad f(x, y) = \sqrt{|xy|}.$$

21. Give an example of a continuous function at $(0,0)$ with directional derivative $D_v f(0,0)$ in all directions v but such that $v \to D_v f(0,0)$ is not lineal.
22. Let $f : \mathbb{R}^n \longrightarrow \mathbb{R}$ of class C^1 with $f(0) = 0$. Prove that there exist continuous $g_i : \mathbb{R}^n \longrightarrow \mathbb{R}$ such that $f(x) = \sum_{i=1}^n x_i g_i(x)$.
23. Let $f(x,y) = \frac{x^3}{x^2+y^2}$. Prove that for all regular arcs $\gamma : (-\varepsilon, \varepsilon) \longrightarrow \mathbb{R}^2, \gamma(0) = (0,0)$ the composition map $f \circ \gamma$ is differentiable at 0, yet f is not differentiable at $(0,0)$.
24. A function $f : \mathbb{R}^n \longrightarrow \mathbb{R}$ is called *homogeneous* of degree m if $f(tx) = t^m f(x), t \in \mathbb{R}, x \in \mathbb{R}^n$. If f is differentiable, prove that

$$mf(x) = \sum_{i=1}^{n} x_i \frac{\partial f}{\partial x_i}(x).$$

25. Let $U = \{(x,y) \in \mathbb{R}^2 : x > 0$ and f differentiable in U such that

$$x\frac{\partial f}{\partial x} + y\frac{\partial f}{\partial y} = 0.$$

Prove that there exists $h : \mathbb{R} \longrightarrow \mathbb{R}$ differentiable such that $f(x,y) = h(\frac{y}{x})$.

26. For the family of functions

$$x, \quad y, \quad x+y+z, \quad xy+yz+zx, \quad xyz,$$

identify the functionally independent subfamilies, precise the domains of independence, and obtain the dependence equations.

27. Prove that $u = x^2 + y + z, v = x^2 + y^2$ are functionally independent in $U = \{(x,y,z) : x,y,z > 0\}$. Find a third function w in U such that (u,v,w) is a coordinate system in U. Prove that a differentiable f in U depends functionally on u,v if and only if

$$x\frac{\partial f}{\partial y} - y\frac{\partial f}{\partial x} + x(2y-1)\frac{\partial f}{\partial z} = 0.$$

28. Prove that $u = x+2y-3z, v = 2x-y+5z$ are functionally independent in $\mathbb{R}^3$ and that a differentiable f depends functionally on u,v if and only if

$$7f_x - 11f_y - 5f_z = 0.$$

29. Let $f : \mathbb{R}^2 \longrightarrow \mathbb{R}$,

$$f(x,y) = \frac{xy(x^2-y^2)}{x^2+y^2}.$$

Study the differentiability properties of f and check that $D_{1,2}f(0,0)$ does not equal $D_{2,1}f(0,0)$.

30. Prove, using a convenient coordinate system, that a twice differentiable function f in $U = \{(x,y) : x, y > 0\}$ is of the form $f(x,y) = \Phi(xy) + \Psi(\frac{y}{x})$, with Φ, Ψ twice differentiable in the real line, if and only if

$$\frac{\partial^2 f}{\partial y^2} - \frac{x^2}{y^2}\frac{\partial^2 f}{\partial x^2} + \frac{1}{y}\frac{\partial f}{\partial y} - \frac{x}{y^2}\frac{\partial f}{\partial x} = 0.$$

31. Find the general solutions $f(x,y), f(x,y,z)$ of

$$yf_y = f, \quad f_x = 2xyf, \quad f_{xx} = 0, \quad f_{xyz} = 0.$$

32. Find solutions with separate variables, $u(x,y) = X(x)Y(y)$, of

$$u_y = yu_x, \quad xu_x = u + yu_y.$$

33. Prove that $u(x,y)$ has the form $f(x)g(y)$ if and only if $uu_{xy} = u_x u_y$.
34. Using the change of variables $u = y + 2x, v = y + 3x$, find the general solution of $f_{xx} - 5f_{xy} + 6f_{yy} = 0$.
35. Analyze the critical points of

    (a) $f(x,y) = 8x^3 - 24xy + y^3$.
    (b) $f(x,y,z) = x^2 + y^2 - z^2 - xy + xz - 2z$.
    (c) $f(x,y) = \log(2 + \sin xy)$.
    (d) $f(x,y) = \sin x \cos y$.
    (e) $f(x,y,z) = xyz(1-x)(1-y)(1-z)$.
    (f) $f(x,y) = 6x^2 - 2x^3 + 3y^2 + 6xy$.
    (g) $f(x,y) = xy(x^2 + y^2 - 1)$.
    (h) $f(x,y) = \sin^2 x + \sin^2 y - \cos^2 x \cos^2 y$.
    (i) $f(x,y) = x^3 + x^2 + 2\alpha xy + y^2 + 2\alpha x + 2y, \alpha > 0$.
    (j) $f(x,y,z) = x^2 + 2y^2 + 2xy + z^2 + 2xyz + 2y^2 z$.

36. Given a cloud of $m > 2$ points $p_j = (x_j, y_y)$ in the plane, find the line $L : y = ax + b$ minimizing $\sum_j d(p_j, L)^2$, where $d(p_j, L)$ denotes the Euclidean distance from p_j to L. Compare the result with the regression line of paragraph 4.4.1.
37. Find the relative extrema of $f(x,y,z) = \frac{x}{2} + \frac{y^2}{2x} + \frac{z^2}{y} + \frac{2}{z}$ in $U = \{x, y, z > 0\}$. Has f absolute extrema in U?
38. Find the absolute extrema of the following functions on the indicated sets:

    (a) $f(x,y) = x^2 + y^2 + 2x$ in $\{(x,y) : x^2 + y^2 \leq 1, y \geq x\}$.

(b) $f(x, y) = x^2 + y^2 - 2x - 2y$ in $\{(x, y) : x^2 + y^2 \leq 4, y \geq 0\}$.
(c) $f(x, y) = 5x^2 + 5y^2 - 8xy$ in $\{(x, y) : x^2 + y^2 - xy \leq 1\}$.
(d) $f(x, y) = x^3 + y^3 - \frac{3}{2}x^2 - 3y^2$ in $x^2 + y^2 \leq 1$.
(e) $f(x, y) = x^2 + y^2 - xy - 4x - 4y$ in the triangle with vertices $(1, 0), (1, 9)$ and $(10, 0)$.
(f) $f(x, y, z) = 11x^2 + 11y^2 + 14z^2 - 2xy - 8xz - 8yz - 24x + 24y$ in the tetrahedron limited by $x = 0, y = 0, z = 0, x - y + z = 3$.
(g) $f(x, y, z) = xy + yz + zx$ on the unit ball

$$K = \{(x, y, z) \in \mathbb{R}^3;\ x^2 + y^2 + z^2 \leq 1\}.$$

39. Prove that $f(x, y) = 3x^4 - 4x^2y + y^2 = (y - 3x^2)(y - x^2)$ has a relative minimum at $(0, 0)$ along every line $y = mx$, but it has no relative minimum at the origin.
40. Find the point on the curve $z = x^2 + y^2, x + 2y + z = 4$ closest to the origin. Same question with $\frac{x^2}{4} + y^2 + \frac{z^2}{4} = 1, x + y + z = 1$.
41. Given a one-variable power series $\sum_k c_k t^k$ with positive radius of convergence R, replace t by an $n \times n$ matrix $X = (x_{ij})$ and consider the matrix-valued power series in the n^2 variables x_{ij} defined by

$$f(X) = \sum_k c_k X^k.$$

Prove that $f(X)$ is a well-defined $n \times n$ matrix for $|X|$ small enough and that for fixed p, q the (p, q)-entry $f_{p,q}(X)$ of $f(X)$ has non-trivial domain of convergence. If $R = +\infty$, $f, f_{p,q}$ are defined everywhere. The most important example is

$$e^X = \sum_k \frac{X^k}{k!}.$$

42. Someone wants to build a cave of dimensions x, y, z with isolated walls. The material to use on each of the three couples of opposite walls has cost A, B, C euros/m2, respectively. If the budget is D euros, which is the largest volume V attainable? If the volume must be V, which is the minimum budget needed?
43. Find the maximum of $f(x_1, \ldots, x_n) = (x_1 \cdots x_n)^2$ on the unit sphere in $\mathbb{R}^n$; prove that for positive a_i,

$$(a_1 a_2 \cdots a_n)^{\frac{1}{n}} \leq \frac{a_1 + \cdots + a_n}{n}.$$

44. Prove Kantorovich's inequality (4.6) using Lagrange multipliers.

45. Find the distance between two lines in space given in parametric form.
46. Find the maximum of $|\det M|$ where M is a square matrix whose column vectors have lengths $L_1, \dots, L_n$.
47. Using Lagrange multipliers, find the semi-axes of the ellipse intersection of $\frac{x^2}{a^2} + \frac{y^2}{b^2} + \frac{z^2}{c^2} = 1$ and $Ax + By + Cz = 0$.
48. Find the point in a triangle minimizing the sum of distances to the vertices. Same question with four points.
49. Consider the system of equations

$$e^x + \alpha y^2 z - z = \beta, \quad x^2 + \alpha y^2 \ln(z - xy) = 0.$$

Find $\alpha, \beta \in \mathbb{R}$ such that the system defines y, z as functions of x around $(0, 1, 1)$. For which values of $\alpha, \beta \in \mathbb{R}$ one has $y'(0) = -1/2, z'(0) = 1$?
50. Prove that

$$\begin{aligned} &2x^3y^3z^2 - 3x^2y^4z^2 - x^4y^2z^2 + x^2y^2z^3 \\ &\quad + 2x^3y^2z^2 - 6x^2y^3z^2 - x^2 - 3y^2 + 2xy + 2x - 6y + z = 0 \end{aligned}$$

defines implicitly z as a function of x, y around $(0, -1, -3)$ and that z has a relative extremum at $(0, -1)$. Prove that

$$x^2 + y^2 + z^2 = 2x + 6y + 4z - 13$$

defines $y = f(x, z)$ as a function of x, z around $(1, 4, 2)$ and that y has a relative minimum at $(1, 2)$.
51. Prove that the system

$$\sin\frac{\pi}{w} = 0, e^{x+u} - 1 = 0, \quad 2x - u + v - w + 1 = 0,$$

defines $u = u(x), v = v(x), w = w(x)$ around $(0, 0, 0, 1)$, and compute the Taylor development of u, v, w at 0 of order two.
52. For a polynomial $p(x) = x^n + a_{n-1}x^{n-1} + \cdots + a_1x + a_0$ with real coefficients, find a local condition ensuring that the roots of p are continuous functions of the coefficients.
53. Prove that the intersection of the level surfaces

$$yz + xz + xy = 5, \quad xyz = 2,$$

is a curve $y = y(x), z = z(x)$ around $(1, 2, 1)$. Compute $y'(1), z'(1)$ and $y''(1), z''(1)$.

54. Prove that there exist differentiable functions $x = x(s,t), y = y(s,t)$, defined around $(2, -1)$ such that

$$xs^2 + yt^2 = 1, \quad x^2 s + y^2 t = xy - 4.$$

How many choices exist? For each one, compute x_s, x_t, y_s, y_t.

55. Let $f : \mathbb{R}^3 \to \mathbb{R}^3$, $f(x,y,z) = (e^x, \sin(x+y), e^z)$. Prove that f has a local differentiable inverse around $(0,0,0)$ but f has no global inverse. Prove that $f(u,v) = (e^u + e^v, e^u - e^v)$ is a diffeomorphism.

56. Let $f : \mathbb{R} \longrightarrow \mathbb{R}$ of class $\mathcal{C}^k$ $(k \geq 1)$ such that

$$|f'(t)| \leq k < 1, \quad t \in \mathbb{R}.$$

Prove that $F : \mathbb{R}^2 \longrightarrow \mathbb{R}^2$ defined by

$$F(x,y) = (x + f(y), y + f(x))$$

is a diffeomorphism of class $\mathcal{C}^k$ of $\mathbb{R}^2$. Is this true for

$$G(x,y,z) = (x + f(y), y + f(z), z + f(x))$$

in $\mathbb{R}^3$?

57. Let S be the surface defined by

$$xy^2z^3 + yz^2x^3 + zx^2y^3 = 3,$$

and f a differentiable function around $p = (1,1,1)$. We know that f has normal derivative 1 at p and that f equals $\sin(xyz)$ on S. Compute $\nabla f(p)$.

58. Let Γ be the curve defined by

$$xyz = 1, \quad x^3 + 2y^3 + z^3 = 4,$$

and f a differentiable function around $p = (1,1,1)$. We know that f equals xy on Γ and that the directional derivative $D_v f(p)$ equals 1 for $v = (0,1,1)$ and $v = (1,0,1)$. Compute $df(p)$.

59. Let M be a k-dimensional sub-manifold in $\mathbb{R}^n$ of class C^1 and f a continuous function on M. Prove that locally around each $p \in M$ there is a continuous extension g of f. If M is compact, use partitions of unity to show that there is a global extension g of f to $\mathbb{R}^n$. If f is differentiable on M, then g can be chosen differentiable, too. In fact, extensions of continuous functions can be defined for arbitrary closed sets.

60. In the half-plane $x > 0$, consider $u(x, y) = -y \log x$. Which equation,
$$A(x, y)f_x + B(x, y)f_y = 0,$$
characterizes the functions $f(x, y) = h(u(x, y))$ for some one-variable function h?

61. Which is the maximal interval of definition of the solutions of $x' = \frac{x^2-1}{2}, x(s) = y$?

62. Find the solutions of
$$x' = 1 + x^2, \quad x(0) = 0,$$
and
$$x' = t + y, \quad y' = t - x^2, \quad (x(0), y(0)) = (2, 1).$$

63. Solve the following ODEs using an integrating factor $\lambda(x)$ or $\lambda(y)$:
$$(x + y^2)\,dx - 2xy\,dy = 0, \quad y(1 + xy)\,dx - x\,dy = 0,$$
$$(x \cos y - y \sin y)\,dy + (x \sin y + y \cos y)\,dx = 0.$$

64. Find, depending on y, the interval of definition of the solution of the Cauchy problem
$$x' = x^3 - x^2, \quad x(s) = y.$$
Same question with
$$x' = t^2 + e^{-x^2}, \quad x(s) = y,$$
and
$$x' = \frac{x^3}{1 + x^2}, \quad x(s) = y.$$

65. Consider the Cauchy problem
$$x' = f(t, x), \quad x(s) = y,$$
where $f : \mathbb{R} \times \mathbb{R}^n \longrightarrow \mathbb{R}^n$ is continuous and $|f(t, x_1) - f(t, x_2)| \leq L(t)|x_1 - x_2|$, with $L(t)$ continuous in $\mathbb{R}$. Prove that there is a unique solution defined in the whole line.

66. Consider the Cauchy problem
$$x' = g(x), \quad x(s) = y,$$
where $g : \mathbb{R} \longrightarrow \mathbb{R}, g(x) > 0$ is locally Lipschitz. Let φ be the maximal solution defined in $I = (a, b)$.

(a) Prove that if $I = \int_s^\infty \frac{1}{g(\tau)}\, d\tau < \infty$, then $b < \infty$, and compute b.
(b) Prove that if $I = \infty$, then $b = +\infty$ and $\lim_{t\to\infty} \varphi(t) = +\infty$.

67. Consider the system

$$x' = x(x-1), \quad y' = -2xy + y.$$

(a) Find the solution $\varphi(t, 0, (x_0, y_0))$, with initial condition $\varphi(0) = (x_0, y_0)$.
(b) Check that $\phi(t) = (1, e^{-t}) = \varphi(t, 0, (1, 1))$.
(c) Compute $\frac{\partial \varphi}{\partial x_0}(t, 0, (1,1))$ and $\frac{\partial \varphi}{\partial y_0}(t, 0, (1,1))$ and check that they are solutions of the first variation equations.

68. Let $\phi = \phi(t, s, y, (a, b))$ be the solution of the problem

$$x' = 2t(ax - bx^2), \quad x(s) = y.$$

(a) Find the solution explicitly.
(b) Check that $\phi(t) = e^{t^2}$ is the solution if $s = 0, y = 1, a = 1, b = 0$.
(c) Check that

$$\frac{\partial \phi}{\partial s}(t, 0, 1, (1, 0)), \quad \frac{\partial \phi}{\partial a}(t, 0, 1, (1, 0)), \quad \frac{\partial \phi}{\partial b}(t, 0, 1, (1, 0))$$

are indeed solutions of the system stated in Theorems 8.5 and 8.6.

69. Find the general solution of the systems

$$\frac{dx}{1+x^2} = \frac{dy}{z - xy} = -\frac{dz}{xz + y}, \quad dz = \frac{dx}{-x(x+y)} = \frac{dy}{y(x+y)},$$

$$\frac{dx}{x(y^2 - z^2)} = \frac{dy}{-y(x^2 + z^2)} = \frac{dz}{z(y^2 + x^2)}, \quad \frac{dx}{ayz} = \frac{dy}{bzx} = \frac{dz}{cxy}.$$

70. Solve the partial differential equations

$$(xy^3 - 2x^4)p + (2y^4 - x^3 y)q = 0, \quad uxp + (3x - 2y)q = 0.$$

71. Find the general solution of

$$ap + bq = \frac{\sqrt{1+u^2}}{u},$$

(with a, b constants), and the solution through the line $y = 0, z = h$.

72. Find the general solution of

$$(xy^3 - 2x^4)p + (2y^4 - x^3y)q = 9(x^3 - y^3)u.$$

73. Find the general solution of

$$-yf_x + xf_y + (1 + z^2)f_z = 3zf,$$

and the ones invariant under rotations with axis z.

74. Find the values of α for which

$$dz = zxy\,dx + z(\alpha x^2 + \tan y)\,dy$$

is completely integrable. Find all solutions for every α.

75. Solve the equation

$$xy^2p + x^2yq = u(x^2 + y^2),$$

and find a family S of solutions satisfying

$$xy(y^2 - x^2)\,dz = yz(y^2 - 3x^2)\,dx + xz(3y^2 - x^2)\,dy.$$

Find the family of curves orthogonal to the family S.

76. Find the family of curves $g(x, y) = c$ orthogonal to the family $x^m + y^n = c$.

77. Prove that there exists a family S of surfaces orthogonal to the system of curves

$$x^2 + 2y^2 = az^2, \quad x^2 + y^2 + z^2 = bz.$$

Find the surfaces orthogonal to the family S and triply orthogonal families containing S.

78. Find the family of curves orthogonal to the family of surfaces

$$S_1 : x(x^2 + y^2 + z^2) + cy^2 = 0.$$

Find a family S_2 of spheres centered at points in the z-axis orthogonal to S_1, and a third family S_3 so that S_1, S_2, S_3 are a triply orthogonal system.

79. Let $f : [0, 1] \times [0, 1] \longrightarrow \mathbb{R}$ be defined by

$$f(x, y) = \begin{cases} 0, & \text{if } x \in \mathbb{R} \setminus \mathbb{Q}; \\ 0, & \text{if } x \in \mathbb{Q} \text{ i } y \in \mathbb{R} \setminus \mathbb{Q}; \\ \dfrac{1}{q}, & \text{if } x = \dfrac{p}{q} \text{ irreducible and } y \in \mathbb{Q}\,. \end{cases}$$

Prove that f is Riemann integrable in $[0, 1] \times [0, 1]$, but for $x \in [0, 1] \cap \mathbb{Q}$ the function $f_x(y) = f(x, y)$ is not Riemann integrable in $[0, 1]$.

80. Prove that

$$C = \left\{ (x,y,z), 0 \le z \le \frac{1}{1+x^2+x^2y^4+2x^2y^2} \right\}$$

has finite volume.

81. Compute the mass and center of mass for

(a) $[-1,1]^2$ with mass density at (x,y) proportional to the distance to the line $2x+2y=1$.

(b) $[-1,1]^3$ with mass density at (x,y,z) proportional to the distance to the plane $2x+2y+2z=1$.

82. Compute the integral over $Q=[0,1]^2$ of

$$f(x,y) = x^2y^2e^{xy}, \quad f(x,y) = xy\sin \pi xy.$$

83. Compute the integral over $Q=[0,1]^3$ of

$$f(x,y,z) = \log(x+y+z), \quad f(x,y,z) = xyz\log(1+xyz).$$

84. Compute the following iterated integrals changing the order of integration:

$\int_0^{\sqrt{2}}(\int_{-\sqrt{4-2y^2}}^{\sqrt{4-2y^2}} y\,dx)\,dy, \quad \int_0^1(\int_{x^3}^{x^2} y^2\,dy)\,dx,$

$\int_0^4(\int_{y/2}^2 e^{x^2}\,dx)\,dy, \qquad \int_{-1}^0(\int_0^{x+1} e^{x+y}\,dy)\,dx + \int_0^1(\int_0^{1-x} e^{x+y}\,dy)\,dx,$

$\int_0^4(\int_{y/2}^{\sqrt{y}} x^2y^2\,dx)\,dy, \qquad \int_0^\pi(\int_x^\pi \frac{\sin y}{y}\,dy)\,dx.$

85. Compute $\int_A f\,dm$ for:

(a) $f(x,y) = \frac{y}{x^2+y^2}$ and A the triangle limited by $y=x$, $y=2x$, $x=2$.

(b) $f(x,y) = xy$ and A the sector in the first quadrant limited by $x^2+y^2=25, y=0, 3x-4y=0$.

86. Compute the volume of A for:

(a) A is a cylinder with base the plane region limited by $y=4-x^2$, $y=3x$ and bounded above by $z=x+4$.

(b) A is the part of the first octant limited by $z=4-x^2-y$.

(c) A is the part of the first octant limited by $z=1-y^2, x+y=1, x+y=3$.

(d) A is limited by $z=x^2+y^2, x^2+y^2+z^2=6$.

(e) A is defined by $0 \le z \le x^2 + y^2, x + y \le 1, 0 \le x, y$.
(f) A is defined by $0 \le z \le \sqrt{x^2 + y^2}, x^2 + y^2 \le 2y$.
(g) A is defined by $\sqrt{x} + \sqrt{y} \le 1, 0 \le z \le \sqrt{xy}$.
(h) A is defined by $\sqrt{x} \le y \le 2\sqrt{x}, 0 \le z \le 9 - x$.

87. Analyze which of the following integrals are finite:
(a) $\int_{\mathbb{R}^2} |x|^\alpha |y|^\beta \log |x + y| \, dm$.
(b) $\int_{\mathbb{R}^2} \frac{\sin xy}{(|x|+|y|)^\alpha} \, dm$.
(c) $\int_{\mathbb{R}^3} \frac{|x|+|y|+|z|}{1+(|xy|+|yz|+|zx|)^\alpha} \, dm$.

88. The following expressions are volumes of solids. Draw them and compute their volume changing the order of integration.
(a) $\int_{-1}^{1} \left(\int_0^{1-x^2} \left(\int_{-\sqrt{1-x^2-z}}^{\sqrt{1-x^2-z}} dy \right) dz \right) dx$.
(b) $\int_0^2 \left(\int_{2x}^4 \left(\int_0^{\sqrt{y^2-4x^2}} dz \right) dy \right) dx$.
(c) $\int_0^6 \left(\int_{z/2}^3 \left(\int_{z/2}^y dx \right) dy \right) dz + \int_0^6 \left(\int_3^{(12-z)/2} \left(\int_{z/2}^{6-y} dx \right) dy \right) dz$.

89. Consider the polynomial $P = ax^2 + bx + c, a, b, c \in [0, 1]$. What event is more probable: P has two real roots or P has two non-real roots?

90. Compute the convolution $1_I * 1_J$ for two intervals I, J and $1_I * 1_J * 1_K$ with a third interval.

91. Prove that if f is smooth in $\mathbb{R}^n$, there is a sequence of polynomials P_m such that $D^\alpha P_m \to D^\alpha f$ uniformly on compacts for each muti-index α.

92. Compute the area of the flower given in polar coordinates by $r = \cos m\theta$.

93. Compute the integral of $(1+x^2+y^2)^{-2}$ in the interior of the lemniscata $(x^2 + y^2)^2 = x^2 - y^2$.

94. Compute the integral of $(x + y + z)x^2y^2z^2$ in the tetrahedron $x + y + z \le 1, x, y, z \ge 0$.

95. Compute the volume of the solid generated by the cardioid $r = 1+\cos\theta$ when rotating around the x axis, and the area of its boundary.

96. Use suitable coordinates to find the volume of the solid defined by:
(a) $z \le \sqrt{16 - x^2 - y^2}$, $x^2 + y^2 \le 4x$.
(b) $z \le x^2 + y^2 + 3$, $z \ge 0$, $x^2 + y^2 \le 1$.
(c) $z \le \sqrt{16 - x^2 - y^2}$, $x^2 + y^2 \ge 1$.
(d) $x^2 + y^2 \le 1$, $x^2 + z^2 \le 1$.
(e) $x^2 + y^2 \le 1$, $y^2 + z^2 \le 1$, $z^2 + x^2 \le 1$.

(f) The intersection of the ellipsoid $E : \frac{x^2}{a^2} + \frac{y^2}{b^2} + +\frac{z^2}{c^2} \leq 1$ and the translated ellipsoid $E + (a, 0, 0)$.

97. Using the variables $u = \sqrt{xy}, v = \sqrt{\frac{y}{x}}$, compute the area of the region bounded by the curves $xy = 1, xy = 9, y = x, y = 4x$.
98. Compute the area of the region bounded by the curves $xy = a$, $xy = b, x^2 - y^2 = c, x^2 - y^2 = d$.
99. Compute the mass and center of mass of:

(a) The interior of $4x^2+4y^2+z^2 = 16, z \geq 0$ with density proportional to the distance to the plane xy.

(b) A cone with basis of radius R and height H with density proportional to the distance to the vertex.

100. Prove Newton's result stating that the potential created by a uniform mass distribution in a ball B outside B

$$U(y) = k \int_B \frac{1}{|x - y|}\, dm(x), \quad y \notin B,$$

is the same as if all the mass is concentrated at the center of B.

101. Prove that if A has finite Lebesgue measure and f is bounded measurable on A,

$$\lim_{p \to +\infty} \left(\int_A |f|^p\, dm \right)^{\frac{1}{p}} = \sup_{x \in A} |f(x)|.$$

102. Prove *Jensen's inequality*: for φ convex

$$\varphi\left(\frac{1}{m(A)} \int_A f(x)\, dm \right) \leq \frac{1}{m(A)} \int_A \varphi(f(x))\, dm).$$

103. This exercise proposes a derivation of Euler's formula

$$\sum_{n=1}^{\infty} 1/n^2 = \pi^2/6,$$

different from Euler's original beautiful proof.

(a) Expand $\frac{1}{1-x^2y^2}$ in a power series in xy and justify the identity

$$I = \int_{[0,1]\times[0,1]} \frac{1}{1 - x^2y^2}\, dm = \sum_{n=0}^{\infty} \frac{1}{(2n+1)^2}.$$

(b) Check that the change of variables

$$x = \frac{\sin u}{\cos v}, \quad y = \frac{\sin v}{\cos u},$$

transforms the triangle limited by $u = 0, v = 0, u + v = \pi/2$ in $[0,1] \times [0,1]$.

(c) Prove that $I = \frac{\pi^2}{8}$ and Euler's formula as a consequence.

104. For the unit ball B, compute

$$\int_B \frac{1}{x^2 + y^2 + (z-2)^2} \, dm.$$

105. Compute the length of the following arcs in the plane:

(a) $r = e^{-\theta}$, (b) $r^2 = \cos 2\theta$, (c) $r = \sin 3\theta$,
(d) $r = 1 + \cos\theta$, (e) $r = |\sin 2\theta|$, (f) $r = 1 + \cos\frac{\theta}{2}$.

106. Draw and compute the length of the graphic $y = -\log(\cos x), 0 \leq x \leq \frac{\pi}{4}$ and the catenary $y = a\cosh\frac{x}{a}, -a \leq x \leq a$.

107. The *cycloid* is the path described by a fixed point P of a bike wheel. Find, in terms of the wheel radius, a parametrization and the length of the cycloid between two consecutive points of contact of P with the road (assume the road is flat and smooth).

108. A wire has the shape of the curve

$$x^2 + y^2 + z^2 = 4, \quad x + z = 2,$$

and mass density proportional to the distance to the z axis. Compute the mass and center of mass.

109. Compute the area of the surfaces S and the center of mass if the density is constant:

(a) The piece of $z = x^2 + y^2$ interior to $x^2 + y^2 = 1$.
(b) The part of a sphere of radius R between $z = a$ and $z = b$, $|a|, |b| < R$.
(c) The piece of $z = xy$ interior to $x^2 + y^2 = a^2$.

110. Compute the area of the *Viviani cupola*, the part of the sphere $x^2 + y^2 + z^2 = R^2$ interior to the cylinder $x^2 + y^2 = Rx$. Compute the volume of the interior of the cylinder below the cupola.

111. Compute $\int_\Gamma \langle F, T \rangle \, ds$ for:

(a) $F(x,y) = (y^2, -2x)$, Γ the triangle with vertices $(0,0), (1,0), (1,1)$, oriented clockwise.

(b) $F(x,y) = (x^2y, x^3y^2)$, Γ the closed curve determined by $y = 4$, $y = x^2$, oriented counterclockwise.

(c) $F(x,y) = (y^m, x^m), m = 0, 1, 2, \ldots$, and Γ the circle of radius r centered at the origin, oriented counterclockwise.

(d) $F(x,y) = (x^2e^x + y - \log(1+x^2), 8x - \sin y)$, Γ the unit circle oriented counterclockwise.

(e) $F = (xy + x + y, xy + x - y)$, Γ the ellipse $\frac{x^2}{a^2} + \frac{y^2}{b^2} = 1$ oriented counterclockwise.

112. Compute the line integral $\int_\Gamma y^2\,dx + x^2\,dy$ where Γ is the upper part of the ellipse $\frac{x^2}{a^2} + \frac{y^2}{b^2} = 1$, oriented clockwise. Same question with

$$\int_\Gamma \frac{x\,dx + y\,dy}{\sqrt{1+x^2+y^2}},$$

where Γ is the part of the same ellipse in the first quadrant, oriented clockwise.

113. Prove that a radial continuous field $F = f(r)(x,y)$, with f continuous, is conservative.

114. Compute the line integral of

$$F(x,y) = \frac{1}{x^2+y^2}\left(2x^3 + 2xy^2 + y, -(2x^2y + 2y^3 + x)\right),$$

along the upper part of the ellipse $x^2 + 4y^2 = 1$, $y \geq 0$ oriented counterclockwise. Same question with

$$F(x,y,z) = (x^2 - yz, y^2 - xz, -xy),$$

along the curve $x = t, y = \frac{t^2}{2}, z = \frac{t^3}{6}, 0 \leq t \leq 1$.

115. Compute the line integrals in space:

(a) $\int_\Gamma (y^2 - z^2)\,dx + (z^2 - x^2)\,dy + (x^2 - y^2)\,dz, \Gamma$ the helix $x = \cos t$, $y = \sin t, z = t, 0 \leq t \leq 2\pi$.

(b) $\int_\Gamma xy\,dx + yz\,dy + zx\,dz, \Gamma$ the arc of the circle $x^2 + y^2 + z^2 = 2x, z = x$, where $y > 0$.

116. Check that the following 1-forms ω are exact and compute their line integral along a path from p to q in the domain of definition of ω:

(a) $\omega = \frac{1}{y^2}(y\,dx - x\,dy), p = (1,2), q = (2,1)$.

(b) $\omega = \frac{1}{(x-y)^2}(x\,dy - y\,dx), p = (0,2), q = (1,3)$.

(c) $\omega = \frac{yz\,dx + zx\,dy + xy\,dz}{xyz}, p = (1,1,1), q = (2,3,4)$.

117. Check Green's formula for:
    (a) $\int_\Gamma xy^2\,dy - x^2y\,dx$, Γ the unit circle oriented clockwise.
    (b) $\int_\Gamma 2(x^2+y^2)\,dx + (x+y)^2\,dy$, Γ the boundary of the triangle with vertices $(1,1),(1,2),(2,3)$.
118. Compute the flux of:
    (a) $F = (x^3, y^3, z^3)$ across the unit sphere.
    (b) $F = (yz, xz, xy)$ across the tetrahedron bounded by the planes $x = 0, y = 0, z = 0, x + y + z = a$.
    (c) $F = (x, y, z)$ across the ellipsoid $\frac{x^2}{a^2} + \frac{y^2}{b^2} + \frac{z^2}{c^2} = 1$.
119. Use Stokes' theorem to compute the circulation of:
    (a) $F = (y+z, z+x, x+y)$ along the circle $x^2+y^2+z^2 = 1, x+y+z = 0$.
    (b) $F = (y - z, z - x, x - y)$ along the ellipse $x^2 + y^2 = 1, x + z = 1$.
    (c) $F = (x, x + y, x + y + z)$ along the curve $x = a\cos t, y = a\sin t, z = a(\sin t + \cos t), 0 \leq t \leq 2\pi$.
120. Check Gauss' theorem for:
    (a) $F = (x^m.y^m, z^m)$ and the unit cube.
    (b) $F = (x^2, y^2, z^2)$ and the cone $\frac{x^2}{a^2} + \frac{y^2}{a^2} - \frac{z^2}{b^2}, 0 \leq z \leq H$.
121. Prove that $F(p) = f(|p|)p$ is solenoidal if and only if $f(t) = \frac{k}{t^3}$.
122. Assume that u is a C^2- function in $\mathbb{R}^n$ such that its restriction to every line $p + tv$ is linear in t. Prove that u is a linear function.

Bibliography

[1] L. Ahlfors, *Complex Analysis*, Mc Graw-Hill, New York, 1978.

[2] T. Bonnessen, and W. Fenchel, *Theorie der konvexen Körper*, Springer, Berlin, 1934.

[3] J. Bruna and J. Cufí, *Complex Analysis, EMS Textbooks in Mathematics, No. 13*, EMS Publishing House, Zurich, 2013.

[4] E. DiBenedetto, *Partial Differential Equations*, Birkhauser, Basel-Berlin-Boston, 1995.

[5] G. Darboux, *Leçons sur les systèmes orthogonaux et les coordonnées curvilignes*, Gauthier-Villars, Paris, 1910.

[6] M. P. Do, *Differential Geometry of Curves and Surfaces*, Prentice-Hall, Englewood Cliffs NJ, 1976.

[7] H. Federer, *Geometric Measure Theory*, Springer, New York, 1969.

[8] P. Lax, A short path to the shortest path, *Amer. Math. Monthly*, 102(2) (1995), 158–159.

[9] J. M. Lee, *Introduction to Smooth Manifolds*, Springer, New York, 2002.

[10] S. Lojasiewicz, *An Introduction to the Theory of Real Functions*, Wiley-Interscience, New York, Hoboken NJ, 1988.

[11] Encyclopédie des formes mathématiques remarquables, https://mathcurve.com.

[12] J. Milnor, *Topology from a Differential Viewpoint*, University of Virginia Press, Charlottesville, 1965.

[13] M. A. Naimark, *Linear Representations of the Lorentz Group*, Pergamon Press, Oxford, 1964.

[14] W. Rudin, *Real and Complex Analysis*, Mc Graw-Hill, New York, 1987.

[15] L. Santaló, *Integral Geometry and Geometric Probability*, 2nd edition, Cambridge Univ. Press, Cambridge, 2009.

[16] G.E. Schwarz, A pretender to the title "A canonical Möbius strip", *Pacific J. Math.* 143(1990).

[17] J. Sotomayor, *Lições de Equações Diferenciais Ordinárias*, IMPA, Rio de Janeiro, 1979.

[18] E. H. Spanier, *Algebraic Topology*, Springer-Verlag, New York, 1966.
[19] M. A. Sadybekov, B. T. Torebek, and B. Kh. Turmetov, Construction of Green's function of the Neumann problem in a ball, *Eur. Math. J.*, V 7(2), (2016), 100–105.
[20] T. Tao, *An Introduction to Measure Theory*, AMS Graduate Studies in Mathematics, Providence RI, 2011.

Index

D

E

F

G

H

Q

R

S

www.ingramcontent.com/pod-product-compliance
Ingram Content Group UK Ltd.
Pitfield, Milton Keynes, MK11 3LW, UK
UKHW020026130625
459608UK00002B/5

9 781800 611719